T. Schrader, A. D. Hamilton (Eds.)
Functional Synthetic Receptors

Further Titles of interest:

W. R. Briggs, J. L. Spudich (Eds.)

Handbook of Photosensory Receptors

2005
ISBN 3-527-31019-3

S. Frings, J. Bradley (Eds.)

Transduction Channels in Sensory Cells

2004
ISBN 3-527-30836-9

I. Willner, E. Katz (Eds.)

Bioelectronics
From Theory to Applications

2005
ISBN 3-527-30690-0

C. M. Niemeyer, C. A. Mirkin (Eds.)

Nanobiotechnology
Concepts, Applications and Perspectives

2004
ISBN 3-527-30658-7

G. Krauss

Biochemistry of Signal Transduction and Regulation
Third, Completely Revised Edition

2003
ISBN 3-527-30591-2

Thomas Schrader, Andrew D. Hamilton (Eds.)

Functional Synthetic Receptors

**WILEY-
VCH**

WILEY-VCH Verlag GmbH & Co. KGaA

Editors

Prof. Dr. Thomas Schrader
Philipps-Universität Marburg
Fachbereich Chemie
Hans-Meerwein-Strasse
35032 Marburg
Germany

Prof. Dr. Andrew D. Hamilton
Yale University
Department of Chemistry
Sterling Chemistry Laboratory
225 Prospect Street
New Haven, CT 06520
USA

■ All books published by Wiley-VCH are carefully produced. Nevertheless, authors, editors, and publisher
do not warrant the information contained in these
books, including this book, to be free of errors.
Readers are advised to keep in mind that statements,
data, illustrations, procedural details or other items
may inadvertently be inaccurate.

Library of Congress Card No.: Applied for

British Library Cataloging-in-Publication Data:
A catalogue record for this book is available from the
British Library

Bibliographic information published by
Die Deutsche Bibliothek
Die Deutsche Bibliothek lists this publication in the
Deutsche Nationalbibliografie; detailed bibliographic
data is available in the
Internet at <http://dnb.ddb.de>.

Printed in the Federal Republic of Germany

Printed on acid-free paper

Typesetting   TypoDesign Hecker GmbH, Leimen
Printing   betz-druck gmbH, Darmstadt
Bookbinding   Litges & Dopf Buchbinderei GmbH,
Heppenheim

ISBN-10   3-527-30655-2
ISBN-13   978-3-527-30655-8

Table of Contents

Functional Synthetic Receptors. Edited by T. Schrader, A. D. Hamilton
Copyright © 2005 WILEY-VCH Verlag GmbH & Co. KGaA, Weinheim
ISBN: 3-527-30655-2

Preface

Although numerous books appeared on the general topic of Supramolecular Chemistry, the combination of an overview about specific receptor structures for important compound classes with a broad description of their potential functions and applications is not found today. Either the supramolecular principles are emphasized (noncovalent interactions, enthalpy-entropy calculations, selectivity etc.) or the (often narrow) range of applications is detailed from a certain view point, e.g. in the field of sensors, electrochemical devices etc. In addition, several reviews or books which were already written on certain chapters of this book, appeared in the 90's, so that a fresh compilation of recent advances in the field during the past 4-5 years seems necessary.

This book comprises a timely overview about receptor molecules for the most important classes of compounds with a concentration on recent literature (1998-2004). Special emphasis is placed on potential applications. As a consequence, a strong interdisciplinary touch reaches out to other fields like Materials Science (devices), Bioorganic Chemistry (enzyme mechanisms, model compounds), Medicinal Chemistry (prevention of pathological processes with synthetic receptors) and Organic Synthesis (catalysts, mechanistic elucidations), to name just a few. Especially in the biological context challenging solvents are preferred: which receptor systems work well in water or even better under physiological conditions? Any overlap with other books in the neighbourhood of supramolecular chemistry is avoided, unless the latest review is already more than 5-6 years old.

The level of presentation aims at the advanced readership, i. e., graduate students and specialists in the field. Introductory remarks are restricted to a minimum. The topics dealt within will be interesting for pharmaceutical companies dealing with drug design as well as chemical companies with a polymer branch or a nanotechnology group. Firms working in the field of molecular biology and biotechnology will also benefit from the chapters with biologic content. Finally analytical companies working with or producing the advanced analytical equipment mentioned in this book will find interesting new fields of applications of their technologies.

The editors have asked leading authors in their fields to summarize the most important developments of the recent past, and place their emphasis on the most fascinating and promising applications of these new receptor molecules. Unfortunately, this compilation must be a (subjective) selection, leaving out many excellent contributions – we apologize to all those at the outset.

Functional Synthetic Receptors. Edited by T. Schrader, A. D. Hamilton
Copyright © 2005 WILEY-VCH Verlag GmbH & Co. KGaA, Weinheim
ISBN: 3-527-30655-2

We are indebted to all authors for their invaluable contributions to this book. Dr. Gudrun Walter and Dr. Rainer Münz of Wiley-VCH did a great job in shaping the concept and assembling the individual items into a complete opus.

Marburg and New Haven,
January 2005

Thomas Schrader and Andrew D. Hamilton

List of Contributors

Sergey Antsypovich
Chair of Organic Chemistry I
Ruhr-University Bochum
Universitätsstraße 150
44780 Bochum (Germany)

Won-Seob Cho
Department of Chemistry
and Biochemistry
Institute for Cellular and
Molecular Biology
1 University Station – A5300
The University of Texas at Austin
Austin, TX 78712-0165 (USA)

Anthony P. Davis
School of Chemistry
University of Bristol
Cantock's Close
Bristol BS8 1TS (UK)

Lars Eckardt
Chair of Organic Chemistry I
Ruhr-University Bochum NC2/133
Universitätsstraße 150
44780 Bochum (Germany)

Andrew D. Hamilton
Department of Chemistry
Yale University
225 Prospect Street
P.O. Box 208 107
New Haven, CT 06520 (USA)

Tony D. James
Department of Chemistry
University of Bath
Bath BA2 7AY (UK)

Euan R. Kay
School of Chemistry
University of Edinburgh
The King's Buildings
West Mains Road
Edinburgh EH9 3JJ (UK)

Günter von Kiedrowski
Chair of Organic Chemistry I
Ruhr-University Bochum NC2/171
Universitätsstraße 150
44780 Bochum (Germany)

David A. Leigh
School of Chemistry
University of Edinburgh
The King's Buildings
West Mains Road
Edinburgh EH9 3JJ (UK)

Michael Maue
Department of Chemistry
Marburg University
Hans-Meerwein-Straße
35032 Marburg (Germany)

Functional Synthetic Receptors. Edited by T. Schrader, A. D. Hamilton
Copyright © 2005 WILEY-VCH Verlag GmbH & Co. KGaA, Weinheim
ISBN: 3-527-30655-2

Sven Mönninghoff
Chair of Organic Chemistry I
Ruhr-University Bochum NC2/132
Universitätsstraße 150
44780 Bochum (Germany)

Wolf-Matthias Pankau
Chair of Organic Chemistry I
Ruhr-University Bochum NC2/132
Universitätsstraße 150
44780 Bochum (Germany)

Leonard J. Prins
Department of Chemical Sciences
University of Padova
Via F. Marzolo 1
35131 Padova (Italy)

Maya Radeva
Chair of Organic Chemistry I
Ruhr-University Bochum NC2/126
Universitätsstraße 150
44780 Bochum (Germany)

Dmitry M. Rudkevich
Department of Chemistry
and Biochemistry
University of Texas at Arlington
Arlington, TX 76019-0065 (USA)

Thomas Schrader
Department of Chemistry
Marburg University
Hans-Meerwein-Straße
35032 Marburg (Germany)

Paolo Scrimin
Department of Chemical Sciences
University of Padova
Via Marzolo 1
35131 Padova (Italy)

Jonathan L. Sessler
Department of Chemistry
and Biochemistry
Institute for Cellular and
Molecular Biology
1 University Station – A5300
The University of Texas at Austin
Austin, TX 78712-0165 (USA)

K. Ingrid Sprinz
Department of Chemistry
Yale University
PO Box 208107
New Haven, CT 06520 (USA)

Johanna Stankiewicz
Chair of Organic Chemistry I
Ruhr-University Bochum NC2/164
Universitätsstraße 150
44780 Bochum (Germany)

Debarati M. Tagore
Department of Chemistry
Yale University
PO Box 208107
New Haven, CT 06520 (USA)

Jan Zimmermann
Chair of Organic Chemistry I
Ruhr-University Bochum NC2/133
Universitätsstraße 150
44780 Bochum (Germany)

1
Artificial (Pseudo)peptides for Molecular Recognition and Catalysis

Leonard J. Prins and Paolo Scrimin

1.1
Introduction

This chapter focuses on recognition and catalytic processes in which artificial (pseudo)peptide sequences, which can be very short, play a decisive role. The enormous amount of literature related to this topic is far beyond the scope of a single chapter, and, therefore, we intend to emphasize concepts and breakthroughs by using representative examples. Obviously, the reason for the interest in the role of (pseudo)peptides in molecular recognition and catalysis is the fact that polypeptides, e.g. proteins, play a crucial role in practically all biologically relevant processes. An incredible number of recognition events is of key importance for the occurrence of life. The origin of biological recognition is the tertiary structure of proteins, which is marvelously determined by conformationally well-defined secondary structures such as α-helices, β-sheets, coiled coils, etc. These locally structured units give order to the overall system, positioning functional groups precisely in three-dimensional space, thus creating an active site where recognition takes place. Molecular recognition is especially crucial in the functioning of enzymes. To accomplish its powerful tasks an enzyme first needs to recognize the substrate and, subsequently, in the course of its chemical transformation, also the intermediate transition state that lies on the reaction pathway toward the product. These impressive results in Nature form an almost infinite source of inspiration for the chemist, not only to mimic natural functions but also to modify them and apply them in unnatural situations.

In this chapter we will discuss advances that have been made in our learning process from Nature and, more specifically, show how chemists are able to mimic natural functions using artificial synthetic molecules. First, we will focus on the biomolecular recognition of oligonucleotides (DNA/RNA) and protein surfaces by artificial oligopeptides. Next, we will show that chemists have learned to control the secondary structure of (pseudo)peptides and that specific (catalytic) functions can be introduced at will. Finally, we will conclude with a brief overview of the selection of (pseudo)peptide catalysts by a combinatorial approach.

Functional Synthetic Receptors. Edited by T. Schrader, A. D. Hamilton

ISBN: 3-527-30655-2

1.2
Recognition of Biological Targets by Pseudo-peptides

1.2.1
Introduction

In all organisms, nucleic acids are responsible for the storage and transfer of genetic information. With the aim of curing gene-originated diseases, artificial molecules that can interact with DNA and RNA are of utmost interest. In this section we will discuss the current state of two major classes of pseudo-peptides that are currently under intense investigation – polyamides that bind in the minor groove of DNA and peptide nucleic acids (PNA). Both classes of compounds are inspired by naturally occurring analogs. The high synthetic accessibility and the ease with which chemical functionality can be introduced illustrate the high potential of artificial pseudo-peptides. In addition, their high biostability has enabled successful applications in both *in-vitro* and *in-vivo* studies. The limiting properties of these compounds will also be addressed.

Another way of interfering with biological processes is to obstruct the activity of proteins themselves. Pseudo-peptides that inhibit the formation of protein–protein complexes via competitive binding to the dimerization interface will be discussed. Selected examples will be given that clearly illustrate the strong increase in activity when amino acids present in a wild-type peptide sequence are replaced by artificial amino acids.

1.2.2
Polyamides as Sequence-specific DNA-minor-groove Binders

The discovery of the mode of interaction between the natural compounds distamycin and netropsin and the minor groove of DNA has been the impetus for the development of a set of chemical rules that determine how the minor groove of DNA can be addressed sequence-specifically [1]. NMR and X-ray spectroscopy showed that distamycin binds to A,T-tracts 4 to 5 base pairs in length either in a 1:1 or 2:1 fashion, depending on the concentration (Fig. 1.1) [2, 3]. It was then immediately realized by the groups of Dickerson, Lown, and Dervan that the minor groove of DNA is chemically addressable and, importantly, that chemical modifications of the natural compounds should, in theory, provide an entry to complementary molecules for each desirable sequence [4, 5].

1.2.2.1 Pairing Rules
The minor groove of DNA is chemically characterized by several properties. First, the specific positions of hydrogen-bond donor and acceptor sites on each Watson–Crick base pair, as depicted schematically in Fig. 1.2. Next, the molecular shape of the minor groove in terms of specific steric size, such as the exocyclic NH_2 guanine. Finally, an important property is the curvature of the double stranded DNA helix. Having

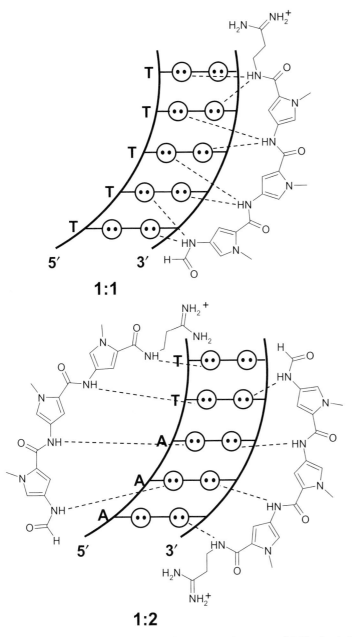

1:1

1:2

Figure 1.1 Observed binding modes of distamycin to DNA (1:1 and 1:2 complexes). The dotted circles represent the lone pairs of N(3) of purines and O(2) of pyrimidines in the minor groove. The dotted lines represent hydrogen bonds between distamycin and DNA.

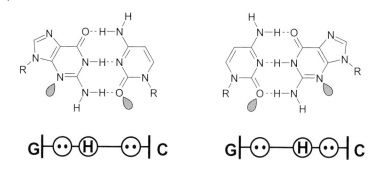

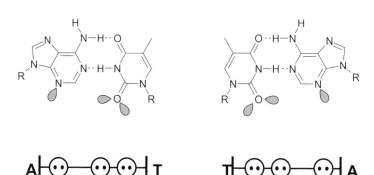

Figure 1.2 Hydrogen-bond donors and acceptors present in the minor groove of ds DNA for each of the four Watson–Crick base pairs. Lone pairs are indicated by shaded orbitals and R represents the sugar–phosphate backbone. In the schematic representations circles with dots are hydrogen-bond acceptors and circles with H are hydrogen-bond donors.

these properties as a guideline, Dervan and coworkers have developed a series of five-membered heterocycles that pairwise can recognize each of the four base pairs [6, 7]. These couples and their binding modes are schematically depicted in Fig. 1.3. To gain selectivity for a G,C over an A,T base pair, the pyrrole ring (Py) was substituted by an imidazole (Im), which forms an additional hydrogen-bond with the exocyclic NH_2 of guanine, as confirmed by crystal structure analysis. In addition, replacement of the pyrrole CH for an N eliminates the steric clash of pyrrole and the exocyclic NH_2 of guanine. The presence of an additional hydrogen-bond acceptor on thymine residues stimulated the synthesis of the *N*-methyl-3-hydroxypyrrole (Hp) monomer, which contains an additional hydrogen-bond donor. Also, in this case, the complementary molecular shape between the cleft imposed by the thymine-O2 and the adenine-C2 and the bumpy –OH are important.

The selective binding to T,A over A,T base pairs (and, similarly, G,C over C,G) originates from the *antiparallel* binding of *two* polyamide strands in the minor groove of DNA. A key NMR spectroscopy study confirmed that an ImPyPy polyamide bound

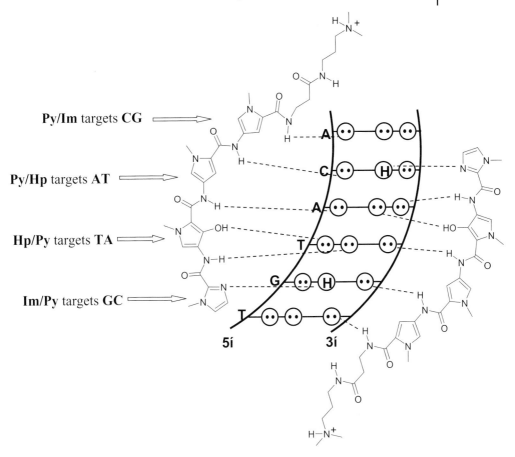

Figure 1.3 Pairing rules for dsDNA recognition by polyamides.
Py: pyrrole, Im: imidazole, Hp: hydroxypyrrole.

antiparallel in a 2:1 fashion to a 5'-WGWCW-3' sequence (W = A or T) with the polyamide oriented N→C with respect to the 5'→3' direction of the adjacent DNA strand.

Recently, the repertoire of the heterocycles used (Py, Im, and Hp) has been expanded to novel structures based on pyrazole, thiophene, and furan, to increase binding specificity and stability (the Hp monomer has limited stability in the presence of free acid or radicals) [8]. In addition, benzimidazole-based monomers (Ip and Hz) were incorporated in polyamides as alternatives for the dimeric subunits PyIm and PyHp, respectively [9, 10]. DNase I footprinting revealed functionally similar behavior with regard to the parent compounds containing exclusively Py, Im, and Hp monomers. An important advantage is the chemical robustness of the benzimidazole monomer Hz relative to Hp.

1.2.2.2 Binding Affinity and Selectivity

The ternary complex composed of two three-ring structures, such as distamycin, and DNA is rather modest, for entropic reasons and because of the low number of hydrogen bonds involved. In an important step forward towards artificial DNA binders that can effectively compete with DNA-binding proteins, the carboxyl and amino termini of two polyamide chains were covalently connected via a γ-aminobutyric acid linker (Fig. 1.4a) [11]. A so-called hairpin polyamide composed of eight heterocycles was shown to bind to the complementary six-base-pair DNA sequence with an affinity constant of the order of 10^{10} M^{-1}. A single base-pair mismatch site induced a 10–100-fold drop in affinity. Importantly, the N→C orientation with respect to the 5'→3' direction of DNA is generally retained for these compounds. An additional tenfold increase in affinity was observed for a cyclic polyamide in which the two strands were covalently connected at both termini (Fig. 1.4b) [12]. The γ-turn has a preference for an A,T over a G,C base pair, presumably because of a steric clash between the aliphatic turn and the exocyclic amine of guanine. New polyamide structures that are covalently bridged via the heterocycle nitrogen atoms, either at the center (H-pin, Fig. 1.4c) or terminus (U-pin, Fig. 1.4d) have recently been prepared [13, 14]. The U-pins resulted in a loss in affinity, because of the removal of two hydrogen bond donors, but were insensitive to the base pair adjacent to the turn. Cleverly, the H-pin polyamides were synthesized on a solid support using the Ru-catalyzed alkene metathesis reaction to connect the different polyamides. This approach enabled the rapid synthesis and screening of a series of polyamides with alkyl bridges differing in size ($(CH_2)_n$, with n ranging from 4 to 8); the optimum affinity and specificity was obtained for $n = 6$.

In the gigabase-sized DNA database it is desirable to address sequences of 10–16 base pairs, because these occur much less frequently. Increasing the number of heterocycles in polyamides increases the sequence size that can be targeted, but only up to a certain limit. Studies revealed that the binding affinity is maximized at a contiguous ring number of 5. For longer systems affinity drops because the different curvatures of polyamides and B-DNA starts to give energetically strongly unfavorable interactions. These problems can be partially overcome by replacing one (or more) of the pyrrole units by a more flexible β-alanine unit. In this way polyamides have been prepared that bind sequences as long as 11 base pairs with subnanomolar affinities [15]. Alternatively, two hairpin polyamides have been covalently connected either turn-to-turn or turn-to-tail and were shown to bind ten-base-pair sequences with impressive affinities in the order of 10^{12} M^{-1} [16, 17]. It should be noted, however, that these high affinities come with rather low selectivity.

The potential of these molecules in controlling gene expression is extremely important; here are examples that illustrate this point. Because excellent reviews have appeared that cover in great detail all results obtained, we will limit ourselves to recent examples that illustrate well the different concepts.

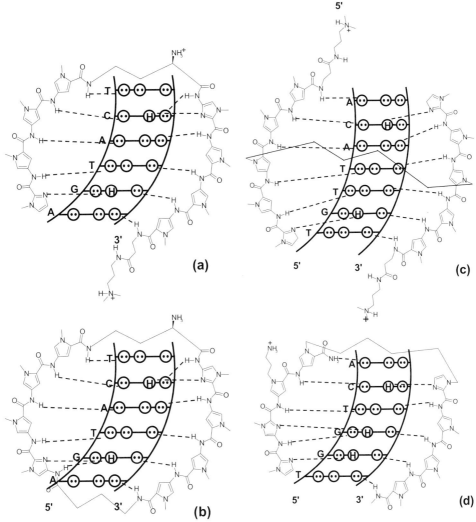

Figure 1.4 Strategies in polyamide design: (a) hairpin ($K_a \approx 10^{11}$ M^{-1});
(b) cycle ($K_a \approx 10^{11}$ M^{-1}); (c) H-pin ($K_a \approx 10^{10}$ M^{-1}); (d) U-pin
($K_a \approx 10^9$ M^{-1}).

1.2.2.3 DNA Detection

The ability to detect double stranded DNA sequences, *and* single base-pair mismatches, is an extremely useful tool in the field of genetics. Most methods involve hybridization of single-stranded DNA by a complementary oligonucleotide probe, which carries a signaling moiety. These techniques, however, require denaturation of DNA. On the other hand, double-stranded DNA can be detected by dyes such as ethidium bromide and thiazole orange, but binding is unspecific, making these dyes useful solely as quantitative tools for DNA detection. Dervan and coworkers prepared

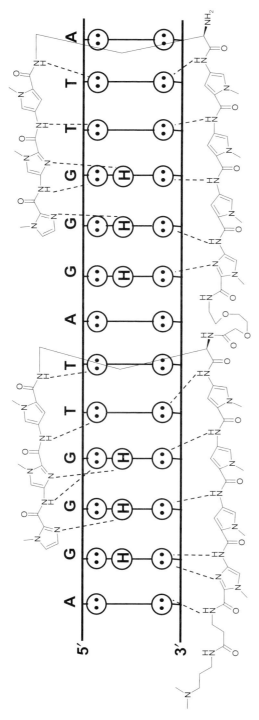

Figure 1.5 Tandem polyamide designed to target the repetitive sequence TTAGG present in insect telomers.

a series of eight-ring hairpin polyamides with tetramethyl rhodamine (TMR) attached to internal pyrrole rings and studied the fluorescence in the presence and absence of 17-mer duplex DNA [18]. In the absence of DNA, the fluorescence of the conjugates was strongly diminished compared with that of the free dye. This was hypothesized to result from the short linker separating the polyamide and the dye, which enables nonradiative decay of the excited state. The addition of increasing amounts of duplex DNA with a match sequence resulted in an increase in fluorescence until 1:1 DNA:conjugate stoichiometry was reached. Binding of the polyamide fragment to the minor groove of DNA results in forced spacing between polyamide and dye, thus diminishing any quenching effect.

In an impressive study by Laemmli and coworkers the ability of polyamide–dye conjugates to function in a genomic context was demonstrated [19]. A series of tandem polyamides was synthesized that interact specifically with two consecutive insect-type telomeric repeat sequences (TTAGG) (Fig. 1.5). The dissociation constant for the best polyamide was 0.5 nM, as determined by DNase I footprinting. Epifluorescence microscopy studies using Texas Red-conjugated analogs of these polyamides showed a very strong staining of both insect and vertebrate telomeres of chromosomes and nuclei. Convincingly, the telomere-specific polyamide signals of HeLa chromosomes colocalize with the immunofluorescence signals of the telomere-binding protein TRF1. Studies in live Sf9 cells seem to suggest rapid uptake of the conjugates, thus enlarging the potential of these compounds as human medicine. These results should be interpreted with caution, however, because the fluorescence studies were performed after fixation of the cells, which is known to dramatically increase the membrane permeability of cells [20].

In a related approach, Trask et al. targeted the TTCCA motif repeated in the heterochromatic regions of human chromosomes 9, Y, and 1, using polyamides tagged with fluorescein [21]. Staining of the targeted regions was similar to that with the conventional technique (FISH), which employs hybridization of fluorescent complementary sequences. In sharp contrast, however, polyamide–dye conjugates do not require denaturation of the chromosomes.

1.2.2.4 Gene Inhibition

Gene expression requires recruitment of the transcription machinery to the promoter region, after which transcription of the coding region into mRNA can start. Inhibition of this process by polyamides can occur in either the promoter or coding region of a gene. The latter is more difficult to achieve, because any molecule noncovalently bound to the double helix will be expelled by RNA polymerases during the transcription of DNA. To address this issue polyamides have been tagged with alkylating agents such as chlorambucil and *seco*-CBI [22]. Indeed it was observed that alkylation occurs specifically at base pairs flanking the binding site of the polyamide. Whether this strategy enables effective inhibition of RNA polymerases has not yet been reported.

Most attention has been paid toward polyamides that act in the promoter region of a gene as competitors for the binding of transcription factors to DNA. Inhibition by polyamides can occur for a variety of reasons. A minor-groove-binding protein can be inhibited by a minor-groove polyamide because of steric hindrance. Binding of a protein with major-groove/minor-groove contacts can be similarly inhibited when the polyamide is crucially located in the minor groove. Alternatively, polyamides can also function as allosteric effectors that rigidify the shape of B-DNA, thus competing with a major-groove binding protein that requires helical distortion. Examples of protein–DNA complexes that have been inhibited by polyamides are TBP, LEF-1, Ets-1, and Zif268. In a key study, the viral HIV-1 gene was targeted (Fig. 1.6) [23]. The HIV-1 enhancer/promoter region contains binding sites for multiple transcription factors, among them Ets-1, TBP, and LEF-1. Two different polyamides were designed to target DNA sequences immediately adjacent to the binding sites of these transcription factors. Cell-free assays showed that these ligands specifically inhibited binding of the transcription factors to DNA and consequently repressed HIV-1 transcription. In isolated human peripheral blood cells, incubation with a combination of these two polyamides resulted in 99% inhibition of viral replication, with no obvious decrease in cell viability. RNase protection assays indicated that the transcript levels of some other genes were not affected, suggesting that the polyamides indeed affect transcription directly.

However, despite the success in inhibiting binding of a large variety of proteins to DNA, problems remain with the class of major-groove-binding proteins that are not affected by the presence of ligands in the minor groove. Recent studies have been aimed at a generic solution that would inhibit binding of any sort of transcription factor [24]. Very promisingly, it was observed that attachment of an acridine intercalator to a polyamide locally extended and unwound the double helix and thus acted as an allosteric inhibitor for the major-groove binding of the GCN4 bZip protein.

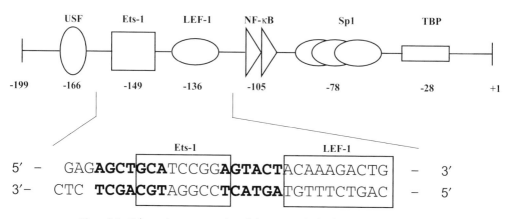

Figure 1.6 Schematic representation of the enhancer/promoter-region of the HIV-1 gene, with indications of the binding sites of the various transcription factors. In the enlargement, the binding sites of Ets-1 and LEF-1 are shown together with the binding sites of the polyamides (bold).

1.2.2.5 Gene Activation

Gene activation can occur either by inhibiting the binding of a repressor protein or via recruitment of the transcription machinery. The first method is conceptually identical to the examples given in the previous section. For instance, polyamides have been successfully applied as an upregulator for transcription of the human cytomegalovirus MIEP [25]. The second method requires an entirely different role of the polyamide. Eukaryotic transcription factors are minimally composed of a DNA-binding and an activation domain. Most activator proteins also contain a dimerization element. Generally, gene transcription starts with binding of a transcription factor to the promoter-region, which induces recruitment of a series of other transcription factors to nearby promoter sites, and finally transcription is initiated. Dervan and coworkers showed that it is possible to replace an activator protein by an artificial minimum system composed only of short oligopeptides (Fig. 1.7) [26, 27]. In the mimic of the Gal4 yeast activator the DNA-binding domain was replaced with an eight-ring hairpin polyamide and the activation domain with VP2, a 16-amino-acid residue oligopeptide derived from the viral activator VP16. The two modules were linked with an eight-atom spacer giving a polyamide–peptide conjugate with a size not exceeding 3.2 kDa. Cell-free activation assays revealed an upregulation of almost 15 fold at concentrations that caused full occupancy of the binding sites. In a recent study, the influence of the linker on activation activity was investigated by examining a series of polyamide–peptide conjugates linked via rigid oligoproline sequences varying in size between 18 and 45 Å [28]. Optimum activity was observed for a Pro_{12} linker, about 36 Å in length.

The Hox proteins belong to a family of transcriptional regulators that bear the "home domain" – a trihelical DNA-binding domain that is conserved across vast evo-

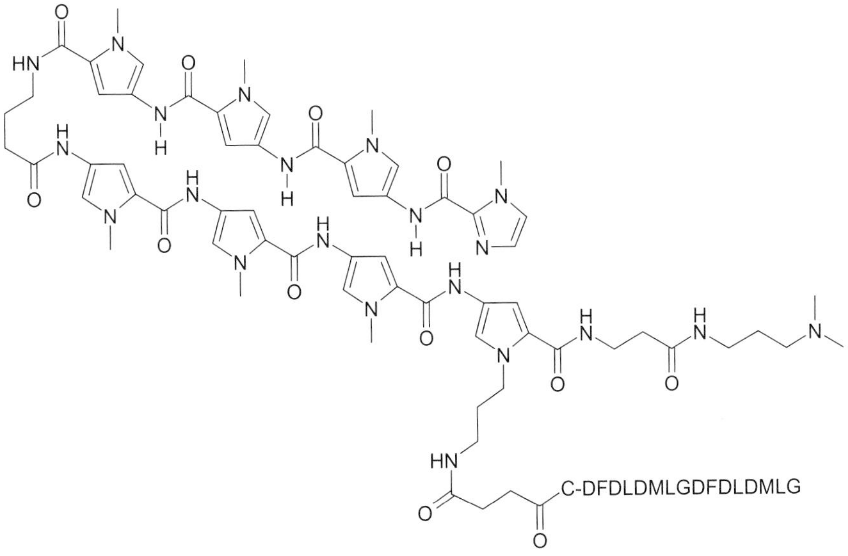

Figure 1.7 A polyamide-based artificial transcription factor for Gal4.

lutionary distances. Generally, Hox proteins by themselves bind DNA with low affinity and selectivity, but recently it has been suggested that heterodimer formation with the TALE class of homeoprotein causes high affinity and selectivity. Interaction between a *Drosophila* Hox (Ubx) and TALE (Exd) protein occurs via a short YPWM peptide. This conserved peptide is a feature of all Hox proteins. A polyamide was conjugated to this small oligopeptide to serve as a Hox-mimic [29]. Electrophoretic mobility shift assays revealed positive cooperative interactions between Exd and the conjugate in the formation of a ternary complex with DNA. It was, in fact, shown that the artificial activator was more effective than its naturally occurring equivalent, clearly illustrating the strong potential of this class of conjugates.

1.2.2.6 Future Perspective

The examples discussed in this section demonstrate the enormous achievements made in using polyamides, and their large potential as medicine for genetic diseases. However, some important hurdles still have to be passed to reach this ultimate goal. The main problem, as for many man-made synthetic structures, is the problem of delivery of the compounds to where they are needed – in the cellular nuclei in an organism. Evolution has decided, and for good reasons, that cells should not be a social room in which everybody can enter and leave as desired. Most polyamide studies have been performed on cell-free assays, clearly showing that the principles work. Also some encouraging results have been obtained in a cellular context, most notably inhibition of HIV-1 expression in human blood lymphocytes and also the induction of specific gain- and loss-of-function phenotypes in *Drosophila* embryos [30]. Recent studies, however, have revealed that polyamide uptake is largely cell-dependent and that in many cell lines polyamides are excluded from the nucleus [20, 31]. The next research phase for these compounds will reveal whether a generic solution to the cellular uptake problem is feasible, or if each target will require different chemical modifications of the polyamides.

1.2.3
Peptide Nucleic Acids

The second large class of (pseudo)peptides renowned for their ability to interact specifically with nucleic acids (both RNA and DNA) are the peptide nucleic acids (PNA). This research area also has been covered extensively in reviews [32–35] and we will therefore limit ourselves to illustrative and, where possible, recent examples.

1.2.3.1 Chemistry and Interaction with DNA/RNA

PNA is a DNA structural mimic in which the DNA backbone is replaced by a (pseudo)peptide, to which the nucleobases are connected by methylenecarbonyl linkages (Fig. 1.8) [36]. The structural simplicity and high accessibility of PNA has stimulated many scientists to study its properties in DNA/RNA recognition and to devise alternative structures [37, 38]. The strong binding properties of PNA to complementary ss DNA and RNA strands are best illustrated by the much higher melting points of the PNA–DNA and PNA–RNA dimers compared with the corresponding ds DNA and

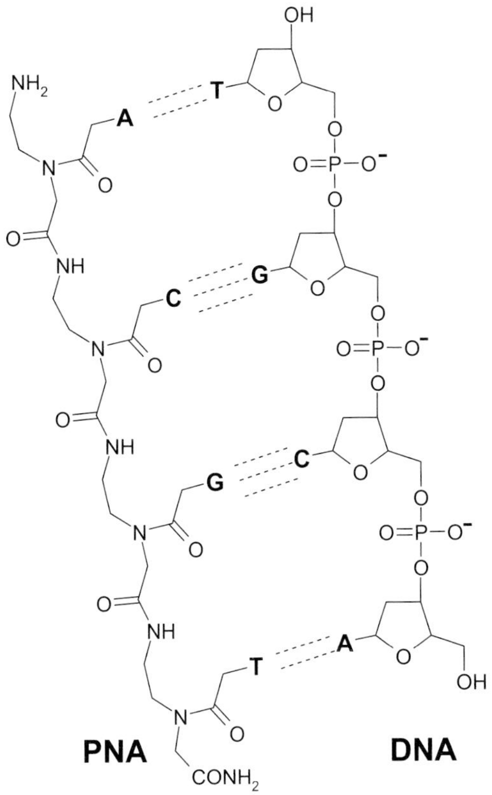

Figure 1.8. Chemical structures of PNA and DNA.

DNA–RNA duplexes (69.5 and 72.3 compared with 53.3 and 50.6 °C, respectively) [39]. In addition, the PNA–DNA dimer proved much less tolerant of single base pair mismatches than the native DNA dimer. The increased stability of the PNA-containing dimers is primarily ascribed to the lack of repulsive electrostatic interactions between the two strands.

Originally designed in the early 1990s to serve as the third strand in triple helix DNA, the original *N*-(2-aminoethylglycine) PNA turned out to behave differently than expected. Addition of a 10mer homothymine PNA to complementary ds DNA resulted in helix invasion rather than triple-helix formation (Fig. 1.9) [36, 40]. In the resulting complex one PNA strand binds via Watson–Crick base-pairing antiparallel to the complementary DNA strands. A second PNA binds in a parallel fashion via Hoogsteen base-pairing. Triplex invasion is limited to homopurine tracks, similar to the behavior observed for triple-helix-forming oligonucleotides. Interestingly, the PNA$_2$–DNA complexes are kinetically very stable with half-lives in the order of hours [41]. The stability is highly dependent on the ionic strength of the solution, however, with even physiologically relevant levels having a detrimental effect on binding. On the other hand, binding is greatly facilitated in cases where DNA is (tran-

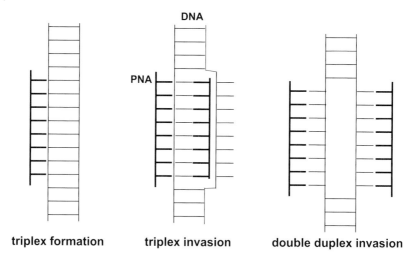

triplex formation   **triplex invasion**   **double duplex invasion**

Figure 1.9   Different modes of PNA-interaction with dsDNA.
PNA strands are depicted in bold.

siently) unwound, such as negative DNA supercoiling or by a passing RNA polymerase [42, 43]. Other factors beneficial for complex stability are a covalent connection of the two PNA strands and replacement of the cytosines in the Hoogsteen strand
for pseudoisocytosines, which do not require a low pH for protonation of the N3 position. Alternatively, it was recently shown that conjugation of a DNA-intercalator
such as 9-aminoacridine to a PNA significantly increases binding affinity and enables helix invasion even at physiologically relevant ionic strength [44].

Also, sequences other than homopurine tracks can be targeted when pseudo-com-
plementary PNA are used. Pseudo-complementarity means that the AT nucleobases
in the PNA strands are replaced by a 2,6-diaminopurine/2-thiouracil pair [45]. This
artificial base-pair is very unstable, because of steric hindrance. Ds DNA recognition
in this case occurs via so-called double-duplex invasion, in which each PNA binds to
the complementary DNA strand, thus relieving the steric repulsion present in the
PNA duplex (Fig. 1.9).

Other attractive features of PNA that increase its potential as a therapeutic agent include a high chemical and enzymatic stability and low toxicity. A limiting factor, on the
other hand, is the poor cellular (nuclear) delivery of this class of compounds [46]. These
issues will be addressed in more detail in the selective examples discussed below.

1.2.3.2   PNA as a Regulator of Gene Expression
In-vitro studies showed that formation of the triple helix complex $(PNA)_2$–DNA can
inhibit gene transcription by RNA polymerase [47]. Using an *in-vitro* replication run-
off assay under physiological conditions, 14-mer PNA inhibited replication of mitochondrial DNA mutant templates by more than 80%, whereas no inhibition of wild-
type template replication was observed. From a medicinal perspective this is an important observation, because numerous genetic diseases result from single nu-

cleotide polymorphism (SNP). Similarly, binding of either a 10-mer homothymidine PNA or a 15-mer PNA with mixed sequence to a plasmid (pBSA10) containing a complementary target sequence caused 90 to 100 % site-specific termination of pol II transcription elongation [48].

Interestingly, transcription activation using PNA has also been achieved via two conceptually different approaches. Triplex invasion causes the expelled DNA strand to form a stable D-loop structure at the DNA-binding site, which is known to induce transcription. A series of C,T-PNA with different lengths (8mer – 20mer) was designed to bind to a 20 base-pair DNA homopurine sequence cloned into two promoter reporter vectors – pGL3-Basic, which carries a promoterless luciferase gene, and pEGFP-1, which carries a promoterless green fluorescent protein (GFP) gene [49]. In this setup, fluorescence readout is a direct reporter of the extent of transcription activation. Gel-mobility shift assays showed the strongest binding for the 16mer PNA, even at 1×10^{-7} M concentrations. Transcription activation was performed both in HeLa nuclear extracts and in human NF cells. Transfection of the cells in the *in-vivo* experiment was performed by preloading the plasmid with PNA and subsequent transfection using cationic liposomes. Both experiments resulted in high expression levels of the fluorescent protein, with an optimum for the 16- and 18-mer PNA.

The second approach toward gene activation is conceptually identical to the strategy employed with the minor-groove-binding polyamides already discussed [50]. A synthetic activator was formed by connecting a twenty residue Gal80-binding peptide sequence to a bis-PNA, i.e. two homopyrimidine PNA "clamped" via a poly(ethylene glycol) linker. The ability of this artificial activator to recruit Gal80 was assessed by binding of the PNA–DNA complex to Gal80 attached to agarose beads. Large amounts of the complex were retained for the DNA duplex containing five potential PNA binding sites. DNA retention was practically absent in a series of control experiments. The results of an *in-vitro* study on HeLa nuclear extracts were recently reported [51]. It was shown that promoter-targeted PNA alone acts as a strong inhibitor of basal transcription, for reasons already discussed, but that conjugation of the Gal80-binding peptide reactivates transcription.

Finally, in a very exciting and controversial study Richelson and coworkers have claimed that antigene and antisense PNA can pass the blood–brain barrier and can bind *in-vivo* to the neurotensin receptor (NTR1) in rats [52, 53]. Specifically, intraperitoneal injection of PNA inhibited the hypothermic and antinociceptive activities of neurotensin microinjected in the brain. Controversy has arisen because of the lack of evidence that the unmodified PNA used can indeed enter cells, not to mention nuclei. A second, more fundamental, problem lies in the design of the PNA used – all current information known about this class of compound suggests that it is unlikely that short hetero purine/pyrimidine PNA sequences can effectively bind to their complementary DNA strands.

1.2.3.3 Antisense Properties of PNA
Principally, oligodeoxynucleotides (ODN) and analogs can inhibit translation of mRNA in two ways. The first is activation of ribonuclease H (RNase H) toward cleaving of the mRNA strand in the ODN–RNA complex. Although this property has been

observed for several ODN analogs, for example phosphorothioates, PNA–RNA complexes do not seem to be substrates for RNase H. It has, however, been shown that antisense properties of PNA can arise from the second mode, in which either formation of the initial ribosomal translation machinery is inhibited or ribosome elongation in the coding region of mRNA is arrested. The efficiency depends on the type of complex formed [54]. Triplex-forming 10mer PNA were effective both when the AUG start codon was targeted, but also when bound to a downstream sequence in the coding region [55]. Duplex-forming PNA, however, seems effective in inhibiting formation of the translation machinery only; except in some special cases it is unable to stop translation once started. This difference is ascribed to the higher stability of the $(PNA)_2$/RNA triplex and its tighter binding. Presumably, the "looser" binding of the duplex at the termini makes it easier for the elongating ribosome to expel it. Similar behavior has also been observed for minor-groove-binding polyamides. Also processes involving RNA, other than translation, have seemed to be sensitive to interactions with PNA. Examples include reverse transcription, telomerase activity, and RNA splicing.

Although most studies have been performed under cell-free *in-vitro* conditions, results *in vivo* have also been obtained. Most notably, Kole and coworkers reported on the antisense activity of PNA in a transgenic mouse containing the gene EGFP-654 encoding for the enhanced green fluorescent protein (EGFP) [56]. This gene was, however, interrupt by an aberrantly spliced mutated intron of the human β-globin gene. Consequently, in this model EGFP is only expressed in tissues in which an antisense oligomer (such as PNA) has restored the correct splicing. PNA was administered for one to four days via intraperitoneal injection, after which various organs were examined for the presence of EGFP. High activity was observed in the kidney, liver, and small intestine. Importantly, PNA activity was only observed for PNA that had four lysines connected to the C-terminus. A PNA containing only one lysine residue was completely inactive.

1.2.3.4 Future Perspective

Minor-groove binding polyamides and peptide nucleic acids are impressive examples of the progress made starting from a chemical design towards biologically active compounds. *In-vitro* studies have revealed the chemical robustness (and limitations) of these compounds as antigene and antisense agents, and it is striking that both classes are now more or less at the same stage of development, with the hurdle of cellular uptake preventing a wide applicability *in vivo*. In both instances, isolated examples have illustrated their potential *in vivo*, but a general solution has not yet been found. It should be remembered, however, that development of these classes of compound was started no more than approximately 15 years ago. The impressive progress made in this relatively short time and their wide use throughout the (bio)chemical community should evoke optimism for their future use as gene medicine.

1.2.4
Protein Recognition by (Pseudo)peptides

DNA and RNA comprise a relatively small set of components and, in most cases, the three dimensional (sub)structure of the targeted site is well-known or can be easily deduced. Consequently, as we have seen in the previous section, highly specific artificial ligands that bind practically any desirable sequence with high affinity can be prepared by following a rather limited set of chemical rules. For proteins this is entirely different. The large number of amino acid residues, placed in a random order, forming a fascinating wealth of three dimensional structures with practically all biochemical functions possible, renders a rational approach to the design of artificial ligands an enormous challenge. Some emerging approaches will be discussed next. In accordance with the theme of this chapter, we will limit ourselves to a discussion of small (pseudo)peptides and, more specifically, their ability to recognize protein surfaces. This choice is motivated by the strong current interest in protein–protein interactions as part of the emerging functional proteomic era. Also, this avoids treatment of small molecular inhibitors of protein activity, which could easily fill a chapter on its own. The same is true for the sequence-selective recognition of short peptide fragments and isolated secondary structure elements, such as α-helices and β-sheets, using artificial – not necessarily peptidic – compounds.

Many proteins are active only when part of a larger complex with other proteins. Small molecules that either inhibit or enhance protein assembly are important targets for pharmaceutical purposes, because they hold promise for the treatment of human diseases [57–59]. Complicating matters are not only the enormous structural diversity already mentioned, but also the large surface area involved in dimerization, which often exceeds 1100 Å^2. Importantly, however, it was found that the major contribution to the free energy of binding results from so-called "hot spots" on the protein surfaces about 600 Å^2 in size [60, 61]. Mutation of the amino acid residues that constitute these hot spots significantly reduce the binding affinity. The hot-spots are generally located in the center of the dimerization surface. Current views are that in this way a watertight seal can be generated around the energetically favorable interactions.

1.2.4.1 Dimerization Inhibitors of HIV-1 Protease
HIV-1 protease is a dimeric enzyme whose active site is composed of residues from both subunits [62]. Because the monomeric forms are completely inactive, agents which can inhibit dimerization are under intense investigation. In addition, whereas the HIV-1 virus continuously mutates and renders inhibitors ineffective, it seems that the dimerization domains are highly conserved. The dimerization interface of HIV-1 protease is a four-stranded, antiparallel β-sheet composed of the interdigitated N- and C-termini of the monomers. Chmielewski and coworkers prepared a dimerization inhibitor in which the peptide sequence derived from the N-terminus was coupled, via a linker, to a peptide sequence derived from the C-terminus (Fig. 1.10) [63]. HIV-1 protease was inhibited with an IC$_{50}$ of 350 nM. In subsequent studies the influence of variations in both the peptide sequences and the linker were examined. Mutations in the peptide sequences led to a new inhibitor with a slightly

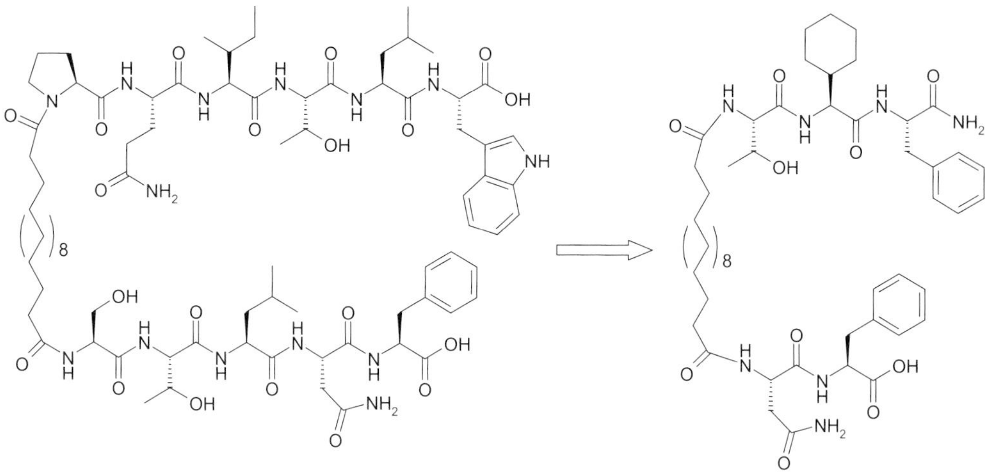

IC$_{50}$ = 350 nM

IC$_{50}$ = 680 nM

Figure 1.10 HIV-1 protease dimerization inhibitors.

higher IC$_{50}$ of 680 nM, but with a significantly lower molecular weight (933 g mol^{-1} compared with 1586 g mol^{-1}) [64]. Interestingly, in this sequence a nonnatural amino acid with an appending cyclohexyl group was used. Linker studies revealed that rigidification of the spacer using naphthyl or dibenzofuran groups resulted in reduced activity of the inhibitor, even although models suggested that these linkers should position the two peptide sequences at the required 10-Å distance.

Intrigued by these studies, we reasoned that the reversible control of the distance between the two tethered peptides in this kind of inhibitor would lead to modulation of its activity. For this purpose we connected the two peptides initially employed by Chmielewski et al. to the Tren template via a rigid aromatic spacer [65]. Molecular modeling predicted that the distance between the two N-termini would be close to 10 Å (presumably required for optimum activity) only in the presence of this spacer and when a Zn^{II} metal ion was complexed by the Tren template. In the design of the system the third arm, necessary to impart rigidity to the Zn^{II} complex but not used for enzyme recognition, was functionalized with a naphthalene unit to enable fluorescent readout for the protonation and/or Zn^{II} complexation of the Tren template. We were able to verify our hypothesis using a fluorogenic substrate assay, which revealed enhancement of the inhibition of HIV-1 protease activity on the addition of Zn^{II}. Control experiments revealed that addition of EDTA (ethylenediaminetetraacetate) reversed this effect, because of depletion of the Zn^{II} ion from the Tren template.

1.2.4.2 Inhibitors of p53/Hdm2 Interactions
The p53 tumor suppressor is a multifunctional protein that regulates cell proliferation by induction of growth arrest or apoptosis in response to DNA damage and/or stress stimuli. The p53 protein is inactivated in most human tumors, which in more than 30% of soft tissue sarcomas can be linked to amplified expression of the gene

coding for the oncoprotein human double minute 2 (hdm2). Hdm2 affects p53 in two ways: first, it inhibits p53 expression by binding to the transcriptional activation domain of the p53 gene and, second, a protein complex involving hdm2 causes nuclear export of p53 and subsequent cytoplasmic degradation. Small molecules that inhibit formation of this complex might help restoring p53 levels in cancer cells. The X-ray crystal structure of the N-terminal domain of hdm2 bound to a 15-mer p53-derived peptide revealed a deep hydrophobic cleft to which the peptide binds as an amphipathic α-helix [66]. On the basis of this knowledge a dodecapeptide was designed that inhibited hdm2/p53 complex formation with an IC_{50} of around 8.7 μM, which served as a starting point for further optimization (Fig. 1.11) [67]. A jump forward was achieved when artificial amino acids were used. Among these were the α,α-disubstituted amino acids Aib (α-aminoisobutyric acid), which increased the propensity of the peptide sequence to adopt an helical structure and Pmp (phosphonomethylphenylalanine), which stabilized a salt bridge with an Hdm2 lysine residue. The resulting optimized eight-amino-acid pseudo-peptide had an IC_{50} value of 5 nM and in a cellular assay caused p53-dependent cell-cycle arrest in an Hdm2-overexpressing tumor cell line.

peptide sequence	IC_{50} (nM)
Ac-Gln-Glu-Thr-Phe-Ser-Asp-Leu-Trp-Lys-Leu-Leu-Pro-NH$_2$	8673 ± 164
Ac-Phe-Met-Aib-Tyr-Trp-Glu-Ac$_3$c-Leu-NH$_2$	8949 ± 588
Ac-Phe-Met-Aib-Pmp-Trp-Glu-Ac$_3$c-Leu-NH$_2$	314 ± 88
Ac-Phe-Met-Aib-Pmp-FTrp-Glu-Ac$_3$c-Leu-NH$_2$	14 ± 1
Ac-Phe-Met-Aib-Pmp-ClTrp-Glu-Ac$_3$c-Leu-NH$_2$	5 ± 1

Figure 1.11 The use of nonnatural amino acids to optimize an oligopeptide that inhibits the binding of human wild-type p53 to GST-hdm2. Abbreviations for the nonnatural amino acids: Aib, α-aminoisobutyric acid; Pmp, phosphonomethylphenylalanine; ClTrp, 6-chlorotryptophan; FTrp, 6-fluorotryptophan; Ac$_3$c, 1-aminocyclopropanecarboxylic acid.

1.2.4.3 Artificial Antibody Mimics for Protein Recognition

Inspired by the impressive achievement of the immune system to develop antibodies that bind nonendogenous protein surfaces with high selectivity and affinity, Hamilton and coworkers took up the challenge to develop an equivalent synthetic system. Structural diversity in antibodies arises from the presence of six hypervariable loops in the fragment antigen binding (FAB) region. Four of these loops generally take up a hairpin conformation and the remaining two form more extended loops. As a mimic, calix[4]arene was used as a rigid cyclic scaffold and the upper rim was functionalized with cyclic peptides containing an artificial 3-aminomethylbenzoyl amino acid to facilitate coupling to the platform (Fig. 1.12) [68]. Cytochrome c was set as an initial target, because of its well-characterized structure and its positively charged sur-

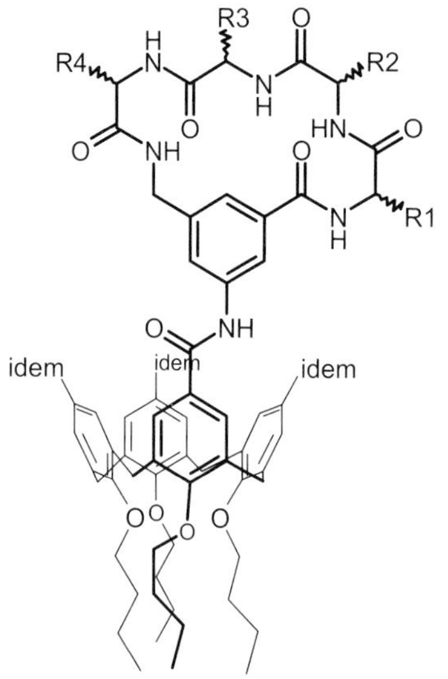

Figure 1.12 Design of an artificial receptor for chymotrypsin surface recognition.

face originating from the presence of several lysine and arginine residues. To introduce complementary charge in the antibody mimic the negatively charged sequence GlyAspGlyAsp was used in the loops. Protein binding was confirmed by gel-permeation chromatography, ^{1}H NMR spectroscopy, and kinetic assays on the ascorbate reduction of Fe^{III}-cyt c. A follow-up study revealed that this compound effectively competes with cytochrome *c* peroxidase, a natural high affinity partner of cytochrome *c*, forming a 1:1 complex with a binding affinity in the order of 10^8 M^{-1} [69]. It was recently shown that the same receptor also binds to the protein surface of chymotrypsin (ChT) and blocks the protein–protein interaction of ChT with a variety of proteinaceous inhibitors [70].

1.3
Synthetic (Pseudo)peptide-based Supermolecules: From Structure to Function

Although amino acid-based superstructures are almost the rule in the biological world, the number of synthetic supermolecules based on oligopeptides is relatively small. The main obstacles are the difficulty of predicting and controlling the conformation of amino acid sequences and, at the same time, introducing (un)natural functional groups in appropriate positions where they can perform recognition or catalytic tasks. Early examples in this field were extensively reviewed by Voyer a few years ago [71]. More recently, scientists have started to focus on the development of new foldamers [72] by introducing key structural elements in an artificial sequence that imposes conformational restriction on the oligopeptide. In this context, a foldamer is defined as an oligomer with a characteristic tendency to fold into a specific structure in solution, which is stabilized by noncovalent interactions between nonadjacent subunits. In recent years large progress has been made in the *de-novo* design of short peptide sequences (<100 residues) that adopt a well-defined conformation. Several reviews have recently appeared that summarize the achievements in this field [73]. Recent accomplishments include metal-peptide models which, with remarkable success, control the placement of metals within designed secondary structures [74, 75].

1.3.1
Catalytic (Pseudo)peptides

One of the earliest examples of a short (14 amino acids), catalytically active sequence was reported by Benner et al. more than ten years ago [76]. The oligopeptide H_2N–Leu–Ala–Lys–Leu–Leu–Lys–Ala–Leu–Ala–Lys–Leu–Leu–Lys–Lys–$CONH_2$ is not highly organized at infinite dilution in aqueous buffer (pH 7), having an α-helix content of only 18%. Increasing the concentration, however, induces aggregation into a four-helix bundle characterized by a much greater content of helix. The oligopeptide catalyzes the decarboxylation of oxaloacetate via an imine intermediate with three orders of magnitude rate acceleration relative to the uncatalyzed reaction. The catalytic activity is because of two factors. Placement of the reacting amine at the terminus of the helix causes a critical decrease of the pK_a ($\Delta pK_a = 0.6$). Second, the anionic transition state is stabilized by electrostatic interactions with the cationic lysine residues present on the peptide surface. As a consequence, imine formation is no longer the rate-determining step, which is at variance with the natural enzyme oxaloacetate decarboxylase.

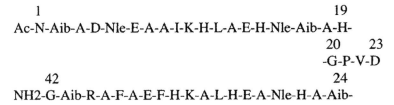

```
   1                                         19
Ac-N-Aib-A-D-Nle-E-A-A-I-K-H-L-A-E-H-Nle-Aib-A-H-
                                          20      23
                                          -G-P-V-D
      42                                    24
NH2-G-Aib-R-A-F-A-E-F-H-K-A-L-H-E-A-Nle-H-A-Aib-
```

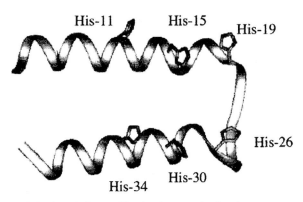

Figure 1.13 Baltzer's helix–loop–helix motif for the cleavage of carboxylate esters.

Baltzer's group, exploiting the conformational preference of designed 42 amino acid sequences for a helix–loop–helix conformation [77], has systematically modified specific residues in key positions of the oligopeptide to catalyze the hydrolysis and transesterification reactions of *p*-nitrophenyl esters (Fig. 1.13) [78]. Imidazole-functionalized peptides obtained by introducing several histidines in the sequence were able to provide substrate recognition, and acceleration exceeding three orders of magnitude compared with *N*-methylimidazole. For example, the sequence depicted in Fig. 1.13 hydrolyzes 2,4-dinitrophenyl acetate with a second-order rate constant of $0.18 \text{ M}^{-1} \text{ s}^{-1}$ at pH 3.1 compared with $9.9 \times 10^{-5} \text{ M}^{-1} \text{ s}^{-1}$ for *N*-methylimidazole.

Interestingly, it has been demonstrated that the reaction mechanism takes advantage of the cooperativity of two adjacent histidines, one acting as the nucleophile and the other as a general base. For flanking His–Lys sequences the unprotonated form of the histidine attacks the ester in the rate-determining step of the process followed by subsequent transacylation of the lysine. If several lysine residues are present, at low pH, only those that flank the His are acylated. This leads to site-selective incorporation of an acyl residue in a natural sequence [79]. Direct alkylation of lysines occurs in sequences devoid of histidines. In this case the nucleophilicity of the lysines could be controlled by site-selective pK_a depression [80]. As stated by the authors, the elucidation of the principles that control reactivity in such elementary processes is the first step toward the construction of biomimetic catalysts that accelerate reactions for which naturally occurring enzymes do not exist.

A pyridoxamine coenzyme amino acid chimera was introduced by Imperiali into designed $\beta\beta\alpha$ motif peptides (23-amino-acid sequences) to obtain a transamination catalyst (Fig. 1.14) [81]. The uncatalyzed transamination of α-keto acids to α-amino acids is a difficult reaction which is catalyzed efficiently by the coenzymes pyridoxamine phosphate and pyridoxal phosphate. A series of 18 peptides was synthesized

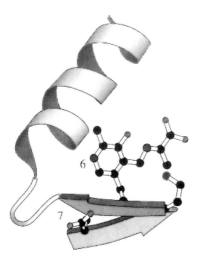

Figure 1.14 A pyridoxamine coenzyme amino acid chimera was introduced by Imperiali into designed $\beta\beta\alpha$ motif peptides (23-amino-acids sequences) to obtain a transamination catalyst.

to create different microenvironments for the introduced pyridoxamine functionality. This led, in the best case, not only to acceleration of the transamination rate but also to promising enantioselectivity in the production of L-alanine (up to 27% ee).

Barton and her group [82] reported the activity of zinc-binding peptides tethered to rhodium intercalators. In these systems the intercalator provides DNA binding affinity and the metal-binding peptide contributes the reactivity. The strategy seems to be rather general because Zn^{II}-promoted DNA cleavage was observed for two widely different tethered metallopeptides. The structural features of one of the two are such that they place two imidazole units of histidines facing each other in a α-helix conformation (Fig. 1.15). In this way they are directly involved in the binding of the metal ion. The other was modeled on the active site of the *Bam*HI endonuclease. Specifically, a short β hairpin was excised from *Bam*HI restriction endonuclease and the sequence was such that it clustered three carboxylate groups (one from aspartic acid and two from glutamic acid) for metal-ion binding. The native protein is known to bind to DNA as a dimer and cleaves at 5'-GGATCC-3' palindromic sites to give four base 5' staggered ends. Barton has also studied other redox-active metals, for example copper, to induce oxidative cleavage. The results indicate that one of the critical issues in the design of an artificial nuclease is selection of the intercalator. Not all metallointercalators studied orient a metallopeptide for the initiation of DNA hydrolysis, and it might be necessary to vary linkers and ancillary ligands to optimize activity. The work done by Barton and others have also shown how extreme caution should be taken in evaluating the mechanism of action of a metallocatalyst when the metal ion can be involved in redox chemistry. Quite often what is assumed to be an hydrolytic mechanism eventually turns out to be oxidative. This can be observed for metals like copper, cobalt and, in selected cases, nickel also.

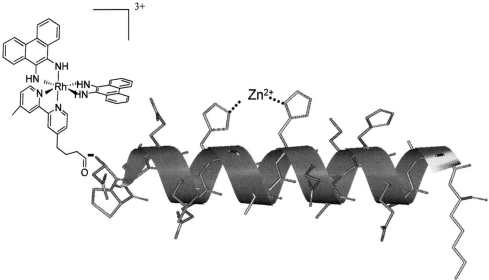

Figure 1.15 Zinc-binding peptide tethered to a rhodium intercalator reported by Barton.

An interesting approach to the engineering of functional peptides is the introduction of synthetic unnatural amino acids in the sequence. The scope can be twofold – synthetic amino acids can be used to limit the conformational freedom of the oligo or, alternatively, to introduce functional groups not available in the biological world [83]. Our contribution to this area has been the synthesis of the artificial amino acid ATANP [84] and its incorporation in a wide variety of oligopeptides. ATANP is characterized by the presence of an appended triazacyclononane which forms 1:1 complexes with metal ions such as Cu^{II} and Zn^{II} with strong affinities. Because these metals play an essential role in the hydrolysis of DNA and RNA, our aim has been to use this amino acid for the development of multinuclear transphosphorylation metallocatalysts. In collaboration with Baltzer's group 42-mer peptides analogous to that reported in Fig. 1.13 were synthesized, with the difference that these new sequences incorporated up to four copies of ATANP (Fig. 1.16) [85].

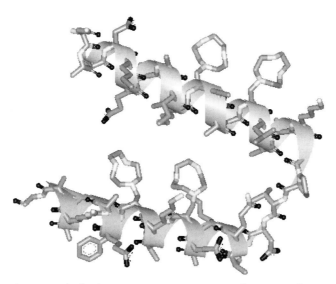

Figure 1.16 Artificial 42-mer sequence incorporating four copies of the unnatural amino acid ATANP used, as Zn^{II} complex, in the cleavage of a RNA model substrate.

These new peptides also form helix–loop–helix motifs and bind Zn^{II} ions with the triazacyclononane subunits present in the lateral arms of ATANP. It was observed that metal complexation causes a decrease in the helix content of the peptide. However, even on partial unfolding of the structure, acceleration of the cleavage of HPNP (2-hydroxypropyl-*p*-nitrophenyl phosphate, an RNA model substrate) was observed. Incorporation of ATANP in more structured peptide sequences controls the three-dimensional positioning of the metals even better and, consequently, significantly improves their catalytic activity. Thus the secondary structure of a 3$_{10}$-helix can be used to position functional groups present in the amino acids at precise reciprocal distance. In this conformation, when placed in positions i and $i + 3$, they will face each other with a separation of ca 6.2 Å, which is the pitch of the 3_{10}-helix. An example is heptapeptide Ac–Aib–L-ATANP–(Aib)$_2$–L-ATANP–(Aib)$_2$–OMe, which was shown to fold into a 3_{10}-helix in water. The two ATANP residues not only aid water solubility but can also form strong complexes with transition metals such as Zn^{II} and Cu^{II} [86]. The dinuclear Zn^{II} complex turned out to be a good catalyst for the cleavage of both an RNA model substrate (2-hydroxypropyl-*p*-nitrophenyl phosphate, HPNP) [87a] and plasmid DNA [87b]. In the latter, cooperative action between the metal centers was demonstrated by comparison of the activity of the mononuclear and dinuclear complexes. Tentatively, a mechanism is proposed which requires the formation of a supramolecular DNA–peptide complex, as shown in Fig. 1.17. The k_ψ for the cleavage process (~1×10^{-5} s^{-1}) enabled estimation of rate acceleration of approximately ten-millionfold over the uncatalyzed cleavage process.

The attachment of peptide sequences to conformationally flexible templates results in polypeptidic structures with the separate peptide chains adopting a random orientation relative to each other. Of special interest are those templates whose conformation can be rigidified by external stimulus, for example pH or the presence of a metal ion, because this can also induce organization of the peptide chains. One example of such a template is the tetraamine tris(2-aminoethyl)amine (Tren) whose three arms are aligned on complexation of a metal ion, for example Zn^{II}, in the tetradentate binding pocket. An artificial multinuclear metallonuclease subject to allosteric control by Zn^{II} ions taking advantage of this property was recently described by us. A tripodal apopeptide was synthesized by connecting three copies of the heptapeptide H–Iva–Api–Iva–ATANP–Iva–Api–Iva–NHCH$_3$ (where Iva is (*S*)-isovaline and Api is 4-amino-4-carboxypiperidine) to the Tren template. The oligopeptides contain C$^\alpha$-tetrasubstituted amino acids to induce helicity and ATANP to introduce a ligand for metal-binding. The apopeptide can bind up to four metal ions (Cu^{II} or Zn^{II}), one in the Tren subsite and three in the azacyclononane subunits [88]. The binding of metals to the Tren platform induces a change from an open to a closed conformation in which the three helical peptides are aligned in parallel, thus creating a pseudo-cavity with the azacyclonane units pointing inward (Fig. 1.18). In the transphosphorylation of phosphate esters, however, the behavior of this tripodal template was very peculiar. The tetrazinc complex catalyzed the cleavage of 2-hydroxypropyl-*p*-nitrophenyl phosphate (HPNP) whereas the free ligand was a catalyst for cleavage of an oligomeric RNA sequence with selectivity for pyrimidine bases. With HPNP, Zn^{II} was acting as a positive allosteric effector by enhancing the catalytic efficiency of the system. With the polyanionic RNA substrate, Zn^{II} was switching off the activity, thus behaving as a negative allosteric regulator. Our explanation of this different behavior is that the small cavity formed on metal–ligand coordination can accommodate HPNP but not the larger RNA oligomer. Consequently, the tetranuclear complex is a catalyst for cleavage of HPNP but not for cleavage of the oligonucleotide.

Although these examples concern the recognition and subsequent transformation of small substrates, Ghadiri [89] and Chmielewski [90] have described short peptides based on the coiled-coil folding motif [91, 92] that act as templates for the recognition of shorter sequences and, eventually, for their ligation. When the two short peptides

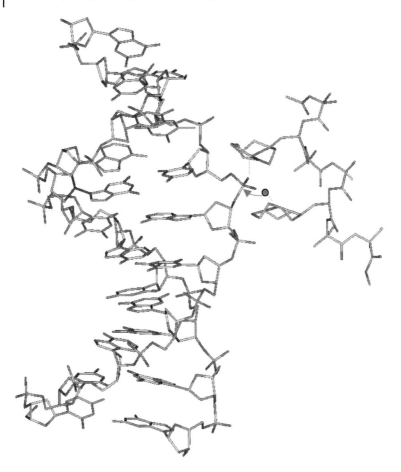

Figure 1.17 Peptide–DNA complex suggested as the active species in the hydrolytic process by Scrimin et al.

constitute each complementary pair of the templating unit, the systems are examples of self-replicating peptides [89c,d, 90]. For instance, a 33-residue polypeptide reported by Ghadiri [89] (Fig. 1.19) accelerates the ligation of a 16- and 17-amino-acid sequence using the coupling chemistry introduced by Kent et al. [93]. Initial rate accelerations were up to 4100 compared with a poorly matching system. In terms of catalytic efficiency this gives a value of $(k_{cat}/K_m)/k_{uncat}$ of 7×10^5, larger than that found for catalytic antibodies [94]. The source of the catalytic acceleration is the specific recognition of the two substrates by the templating peptide which places the reactive termini in close proximity. The recognition selectivity was further exploited for chiroselective amplification of homochiral peptide fragments from a mixture of racemic substrates. The general limitation of the approach is that it is marred by product inhibition contrary to that found with the antibody-based catalyst [94]. However, by destabilizing the coiled-coil structure Chmielewski was able to enhance the catalytic

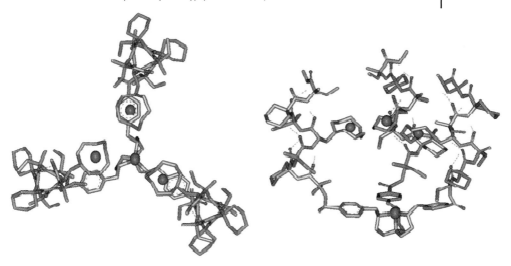

Figure 1.18 The binding of metals to the Tren platform induces a change from an open to a closed conformation in which the three helical peptides are aligned in parallel, thus creating a pseudo-cavity with the azacyclonane units pointing inward.

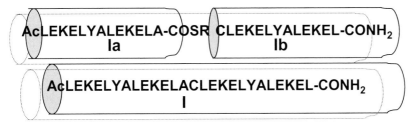

Figure 1.19 A 33-residue polypeptide reported by Ghadiri accelerates the ligation of a 16- and 17-amino acid sequence using the coupling chemistry introduced by Kent et al.

efficiency of the system [90b] and obtained a self-replicating peptide approaching exponential growth [95].

1.3.2
(Pseudo)peptides Altering Membrane Permeability

The amount of work performed on the synthesis of natural sequences able to alter membrane permeability is enormous [96] – a book would probably be needed to address this area properly. We will only provide examples of supramolecular peptide-based structures synthesized to perform this task. The cyclic peptides reported by Ghadiri et al. are excellent examples of conformationally restricted oligopeptides [97]. The peptides are composed of an even number of alternating D and L amino acids and are designed such that a flat conformation of the sequence is energetically favored.

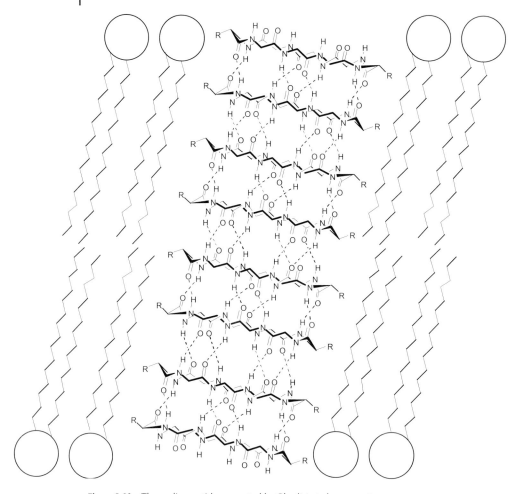

Figure 1.20 The cyclic peptides reported by Ghadiri et al. aggregate in the lipid membrane of bacteria and, by increasing its permeability, exert antibacterial activity.

In this conformation the amide functionalities of the backbone lie perpendicular to the plane of the ring structure, which consequently enables stacking of the cycles into β-sheet like tubular aggregates (Fig. 1.20). It was shown that the peptides aggregate in the lipid membrane of bacteria and, by increasing its permeability, exert antibacterial activity [98]. It was also demonstrated that they could mediate transmembrane transport of glutamic acid [99].

Crown ether-functionalized unnatural amino acids have been introduced in peptide sequences by Voyer and coworkers [100]. Incorporation of six crown-ether-functionalized amino acids in oligopeptides with a high propensity to form an α-helix, resulted in the formation of artificial ion channels [100a,b,e,f]. A helical 14-residue peptide containing four polar but uncharged, benzo-21-crown-7 side-chains aligned along one face induced significantly more membrane leakage than analogous 21-mer or 7-mer peptides [100b]

Matile and coworkers [101] have developed synthetic β-barrels that bypass folding problems with preorganizing "non-peptide" staves but maintain the functional plasticity provided by β-strands. They have shown that rigid-rod β-barrels such as that depicted in Fig. 1.21 are indeed suitable for many tasks such as molecular recognition, transformation of organic molecules, and, relevant in this context, translocation. Their design is simple – short peptide strands are attached to each arene of a *p*-octaphenyl rod. Self-assembly or programmed assembly of complementary rods is driven by stave rigidity and formation of consecutive, interdigitating, intermolecular, antiparallel β-sheets between adjacent rods (i.e., minimized entropy losses and maximized enthalpy gains). The curvature needed to obtain barrels instead of precipitating amphiphilic tapes is initiated by arene–arene torsion angles 180° in the *p*-octaphenyl scaffold and propagated by peripheral crowding with bulky N- and C-terminal amino acid residues. Suprastructural plasticity of *p*-octaphenyl β-barrels with "ideal" stability was demonstrated with regard to barrel truncation, elongation, contraction and expansion in water, and bilayer membranes.

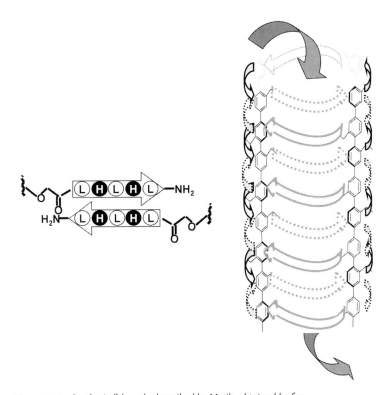

Figure 1.21 Synthetic β-barrels described by Matile obtained by formation of consecutive, interdigitating, intermolecular, antiparallel β-sheets between adjacent rods.

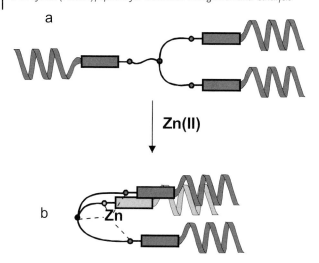

Figure 1.22 The change of conformation upon metal-ion complexation regulates the membrane activity of this tripodal polypeptide template.

We have conjugated peptide chains to the Tren template for synthesis of a system able to affect membrane permeability and subject to allosteric control. Thus three copies of peptide sequences from the peptaibol family, a class of natural antibiotic known to act on cell membranes by channel formation, were coupled to the Tren template. Experiments on leakage of trapped fluorescent dyes from unilamellar vesicles showed that a minimum of five amino acid residues per peptide chain was required for the formation of an active species. More important, it was observed that the tripodal apopeptide was far more effective than its Zn^{II} complex. A mechanism was proposed in which the Zn^{II} ion causes a change in the conformation from extended to globular (Fig. 1.22). Molecular modeling indicated that in the extended form the apopeptide can span the lipid bilayer, thus enabling pore formation via clustering. In this example, binding of Zn^{II} has an inhibitory effect on the activity of the peptide [102].

1.3.3
Nanoparticle- and Dendrimer-based Functional (Pseudo)peptides

Nanosize clusters of several metals can act as templates for self–assembly of organic molecules. For example, organic thiols passivate the surface of gold nanoclusters forming a monolayer on their surface. This prevents growth of the particles and alters their solubility, depending on the functional groups present on the protecting monolayer. Synthetic procedures are available for the synthesis of these monolayer-protected gold nanoparticles (Au-MPC) both in apolar [103] and polar solvents [104]. These materials are characterized by the properties of the organic molecules that constitute the monolayer but also by the properties of the metal cluster core. Multivalency [105], which is at the basis of the amplification of the binding events with functional nanoparticles, has been exploited for the preparation of catalytic systems in which a cluster of functional groups led to catalytic activity which was not just the sum of the individual contributions [106]. We prepared a thiol-functionalized dipeptide by N-acylation of HisPhe speculating that, in this system, the free carboxylate/carboxylic acid of phenylalanine could constitute a simple function for catalysis providing, in conjunction with the imidazole of histidine, either base/nucleophilic and/or acid catalytic contributions. The nanoparticles were prepared by exchange reaction starting from polyether-functionalized water-soluble nanoparticles [104] thus affording the Au-MPC shown in Fig. 1.23 [107]. This new system could be studied in an aqueous solution in which hydrophobic interactions can be exploited for recognition of a substrate. Indeed these nanoparticles bind the lipophilic *p*-nitrophenyl ester of Z-protected L-leucine with a binding constant, K_b, of 2×10^4 M^{-1}. Very interestingly, for the dipeptide-based catalyst there was no evidence of cooperativity between two imidazoles but, rather, the nucleophilic contribution of both

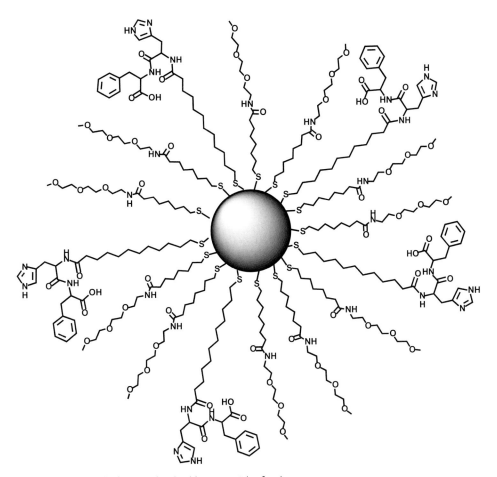

Figure 1.23 Dipeptide-functionalized gold nanoparticles for cleavage of carboxylate esters.

an imidazole and a carboxylate in the cleavage of 2,4-dinitrophenylbutanoate (DNPB). The carboxylate contribution provided more than two orders of magnitude rate acceleration relative to the monomer. Consequently, these functional Au-MPC might be considered as models of an aspartic esterase and, as the real enzyme, have maximum efficiency below pH 5.

The ultimate goal we had in mind was mimicry of an esterolytic enzyme not just by introducing minimal functional groups at the periphery of the nanoparticles but also by providing the appropriate environment (including chirality), as in the catalytic site of a protein. What we were pursuing was the self assembly of a synthetic protein by grafting short but conformationally constrained peptides on the surface of a gold nanopar-

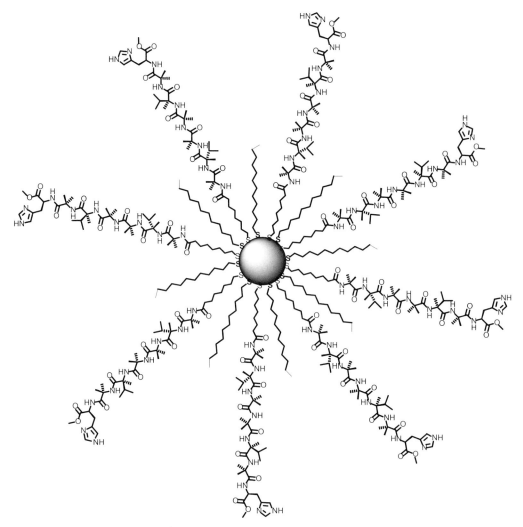

Figure 1.24 Nanoparticle covered with a monolayer comprising a 3_{10}-helical heptapeptide.

ticle. Very naively we made our first approach by using the rather lipophilic (although highly structured [108] in a helical conformation) peptide shown in Fig. 1.24 [109] assuming that the imidazole would be sufficient to ensure solubility in water. Although the exchange reaction with unfunctional MPC was successful, the resulting system proved poorly soluble not only in a aqueous solution but also in a variety of solvents of different polarity. The probable explanation we proposed was that the packing of the peptide in the monolayer was too tight, resulting in an extended network of intra-monolayer hydrogen bonds preventing solution. With these studies we learned two lessons – first, that water-soluble nanoparticles had to be used for the exchange reaction and, second, that exchange of the thiols had to be limited to avoid overcrowding in the monolayer. By applying these rules of thumb we succeeded in preparing nanoparticles covered by a monolayer comprising the dodecapeptide shown in Fig. 1.25. The se-

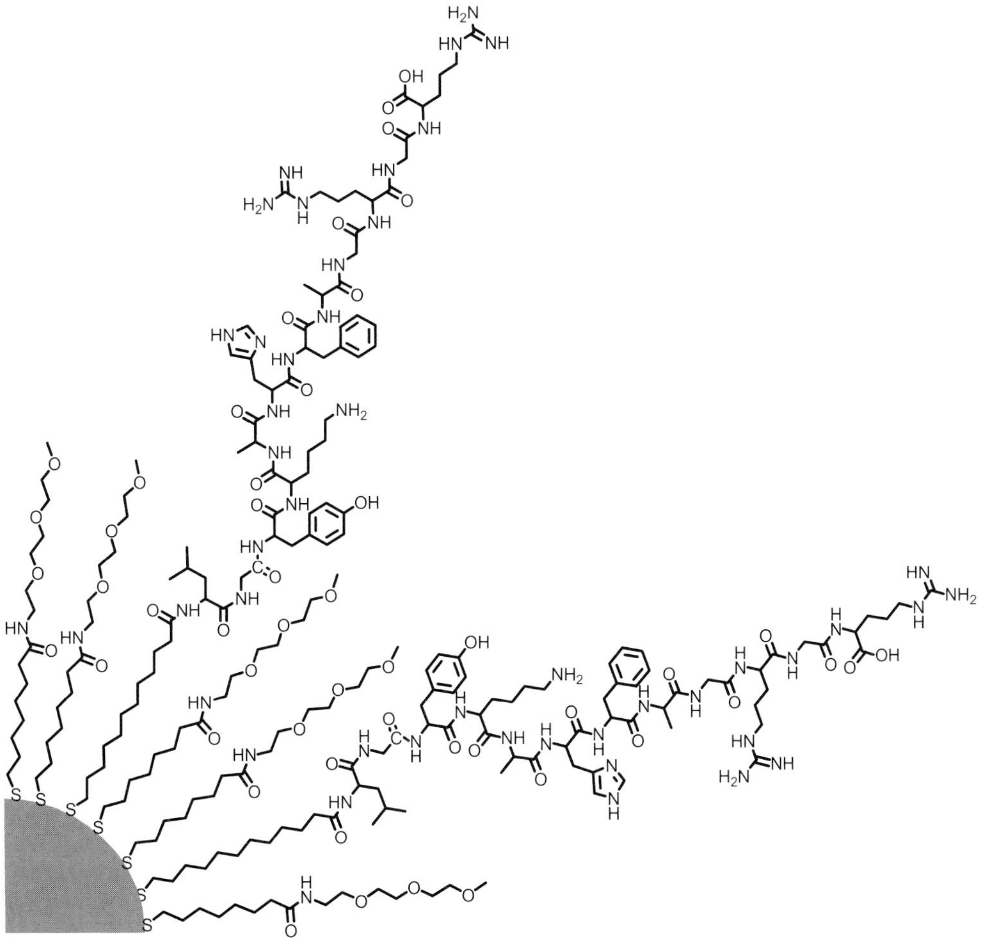

Figure 1.25 The functional groups present in this peptide sequence covering a gold nanocluster catalyze the hydrolytic cleavage of a carboxylate ester.

quence terminates with the free carboxylate of arginine and has several functional groups potentially active for catalysis of the cleavage of a carboxylate ester – one histidine, flanked by hydrophobic amino acids, one lysine, and one tyrosine. These nanoparticles were fully soluble in water in a wide pH range and had remarkable activity in the cleavage of DNPB. However, whereas the dipeptide-functionalized Au-MPC were particularly efficient at low pH, these new nanoparticles proved particularly effective at pH > 8, a clear indication that they were working preferentially with a base-catalyzed or nucleophilic mechanism involving high-pK_a species. Complete analysis of the system is still in progress but we know already that the rate acceleration at pH 10 is ca. two order of magnitude larger than that determined with the dipeptide-functionalized nanoparticles whereas at low pH the acceleration is not so different.

Peptide dendrimers were prepared by solid-phase peptide synthesis [110]. Monomeric dendrimers were first obtained by assembly of a hexapeptide sequence containing alternate standard α-amino acids with diamino acids as branching units. The monomeric dendrimers were then dimerized by disulfide-bridge formation at the core cysteine. Peptide dendrimers composed of the catalytic triad amino acids histidine, aspartate, and serine catalyzed the hydrolysis of *N*-methylquinolinium salts when the histidine residues were placed at the outermost position. The dendrimer-catalyzed hydrolysis of 7-isobutyryl-*N*-methylquinolinium followed saturation kinetics with a rate acceleration (k_{cat}/k_{uncat}) of 3350 and a second order rate constant (k_{cat}/K_M) 350-fold larger than that of the 4-methylimidazole-catalyzed reaction; this corresponds to a 40-fold rate enhancement per histidine side chain. Catalysis was attributed to the presence of histidine residues at the surface of the dendrimers.

Ghosh et al. [111] demonstrated for the first time that discrete molecules comprising leucine zippers tethered to a core dendrimer provide the necessary molecular framework for constructing both monodisperse and fibrillar supramolecular assemblies that span the nanometer to micrometer range. The ability to integrate protein/protein self-assembly principles with dendrimer architecture enabled the facile self-assembly of hybrid dendrimer–protein complexes potentially useful for protein, DNA, or RNA recognition. These novel leucine-zipper dendrimers might provide a rich platform for constructing and utilizing a new class of programmable biomaterial with helical secondary structure. Ghosh's group design entailed choosing a dendrimer scaffold that did not sterically occlude peptide assembly and a series of peptides capable of noncovalent assembly. For the dendrimer, they used the zero generation PAMAM [112] core-functionalized with maleimides, thus enabling chemoselective covalent tethering to cysteine-containing peptides. The peptides consisted of the pH-sensitive leucine zippers with hydrophobic cores and polar exteriors that can form tetramers, which were further modified with a Cys linker at the C-terminus.

1.4
Combinatorial Selection of Functional (Pseudo)peptides

One of the important advantages of a peptide sequence over other building blocks is that solid-phase synthesis and, hence, a combinatorial approach to the preparation of libraries [113] of molecules can be accomplished with ease. The challenge in these cases is to devise methods for screening of these libraries. This means the possibility of estimating the binding strength for a specific target and, depending on the final use of the systems, evaluation also of their activity as catalysts or sensing devices. In contrast with the examples described above, which have in common the rational design of the functional (pseudo)peptide, the combinatorial approach is, by definition, an (almost) totally random selection approach. For example, Miller et al. [114c,d] have described the combinatorial synthesis of a 39-member library of five-amino-acid peptides. Screening of the catalytic activity for converting *myo*-inositol into D-*myo*-inositol-1-phosphate revealed one of the members to be an extremely efficient kinase mimic (Fig. 1.26). Delivery of a phosphoryl group to one of the three hydroxyl sites of the substrate exclusively was observed with ee >98%. Further focusing of the library led to even more remarkable results. Although previous work by the same group using conformationally restrained short sequences in asymmetric transacylation reactions [115] led to promising results with high enantioselectivity factors (>50) this peptide does not seem to be at all organized. Clearly more research is needed to understand the formation and structure of the catalyst–substrate complexes and propose a rationale for these exciting results.

Functional metal peptides are appealing targets. Novel peptide-based catalysts for alkene epoxidation from metal-binding combinatorial libraries were obtained by Francis and Jacobsen [116] A combinatorial approach to hydrolytic metallopeptides for cleavage of phosphates was reported by Berkessel and Herault [117] who prepared a 625-member library of undecapeptides. Their strategy enabled direct on-beads screening of the library with a chromogenic model phosphate. In this way they were able to find active hits whose efficiency was also tested in homogeneous solution. Combinatorial optimization of the tripeptide Xaa–Xaa–His for complexation to Ni[II]

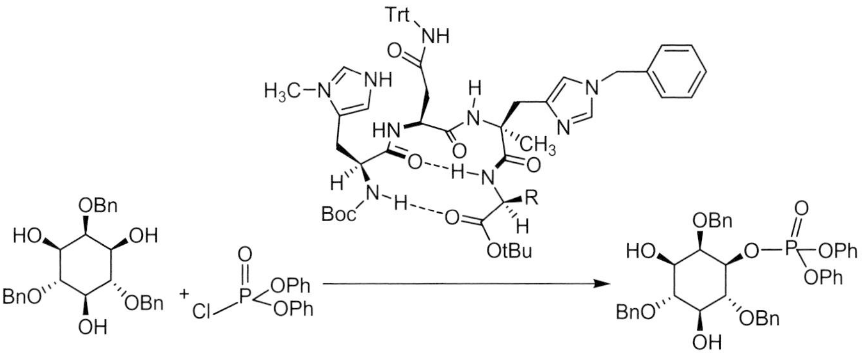

Figure 1.26 Miller's peptide catalyzes the conversion of *myo*-inositol into D-*myo*-inositol-1-phosphate.

and deoxyribose-based (oxidative) cleavage of B-form DNA was also performed by Long and his coworkers [118]. The procedure they employed was to generate two libraries (by using 18 naturally occurring amino acids and excluding Cys and Trp) in which the first and second positions of the ligand were varied. The optimized metallopeptide Ni^{II}–Pro–Lys–His was found to cleave DNA an order of magnitude better than Ni^{2+}–Gly–Gly–His, the reference compound used as a starting point for the selection process. Interestingly, metal complexation and the T/A-rich site selectivity of the optimized metallopeptide were not altered and DNA binding affinity was only slightly increased. Thus the increased activity observed is mostly related to the geometry of binding, as is suggested by molecular models.

The stereoselective catalysis of a series of rather different reactions was performed by Hoveyda's group using metallopeptide complexes. In this work the highly crucial identification of the most effective chiral ligand/metal salt couple for a specific process was performed via preliminary parallel screening of different metal precursors employing readily modifiable (modular) ligands. In all instances peptide-based Schiff bases with different O, N or P donors were used in the presence of both early and late transition metal ions. The development of efficient and highly stereoselective C–C bond-forming processes such as Ti^{IV}-catalyzed addition of CN to epoxides [119] and imines [120], Al^{III}-catalyzed addition of CN to ketones [121], Zr^{IV}-catalyzed dialkyl addition of zinc to imines [122], and Cu-catalyzed conjugate addition to enones [123] and allylic substitution [124] have been achieved. As an example, pyridinyl peptidic derivatives have been found to be effective chiral ligands for the Cu-catalyzed allylic substitution of di- and trisubstituted alkenes. The newly catalytic system enabled the synthesis of (R)-(–)-sporochnol with 82% *ee* and 82% overall yield [124]. The strategy for selection of the optimum reaction conditions was based on a series of repeated steps. Similar enantiomeric excesses were obtained by using a nonrational approach based on screening of parallel libraries of substrates, solvents, and metal ions [120a, 122]. This might indicate that small, focused libraries might provide results as good as a rational but more time-consuming approach in the selection of a catalytic system.

An important aspect was raised by Gilbertson in selecting catalysts from peptide-based phosphine ligands for late transition metal ions [125, 126] – that of the tuning of the best chiral ligand/substrate couple. This point emphasizes the quest for catalysts which are rather specific for a substrate, a property typical of enzymes. By studying the Pd^{II}-catalyzed addition of dimethyl malonate to cyclopentadienyl acetate with catalysts selected from two series of 96-and 40-member peptide libraries, this group also addressed the relevance of the secondary conformation of the peptide to the stereoselectivity of the catalytic process. They found that β-turns are better than helical conformations, and that occasionally the direct screening of the catalysts anchored on the solid support where they have been synthesized is possible with results comparable with those obtained in homogeneous solution.

If catalysis has proven an excellent proving ground for testing the potential of (pseudo)peptides a newly emerging application as fluorescent markers has been reported by Imperiali's group. Previously they had focused their research on peptide-based fluorescent chemosensors for Zn^{II} [127]. Now the scope of their research has widened to the discovery of an Ln^{III}-binding peptide for rapid fluorescence detection of tagged proteins. The development of fluorescent proteins as molecular tags may enable complex biochemical processes to be correlated with the functioning in living cells [128]. The work by Imperiali is based on previous studies by other laboratories that had focused on sequences based on specific loops of calcium-binding proteins [129] for similarities in ionic radii and coordination preferences of Ca^{2+} and Ln^{3+} ions. This previous work [130] had pinpointed hot positions relevant both for binding and fluorescence enhancement in 14-mer peptides. Imperiali's screening [131] of a large (up to 500,000 member) library of peptides of the general sequence: Ac–Gly–Xaa–Zaa–Xaa–Zaa–Xaa–Gly–Trp–Zaa–Glu–Zaa–Zaa–Glu–Leu (where Xaa were varied between potential metal-binding residues Asp, Asn, Ser, or Glu and Zaa were hydrophobic amino acids) led to the discovery of a peptide with K_D for Tb^{3+} ions as low as 0.22 μM. They suggested that these lanthanide-binding tags (LBT) might constitute a new alternative for expressing fluorescent fusion proteins by routine molecular biological techniques. To find sequences with even better binding ability for this lanthanide ion they also devised a powerful combinatorial screening [132] on beads. The methodology utilizes solid-phase split-and-pool combinatorial peptide synthesis where orthogonally cleavable linkers provide the handle for an efficient two-step screening procedure (Fig. 1.27). The initial screening avoids the interference caused by on-bead screening by photochemically releasing a portion of the peptides into an agarose matrix for evaluation. The secondary screening further characterizes each hit in a defined aqueous solution. It is worth mentioning that this procedure

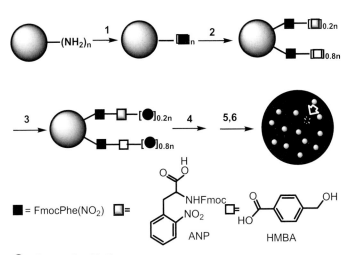

$\blacksquare$ = FmocPhe(NO$_2$) $\square$ = ANP $\square$= HMBA

$\bullet$ = βAlaArgProGly-Fmoc

Figure 1.27 Imperiali's procedure for synthesis and screening of Tb^{3+}-binding peptides: 1. Coupling of N-α-Fmoc-4-nitrophenylalanine. 2. Coupling of ANP:HMBA (1:10). 3. Introduction of the spacer sequence. 4. Synthesis of a "split-and-pool" library with a mass spectral ladder. 5, 6. Amino acid side-chain deprotection followed by casting into agarose with 50 μM Tb^{3+} and photolysis. After selection of the beads with luminescent halos and work-up they are titrated with Tb^{3+} and the sequence deconvoluted by MALDI-MS.

avoids interference from the solid support, in contrast with on-bead screening recently reported by Miller [133] and Davis [134] in which they look for products of a reaction produced by a solid-supported catalyst.

Long-lived proteins are susceptible to nonenzymatic chemical reactions and the evolution of fluorescence; little is known about the sequence-dependence of fluorogenesis, however. To address this point Van Vranken et al. [135] synthesized a library of over half a million octapeptides and exposed it to light and air in pH 7.4 buffer to identify fluorogenic peptides that evolve under mild oxidative conditions. The bead-based peptide library was composed of the general sequence $H_2N–Ala–(Xxx)_6–Ala–resin$, where Xxx was one of nine representative amino acids: Asp, Gly, His, Leu, Lys, Pro, Ser, Trp, and Tyr. Next, they selected five highly fluorescent beads from the library and subjected them to microsequencing, revealing the sequence of the unreacted peptide. All five fluorogenic sequences were ionic; lacked Tyr, His, and Leu; and most of the sequences contained only one Trp. Synthesis of the five soluble peptides revealed that one peptide ($H_2N–AlaLysProTrpGlyGlyAspAla–CONH_2$) evolved into a highly fluorescent photoproduct and a nonfluorescent monooxygenated photoproduct. Their work demonstrated that peptide fluorogenesis depends on sequence and not merely on the presence of tryptophan. The potential importance of fluorogenic peptide sequences is twofold. First, fluorogenic sequences that arise by mutation might prove to be hot spots for human aging. Second, fluorogenic sequences, particularly those compatible with intracellular conditions, might serve as fluorescent tags (see above) for proteins or as fluorescent biomaterials.

1.5
Conclusions

We hope that this chapter (and the others in this book) will give the readers the flavor of what is brewing in the pot of scientists using peptides and pseudo-peptides in general for the synthesis of new supramolecules expanding the scope of the systems naturally available. This is an open-ended chapter, because research in this field is developing rapidly and new breakthroughs emerge almost every day. Particularly fascinating is the design of new foldamers because, as we have seen in many examples, conformational control constitutes a tremendous challenge rewarded by improved efficiency. Although the ability to predict conformation using α-amino acids in very short sequences has improved considerably as the result of the seminal work done by Toniolo and his group, scientists are introducing new building blocks (β- and γ-amino acids for instance) [136] thus expanding the repertoire of accessible foldamers. These molecules, in contrast with standard peptides, might not be cleaved by proteins and, hence, are excellent candidates for new drugs.

The most fascinating field is that of the preparation of "smart" materials by conjugation of (pseudo)peptides and nanosystems such as nanoparticles or dendrimers. The possibility of controlling the structure of these systems at the molecular level (referred to as the "bottom up" approach in their preparation) constitutes an entry to complex molecular architectures with the possibility of defining specific functions.

New fresh ideas will certainly contribute to give further momentum to the development of new supramolecules with improved efficiency.

References

1 M.L. Kopka, C. Yoon, D. Goodsell, P. Pjura, R.E. Dickerson *Proc. Natl. Acad. Sci. USA.* **1985**, *82*, 1376–1380.

2 M. Coll, C.A. Frederick, A.H.J. Wang, A. Rich *Proc. Natl. Acad. Sci. USA* **1987**, *84*, 8385–8389.

3 J.G. Pelton, D.E. Wemmer *Proc. Natl. Acad. Sci. USA* **1989**, *86*, 5723–5727.

4 P.B. Dervan *Science* **1986**, *232*, 464–471.

5 For reviews see: P.B. Dervan, B.S. Edelson *Curr. Opin. Struct. Biol.* **2003**, *13*, 284–299. P.B. Dervan *Bioorg. Med. Chem.* **2001**, *9*, 2215–2235.

6 S. White, J.W. Szewczyk, J.M. Turner, E.E. Baird, P.B. Dervan *Nature*, **1998**, *391*, 468–471.

7 C.L. Kielkopf, S. White, J.W. Szewczyk, J.M. Turner, E.E. Baird, P.B. Dervan, D.C. Rees *Science*, **1998**, *282*, 111–115.

8 M.A. Marques, R.M. Doss, A.R. Urbach, P.B. Dervan *Helv. Chim. Acta* **2002**, *85*, 4485–4517.

9 D. Renneberg, P.B. Dervan *J. Am. Chem. Soc.* **2003**, *125*, 5707–5716.

10 C.A. Briehn, P. Weyermann, P.B. Dervan *Chem. Eur. J.* **2003**, *9*, 2110–2122.

11 J.W. Trauger, E.E. Baird, P.B. Dervan *Nature* **1996**, *382*, 559–561.

12 D.M. Herman, J.M. Turner, E.E. Baird, P.B. Dervan *J. Am. Chem. Soc.* **1999**, *121*, 1121–1129.

13 B. Olenyuk, C. Jitianu, P.B. Dervan *J. Am. Chem. Soc.* **2003**, *125*, 4741–4751.

14 A. Heckel, P.B. Dervan *Chem. Eur. J.* **2003**, *9*, 1–14.

15 S.S. Swalley, E.E. Baird, P.B. Dervan *Chem. Eur. J.* **1997**, *3*, 1600–1607.

16 J.W. Trauger, E.E. Baird, P.B. Dervan *Angew. Chem. Int. Ed.* **1998**, *37*, 1421–1423.

17 D.M. Herman, E.E. Baird, P.B. Dervan *Chem. Eur. J.* **1999**, *5*, 975–983.

18 V.C. Rucker, S. Foister, C. Melander, P.B. Dervan *J. Am. Chem. Soc.* **2003**, *125*, 1195–1202.

19 K. Maeshima, S. Janssen, U.K. Laemmli *EMBO J.* **2001**, *20*, 3218–3228.

20 J.M. Belitzky, S.J. Leslie, P.S. Arora, T.A. Beerman, P.B. Dervan *Bioorg. Med. Chem.* **2002**, *10*, 3313–3318.

21 M.P. Gygi, M.D. Ferguson, H.C. Mefford, K.P. Lund, C. O'Day, P. Zhou, C. Friedman, G. Van den Engh, M.L. Stolowitz, B.J. Trask *Nucleic Acids Res.* **2002**, *30*, 2790–2799.

22 N.R. Wurtz, P.B. Dervan *Chem. Biol.* **2000**, *7*, 153–161.

23 L.A. Dickinson, R.J. Gulizia, J.W. Trauger, E.E. Baird, D.E. Mosier, J.M. Gottesfeld, P.B. Dervan *Proc. Natl. Acad. Sci. USA.* **1998**, *95*, 12890–12895.

24 E.J. Fechter, P.B. Dervan *J. Am. Chem. Soc.* **2003**, *125*, 8476–8485.

25 L.A. Dickinson, J.W. Trauger, E.E. Baird, P. Ghazal, P.B. Dervan, J.M. Gottesfeld *Biochemistry*, **1999**, *38*, 10801–10807.

26 A.K. Mapp, A.Z. Ansari, M. Ptashne, P.B. Dervan *Proc. Natl. Acad. Sci. USA* **2000**, *97*, 3930–3935.

27 A.Z. Ansari, A.K. Mapp, D.H. Nguyen, P.B. Dervan *Chem. Biol.* **2001**, *8*, 583–592.

28 P.S. Arora, A.Z. Ansari, T.P. Best, M. Ptashne, P.B. Dervan *J. Am. Chem. Soc.* **2002**, *124*, 13067–13071.

29 H.-D. Arndt, K.E. Hauschild, D.P. Sullivan, K. Lake, P.B. Dervan, A.Z. Ansari *J. Am. Chem. Soc.* **2003**, *125*, 13322–13323.

30 S. Janssen, O. Cuvier, M. Muller, U.K. Laemmli *Mol. Cell.* **2000**, *6*, 1013–1024.

31 B.S. Edelson, T.P. Best, B. Olenyuk, N.G. Nickols, R.M. Doss, S. Foister, A. Heckel, P.B. Dervan *Nucleic Acids Res.* **2004**, *32*, 2802–2818.

32 G. Wang, X.X.S. Xu *Cell. Res.* **2004**, *14*, 111–116.

33 M.C. De Koning, G.A. Van der Marel, M. Overhand *Curr. Opin. Chem. Biol.* **2003**, *7*, 734–740.

34 M. Pooga, T. Land, T. Bartfai, U. Langel *Biomol. Eng.* **2001**, *17*, 183–192.

35 P.E. Nielsen *Acc. Chem. Res.* **1999**, *32*, 624–630.

36 P.E. Nielsen, M. Egholm, R.H. Berg, O. Buchardt *Science*, **1991**, *254*, 1497–1500.

37 S. Jordan, C. Schwemler, W. Kosch, A. Kretschmer, U. Stropp, E. Schwenner, B. Mielke *Bioorg. Med. Chem. Lett.* **1997**, *7*, 687–690.

38 M. D'Costa, V.A. Kumar, K.N. Ganesh *Org. Lett.* **2001**, *3*, 1281–1284.

39 M. Egholm, O. Buchardt, L. Christensen, C. Behrens, S.M. Freier, D.A. Driver, R.H. Berg, S.K. Kim, B. Norden, P.E. Nielsen *Nature*, **1993**, *365*, 566–568.

40 D.Y. Cherny, B.P. Belotserkovskii, M.D. Frank-Kamenetskii, M. Egholm, O. Buchardt, R.H. Berg, P.E. Nielsen *Proc. Natl. Acad. Sci. USA* **1993**, *90*, 1667–1670

41 Y.N. Kosaganov, D.A. Stetsenko, E.N. Lubyako, N.P. Kvitko, Y.S. Lazurkin, P.E. Nielsen *Biochemistry*, **2000**, *39*, 11742–11747.

42 H.J. Larsen, P.E. Nielsen *Nucleic Acids Res.* **1996**, *24*, 458–463.

43 T. Bentin, P.E. Nielsen *Biochemistry* **1996**, *35*, 8863–8869.

44 T. Bentin, P.E. Nielsen *J. Am. Chem. Soc.* **2003**, *125*, 6378–6379.

45 X. Zhang, T. Ishihara, D.R. Corey *Nucleic Acids. Res.* **2000**, *28*, 3332–3338.

46 U. Koppelhus, P.E. Nielsen *Adv. Drug. Deliv. Rev.* **2003**, *55*, 267–280.

47 R.W. Taylor, P.F. Chinnery, D.M. Turnbull, R.N. Lightowlers *Nat. Genet.* **1997**, *15*, 212–215.

48 J.C. Hanvey, N.J. Peffer, J.E. Bisi, S.A. Thomson, R. Cadilla, J.A. Josey *Science*, **1992**, *258*, 1481–1485.

49 G. Wang; K. Jing; R. Balczon; X. Xu *J. Mol. Biol.* **2001**, *313*, 933–940.

50 B. Liu; Y. Han, D.R. Corey, T. Kodadek *J. Am. Chem. Soc.* **2002**, *124*, 1838–1839.

51 B. Liu, Y. Han, A. Ferdous, D.R. Corey, T. Kodadek *Chem. Biol.* **2003**, *10*, 909–916.

52 B.M. Tyler, D.J. McCormick, C.V. Hoshall, C.L. Douglas, K. Jansen, B.W. Lacy, B. Cusack, E. Richelson *FEBS Lett.* **1998**, *421* 280–284.

53 B.M. Tyler, K. Jansen, D.J. McCormick, C.L. Douglas, M. Boules, J.A. Stewart, L. Zhao, B. Lacy, B. Cusack, A. Fauq, E. Richelson *Proc. Natl. Acad. Sci. USA* **1999**, *96*, 7053–7058.

54 J.C. Hanvey, N.J. Peffer, J.E. Bisi, S.A. Thomson, R. Cadilla, J.A. Josey, D.J. Ricca, C.F. Hassman, M.A. Bonham, K.G. Au, S.G. Carter, D.A. Bruckenstein, A.L. Boyd, S.A. Noble, L.E. Babiss *Science*, **1992**, *258*, 1481–1485.

55 H. Knudsen, P.E. Nielsen *Nucleic Acids. Res.* **1996**, *24*, 494–500.

56 P. Sazani, F. Gemignani, S.-H. Kang, M.A. Maier, M. Manoharan, M. Persmark, D. Bortner, R. Kole *Nat. Biotech.* **2002**, *20*, 1228–1233.

57 T. Berg *Angew. Chem. Int. Ed.* **2003**, *42*, 2462–2481.

58 M.W. Peczuh, A.D. Hamilton *Chem. Rev.* **2000**, *100*, 2479–2494.

59 R. Zutshi, M. Brickner, J. Chmielewski *Curr. Opin. Chem. Biol.* **1998**, *2*, 62–66.

60 T. Clackson, J.A. Wells *Science*, **1995**, *267*, 383–386.

61 A.A. Bogan, K.S. Thorn *J. Mol. Biol.* **1998**, *280*, 1–9.

62 A. Wlodawer, A. Miller, M. Jaskolski, B. Sathyanarayana, E. Baldwin, I. Weber, L. Selk, L. Clawson; J. Schneider, S. Kent *Science*, **1989**, *245*, 616–621.

63 R. Zutshi, J. Franciskovich, M. Shultz, B. Schweitzer, P. Bishop, M. Wilson, J. Chmielewski *J. Am. Chem. Soc.* **1997**, *119*, 4841–4845.

64 M.D. Shultz, M.J. Bowman, Y.-W. Ham, X. Zhao, G. Tora, J. Chmielewski *Angew. Chem. Int. Ed.* **2000**, *39*, 2710–2713.

65 S. Valente, M. Gobbo, G. Licini, A. Scarso, P. Scrimin *Angew. Chem. Int. Ed.* **2001**, *40*, 3899–3902.

66 P.H. Kussie, S. Gorina, V. Marechal, B. Elenbaas, J. Moreau, A.J. Levine, N.P. Pavletich *Science*, **1996**, *274*, 948–953.

67 C. García-Echeverría, P. Chène, M.J.J. Blommers, P. Furet *J. Med. Chem.* **2000**, *43*, 3205–3208.

68 Y. Hamuro, M. Crego Calama, H.S. Park, A.D. Hamilton *Angew. Chem. Int. Ed.* **1997**, *36*, 2680–2683.

69 Y. Wei, G.L. McLendon, A.D. Hamilton, M.A. Case, C.B. Purring, Q. Lin, H.S. Park, C.-S. Lee, T. Yu *Chem. Commun.* **2001**, 1580–1581.

70 H.S. Park, Q. Lin, A.D. Hamilton *J. Am. Chem. Soc.* **1999**, *121*, 8–13.

71 (a) N. Voyer *Top. Curr. Chem.* **1996**, *184*, 1–37; (b) N. Voyer, J. Lamothe *Tetrahedron* **1995**, *51*, 9241–9284.

72 (a) S.H. Gellman *Acc. Chem. Res.* **1998**, *31*, 173–180; (b) D.J. Hill, M.J. Mio, R.B. Prince, T.S. Hughes, J.S. Moore *Chem. Rev.* **2001**, *101*, 3893–4011.

73 Recent reviews: (a) L. Baltzer, H. Nilsson, J. Nilsson *Chem. Rev.* **2001**, *101*, 3153–3163; (b) L. Baltzer, J. Nilsson *Curr. Opin. Chem. Biol.* **2001**, *12*, 355–360; (c) R.B. Hill, D.P. Raleigh, A. Lombardi, W.F. DeGrado *Acc. Chem. Res.* **2000**, *33*, 745; (d) L. Baltzer, K.S. Broo *Biopolymers*, **1998**, 31–40.

74 (a) R.G. Daugherty, T. Wasowicz, B.R. Gibney, V.J. DeRose *Inorg. Chem.* **2002**, *41*, 2623–2632; (b) G. Xing, V.J. DeRose *Curr. Opin. Chem. Biol.* **2001**, *5*, 196–200.

75 (a) J.R. Calhoum, H. Kono, S. Lahr, W. Wang, W.F. DeGrado, J.G. Saven *J. Mol. Biol.* **2003**, *334*, 1101–1115; (b) G. Maglio, F. Nastri, V. Pavone, A. Lombardi, W.F. DeGrado *Proc. Natl. Acad. Sci USA* **2003**, *100*, 3772–3777; (c) W.F. DeGrado, L. Di Costanzo, S. Geremia, A. Lombardi, V. Pavone, L. Randaccio *Angew. Chem. Int. Ed.* **2003**, *42*, 417–422; (d) E.N.G. Marsh, W.F. DeGrado *Proc. Natl. Acad. Sci USA* **2002**, *99*, 5150–5154; (e) L. Di Costanzo, H. Wade, S. Geremia, L. Randaccio, V. Pavone, W.F. DeGrado, A. Lombardi *J. Am. Chem. Soc.* **2001**, *123*, 12749–12757; (f) J. Venkatraman, G.A. Naganawda, R. Sudha, P. Balaram *Chem. Commun.* **2001**, 2660–2661.

76 K. Jonsson, R.K. Allemann, H. Widmer, S.A. Benner *Nature* **1993**, *365*, 530–532.

77 S. Olofsson, G. Johansson, L. Baltzer *J. Chem. Soc. , Perkin Trans 2* **1995**, 2047–2056.

78 (a) J. Nilsson, L. Baltzer *Chem. Eur. J.* **2000**, *6*, 2214–2221; (b) L. Baltzer, K.S. Broo, H. Nilsson, J. Nilsson *Bioorg. Med. Chem.* **1999**, *7*, 83–91; (c) K.S. Broo, H. Nilsson, J. Nilsson, L. Baltzer *J. Am. Chem. Soc.* **1998**, *120*, 10287–10295.

79 (a) L.K. Anderson, G.T. Dolphin, L. Baltzer *ChemBioChem* **2002**, *3*, 741–751; (b) L.K. Andersson, G.T. Dolphin, J. Kihlberg, L. Baltzer *J. Chem. Soc. Perkin Trans. 2* **2000**, 459–464.

80 L.K. Andersson, M. Caspersson, L. Baltzer *Chem. Eur. J.* **2002**, *8*, 3687–3697.

81 M.A. Shogren-Knaak, B. Imperiali *Bioorg. Med. Chem.* **1999**, *7*, 1993–2002.

82 (a) M.P. Fitzsimons, J.K. Barton, *J. Am. Chem. Soc.* **1997**, *119*, 3379–3380; (b) K.D. Copeland, M.P. Fitzimons, R.P. Houser, J.K. Barton, *Biochemistry* **2002**, *41*, 343–356.

83 H. Ishida, Y. Inoue *Rev. Heteroatom Chem.* **1999**, *19*, 79–142.

84 P. Rossi, F. Felluga, P. Scrimin, *Tetrahedron Lett.* **1998**, *39*, 7159–7162.

85 P. Rossi, F. Felluga, P. Tecilla, L. Baltzer, P. Scrimin, *Chem. Eur. J. ,* **2004**, *10*, 4163–4170.

86 P. Rossi, F. Felluga, P. Tecilla, F. Formaggio, M. Crisma, C. Toniolo, P. Scrimin *Biopolymers* **2000**, *55*, 496–501.

87 (a) P. Rossi, F. Felluga, P. Tecilla, F. Formaggio, M. Crisma, C. Toniolo, P. Scrimin *J. Am. Chem. Soc.* **1999**, *121*, 6948–6949; (b) C. Sissi, P. Rossi, F. Felluga, F. Formaggio, M. Palumbo, P. Tecilla, C. Toniolo, P. Scrimin *J. Am. Chem. Soc.* **2001**, *123*, 3169–3170.

88 A. Scarso, U. Scheffer, M. Göbel, Q.B. Broxterman, B. Kaptein, F. Formaggio, C. Toniolo, P. Scrimin *Proc. Natl. Acad. Sci. USA* **2002**, *99*, 5144–5149.

89 (a) A.J. Kennan, V. Haridas, K. Severin, D.H. Lee, M.R. Ghadiri *J. Am. Chem. Soc.* **2001**, *123*, 1797–1803; (b) A. Sghatelian, Y. Yokobayashi, K. Soltani, M.R. Ghadiri *Nature*, **2001**, *409*, 797–801; (c) K. Severin, D.H. Lee, J.A. Martinez, M. Veithm M.R. Ghadiri *Angew. Chem. Int. Ed. Engl.* **1998**, *37*, 126–128; (d) D.H. Lee, K. Severin, Y. Yokobayashi, M.R. Ghadiri *Nature* **1997**, *390*, 591–594; (e) D.H. Lee, J.R. Granja, J.A. Martinez, K. Severin, M.R. Ghadiri *Nature* **1996**, 382, 525–528.

90 (a) X.Q. Li, J. Chmielewski *J. Am. Chem. Soc.* **2003**, *125*, 11820–11821; (b) X.Q. Li, J. Chmielewski *Org. Biomol. Chem.* **2003**, *1*, 901–904; (c) R. Issac, J. Chmielewski *J. Am. Chem. Soc.* **2002**, *124*, 6808–6809; (d) R. Issac, Y.W. Ham, J. Chmielewski *Curr. Opin. Struct. Biol.* **2001**, *11*, 458–463; (e) S. Yao, I. Ghosh, R. Zutshi, J. Chmielewski *Nature* **1998**, *396*, 447–450; (f) S. Yao, I. Ghosh, R. Zutshi, J. Chmielewski *Angew. Chem. Int. Ed.* **1998**, *37*, 478–481; (g) S. Yao, I. Ghosh, R. Zutshi, J. Chmielewski *J. Am. Chem. Soc.* **1997**, *119*, 10559–10560.

91 F.H. Crick *Acta Crystallogr.* **1953**, *6*, 689–695.

92 P.B. Harbury, J.J. Plecs, B. Tidor, T. Alber, P.S. Kim *Science* **1998**, *282*, 1462–1467.

93 P.E. Dawson, T.W. Muir, I. Clark-Lewis, S.B.H. Kent *Science* **1994**, *266*, 776–779.

94 D.B. Smithrud, P.A. Benkovic, S.J. Benkovic, M. Taylor, K.M. Yager, J. Witherington, B.W. Philips, P.A. Spengeler, A.B. Smith III, R. Hirschmann *Proc. Natl. Acad. Sci USA* **2000**, *97*, 1953–1958.

95 The "guru" of self-replication and of the treatment of the data of self-replicating systems is von Kiedrowski: G. von Kiedrowski *Bioorg. Chem. Front* **1993**, *3*, 113–118; A. Luther, R. Brandsch, G. von Kiedrowski *Nature* **1998**, *396*, 245–248.

96 A single recent review: A. Tossi, L. Sandri, A. Giangaspero *Biopolymers* **2000**, *55*, 4–30.

97 (a) M.R. Ghadiri *Biopolymers* **2003**, *71*, L21; (b) J. Sanchez-Ouesada, M.P. Isler, M.R. Ghadiri *J. Am. Chem. Soc.* **2002**, *124*, 10004–10005; (c) D.T. Bong, M.R. Ghadiri *Angew. Chem. Int. Ed.* **2001**, *40*, 2163–2166; (d) D.T. Bong J.D. Hartgerink, T.D. Clark, M.R. Ghadiri *Chem. Eur. J.* **1998**, *4*, 1367–1372; (e) T.D. Clark, J.M. Buriak, K. Kobayashi, M.P. Isler, D.E. McRee, M.R. Ghadiri *J. Am. Chem. Soc.* **1998**, *120*, 8949–8962; (f) H.S. Kim, J.D. Hartgerink, M.R. Ghadiri *J. Am. Chem. Soc.* **1998**, *120*, 4417–4424; (g) J.D. Hartgerink, J.R. Granja, R.A. Milligan, M.R. Ghadiri *J. Am. Chem. Soc.* **1996**, *118*, 43–50.

98 S. Fernandez-Lopez, H.S. Kim, E.C. Choi, M. Delgado, J.R. Granja, A. Khasanov, K. Ktaehenbuehl, G. Long, D.A. Weinberger, K.M. Wilcoxen, M.R. Ghadiri *Nature* **2001**, *412*, 452–455

99 J. Sanchez-Quesada, H.S. Kim, M.R. Ghadiri *Angew. Chem. Int. Ed.* **2001**, *40*, 2503–2507.

100 (a) E. Biron, F. Otis, J.-C. Meillon, M. Robitaille, J. Lamothe, P. Van Hove, M.-E. Cormier, N. Voyer *Bioorg. Med. Chem.* **2004**, *12*, 1279–1290; (b) Y.R. Vandenburg, B.D. Smith, E. Biron, N. Voyer *Chem. Commun.* **2002**, 1694–1695; (c) J.-C. Meillon, N. Voyer, E. Biron, F. Sanschagrin, J.F. Stoddart *Angew. Chem. Int. Ed.* **2000**, *39*, 143–147; (d) N. Voyer, B. Guerin *Chem. Commun.* **1997**, 2329–2331; (e) N. Voyer, M. Robitaille *J. Am. Chem. Soc.* **1995**, *117*, 6599–6600; (f) Voyer, N. *J. Am. Chem. Soc.* **1991**, *113*, 1818–1821.

101 (a) B. Baumeister and S. Matile, *Chem. Eur. J.*, **2000**, *6*, 1739–1749; (b) B. Baumeister, N. Sakai and S. Matile, *Angew. Chem. Int. Ed.*, **2000**, *39*, 1955–1959;

(c) N. Sakai, B. Baumeister and S. Matile, *ChemBioChem*, **2000**, *1*, 123–125.

102 (a) P. Scrimin, A. Veronese, P. Tecilla, U. Tonellato, V. Monaco, F. Formaggio, M. Crisma, C. Toniolo *J. Am. Chem. Soc.* **1996**, *118*, 2505–2506; (b) P. Scrimin, P. Tecilla, U. Tonellato, A. Veronese, M. Crisma, F. Formaggio, C. Toniolo *Chem. Eur. J.* **2002**, *8*, 2753–2763.

103 M. Brust, M. Walker, D. Bethell, D. Schiffrin, R. Whyman, R. *J. Chem. Soc., Chem. Commun.* **1994**, 801–802.

104 P. Pengo, S. Polizzi, M. Battagliarin, L. Pasquato, P. Scrimin, P. *J. Mater. Chem.* **2003**, *13*, 2471–2478.

105 M. Mammen, S.-K. Choi, G.M. Whitesides *Angew. Chem. Int. Ed.* **1998**, *37*, 2754–2794.

106 F. Manea, F. Bodar Houillon, L. Pasquato, P Scrimin *Angew. Chem. Int. Ed.* **2004**, *43*, 6165–6169.

107 P. Pengo, L. Pasquato, P. Scrimin, P. submitted for publication.

108 P. Pengo, L. Pasquato, S. Moro, A. Brigo, F. Fogolari, Q.B. Broxterman, B. Kaptein, P. Scrimin *Angew. Chem. Int. Ed.* **2003**, *42*, 3388–3392.

109 P. Pengo, Q.B. Broxterman, B. Kaptein, L. Pasquato, P. Scrimin *Langmuir* **2003**, *19*, 2521–2524.

110 (a) A. Esposito, E. Delort, D. Lagnoux, F. Djojo, J.-L. Reymond *Angew. Chem. Int. Ed.* **2003**, *42*, 1381–1383; (b) D. Lagnoux, E. Delort, C. Douat-Casassus, A. Esposito, J.-L. Reymond *Chem. Eur. J.* **2004**, *10*, 1215–1226.

111 M. Zhou, D. Bentley, I. Ghosh *J. Am. Chem. Soc.* **2004**, *126*, 734–735.

112 D.A. Tomalia *Adv. Mater.* **1994**, *6*, 529–539.

113 B.E. Turk, L.C. Cantley, *Curr. Opin. Chem. Biol.* **2003**, *7*, 84–90.

114 (a) K.S. Griswold, S.J. Miller *Tetrahedron* **2003**, *59*, 8869–8875; (b) B.R. Sculimbrene, A.J. Morgan, S.J. Miller *Chem. Commun.* **2003**, 1781–1785; (c) B.R. Sculimbrene, A.J. Morgan, S.J. Miller *J. Am. Chem. Soc.* **2002**, *124*, 11653–11656; (d) B.R. Sculimbrene, S.J. Miller *J. Am. Chem. Soc.* **2001**, *123*, 10125–10126.

115 (a) E.R. Jarvo, S.J. Miller *Tetrahedron* **2002**, *58*, 2481–2485; (b) G.T. Copeland, S.J. Miller *J. Am. Chem. Soc.* **2001**, *123*, 6496–6502; (c) E.R. Jorvo, G.T.

Copeland, N. Papaioannou, P.J. Bonitate-bus, Jr.; S.J. Miller *J. Am. Chem. Soc.* **1999**, *121*, 11638–11643; (d) S.J. Miller, G.T. Copeland, N. Papaioannou, T.E. Horstmann, E.M. Ruel *J. Am. Chem. Soc.* **1998**, *120*, 1629–1630.

116 M.B. Francis, E.N. Jacobsen, *Angew. Chem. Int. Ed.* **1999**, *38*, 937–941.

117 A. Berkessel, D.A. Hérault, *Angew. Chem. Int. Ed.* **1999**, *38*, 102–105.

118 X. Huang, M.E. Pieczko, E.C. Long, *Biochemistry* **1999**, *38*, 2160–2166.

119 K.D. Shimizu, B.M. Cole, C.A. Krueger, K.W. Kuntz, M.L. Snapper, A.H. Hoveyda, . *Angew. Chem. Int. Ed. Engl.* **1997**, *36*, 1703–1707 and references therein.

120 (a) C.A. Krueger, K.W. Kruntz, C.D. Dzierba, W.G. Wirschum, J.D. Gleason, M.L. Snapper, A.H. Hoveyda *J. Am. Chem. Soc.* **1999**, *121*, 4284–4285; (b) J.R. Porter, W.G. Wirschun, K.W. Kuntz, M.L. Snapper, A.H. Hoveyda *J. Am. Chem. Soc.* **2000**, *122*, 2657–2658.

121 H. Deng, M.P. Isler, M. Snapper, A.H. Hoveyda, *Angew. Chem. Int. Ed.* **2002**, *41*, 1009–1012.

122 J.R. Porter, J.F. Traverse, A.H. Hoveyda, M.L. Snapper *J. Am. Chem. Soc.* **2001**, *123*, 984–985.

123 (a) C.A. Luchaco-Cullis, H. Mizutani, K.E. Murphy, A.H. Hoveyda, *Angew. Chem. Int. Ed. Engl.* **2001**, *40*, 1456–1460; (b) S.J. Degrado, H. Mizutani, A.H. Hoveyda, *J. Am. Chem. Soc.* **2002**, *124*, 13362–13363.

124 S.J. Degrado, H. Mizutani, A.H. Hoveyda *J. Am. Chem. Soc.* **2001**, *123*, 755–756.

125 S.R. Gilbertson, S.E. Collibee, A. Agarkov *J. Am. Chem. Soc.* **2000**, *122*, 6522–6523.

126 S.R. Gilbertson, X. Wang *Tetrahedron* **1999**, *55*, 11609–11618 and references therein.

127 G.K. Walkup, B. Imperiali, *J. Am. Chem. Soc.* **1997**, *119*, 3443–3450.

128 J. Lippincott-Schwartz, G.H. Patterson, *Science* **2003**, *300*, 87–91.

129 For the use of EF-hand sequences as Ln(III)-based metallonucleases see: J.T. Welch, W.R. Kearney, S.J. Franklin *Proc. Natl. Acad. Sci USA* **2003**, *100*, 3725–3730 and references therein.

130 J.P. MacManus, C.W. Hogue, B.J. Marsden, M. Sikorska, A.G. Szabo, *J. Biol. Chem.* **1990**, 10358–10366.

131 K.J. Franz, M. Nitz, B. Imperiali, *ChemBioChem* **2003**, *4*, 265–271.

132 M. Nitz, K.J. Franz, R.L. Maglathlin, B. Imperiali, *ChemBioChem* **2003**, *4*, 272–276.

133 R.F. Harris, A.J. Nation, G.T. Copeland, S.J. Miller, *J. Am. Chem. Soc.* **2000**, *122*, 11270–11271.

134 M. Müller, T.W. Mathers, A.P. Davis, *Angew. Chem. Int. Ed.* **2001**, *40*, 3813–3815.

135 G.L. Juskowiak, S.J. Stachel, P. Tivitmahaisoon, D.L. Van Vranken *J. Am. Chem. Soc.*, **2004**, *126*, 550–556.

136 R.P. Cheng, S.H. Gellman, W.F. DeGrado *Chem. Rev.* **2001**, *101*, 3219–3232.

2
Carbohydrate Receptors

Anthony P. Davis and Tony D. James

2.1
Introduction

Carbohydrates are the most abundant organic molecules on the surface of the Earth. Formed by photosynthesis from carbon dioxide and water, they serve as starting materials and fuels for biochemical processes and, in the form of cellulose, as the major biological building material. They also play more subtle roles. Combined in oligosaccharides, they are well-suited to conveying information by variation of sequence, linkage mode, ring size, and branching. Indeed whereas there are 4096 hexanucleotide combinations and 6.4×10^7 hexapeptides, there are $>1.05 \times 10^{12}$ isomeric hexasaccharide structures [1]. Unsurprisingly, this potential is exploited by Nature. Proteins and cell surfaces are decorated with oligosaccharides which control many properties and affect behavior. Protein folding and trafficking, cell–cell recognition, infection by pathogens, tumor metastasis, and many aspects of the immune response, are mediated by carbohydrate recognition events [2].

In this context, carbohydrates are attractive targets for supramolecular chemists. Carbohydrate receptors can fulfill theoretical functions, throwing light on natural saccharide recognition. Potentially they could have useful biological effects, binding to oligosaccharides and thus interrupting unwanted processes (e.g. cell infection). They could also act by transporting carbohydrates, or carbohydrate-mimetic drug molecules, across cell membranes. "Synthetic antibodies" aimed at specific saccharide epitopes, could also be envisaged. Although such applications are admittedly distant, for reasons which will be discussed later in this chapter, an immediate and realistic medical goal is the recognition and sensing of glucose in biological fluids, potentially of great benefit for the management of diabetes [3]. Other functions are the separation of carbohydrates, for example by large-scale transport processes, and the catalysis of carbohydrate reactions.

A further motivation is the unusual challenge presented by carbohydrate recognition. Saccharides are structurally complex, and intrinsically difficult to bind from aqueous solution. Their dominant functionality (hydroxyl) bears a close similarity to the water molecules which must be expelled from a binding site. It is only the precise

Functional Synthetic Receptors. Edited by T. Schrader, A. D. Hamilton
Copyright © 2005 WILEY-VCH Verlag GmbH & Co. KGaA, Weinheim
ISBN: 3-527-30655-2

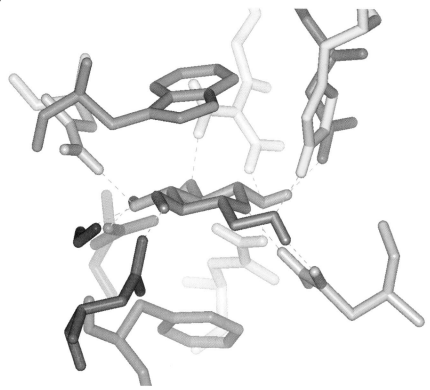

Fig. 2.1 A glucose molecule in the binding site of the *E. coli* galactose chemoreceptor protein, as revealed by crystallography (Ref. [7a]). The substrate makes contact with two apolar residues (phenylalanine and tryptophan, shown as light green) a water molecule (dark blue), and eight polar residues (3 × aspartate, red; 3 × asparagine, pink; 1 × histidine, yellow; and 1 × arginine, light blue). The water molecule and polar residues make a total of 13 hydrogen bonds with the carbohydrate.

arrangement of the OH groups, the occasional presence of more "helpful" functionality (ammonium, carboxylate, etc.), and the small patches of apolar surface which enable recognition to occur. The problem is emphasized by studies on natural carbohydrate recognition. Aside from antibodies, this function is performed mainly by a family of proteins termed lectins [4]. A notable feature of lectins is their low affinity for their targets. Typical lectin–oligosaccharide binding constants are in the range 10^3–10^4 M^{-1}, exceptionally small for biological molecular recognition. Nature compensates partly through multivalency, whereby multiple interactions are required to elicit an effect (hence the widespread interest in the synthesis of "glycoside clusters") [5]. Although higher binding constants are possible among the smaller family of carbohydrate-binding bacterial periplasmic proteins involved in membrane transport and chemotaxis (motion up a concentration gradient) [6], even these perform modestly by biological standards, with K_a for monosaccharides in the range 10^6–10^7 M^{-1}. Moreover, these proteins succeed by surrounding their targets and making a comprehensive array of H-bonding and apolar contacts (Fig. 2.1) [7]. It seems likely they are fully optimized for strong binding, representing the limit achievable by natural selection.

Two general strategies have been employed for the design of carbohydrate receptors. The first involves exploitation of noncovalent forces, the second relies on the reversible formation of covalent B–O bonds from diol units and boronic acids. The first approach, discussed in Sect. 2.2, is biomimetic, and might be directly related to natural saccharide recognition. The second, addressed in Sect. 2.3, is less biologically relevant, but has thus far proved the more effective (at least in aqueous solution). Reviews on both appeared in the mid-late 1990s [8, 9] so this chapter, while attempting a balanced overall view, will concentrate especially on the more recent developments in each area.

2.2
Carbohydrate Receptors Employing Noncovalent Interactions

The noncovalent interactions available for binding carbohydrates are hydrogen-bonding, CH–π interactions, oxygen–metal ion coordination, occasionally electrostatic interactions, and solvophobic effects. In nonpolar solvents polar groups will be poorly solvated, so hydrogen bonding and other polar interactions will be effective drivers for complex formation. Receptors designed to operate in such media should thus possess arrays of polar groups, pre-organized to bind the carbohydrate and to avoid counterproductive intramolecular contacts. Aromatic surfaces can also be used to form CH–π interactions with the substrate, but these are weaker and likely to make only secondary contributions to binding energies. Solvophobic interactions should be relatively unimportant, given the low cohesive energies of nonpolar solvents.

In water the situation is different and somewhat less clear [10]. In principle, guidance can be obtained from natural carbohydrate recognition, but in practice this phenomenon is not well understood. In particular, the driving force for protein–carbohydrate recognition is a matter of debate. The extensive hydrogen bonding observed in protein crystal structures (cf. Fig. 2.1) might be thought to suggest that polar interactions are dominant [6]. On the other hand, the polar groups are well solvated before binding occurs. The net energy change on replacing, for example, $NH\cdots OH_2$ with $NH\cdots OHR$ is surely minimal. An alternative view is that the driving force for binding in water is primarily hydrophobic, and dependent on the release of water from apolar surfaces in receptor and substrate [11]. The apolar surfaces in saccharides are, however, quite small. The discussion has been complicated by calorimetric studies showing that protein–carbohydrate binding is usually enthalpy-driven, with negative (unfavorable) changes in entropy [12, 13]. The classical hydrophobic effect is entropic [14], so this might be taken as evidence against a solvophobic driving force. More recently, however, evidence has emerged for an enthalpically-driven "non-classical" hydrophobic effect [15]. Desolvation is thus a credible driver for carbohydrate recognition. Currently the safest assumption is that both effects are important, i.e. complexation occurs partly because of optimum polar interactions between protein and carbohydrate and partly because of unfavorable interactions between water and the binding partners. Receptors should thus be designed to make both polar and apolar/hydrophobic contact with their substrates.

The solvent in which recognition occurs affects the potential for function and the biological relevance. Carbohydrates occur naturally in water, so this medium is clearly the most significant. Phase transfer from aqueous to organic solvents is also important, because it models transport across cell membranes and can be employed in carbohydrate sensors and separation methods. A similar case can be made for saccharide recognition at aqueous–solid interfaces. Although studies in organic media have less relevance and potential, they are easier to perform and can be used to develop architectural ideas. In the remainder of this section, the available carbohydrate-binding systems will be discussed according to the medium employed, beginning with organic solvents and working via two-phase and surface studies toward homogeneous aqueous solution.

2.2.1
Recognition in Organic Solvents

As described above, the key to saccharide binding in organic media is the provision of pre-organized polar functionality. Hydrogen-bonding groups dominate, although metal centers have been used occasionally. Carbohydrate molecules are quite large compared with other substrates addressed in supramolecular chemistry, so the receptor frameworks also tend to be substantial (especially if an attempt is made to surround the carbohydrate). Conformational control is important, as in most areas of supramolecular chemistry, but especially so when receptors are complex, extended, and likely to contain self-complementary groups.

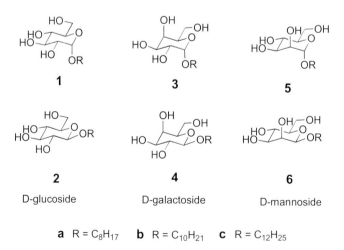

Fig. 2.2 Lipophilic glycosides for studies of carbohydrate binding in nonpolar solvents.

The choice of substrate is a non-trivial issue. Unprotected carbohydrates are insoluble in all but the most polar organic solvents (and these, of course, do not favor polar interactions). On the other hand, protected saccharides are not usually good models for the natural species. The solution was found in lipophilic glycosides such as **1–6** (Fig. 2.2). These compounds have "biorelevant" head-groups but are usually chloroform-soluble. Some examples, such as octyl α- and β-glucosides **1/2a**, and β-galactoside **4a**, are commercially available. Unsurprisingly they aggregate at higher concentrations in nonpolar solvents [16, 17], but meaningful binding studies can often still be performed.

The receptors used for recognition in organic solvents fall naturally into categories depending on their molecular frameworks and the types of functionality employed. These designs are grouped accordingly in the following pages.

2.2.1.1 Calixarenes

An account of synthetic carbohydrate receptors should begin with the octahydroxy calix[4]arene **7**, reported in 1988 by Aoyama and coworkers [18]. This was the first synthetic carbohydrate receptor, and is still remarkable for its accessibility; it is formed in 70 % yield in a single step from resorcinol and dodecanal. The calixarene framework is not large enough to encapsulate the carbohydrate, but **7** can interact effectively with a variety of substrates in "face-to-face" geometry. It has been used largely in two-phase systems, so more will be said in Sect. 2.2.2, but studies have also been performed with **1a** and **2a** in CDCl$_3$. Complexation was followed by NMR and CD analysis, which revealed unusual 1:4 host:guest stoichiometry [16]. The use of CD is notable, because this technique is especially suitable for studying carbohydrate recognition [19]. Saccharides are chiral but have no chromophore, whereas many receptors are achiral and UV-absorbing. Only when substrate associates with receptor, imposing a chiral conformation, does a CD signal emerge.

More recently, two other groups have employed the calixarene scaffold for constructing carbohydrate receptors. Ungaro and coworkers synthesized the bridged system **8**, which does have the potential to surround a substrate [20]. NMR titrations of **8a** in CDCl$_3$ against glucosides **1a** and **2a** showed significant 1:1 binding constants of

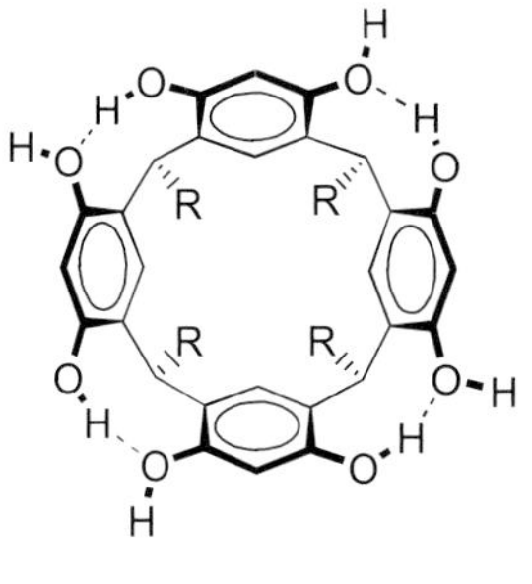

7 R = C$_{11}$H$_{23}$

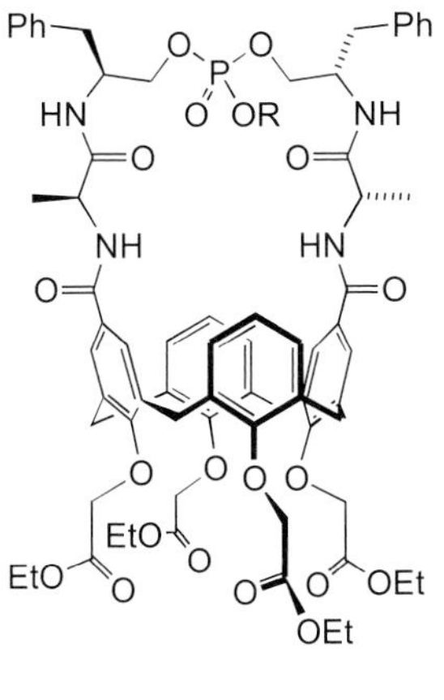

8a R = Me
8b R = -

ca. 500 M^{-1}, whereas the anionic **8b** was more powerful (K_a = 2700 M^{-1} for **2a**) and selective (3:1 for **2a** compared with **1a**). Phosphate units have been widely used in studies of carbohydrate recognition, as discussed further in Section 2.2.1.5. Calixarenes were also used in remarkable self-assembling receptors composed of 15 molecular components, held together by formation of melamine–barbituric acid "rosettes" [21]. The system bound **2a** rather weakly (~20 M^{-1} in CDCl₃) but, in a chiral variant, selectivity was observed between **2a** and its enantiomer.

2.2.1.2 Frameworks Based on Cholic Acid

Cholic acid **9** is a steroidal natural product which is available cheaply in bulk quantities. It is an attractive starting material for supramolecular chemistry because of its size, rigidity, and well-spaced array of functional groups groups [22]. It is especially suitable for preparation of molecules with quite large clefts or cavities and inward-directed polar functionality, as required for carbohydrate recognition. For example, replacement of the 3α-OH with an aromatic spacer, followed by cyclodimerization, gives the "cholaphane" **10**. Reported in 1989/90, this was the first carbohydrate receptor to be capable of fully surrounding a monosaccharide substrate [23]. It was also the first with enantioselectivity, binding D-glucoside **2a** with K_a = 3100 M^{-1} in CDCl₃, but L-glucoside **ent-2a** with K_a = 1000 M^{-1} [24]. Receptor **10** was followed by a series of related compounds such as **11** and **12** [25]. Although some of these had increased rigidity, and seemed to be well pre-organized for monosaccharide binding, none was able to match the original system.

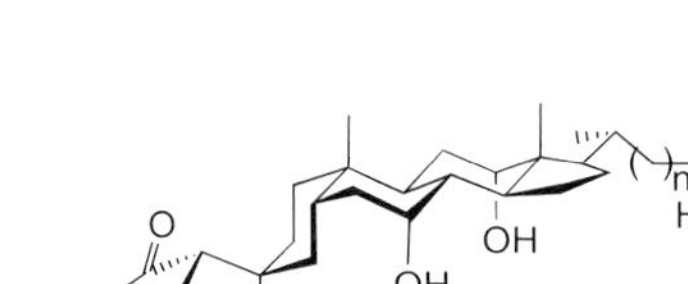

9

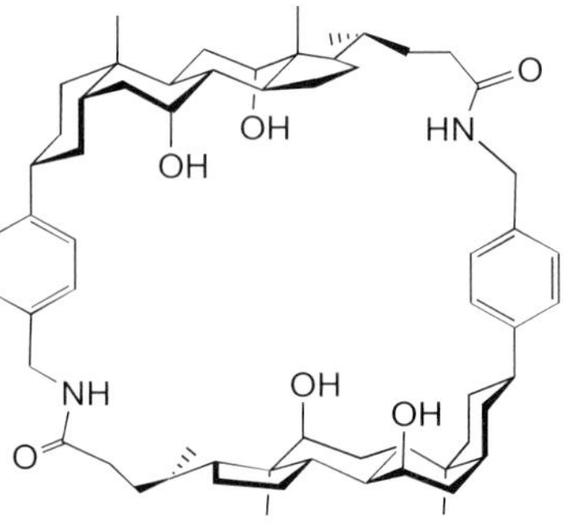

10

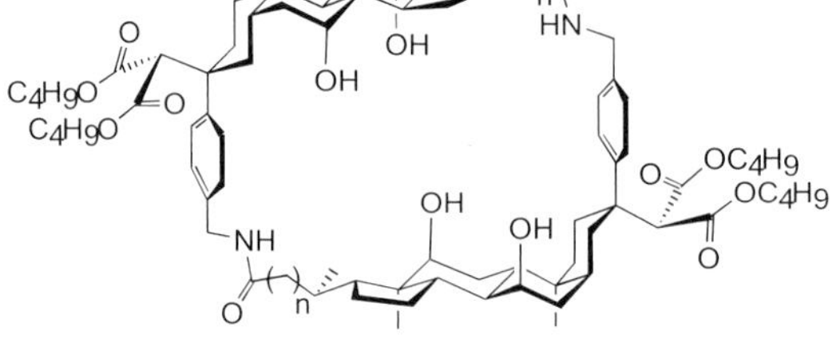

11a n = 0
11b n = 1
11c n = 2

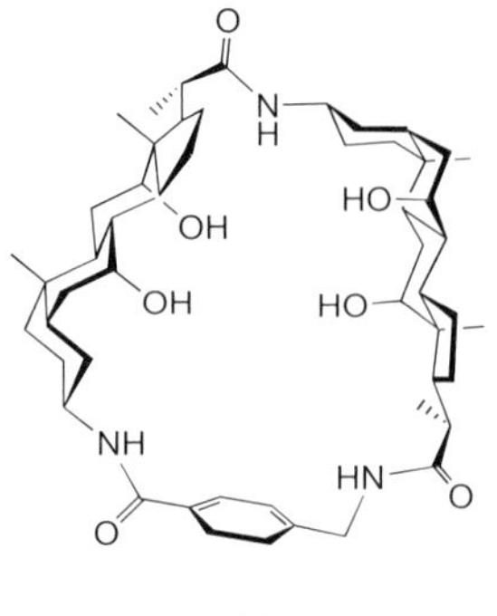

12

Cholic acid was also combined with a porphyrin unit in **13**, capable of forming both hydrogen and Zn···O coordination bonds with carbohydrate substrates [26]. The involvement of a metal center mimics the role of calcium ions in "C-type lectins" [4c]. This study, by Bonar-Law and Sanders, was notable for considering a wide range of glycosyl head-groups (Tab. 2.1). Like **10**, receptor **13** was appreciably enantioselective (**1a** vs **ent-1a**, factor of 3.5). Substantial diastereoselectivity was also observed, but the authors note that this may result largely from the natures of the substrates. Indeed, evidence was presented that the saccharides vary in "stickiness" in the order α-mannoside > β-mannoside > β-glucoside > α-glucoside $\approx$ β-galactoside > α-galactoside. This sequence is not mirrored precisely in Tab. 2.1, but there are clear similarities. Also significant was the observation that small amounts of water increased the binding constants. It is known that immobilized water molecules can participate in protein–carbohydrate recognition [4c] (an example is given in Fig. 2.1), and **13** is assumed to model this process.

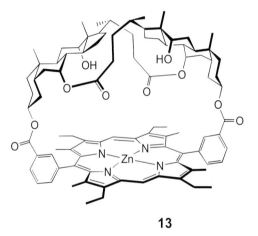

13

Tab. 2.1 Association constants (K_a, M^{-1}) between receptor **13** and organic-soluble glycosides in dry and wet CH_2Cl_2.[a]

	CH_2Cl_2	CH_2Cl_2 + H_2O[b]
α-D-Glucoside **1a**	910	
α-L-Glucoside **ent-1a**	260	
β-D-Glucoside **2a**	1480	6700
α-D-Galactoside **3b**	490	
β-D-Galactoside **4b**	200	
α-D-Mannoside **5b**	7000	12,300
β-D-Mannoside **6b**	5200	18,500

[a] Determined by UV and/or ^1H NMR titration at 295 K
[b] ca 0.09 mol L^{-1} water added to titration solvent

2.2.1.3 Porphyrins

The porphyrin nucleus can be used on its own to pre-organize polar functionality. Some very effective systems have been based on this strategy. Mizutani and coworkers have studied several Zn-porphyrins; of these the quinolyl-substituted receptors **14** proved especially powerful [27]. As shown in Tab. 2.2, **14a** bound octyl glycosides in CHCl$_3$ with K_a up to ~60,000 M^{-1}, and very high diastereoselectivity. The oxy-aryl substituents in **14b** increased K_a for **2a** to 74,000 M^{-1}. Affinities were even higher for the tetraureas **15**, studied in parallel by the groups of Bonar-Law and Kim (Tab. 2.2) [28, 29]. This design features four highly polar functional groups which cannot form intramolecular hydrogen bonds and are well-spaced for cooperative binding to a monosaccharide. The sidechains in **15b/c** contain further functionality, derived from amino acids. The K_a of 2×10^7 M^{-1} for **15c** + **2a** in CHCl$_3$ is the highest recorded for monosaccharide binding in this popular solvent. Lower values are reported for **15a/b**, but this might be largely because of the different medium. Both groups found little difference between the binding properties of **15** (M = Zn) and **15** (M = H$_2$), so Zn···O interactions would seem to be unimportant.

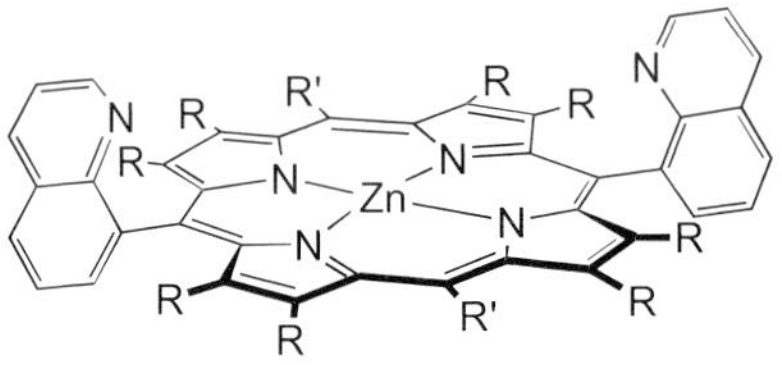

14a R = Et, R' = H   **14b** R = H, R' =

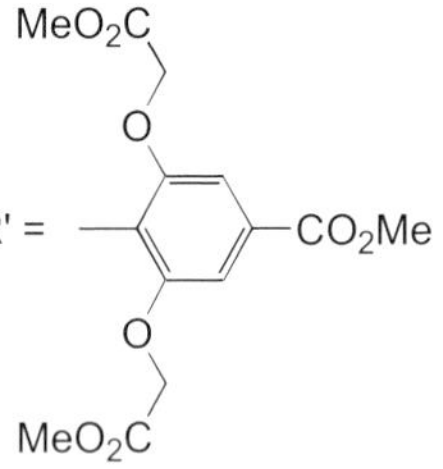

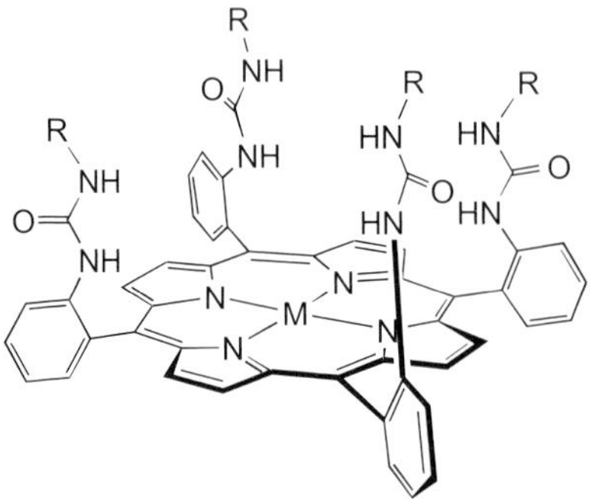

15a R = decyl

15b R =

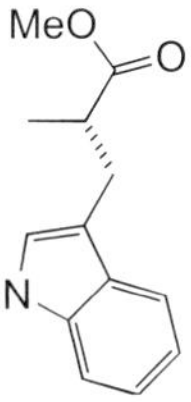

15c R =

M = Zn or H$_2$

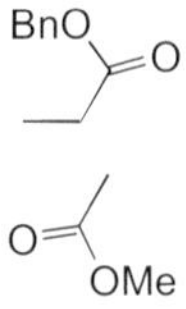

Tab. 2.2 Association constants $(K_a, \text{M}^{-1})^a$ between receptors **14/15** and organic-soluble glycosides[b] in nonpolar solvents.

	14a in CHCl$_3$[c]	15a (M = Zn) in CH$_2$Cl$_2$[d]	15b (M = Zn) in CH$_2$Cl$_2$[d]	15c (M = Zn) in CHCl$_3$[e]
α-D-Glucoside **1**	7570	1.5×10^5	3×10^4	4×10^6
β-D-Glucoside **2**	41,400	3×10^5	5×10^5	2×10^7
α-D-Galactoside **3**	1870			
β-D-Galactoside **4**	3840	$8 \times 10^{5\,f}$	$9 \times 10^{5\,f}$	5×10^6
α-D-Mannoside **5**	15,500			
β-D-Mannoside **6**	61,700	$2 \times 10^{5\,f}$	$1 \times 10^{5\,f}$	

[a] Determined by UV or fluorescence titration
[b] ca 0.09 mol L^{-1} water added to titration solvent; octyl glycosides unless otherwise stated
[c] Ref. [27b]
[d] Ref. [28]
[e] Ref. [29]
[f] Decyl glycoside.

2.2.1.4 Polyaza Clefts and Cavities

A variety of H-bonding arrays can be generated by combining NH and =N– units in rigid frameworks. This has proved a popular approach to carbohydrate recognition, attracting the attention of several groups. One type of design employs "flat", crescent-shaped architecture, as illustrated by **16–20**. These systems are not, in fact, rigidly planar, and will tend to assume twisted conformations on binding to chiral substrates. This means that carbohydrate recognition can often be detected by CD spectroscopy. Other optical effects (e.g. changes in UV or fluorescence spectra) are also likely. Receptors **16** were reported by Anslyn et al. in 1994 [30]. Most of the work involved cyclohexane 1,2-diols and 1,2,3-triols as substrates, but a binding constant of 190 M^{-1} to glucoside **2c** was measured in CDCl$_3$. Receptors **17** were studied more recently by Bell and coworkers [31]. Strong CD and fluorescence effects were observed on complexation of **17a** with octyl glycosides, although affinities were still modest (e.g. K_a = 640 M^{-1} for **2a** in CHCl$_3$) [31a]. **17b** bound amino-sugar salts strongly, even in methanol (e.g. K_a = 2×10^5 M^{-1} for glucoseammonium **21**), although control experiments showed that most of the binding energy was associated with the ammonium group [31b].

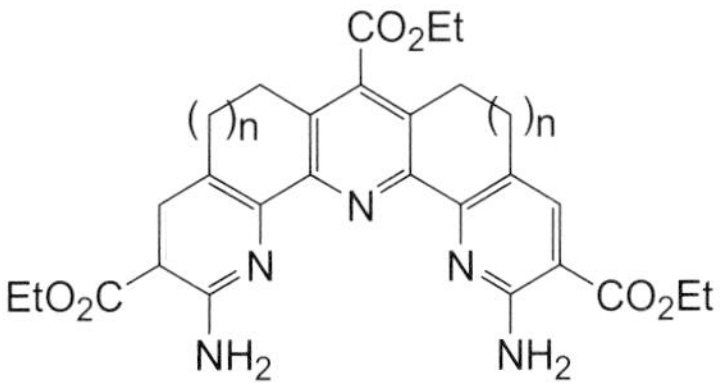

16a n = 1
16b n = 2

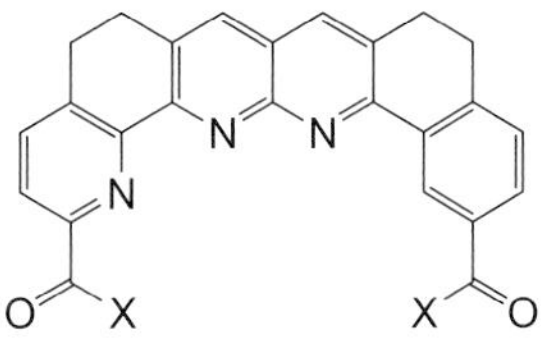

17a X = NHC$_6$H$_{13}$
17b X = O$^-$

The extended bis-ethynyl frameworks in **18–20** are better able to surround a carbohydrate, but are also more subject to twisting. This point is illustrated by **18a/b**, designed by Inouye et al. to bind ribofuranosides [32]. Tested against **22** in CDCl$_3$, acyclic receptor **18a** was poorly effective ($K_a = 30$ M^{-1}) but cyclic analog **18b** performed far better ($K_a = 2400$ M^{-1}). The system was later furnished with a polyoxyethylene bridge, intended to promote complexation of pyranosides [33]. Although receptor **18c** did indeed bind glucoside **2a** quite effectively ($K_a = 5600$ M^{-1} in CDCl$_3$), the affinity for riboside **22** was very similar. Receptor **19a** was designed to bind deoxyribosides, after discovering that **18** is quite poor for these substrates. The "mutation" of =N– to NH produced the desired effect; the association constant to **23** in CDCl$_3$ was increased from 690 to 19,000 M^{-1} [34]. In contrast with **18/19a**, the naphthyridine-based system **20** proved effective in the acyclic form, having good affinity and interesting selectivity (e.g. K_a' values in CDCl$_3$ of 20,000, 6200, and 600 M^{-1} for binding to β-glucoside **2a**, β-galactoside **4a**, and riboside **22**, respectively) [35]. Complex formation was accompanied by large changes in UV–visible and fluorescence spectra, suggesting potential for carbohydrate sensing. The bis-benzo (i.e. indolyl-terminated) analog gave strong CD signals on saccharide binding.

18a R = C$_{11}$H$_{23}$

18b R,R = (H$_2$C)$_2$ ⟨...⟩ (CH$_2$)$_2$

18c R,R =

19a R = i-Bu
19b R = C$_{11}$H$_{23}$

20

The polymer **24**, essentially a recurring version of **18**, also led to strong CD effects in the presence of carbohydrates [36]. The substrates are assumed to act as templates, promoting helical conformations with convergent =N– units. The largest effect was observed for β-Glucosides **2a** and **ent-2a**. Oligomers with $n = 6, 12, 18$ and 24 were also tested, and only the last two gave significant signals. The polymer can form a helix with ca. six units per turn, so this result implies that $\geq +3$ turns are required to produce the effect. A binding study on **2a** + **24** ($n = 24$) in CH$_2$Cl$_2$ gave $K_a = 1200$ M^{-1}.

Clefts with enforced chiral 3D structures have been studied by the Diederich group [37]. Their design is based on a spirobifluorene framework, as in the C$_2$-symmetric **25** [37a]. These receptors are not especially powerful but have interesting selectivity.

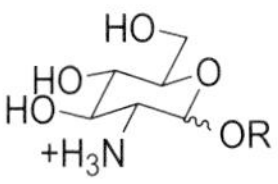

21 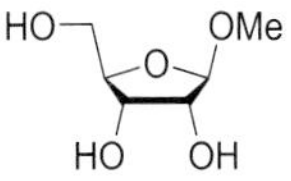   **22** 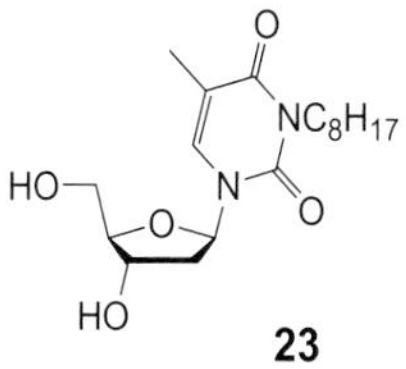   **23**

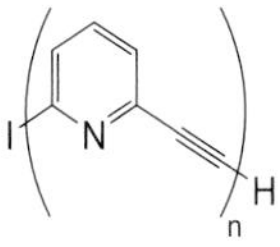

24

Receptor **25a** binds **1a** and **ent-1a** with K_a = 425 and 100 M^{-1}, respectively (CDCl$_3$). Reduced enantioselectivity but increased diastereoselectivity (up to 2.5:1 for **2a** compared with **1a**) is observed for the dendrimeric analogs **25b**. C$_3$-symmetric clefts have also been explored, principally by Mazik and coworkers [38]. A series of general form **26** bound **2a** in CDCl$_3$ with K_a rising to 13,700 M^{-1} (**26a**) and 26,500 (**26b**) [38a,b]. A later example, **27**, resulted in similar affinity but greatly improved diastereoselectivity (26:1 for **2a** compared with **1a**) [38c]. Finally, the C$_3$-symmetric macrocycle **28** has been studied by Lowe, and shown to bind **1a**, **2a**, and **4a** with K_a in the range 3400–7600 M^{-1} [39].

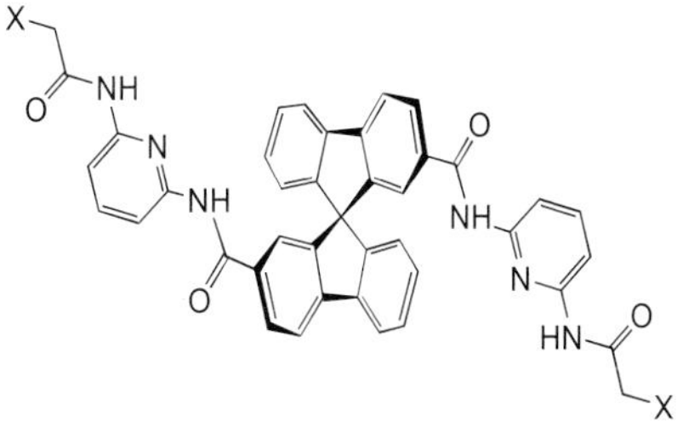

25a   X = NHBoc

25b   X =

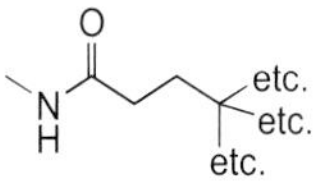

dendrimeric structures, terminated with (OCH$_2$CH$_2$)$_4$OMe

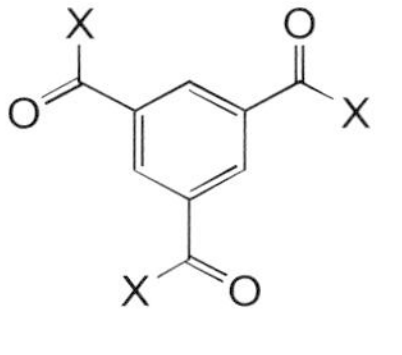

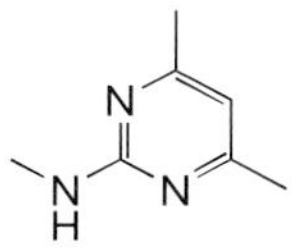

26a

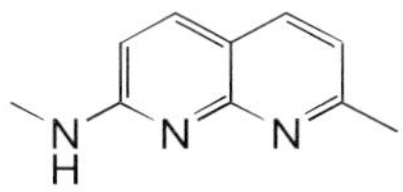

26b

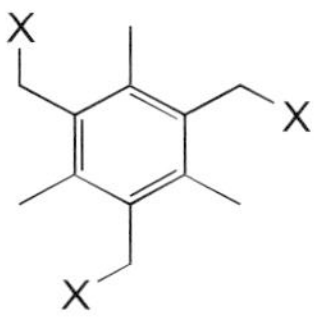

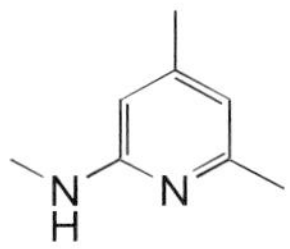

27

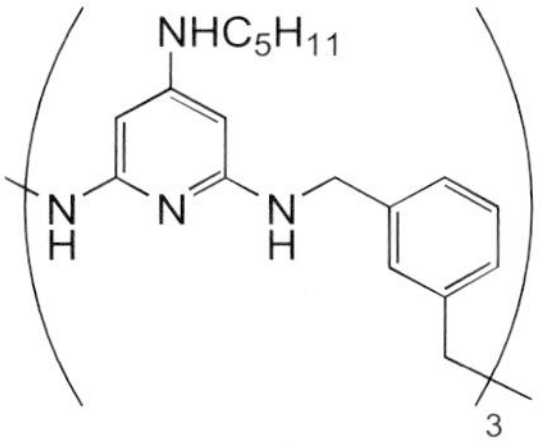

28

2.2.1.5 **Anionic H-bond Acceptors**

Although pre-organization is important in supramolecular chemistry, it is also advantageous to use the strongest possible noncovalent interactions. Anionic centers can be powerful H-bond acceptors, especially if they are relatively basic. For carbohydrates, phosph(on)ate and carboxylate units are especially attractive, because of the potential for anion–diol binding by bidentate H-bonding, as in **29**. Indeed, the groups of Hamilton [40], Diederich [41] and Schneider [42] have all shown that a single PO_2^- unit can associate quite strongly with carbohydrates in organic media. The first two have also developed more sophisticated receptors containing multiple anionic centers. In these, anionic centers have further advantages; they cannot "neutralize" each other by intramolecular hydrogen bonding and, indeed, will tend to repel each other to maintain a pre-organized cleft or cavity. The effectiveness of this strategy is reflected in the fact that studies were generally performed in polar (i.e. competitive) organic solvents such as CH_3CN. The use of such hydrophilic functional groups might not, on the other hand, be favorable for applications in two-phase solvent systems (cf. Sect. 2.2.2).

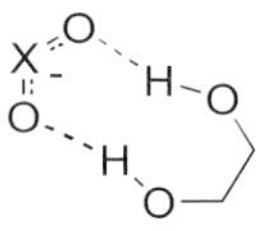

X = C, P

29

The designs employed by Hamilton involve two phosphonate units deployed on rigid spacers. An example is the chiral spirobifluorene **30** [40b]. This compound was not resolved, but tested in racemic form against octyl glycosides in CH_3CN. For β-glucoside **2a** an average K_a of 47,000 M^{-1} was obtained for the two diastereomeric interactions. The ratio of the binding constants (i.e. the enantioselectivity) could also be extracted from the data, and proved to be 5.1:1 This is still (by a small margin) the highest figure recorded for carbohydrate recognition by noncovalent interactions.

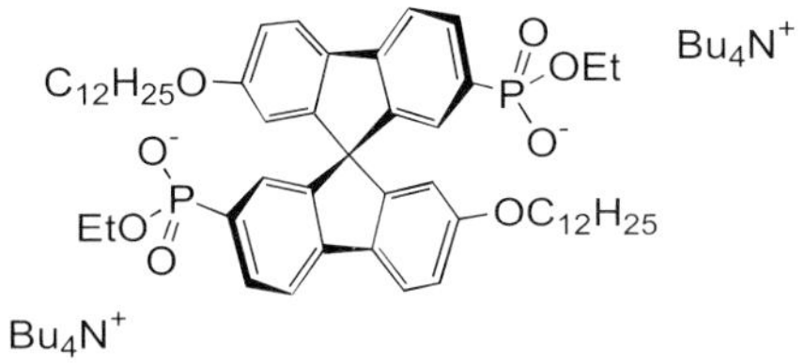

30

The Diederich group have reported a range of receptors based on cyclophosphorylated binaphthol units [41, 43]. Especially interesting are the macrocycles **31** and **32**. Cyclotrimer **31a** is too small to accept a monosaccharide guest but still binds glucoside **2a** with K_a = 3500 in CD_3CN (~20 times more strongly than a monomeric control). Association presumably occurs by face-to-face geometry [43c]. Cyclotetramer **31b** can surround a pyranose and is correspondingly more effective [41a, 43c]. Bind-

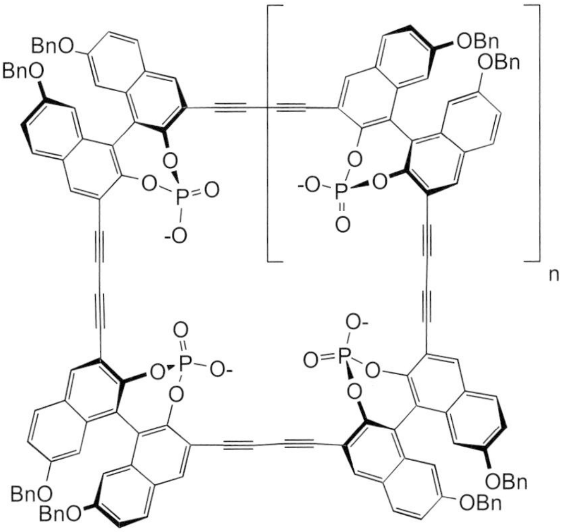

31a n = 0

31b n = 1

ing constants of 4500–5800 M^{-1} were measured for **2a**, **ent-2a**, and **4a** in CD$_3$CN–CD$_3$OD, 98:2 – addition of methanol clearly makes the solvent more competitive. The extended system **32** was intended for complementarity to disaccharides [43a,c]. **32a** was tested against the organic-soluble lactose, maltose, and melibiose derivatives **33–35** (Fig. 2.3), in a solvent system with even more methanol (CD$_3$CN–CD$_3$OD, 88:12). K_a values of 10,700–12,500 M^{-1} were recorded, remarkable considering the polarity of the medium. Although there was negligible selectivity between disaccharides, there was no detectable interaction with monosaccharide derivatives. Receptor **32b** was prepared mainly with a view to introducing carboxyl groups,

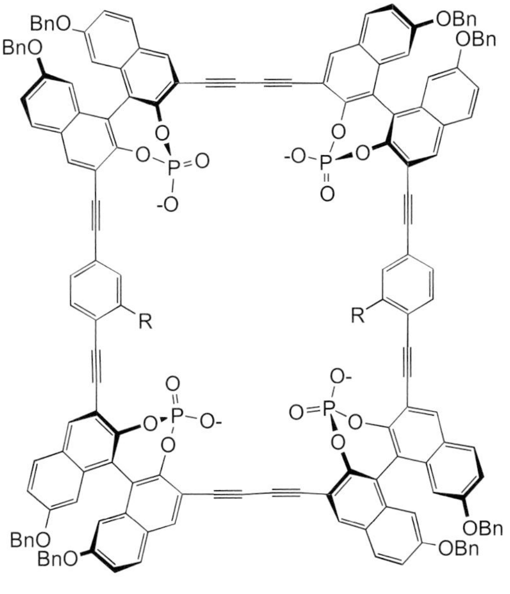

32a R = H

32b R = CO$_2$Me

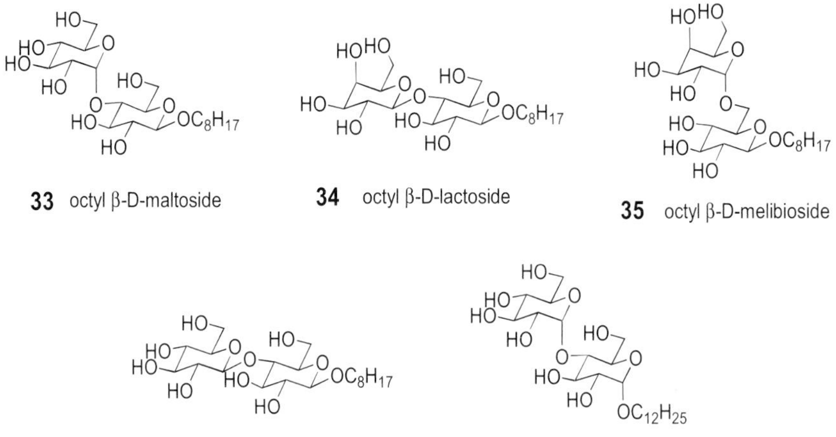

33 octyl β-D-maltoside

34 octyl β-D-lactoside

35 octyl β-D-melibioside

36 octyl β-D-cellobioside

37 dodecyl α-D-maltoside

Fig. 2.3 Lipophilic disaccharide derivatives for binding studies in nonpolar solvents.

with potential for "glycosidase" activity. Catalysis has not yet been demonstrated, but the ester group resulted in a further increase in affinity; melibioside **35** was bound with $K_a = 41,000$ M^{-1} in the above solvent system.

Anionic binding groups are clearly advantageous if the target is cationic. Schrader has shown that the bis-phosphonate **38** binds protonated amino sugars strongly, even in polar DMSO [44]. For example, β-D-glucoseammonium was complexed with $K_a = 92,000$ M^{-1} [44b]. NMR clearly indicated the binding geometry shown, in which the 3- and 1-OH groups contribute to the association. Simple amines were bound a factor of six less strongly [44a] and there were appreciable differences (factors of 2–3) between anomeric and epimeric substrates. Carboxylates were employed by Kubik as

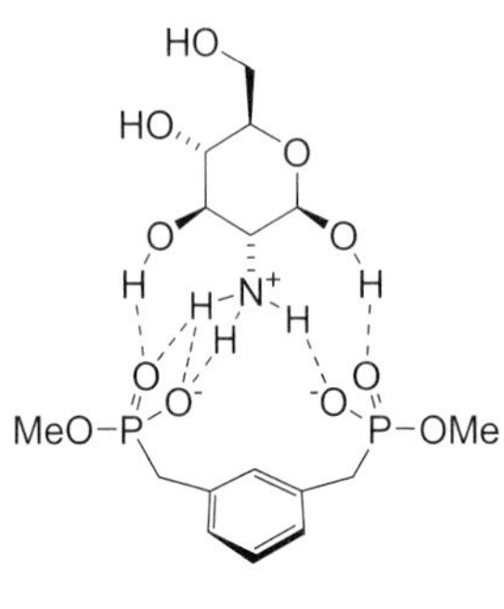

38. β-D-glucoseammonium

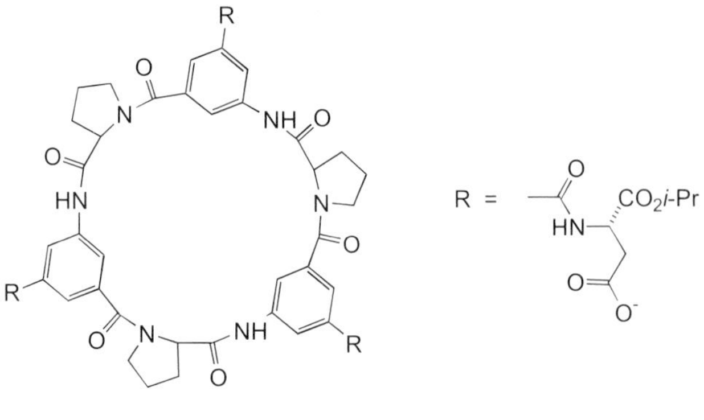

39

binding units in **39** [45]. This receptor was also effective in a relatively polar solvent system. Studies performed in $CDCl_3$–MeOH, 96:4 revealed $K_a = 540$–810 м$^{-1}$ for several alkyl pyranosides.

2.2.1.6 Tricyclic Oligoamides

Of the receptors described above, some can surround a carbohydrate, making contact from several directions. There are, however, few, if any, which can be said to fully encapsulate their substrate with pre-organized binding units. Tricyclic structures provide one approach to this goal. Oligoamide designs have been favored because (1) amide bonds are readily formed, even in high-dilution macrocyclizations, and (2) the secondary amide is a versatile hydrogen-bonding functionality. An early example of this architecture was provided in 1993 by the group of Still [46]. Receptor **40** was designed to bind peptide models rather than carbohydrates, and cannot, in fact, enclose the latter. It illustrates, however, the value of a pre-organized, polycyclic 3D framework. As shown in Tab. 2.3, affinity was moderate but selectivity was occasionally quite impressive.

In contrast, the biphenyl-based octa-amide **41a** was specifically designed to enclose a monosaccharide and was, moreover, targeted at a particular class of substrate [47].

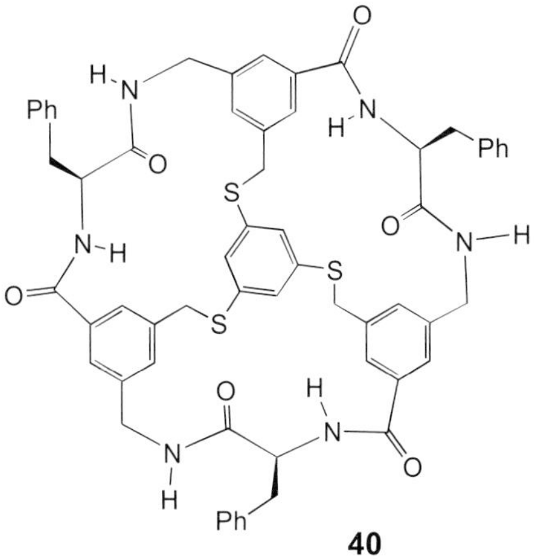

40

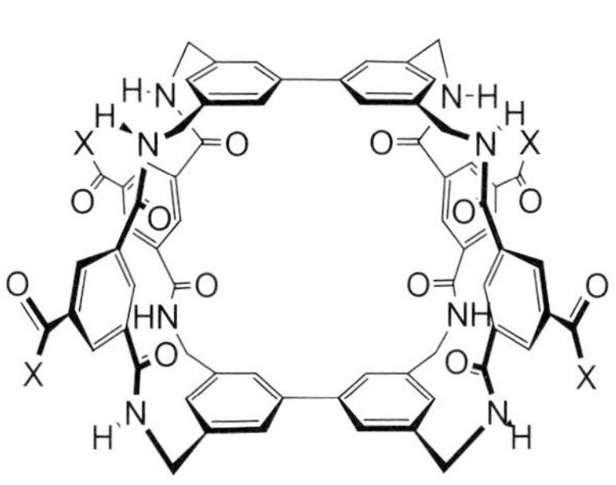

41a X = OC_5H_{11}

41b X = NH

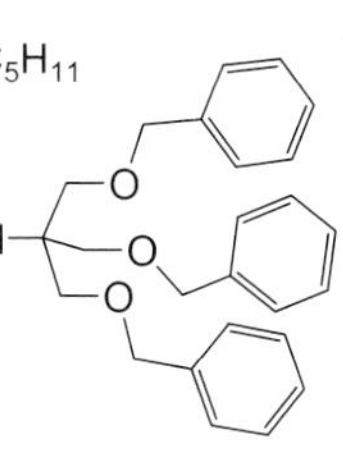

Tab. 2.3   Association constants (K_a, M^{-1}) between receptors **40/41a** and organic-soluble glycosides.

	40 in CDCl$_3$[a]	41a in CDCl$_3$/CD$_3$OH, 92:8[b]	41a in CHCl$_3$[c]
α-D-glucoside **1a**	370	20	13,000
α-L-glucoside **ent-1a**	160		
β-D-glucoside **2a**	1700	980	300,000
β-D-galactoside **ent-2a**	440		
α-D-galactoside **3a**	260		
β-D-galactoside **4a**	610	220	110,000
α-D-mannoside **5a**	5500		
β-D-mannoside **6a**	220		

[a]   Determined by 1H NMR titration; see Ref. [46]
[b]   1H NMR titration; see Ref. [47]
[c]   Fluorescence titration; see Ref. [47]

β-Glucosides **2** are distinguished by their exclusively equatorial array of polar substituents. Crudely, they can be seen as disks with apolar top and bottom surfaces (composed of CH units) and peripheral, externally directed, polar groups (OH and OR; Fig. 2.4). At this level of detail, complementarity can be achieved by means of a cage constructed from two parallel aromatic units joined by rigid, polar spacers. Receptor **41a** conforms to this requirement and, according to modeling, is the appropriate size and shape. It is worth noting that the arrangement of polar and apolar units is quite similar to that employed by the *E. coli* galactose chemoreceptor protein (Fig. 2.1), especially in the use of aromatic surfaces to "sandwich" the substrate. As shown in Tab. 2.3, octyl *β*-D-glucoside **2a** was indeed bound strongly in CHCl$_3$, and significantly in CDCl$_3$/CD$_3$OH, 92:8. As expected, the all-equatorial stereoisomer was preferred. Especially significant was the relative selectivity for *β*- and *α*-glucosides – ca 50:1 in the mixed-solvent system. Receptor **41b**, a more lipophilic analogue of **41a**, is discussed in Section 2.2.2.2.

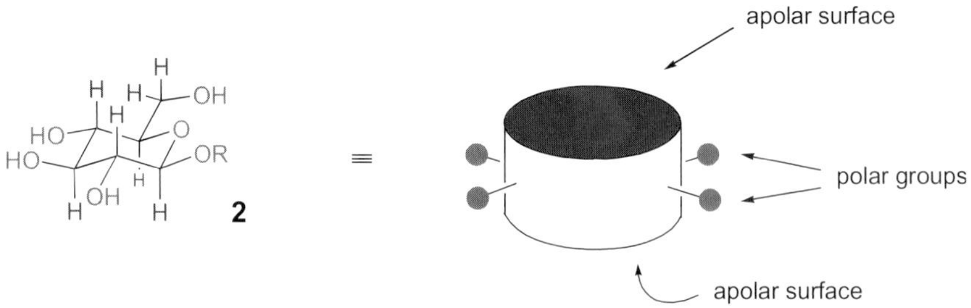

Fig. 2.4   *β*-D-Glucopyranoside **2** interpreted as polar and apolar binding regions.

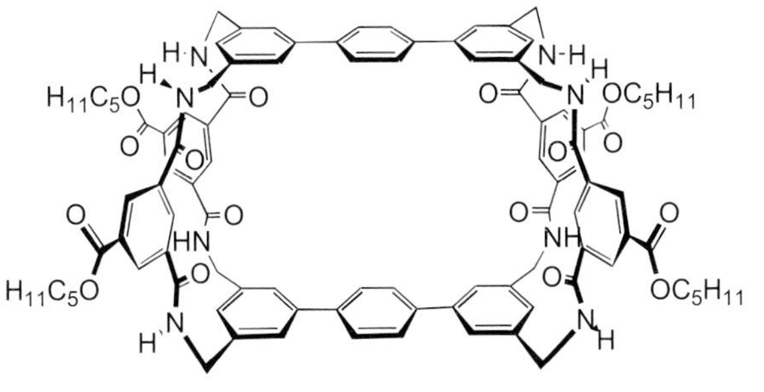

42

An extended version of this system, targeted at disaccharides, has also been re-ported [48]. The terphenyl-based receptor **42** was expected to prefer β-cellobiosides such as **36**, in which all polar substituents are equatorial. Indeed, **36** was bound with $K_a = 7000 \text{ M}^{-1}$ in $CDCl_3/CD_3OH$, 92:8. Glycosides **1a**, **2a**, **33**, **34**, and **37** were also test-ed as substrates but, remarkably, binding was always undetectable. The ability of **42** to distinguish between cellobioside **36** and lactoside **34** is especially notable, because these stereoisomers differ at just 1 of 10 asymmetric centers.

A series of chiral tricyclic hexalactams, exemplified by **43**, has been described by the Diederich group [49]. Binding constants were low, however (e.g. 270 M^{-1} for **43** + **2a** in $CDCl_3$) suggesting that, for these, the carbohydrate might not fully enter the cavity.

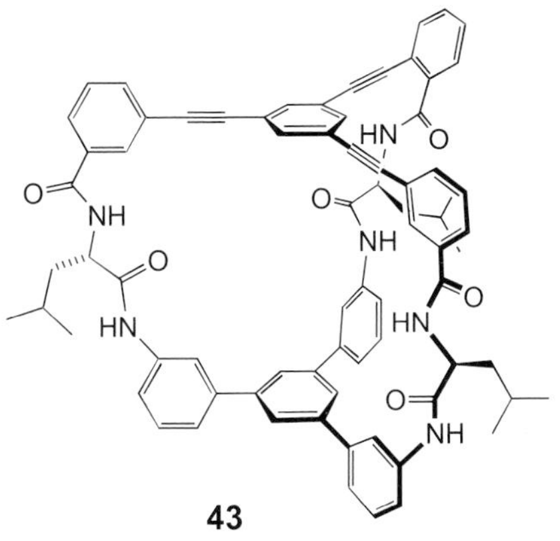

43

2.2.1.7 Other Macrocycles with Polar Cavities

Finally, two other macrocyclic systems fall outside the preceding categories. Receptor **44**, due to Gellman, was an early attempt to target protonated amino-sugars [50]. Binding was dominated by $N^+–H\cdots O=S/P$ interactions, but moderated by $OH\cdots OH$ hydrogen bonding. The hexaol **45**, due to Diederich, had high affinity for octyl glycosides in $CDCl_3$ and notable enantioselectivity (e.g. K_a for **1a** and **ent-1a** of 1100 and 4400 M^{-1}, respectively) [51]. Interestingly, this C_2-symmetric S,S,R-system was much more effective than its C_3-symmetric S,S,S stereoisomer.

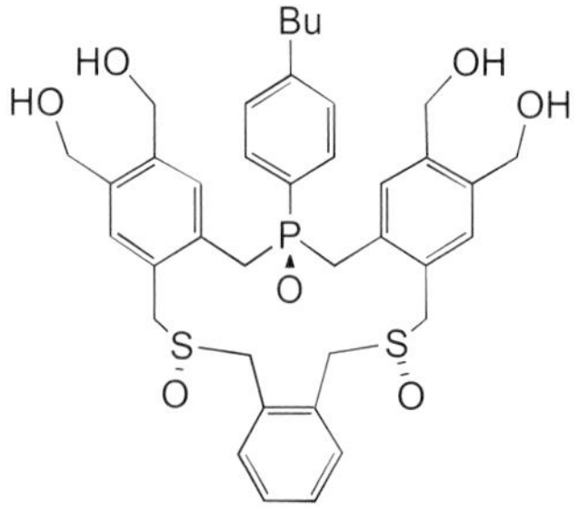

44

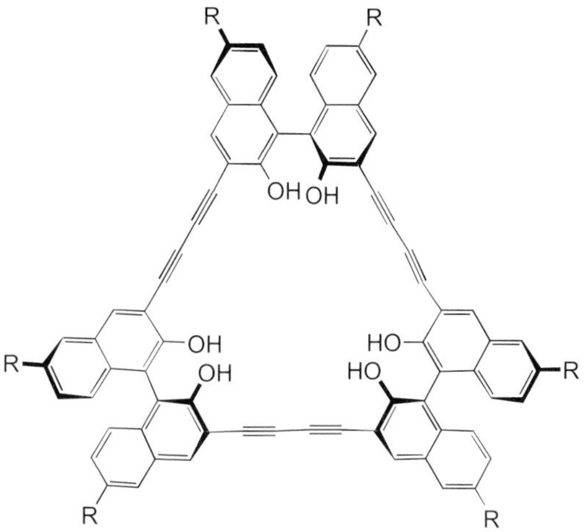

45 R = CH₂CH₂Ph

2.2.2
Recognition in Two-phase Systems

For carbohydrate receptors to be "functional" in the fullest sense it is necessary that they address substrates of real importance. Organic-soluble receptors are thus at a disadvantage; binding studies require specialized lipophilic substrates which are not typical of natural carbohydrates, and might require special synthesis. This problem can be avoided by employing two-phase systems, and showing that natural carbohydrates can be extracted into organic media. The dissolution of solid carbohydrates in nonpolar solvents might be seen as an interesting function, potentially enabling new synthetic processes (although this possibility is still largely unexplored). The extraction of carbohydrates from aqueous into nonpolar media is more challenging but more significant. First, it shows that the receptor can compete for the saccharide with liquid water. Second, it is relevant to carbohydrate transport across cell membranes, and possible applications in the delivery of carbohydrate-like pharmaceuticals. Third, it relates directly to carbohydrate sensing. For example, a receptor capable of glucose extraction could be dissolved in an apolar polymeric membrane, and placed in contact with the bloodstream. Given some method for translating phase transfer into an electronic signal, the system could act as a glucose sensor. Finally, phase transfer can be used in separation systems based on selective transport across apolar barriers. A particular need is enhancement of fructose levels in glucose–fructose mixtures. The resulting "high-fructose syrup" is especially sweet-tasting, and valuable to the food industry [52].

2.2.2.1   Dissolution of Solid Carbohydrates in Apolar Media
Several of the receptors discussed in Sect. 2.2.1 have been shown to solubilize carbohydrates, usually in CH(D)Cl₃ (assumed hereafter, unless otherwise stated). The C_3-

symmetric tris-aminopyridine **27** dissolved ~0.7 equiv. methyl β-D-glucoside [38c] whereas the more powerful tricyclic octalactam **41a** solubilized ~0.9 equiv. glucose [47]. Interestingly, the glucose dissolved by **41a** was predominantly the β-anomer when analyzed by NMR, in accordance with the preference for β-glycosides noted previously.

More extensive studies have been performed by Inouye et al. on receptors **46** (a close relative of **18b**) and **19b**, and on polymer **24**. As shown in Tab. 2.4, receptor **46** was more successful with pentoses **47–51** than with the hexoses **52–54** and **57** [32]. Among the pentoses there seemed a clear preference for ribose **47**, the intended target of this system. Once again, it should be noted that such apparent selectivity might simply reflect the natures of the substrates; some might be more "difficult" than others. Comparison with **19b** (Tab. 2.4) shows, however, that receptor structure has a very significant effect. The latter does succeed with the hexoses, and especially favors deoxyribose **51**, the substrate for which it was designed [34]. Polymer **24** dissolved glucose, mannose, galactose, and fructose in both chloroform and CH_2Cl_2. The amounts were not quantified, but the presence of the carbohydrates could be inferred from induced CD signals [36].

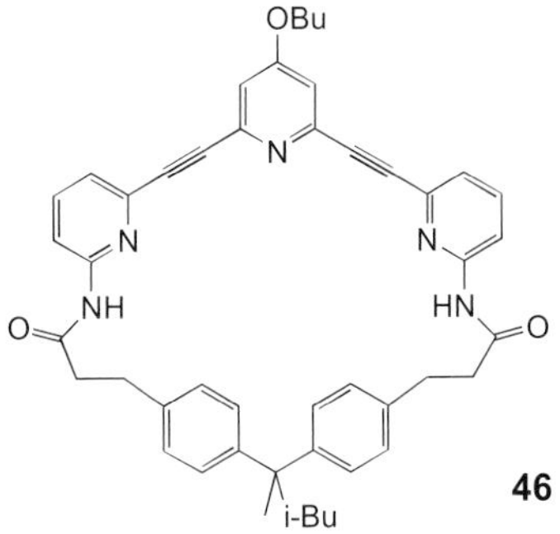

46

Tab. 2.4 Solubilization of sugars in chloroform by **46**, **19b**, and **15c** (M = Zn).

Sugar[a]	Sugar/46[b,c]	Sugar/19b[b,d]	Sugar/15c (M = Zn)[b,e]
D-Ribose **47**	1.0	0.65	0.91
D-Arabinose **48**	0.06	0.53	
D-Xylose **49**	0.5	0.52	2.4
D-Lyxose **50**	0.63	0.66	
2-Deoxy-D-ribose **51**	0.7	1.0	
D-Glucose **52**	_[f]	0.45	0.56
L-Glucose **ent-52**			0.42
D-Galactose **53**	_[f]	0.32	0.44
D-Mannose **54**	_[f]	0.39	0.24
D-Fructose **57**	0.2	0.52	

[a] Fig. 2.5
[b] Molar ratios sugar/receptor occurring in solution
[c] Ref. [32]
[d] Ref. [34]
[e] Ref. [29]
[f] Negligible quantities extracted

The tetrakis-ureido Zn porphyrin **15c** (M = Zn) was also tested against several monosaccharides (Tab. 2.4) [29]. The results suggest general effectiveness roughly equivalent to **19b**, although with somewhat different preferences. It is interesting to compare this system with **41a**, another for which both glucose extraction and homogeneous glucoside binding have been quantified. For the Zn porphyrin the figures are 0.56 (D-glucose extraction, equiv.) and 2×10^7 (β-D-glucoside binding, M^{-1}), whereas for octalactam **41a** they are ~0.9 equiv. and 3×10^5 M^{-1}, respectively. Thus the porphyrin, with its deep binding pocket, seems better suited to glucoside binding whereas the tricyclic framework might be preferable for the native monosaccharide.

Bis-anion **58** dissolved a full equivalent of glucose into CD_3CN, and two equivalents into $CDCl_3$ [41b]. Unfortunately it did not give analyzable data in homogeneous binding experiments. It also failed to extract glucose from aqueous solution. Anionic **59** forms reverse micelles which solubilize glucose monohydrate in the ratio 9:1 anion:carbohydrate [53]. This disorganized system stretches the definition of a receptor, but is one of the more convenient methods for mobilizing carbohydrates in a nonpolar medium.

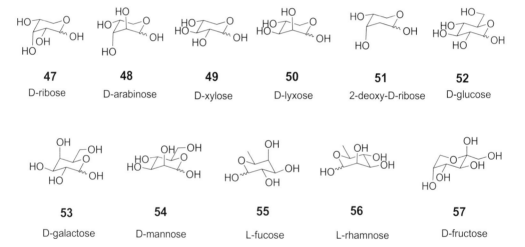

Fig. 2.5 Carbohydrates used for extraction studies, represented as pyranosides. The conformation shown is not necessarily always the most stable.

2.2.2.2 Phase Transfer of Carbohydrates from Aqueous into Organic Solvents

The extraction of saccharides from water was pioneered by Aoyama's group, who used the octahydroxycalixarene **7**. Dissolved in CCl_4, it proved capable of transferring several carbohydrates from quite concentrated aqueous solutions into the organic phase [18, 54]. Some of the results are summarized in Tab. 2.5. The extent of extraction seems to depend on two factors. The first is the hydrophilicity of the substrate. Thus, galactose was barely extracted whereas fucose (6-deoxygalactose) was readily transferred. The second is the relative configurations at C3 and C4 of the substrates.

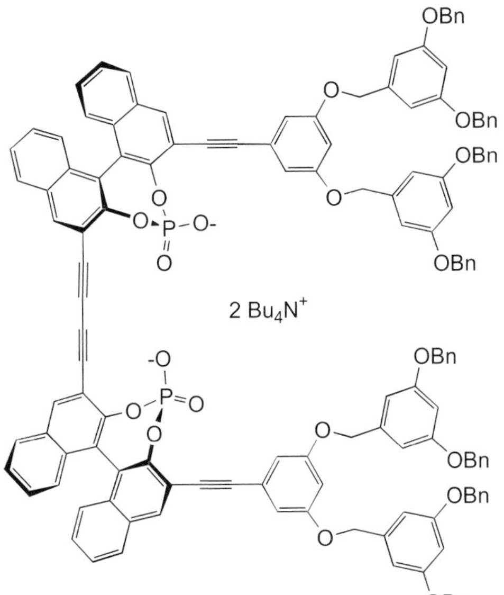

Extraction seems to be favored by a cis arrangement in the Fischer projection (e.g. **47**, **48**, **51**, **55**) which translates to cis in the pyranose form of the carbohydrates. The results suggest face-to-face binding geometry in which OH groups emerging from the same face of the saccharide are best able to form multiple interactions with the receptor. It is notable that ribose **47** was extracted almost exclusively as α-pyranose **60**, whereas in water this form contributes just 20 % to the total isomeric mixture (the remainder being furanose and β-pyranose). This manipulation of the isomeric form

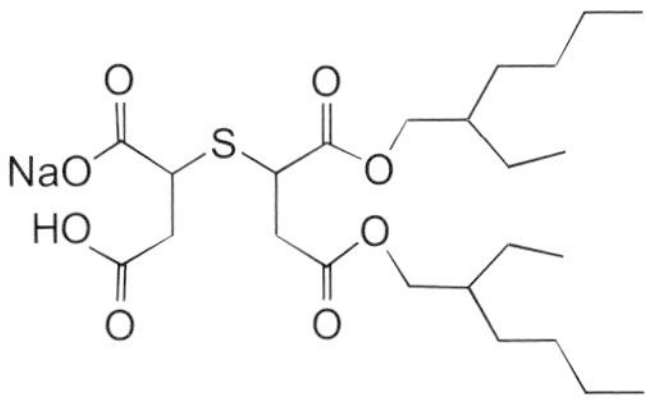

Tab. 2.5 Extraction of sugars from water into CCl_4 by **7**.[a]

Sugar[b]	Sugar/7[c]
D-Ribose **47**	0.5
D-Arabinose **48**	0.1
D-Xylose **49**	vs[d]
D-Lyxose **50**	vs[d]
2-Deoxy-D-ribose **51**	0.8
D-Glucose **52**	vs[d]
D-Galactose **53**	vs[d]
D-Mannose **54**	vs[d]
L-Fucose **55**	1.0
L-Rhamnose **56**	0.1

[a] Data from Ref. [54]; $[\text{Sugar}]_{\text{aq}} = 2.4$ M, $[\text{7}]_{\text{org}} = 0.9 \times 10^{-2}$ M
[b] Fig. 2.5
[c] Molar ratios sugar/7 appearing in the organic phase
[d] Very small; ≤ 0.03

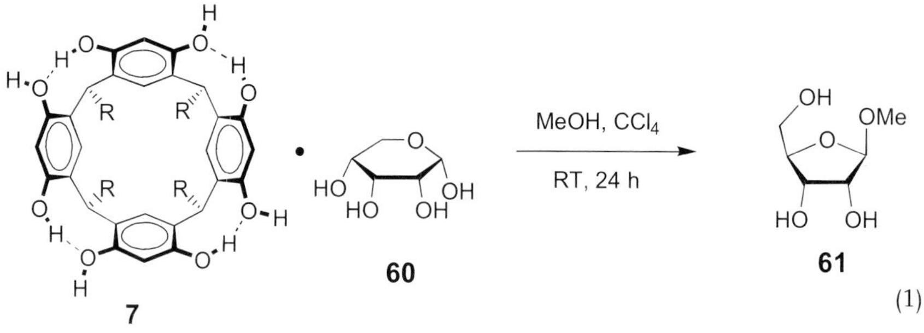

$$(1)$$

might be seen as a functional effect of **7**. A further development involved the promotion and control of glycoside formation. If the complex between **7** and **60** in CCl$_4$ was treated with methanol (10 equiv.), methyl β-D-ribofuranoside **61** was formed with ~100 % conversion and almost complete selectivity (Eq. 1) [55]. The receptor presumably acts as the acid catalyst for glycoside formation, and also controls the stereochemistry; conventional methods are far less stereoselective.

It was later shown that methyl β-D-glucoside **62** could also be extracted by **7** from water into CCl$_4$, despite the absence of *cis*-1,2-diol units [16]. In this instance, however, a 2:1 receptor:substrate ratio suggested a "sandwich" structure in which receptor molecules bind to both faces of the guest. The α anomer **63**, for which the sandwich structure is probably not possible, was very poorly extracted under the same conditions.

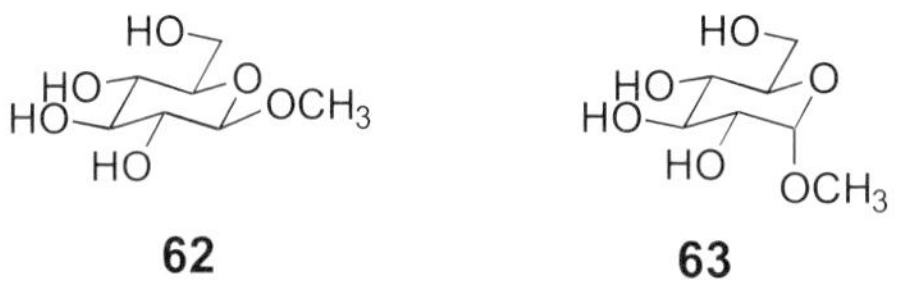

More recently, receptor **7** has been employed as a saccharide carrier in a supported liquid membrane. The membrane was prepared by impregnating a porous PTFE film with a CCl$_4$ solution of **7**. Surprisingly, the transport selectivity was much smaller than might be expected from Tab. 2.5. Pentoses **47–49** and hexoses **52/3** were all transported; the difference between the fastest (arabinose) and slowest (mannose) initial rates was a factor of just 3.8 [56]. Calixarene **7** has also been studied in an electrode–monolayer–water interfacial system. The electrode responded to quite low levels (~10^{-4} M) of monosaccharides although, surprisingly, simple amphiphiles such as octadecanol were also effective [57].

The steroid-based cholaphanes **10** and **11** were tested as phase-transfer agents for D-glucose **52** and methyl β-D-glucoside **62** in water–chloroform [25a]. Results for glucose were negative but measurable amounts of methyl glucoside were extracted. Receptor, **11a** was the most efficient, transferring ~0.2 equiv. **62** into CHCl$_3$ from a 1.75 M aqueous solution. It was also shown to transport this substrate through a chloroform bulk liquid membrane (BLM) in a "U-tube" apparatus.

Tricyclic octalactam **41a** showed high affinities in homogeneous solution, and was expected to act as a powerful phase-transfer agent. In practice it proved unsuitable; attempted extractions gave precipitates at the chloroform–water interface. The dodecabenzyloxy analog **41b**, however, was more successful [58]. This highly lipophilic derivative remained in the chloroform phase and transferred substantial amounts of monosaccharides from aqueous solutions. Selected results are shown in Tab. 2.6. It is notable that, unlike **7** and **10/11**, receptor **41b** can extract the common hexoses glucose, galactose, and mannose. Moreover, in favorable cases such as glucose, very high concentrations are no longer required; extraction is detectable down to 0.1 M aqueous saccharide. Although extraction from physiological levels (~0.005 M), as required for medical glucose sensing, is not yet possible, this goal might not be out of reach in the short–medium term. The selectivity of **41b** is difficult to assess because of uncertainties about the "intrinsic extractabilities" of the different substrates. There seems, however, to be a clear preference for the all-equatorial substitution pattern, as in glucose and xylose. The selectivity for glucose is probably not driven by lower hydrophilicity; mannose is thought to be less well hydrated in aqueous solution [59].

Tab. 2.6 Extractabilities of monosaccharide substrates from water into chloroform by receptor **41b**.[a]

Substrate[b]	Concn substrate in aqueous phase		
	1.0 M	0.5 M	0.1 M
D-Ribose **47**	0.7		
D-Xylose **49**	1.1		
D-Glucose **52**	1.0	0.5	<0.1[c]
D-Galactose **53**	0.2	<0.1[b]	None detectable
D-Mannose **54**	<0.1[b]	<0.1[b]	None detectable
Methyl β-D-glucoside **62**	1.0		
Methyl α-D-glucoside **63**	1.0		

[a] Molar ratios sugar/**41b** appearing in the organic phase. [**41b**]$_{org}$ = 2.9×10^{-4} M; data from Ref. [58]

[b] Fig. 2.5
[c] Carbohydrate detectable, but amounts too small for quantification by NMR integration

The tricyclic framework of **41** can in principle be "tuned" by introducing alternative rigid diacid spacers. Receptors **64**, incorporating pyridyl units, are a first step in this direction [60]. The symmetrical **64a** had low affinity, extracting just 0.15 equiv. glucose from a 1 M aqueous solution into CHCl$_3$. However, the less symmetric **64b** was more useful. This receptor seemed roughly as powerful as **41b** but had complementary selectivity; under the same conditions, glucose, galactose and mannose were extracted to the extent of 0.55, 0.6, and 0.3 equiv., respectively (cf. Tab. 2.6).

There remain a few systems which extract or transport carbohydrates through less well-defined interactions. Aoyama et al. have studied **65**, a macrocyclic but flexible system best seen as a "unimolecular reversed micelle" [61]. Solutions of **65a** in CCl$_4$ mobilize ca. 40 mol equiv. H$_2$O in the organic phase, but extract ribose, glucose, or fructose from 3 M aqueous solutions without co-transfer of water. For glucose, a 1:1

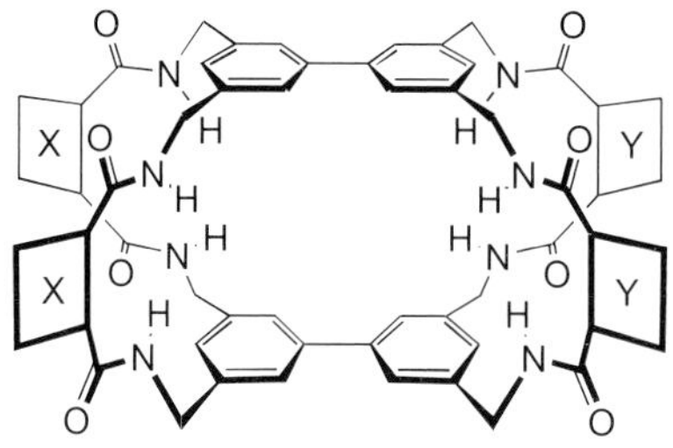

64a   X = Y =

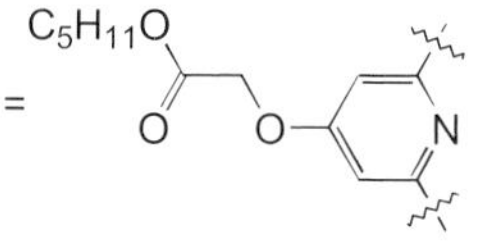

64b   X = 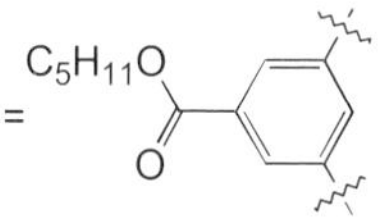   Y =

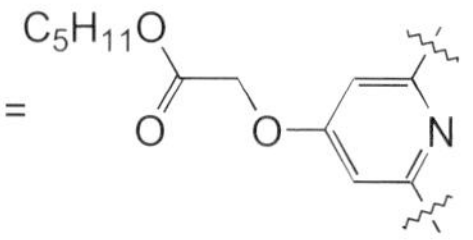

ratio of receptor:substrate was established directly by ^{1}H NMR integration, and also by re-extraction of the sugar back into water. Sugar extraction, and water pool accommodation, was also observed with the acyclic heptakis(dihydroxyalkyl) system **66**. No complexation was observed for **65b** or **67**, however, implying that clustering of the dihydroxyalkyl chains is essential for recognition. The barium complex **68** transported

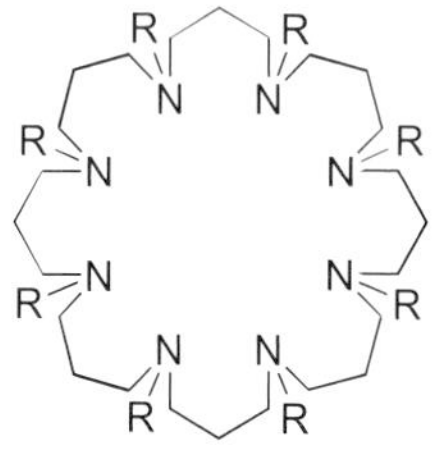

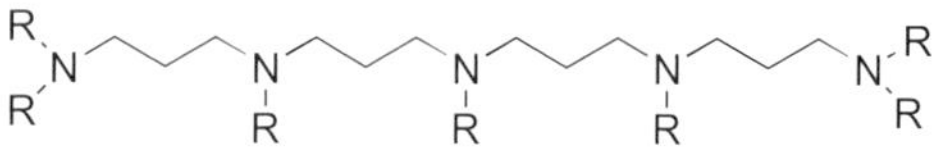

66   R = $CH_2CH(OH)CH(OH)(CH_2)_{10}CH_3$

65a   R = $CH_2CH(OH)CH(OH)(CH_2)_{10}CH_3$

65b   R = $(CH_2)_{11}CH_3$

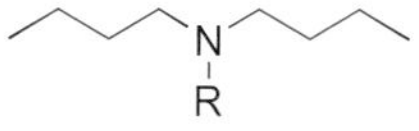

67   R = $CH_2CH(OH)CH(OH)(CH_2)_{10}CH_3$

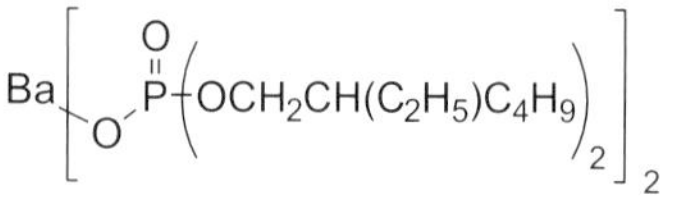

68

several monosaccharides through chloroform bulk liquid membranes, proving especially effective for ribose [62]. The organic-soluble complexes were not fully characterized, but preliminary experiments suggested **68**.(ribose)$_2$ stoichiometry.

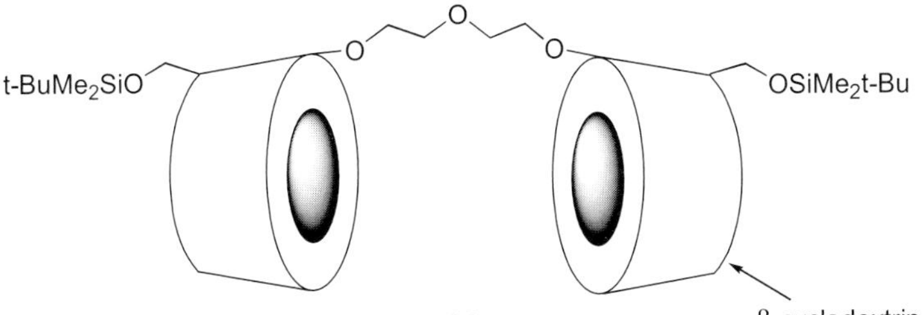

69   β-cyclodextrin

The lipophilic cyclodextrin dimer **69** was active in the transport of ribose, deoxyribose, and methyl α-D-glucoside, methyl α-D-galactoside, and methyl α-D-mannoside through chloroform BLM [63]. Background transport rates in the authors' apparatus were high, however, ranging from 24 to 85 % of the rates in the presence of **69**. Extracted monosaccharide could not be detected by ^{1}H NMR in the chloroform phase. The octacalixarenes **70** and the tetrakis(phosphine oxide) **71** are also reported as carriers [64]. With these more hydrophilic monosaccharides were transported, including glucose and galactose, although there was again no confirmation of measurable carbohydrate extraction. Finally, a surface-bound cyclodextrin has been reported to bind carbohydrates from water, as detected by suppression of ferrocene electro-oxidation [65].

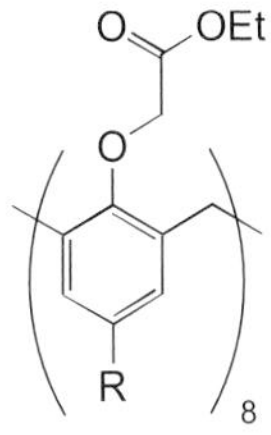

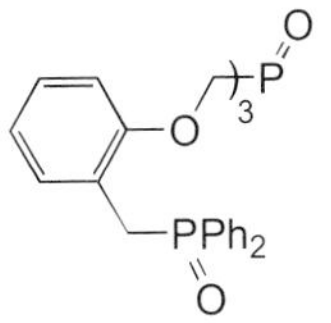

70   R = t-Bu or adamantyl   **71**

2.2.3
Carbohydrate Recognition in Water

Saccharide binding in water by noncovalent interactions is a key goal in the study of carbohydrate recognition. Only in homogeneous aqueous solution can receptors achieve their intellectual function, to furnish information on the mechanisms and driving forces underlying natural saccharide binding. Most applications in medicine and biology require that recognition occurs in water. Cell-surface oligosaccharides are presented to the aqueous extracellular medium and water-soluble receptors (although probably polymer-bound) are likely to provide the most satisfactory solutions for carbohydrate sensing. Furthermore, the use of aqueous media enables a much larger range of substrates to be studied; many oligosaccharides cannot realistically be solubilized in nonpolar solvents or studied in two-phase systems.

Unfortunately, progress towards efficient and selective water-soluble carbohydrate receptors has been slow. The challenge of discriminating between carbohydrates and aqueous solvent has already been emphasized. Control of selectivity presents additional difficulties, especially considering the myriad of subtly-different saccharide structures present in biological systems. There are further practical problems. Merely solubilizing receptors in water is non-trivial. Typical receptors, as discussed in previous sections, are large organic molecules with carbon-based frameworks. Polar groups can be added to carry them into water, but the resulting structures will tend to be amphiphilic. They might, therefore, aggregate in water, interfering with the complexation studies.

Complete and reliable characterization of binding phenomena can also be more difficult in water. In nonpolar solvents, ^{1}H NMR has been the standard technique for detecting complexation and for measuring binding constants. This method is advantageous, because spectra are information-rich and readily interpreted. Complex formation generally involves intermolecular H-bonding, which can be detected directly from downfield motions of the corresponding protons. Further evidence comes from the shielding or deshielding of substrate protons by aromatic rings in the receptor, from changes in coupling constants reflecting and/or from intermolecular nuclear Overhauser effects. It is often possible to estimate binding constants independently of the motions of several signals, raising confidence in the binding model and the accuracy of the results. NMR has one weakness – it cannot be used to determine very high binding constants [66]. Even in these circumstances, however, it can be used to infer complex structures and stoichiometry.

In contrast, NMR has been less widely employed for studies of carbohydrate recognition in water. One reason is probably the issue of solubilization, referred to above. It is easier to achieve concentrations of ~10^{-6} M required for UV–visible or fluorescence studies than the ~10^{-3} M needed for NMR. Second, one is less likely to see clear evidence of H-bond formation. H-bonding protons will be solvated by water molecules before association, so the change in chemical shift will be less and might be unpredictable in direction. Exchange of the protons with solvent can, moreover, preclude the observation of their chemical shifts. The upshot is that many studies in aqueous solution have relied on less clearly diagnostic techniques, for example calorimetry or UV–visible–fluorescence spectroscopy. Such experiments provide accurate and valuable quantitative information, but little or no structural data. Without corroboration there is some risk they might be misinterpreted, a consideration which should be borne in mind while reading this section.

2.2.3.1 Water-soluble Calixarenes

Among the first systems to be studied in water were the anionic calixarenes **72a–c** [67]. Relatives of **7**, these compounds provided an early indication of the challenge awaiting workers in this area. ^{1}H NMR measurements showed unambiguous signs

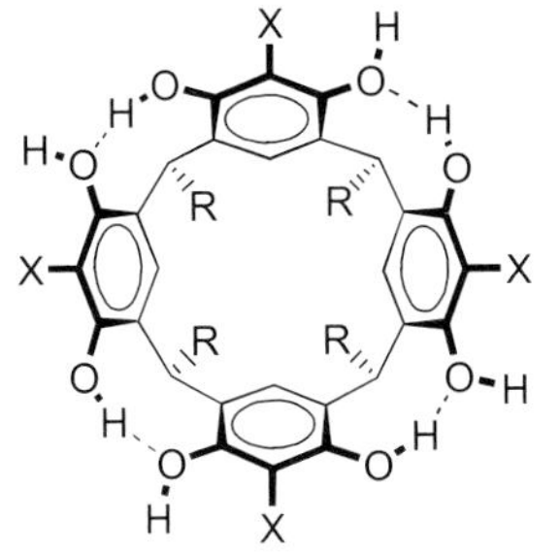

72a   X = H   R = CH$_2$CH$_2$SO$_3$Na

72b   X = Me   R = CH$_2$CH$_2$SO$_3$Na

72c   X = OH   R = CH$_2$CH$_2$SO$_3$Na

72d   X = H   R =   OH OH

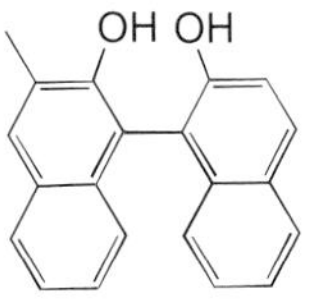

(*R,S*)

of binding to some monosaccharides, but only for quite lipophilic examples such as fucose **55**. Even here, K_a values were very low, as shown in Tab. 2.7. Binding to the common hexoses glucose, galactose, and mannose was undetectable, and even ribose (a favored substrate of **7**) gave $K_a < 1$ M^{-1}. The increase in association constants through the sequence **72a**→**72b**→**72c** suggests a role for CH-π effects, which are maximized by increasing electron density in the π-system [68]. The trend was con-firmed when **72a** was first bis- and then tetra-deprotonated, resulting in further in-creases in affinity (Tab. 2.7) [69]. Recently, the binaphthyl-substituted **72d** was also described [70]. This compound, presumably a mixture of diastereomers, was com-bined with methyl red and titrated against saccharides in water–methanol, 99:1. Changes were observed in the methyl red absorption, implying surprisingly high K_a values (e.g. 1100 M^{-1} for glucose, 4550 M^{-1} for maltose).

Tab. 2.7   Selected association constants (K_a, M^{-1}) for binding of monosaccharides to water-soluble calixarenes **72**.[a]

Substrate[b]	72a	72b	72c	72a – 2H$^+$	72a – 4H$^+$
Arabinose **48**	0.85	2.1	2.5		
2-Deoxyribose **51**	1.2	4.9	3.9		
Fucose **55**	1.8	6.0	8.4	16	26

[a] Determined by ^{1}H NMR titration at 298 K
[b] Fig. 2.5

A second calixarene-type system is the "cyclotetrachromotropylene" **73**, of Poh and coworkers [71]. This receptor is thought to provide a hydrophobic cleft which binds

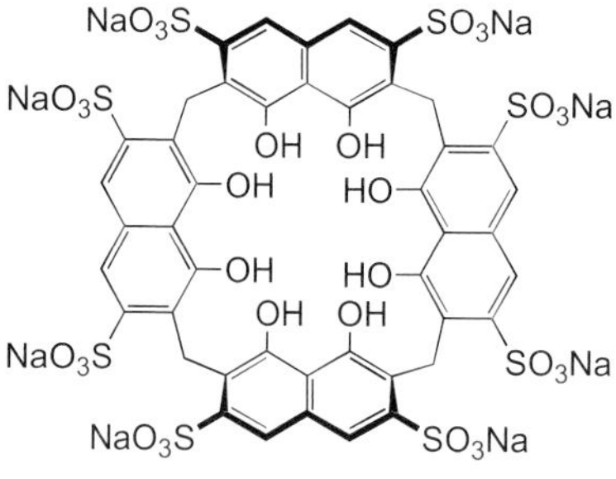

73

simple aliphatic alcohols quite strongly (e.g. K_a = 3000 M^{-1} for *n*-butanol). Again, complex formation was demonstrated by NMR (large upfield shifts for substrate protons). Simple monosaccharides were not bound, but methyl β-D-glucoside **62**, methyl α-D-glucoside **63**, and methyl α-D-mannoside formed complexes with K_a = 6, 28, and 75 M^{-1}, respectively [71a]. **73** was later shown to associate with cyclodextrins, the latter appearing to act, unusually, as the guests in the complexes [71b].

2.2.3.2 Cyclodextrins

In their more familiar role as host molecules, cyclodextrins (cycloamyloses) **74** are capable of binding smaller carbohydrates. Unfortunately, this phenomenon is difficult to investigate, because host and guest tend to give overlapping NMR spectra and there are no structural features which are likely to cause major signal displacements on binding. There are, moreover, no chromophores which might give UV, fluorescence or CD effects. Nonetheless, several groups have published studies. Aoyama [72] and Schneider [73] have both measured binding constants to β-cyclodextrin **74b** using a competition method in which the displacement of a dye from the host is followed by fluorescence spectroscopy. Binding to hexoses was negligible but, as shown in Tab. 2.8, positive and consistent results were obtained for some pentoses. De Namor [74] employed calorimetry for the same host–guest combinations, and deduced somewhat higher affinities (Tab. 2.8).

Tab. 2.8 Association constants (K_a, M^{-1}) for binding of pentoses to β-cyclodextrin **74b** in water.

Substrate[a]	Fluorimetric competition		Microcalorimetry[d]
	b	c	
D-Ribose **47**	5.3	6.3	
D-Arabinose **48**	0.7	1.5	16
D-Xylose **49**	1.0	1.6	17

[a] Fig. 2.5 [c] Ref. [73]
[b] Ref. [72] [d] Ref. [74]

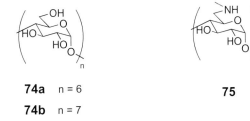

74a n = 6

74b n = 7

75

De Namor also studied α-cyclodextrin **74a**, with the results shown in Tab. 2.9. In this case there seems to be no great difference between affinities to hexoses and pentoses. It is notable that the $\Delta H°$ values are very close to zero, so that binding is mainly entropy-driven. This also implies that the measurements are especially difficult [75], so it would be useful to see these results confirmed by a second technique.

Table 2.9. Association constants (K_a) and thermodynamic data, determined by microcalorimetry, for binding of monosaccharides to α-cyclodextrin **74a** in water.[a]

Monosaccharide[b]	K_a [M^{-1}]	$\Delta G°$ [kJ mol^{-1}]	$\Delta H°$ [kJ mol^{-1}]	$\Delta S°$ [J K^{-1} mol^{-1}]
D-xylose **49**	37	−8.9	−0.09	29.8
L-xylose **ent-49**	117	−11.8	−0.12	39.2
D-glucose **52**	36	−8.9	−0.14	29.4
D-galactose **53**	15	−6.8	−0.32	21.7
D-mannose **54**	59	−10.1	−0.11	33.5
D-fructose **57**	52	−9.8	−0.05	32.8

[a] Data from Ref. [74]; $T = 298.15$ K; errors in $\Delta H°$ estimated
  at ±10–20 %; $K_a{}'$ values uncertain by factors of 1.3–2.0
[b] Fig. 2.5

The modified cyclodextrin **75** was also investigated by Schneider, as a receptor for D-ribose [76]. In this work it was possible to use ^{1}H NMR, following a host C*H* signal, to estimate a K_a of 26 M^{-1}.

2.2.3.3 Receptors Featuring Amide-linked Aromatic Surfaces

The "sandwiching" of saccharides between aromatic surfaces is well known in protein–carbohydrate interactions (cf. Fig. 2.1). Mimicry of this architecture is attractive in being directly relevant to natural carbohydrate recognition. An early attempt was made by Aoyama et al., employing dipeptide **76** [77]. This system was studied with maltose **77a** and maltotriose **77b**, substrates with relatively extended apolar surfaces (Fig. 2.6). Addition of either increased the fluorescence signal from the tryptophan indoles, probably by inhibition of self-quenching. K_a values were small at 1 and 8 M^{-1} for **77a** and **77b** respectively.

More recently, a library of peptides of general form **78** was surveyed for binding to monosaccharides [78]; 62,000 variants were prepared, by split-and-mix solid phase synthesis, and screened by exposure to a yellow sample of D-erythrose **79**. Synthesis

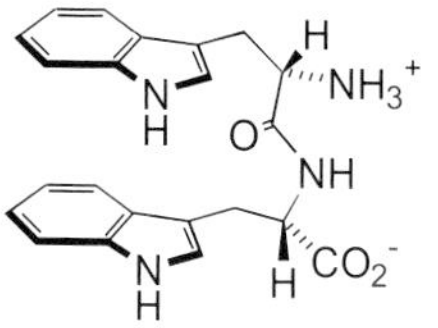

76

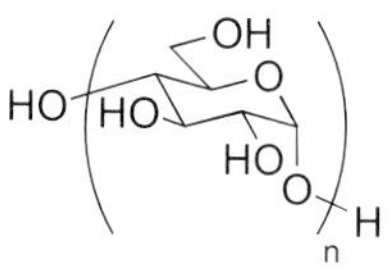

77a n = 2
77b n = 3

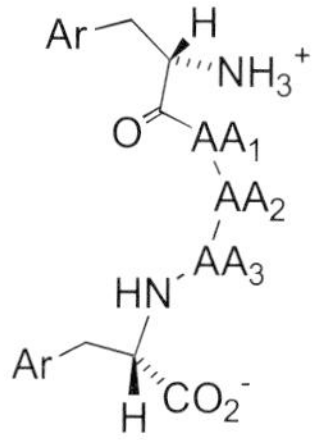

78

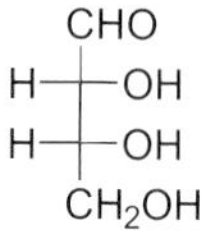

79

beads which acquired the yellow color were identified. The source of this color is unclear, given that **79** has no obvious chromophore. Nonetheless, selected members were then resynthesized in soluble form and tested against erythrose and galactose, using fluorescence spectroscopy in pH 7 aqueous buffer. One variant, with the sequence Trp–Gly–Asp–Glu–Tyr, had remarkable affinity of 350,000 M^{-1} and 52,000 M^{-1} for erythrose and galactose respectively. This apparent success could augur well for biological applications although, given the exceptional results, confirmation by another technique would be welcome.

As discussed in Section 2.2.1.6, the tricyclic cage **41** was designed to provide a more pre-organized version of the "aromatic sandwich". Very recently, the water-soluble dodeca-anionic variant **80** has been synthesized [79]. Binding to a range of saccharides has been studied using both ^{1}H NMR and fluorescence spectroscopy. Results are given in Tab. 2.10. The K_a values are low, but the selectivity is interesting. Unlike **72** and **73** (the most comparable systems, also studied by NMR), receptor **80** gave measurable association constants with the common hexoses. As expected (see Section 2.2.1.6), the β-glucosyl unit was especially favored. Thus cellobiose and methyl β-D-glucoside were the best substrates whereas glucose itself was bound quite strongly. Experiments performed with different anomeric ratios clearly suggested higher affinity for β-D-glucopyranose than for the α anomer. Interestingly, removing hydroxyl groups from the substrate did not usually increase the binding constants; fu-

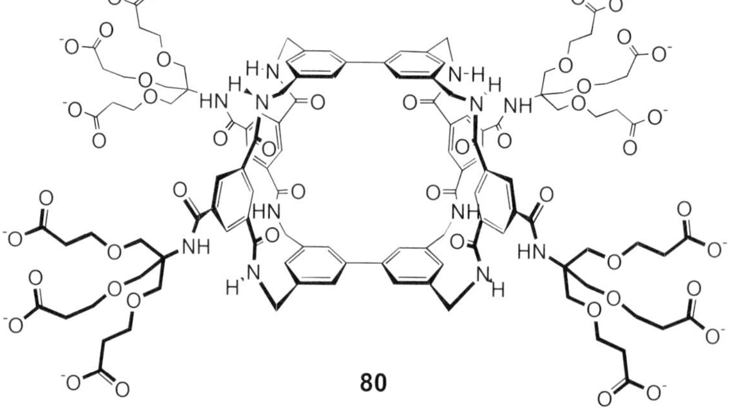

80

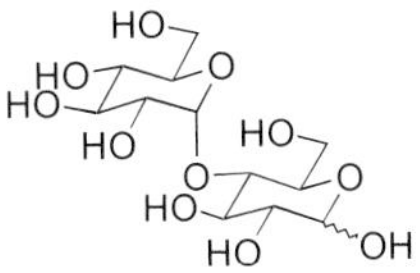

77a

D-maltose

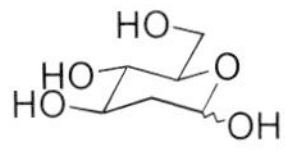

81

2-deoxy-D-glucose

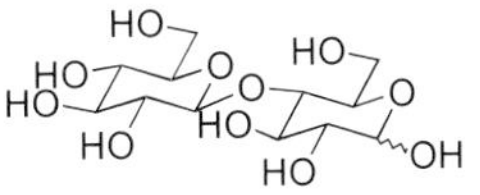

82

D-cellobiose

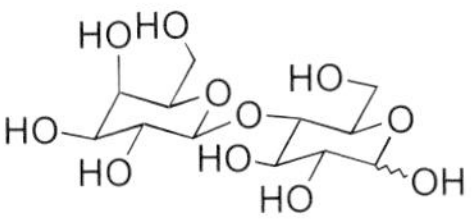

83

D-lactose

Fig. 2.6 Further carbohydrate substrates, shown as pyranosides.

cose, rhamnose, and 2-deoxyglucose were not especially strongly bound. Although the affinities were low, the evidence for complex formation was unambiguous. The K_a values from both titration methods were very similar (Tab. 2.10), and NOESY spectroscopy revealed intermolecular contacts consistent with the proposed binding geometry.

Tab. 2.10 Association constants K_a for binding of carbohydrates to **80** in aqueous solution.[a]

Carbohydrate[b]	K_a [M^{-1}]	Carbohydrate	K_a [M^{-1}]
D-Ribose **47**	3.1	L-Fucose **55**	2.1
D-Arabinose **48**	2.2	L-Rhamnose **56**	_[c]
D-Xylose **49**	4.6 (3.8)	2-Deoxy-D-glucose **81**	7.2
D-Lyxose **50**	_[c]	Methyl β-D-glucoside **62**	27.3 (32)
D-Glucose **52** (α/β = 72/28)[d]	4.6	Methyl α-D-glucoside **63**	6.9 (5.7)
D-Glucose **52** (α/β = 40/60)[e]	9.2 (9.5)	D-Maltose **77a**	_[c]
D-Galactose **53**	2.1	D-Cellobiose **82**	16.6
D-Mannose **54**	_[c]	D-Lactose **83**	_[c]

[a] Unbracketed values were obtained from ^{1}H NMR titration at 296 K, bracketed values from fluorescence titrations

[b] Figs. 2.5 and 2.6

[c] Too small for quantitative estimation

[d] Freshly dissolved glucose was used. The anomeric ratio was determined by NMR, and remained roughly constant throughout the titration

[e] Anomers were allowed to equilibrate.

2.2.3.4 Water-soluble Porphyrins
Král and coworkers have described a range of porphyrins as receptors for carbohydrates in water [70b, 80]. Examples are the series **84**, and the macrocycle **85a** and its

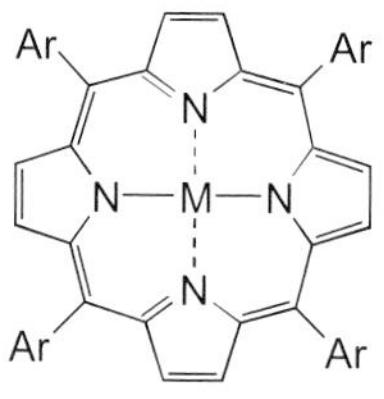

84a M = H$_2$, Ar = ⟶⟨⟩⟶SO$_3$Na

84b M = H$_2$, Ar = ⟶⟨⟩⟶PO$_3$(NH$_4$)$_2$

84c M = H$_2$

84d M = Fe$^+$

Ar =

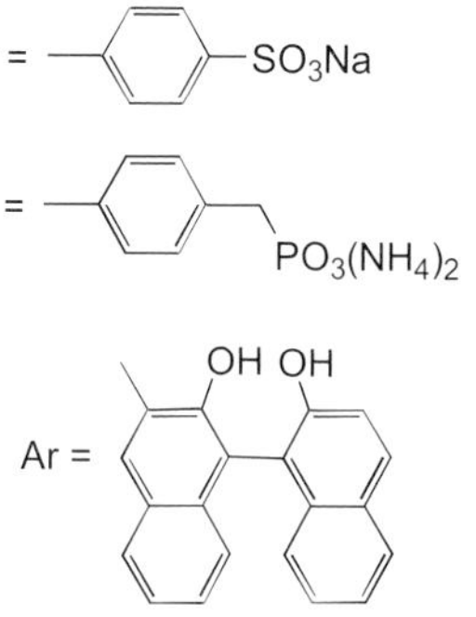

linear analog **85b**. Binding has been studied by UV–visible titrations, although some NMR and fluorescence data have been obtained for nonaqueous solvents. Selected results are shown in Tab. 2.11. Even the simple polyanionic systems **84a/b** seem capable of binding native monosaccharides in an essentially aqueous solvent system (H$_2$O–MeOH, 95:5) [80b,c]. The performance of **84b** is especially remarkable. The nature of the interaction is not obvious; because of the distance between the anionic centers and the porphyrin nuclei it seems unlikely that a monosaccharide could bind simultaneously to both features. The combination of high affinity and optical response is, however, promising for applications in carbohydrate sensing. The cyclodimeric system **85a** was designed to enclose a carbohydrate, binding with two por-

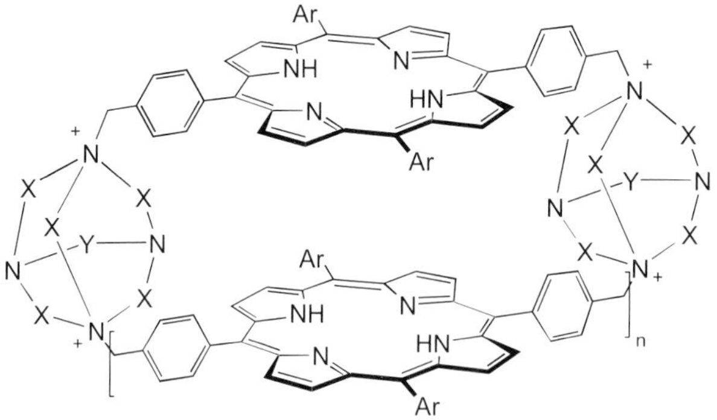

85a n = 1

85b n = 0

Ar = *p*-tolyl
X = (CH$_2$)$_6$, Y = CO(CH$_2$)$_4$CO

phyrin faces [80d]. K_a values are again impressive, although those obtained for linear monomer **85b** are only slightly lower.

Tab. 2.11 Association constants (K_a, M^{-1}) for binding of carbohydrates to porphyrin receptors in aqueous solvent systems.[a]

Substrate[b]	84a[c]	84b[d]	84c[e]	84d[e]	85a[f]	85b[f]
D-Glucose **52**	120	17,600	60	110	4300	1200
D-Galactose **53**	135	19,700	50	100	3300	2100
Methyl β-D-glucoside **62**			<10	20	2300	1100
Methyl α-D-glucoside **63**			20	50	7800	5900
D-Maltose **77a**		24,200	150	230		
D-Lactose **83**	220	25,000			6500	5500

[a] Determined by UV–visible titration
[b] Figs. 2.5 and 2.6
[c] In H_2O–MeOH, 95:5; data from Ref. [80c]
[d] In H_2O–MeOH, 95:5; data from Ref. [80b]
[e] In H_2O–CH_3CN, 50:50; data from Ref. [70b]
[f] In H_2O–MeOH, 95:5; data from Ref. [80d]

2.2.3.5 Metal Complexes

The use of metal ions M^{n+} to coordinate sugar hydroxyls seems a promising approach to carbohydrate recognition in water. It is employed in nature by the "C-type" lectins (for which M = Ca), and is closely related to the highly successful boronate strategy (Sect. 2.3). It has not, however, been widely exploited by supramolecular chemists, possibly because, in simple systems, strong interactions are only observed at high pH [81]. Striegler has studied the binding of carbohydrates to copper(II) complexes [82], with a view to incorporation in molecularly imprinted polymers [83]. At pH 12.4, simple mononuclear complexes such as **86** bound monosaccharides quite strongly ($K_a \approx 5000$ M^{-1}), presumably by deprotonation of the anomeric hydroxyl and chelation to the neighboring OH as shown [82a]. Methyl glycosides did not form complexes, in accord with this interpretation. Affinity was, typically, similar for the dinuclear species **87**, but was far more selective [82b]. K_a values ranged from 360 M^{-1} for glucose to 11,500 M^{-1} for mannose. As shown for the mannose complex, stronger binding was thought to result from coordination of three saccharide oxygen atoms. The analogous glucose complex cannot form for geometrical reasons.

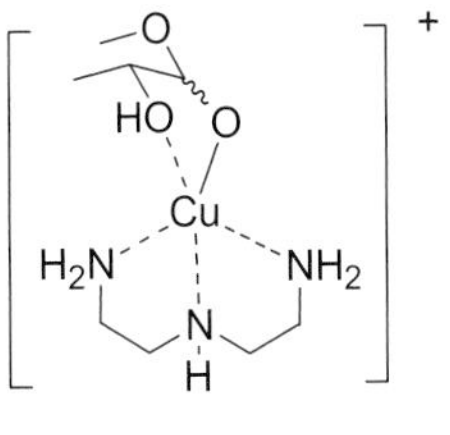

86.monosaccharide

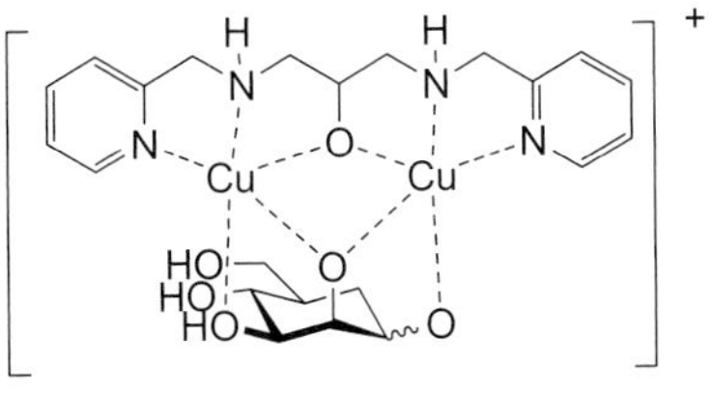

87.mannose

2.2.3.6 Recognition of Lipophilic Saccharides in Aqueous Media

There remain several studies addressing somewhat "easier" targets consisting of carbohydrates conjugated to lipophilic moieties. For example, glycophane **88** [84], cyclophane **89** [85], and anthracene **90** [86] have all been used to bind aryl (phenyl or *p*-nitrophenyl) glycosides. The main driving force is assumed to be hydrophobic contact with the aryl groups, but interactions with the saccharide units moderate the association constants and potentially throw light on carbohydrate recognition in general. Anthracene **90** was also found to bind two targets which are less lipophilic and very biologically relevant, the GM3 derivative **91** and the sialyl Lewis X analog **92**. Complexation was detected by ^{1}H NMR, of the starred galactosyl protons in particular. The signal motions were small (≤ 0.03 ppm) but consistent with 1:1 complexation. It is interesting that these significant substrates can be bound by such a simple and relatively flexible receptor.

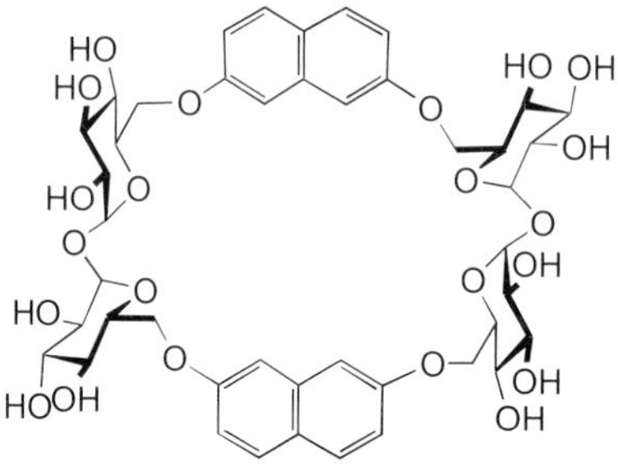

88

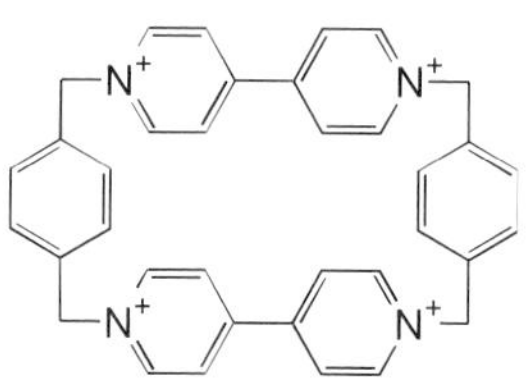

89

Another important lipophilic substrate is Lipid A **93**. This disaccharide is the conserved portion of "bacterial endotoxin", which occurs in the outer membrane of Gram-negative bacteria and is responsible for septic shock. Certain antibiotics (e.g. the polymyxins) bind and neutralize the endotoxin, but pose toxicity problems. The recognition of Lipid A is therefore an attractive goal for carbohydrate receptors. Miller et al. designed the tris-cyclopentane **94** as a receptor for **93** [87]. UV–visible titrations

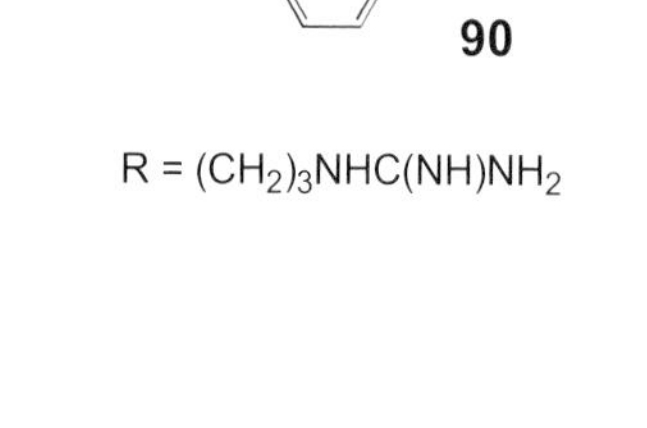

90

R = (CH$_2$)$_3$NHC(NH)NH$_2$

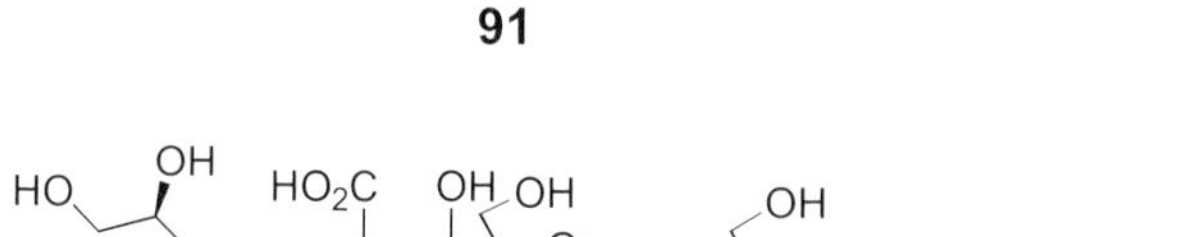

91

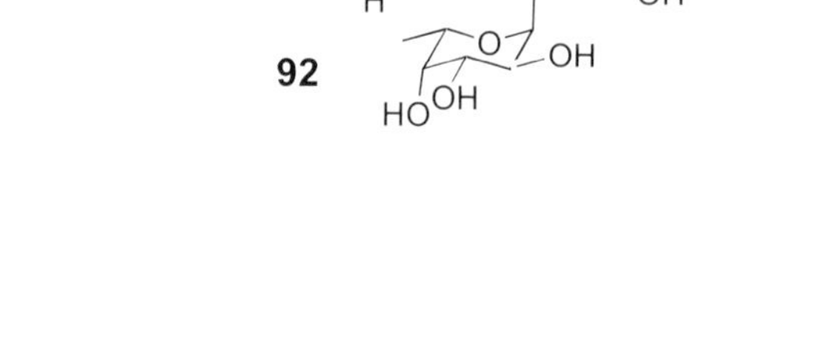

92

revealed an interaction between **93** and **94**, and implied a very high binding constant of 1.7×10^6 M^{-1}. It remains to be seen whether **94** is active *in vivo*, but this work raises hopes of genuine medical applications for carbohydrate receptors.

2.3
Receptors Employing B–O Bond Formation

Stable boronic acid-based saccharide receptors enable the creation of saccharide sensors which, by design, can be selective and sensitive for any chosen saccharide. The recognition of saccharides by boronic acids has a unique place in supramolecular chemistry. The pairwise interaction energy is large enough to enable single-point molecular recognition, and the primary interaction involves the reversible formation of a pair of covalent bonds (rather than noncovalent attractive forces).

Despite a long history – the first structural and quantitative binding constant data was reported in the1950s [88–90] – the structure of the boronic acid–saccharide complexes in aqueous solution continues to be discussed [91–93]. There is general agreement that boronic acids covalently react with 1,2 or 1,3 diols to form five or six membered cyclic esters. The adjacent rigid cis diols of saccharides form stronger cyclic esters than simple acyclic diols such as ethylene glycol. With saccharides the choice of diol used in the formation of a cyclic ester is complicated by the possibility of pyranose-to-furanose isomerization of the saccharide moiety. Lorand and Edwards first determined the selectivity of phenylboronic acid toward saccharides and this selectivity order seems to be retained by all monoboronic acids (D-fructose > D-galactose > D-glucose) [90].

The equilibria involved in the phenylboronate binding of a diol are conventionally summarized as a set of coupled equilibria (Eq. 1). In aqueous solution phenylboronic acid reacts with water to form the boronate anion plus a hydrated proton thereby defining an acidity constant K_a. (Eq. (1) shows an explicit water molecule "coordinated" to the trigonal boronic acids. There is undoubtedly water in rapid exchange on the Lewis acidic boron in the same way that hydrated Lewis acidic metal ions exchange bound water. A good analogy is Zn^{2+} (aq), which ionizes in water to give a pK_a of 8.8, i.e. $Zn\text{-}OH_2 \rightarrow Zn\text{-}OH + H^+$ [94].) The formation of a diol–boronate complex, defined by K_{tet}, formally liberates two equivalents of water, but this stoichiometric factor is usually ignored as a constant in dilute aqueous solution. In a formal sense, phenylboronic acid could also bind diols to form a trigonal complex (K_{trig}), and this species would itself act as an acid according to K_a'. The "acidification" of solutions containing phenylboronic acid and diols is always discussed in terms of the trigonal complex being a stronger acid than the parent phenylboronic acid, i.e. $K_a' > K_a$ [95]. As a result, $K_{tet} > K_{trig}$.

These four coupled equilibria are not the full story, however. Boronic acids readily form stable complexes with buffer conjugate bases (phosphate, citrate, and imidazole) [96]. In fact, both binary boronate–X complexes are formed with Lewis bases (X), as also are ternary boronate–X–saccharide complexes. Occasionally these previously unrecognized species persist into acidic solution and under some stoichiometric conditions they can be the dominant components of the solution. These complexes suppress the boronate and boronic acid concentrations leading to a decrease in the measured apparent formation constants (K_{app}). As a consequence, the scope of the "simple" diol–boronate recognition system is much greater than the simple picture of Eq. (1).

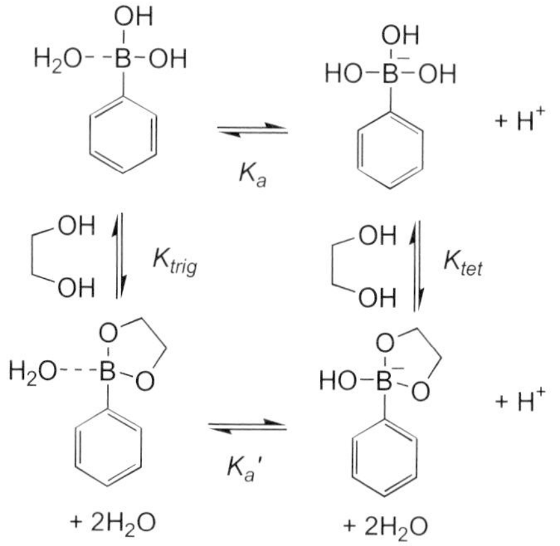

2.3.1
Carbohydrate Recognition in Water

2.3.1.1 Receptors at the Air–Water Interface

Molecular assemblies of monolayers and their properties have been well established [97]. The unique characteristics of Langmuir–Blodgett (LB) films have attracted particular attention. The pressure of an LB film is sensitive to the activity of its individual constituents. The boronic acids **95** and **96** were prepared to demonstrate the effect of saccharide binding at the air/water interface [98]. The reactivity of the boronic acids was tested by solvent extraction methods (solid–solvent, neutral solvent–solvent, and basic solvent–solvent). Extractabilities of both compounds were found in the order D-fructose > D-glucose > D-maltose > D-saccharose. Monolayers formed by **95** were found to be unstable, on the basis of both irreproducible pressure–area (π–A) isotherms and the crystalline nature of the monolayer. The meta isomer, **96**, on the other hand, gave very reproducible results. The π–A isotherm of **96** was affected by introducing a saccharide into the subphase at pH 10. The chiral cholesterylboronic acid derivative **97** behaved similarly [99]. Because of its chiral nature it could selectively recognize chiral isomers of fructose. The effect of quaternized amines on the saccharide binding has also been investigated [100]. Quaternized amines facilitate saccharide detection by the monolayer at neutral pH. Assistance of closely located ammonium cations in the formation of boronate anion is believed to be the source of enhancement. The cooperative binding of saccharides by the diboronic acid derivatives **98a–d** on monolayers has also been investigated and found to be in agreement with its recognition pattern in homogeneous solutions [101, 102]. Molecular recognition in this system also seems to be facilitated by closely located ammonium cations [102].

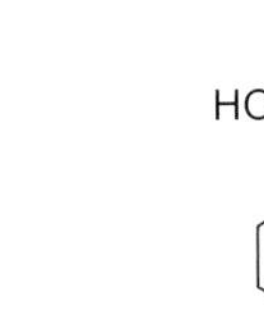

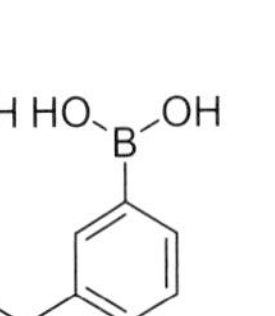

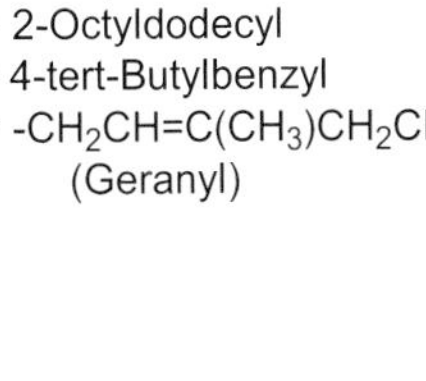

98 a: R = Methyl
b: R = 2-Octyldodecyl
c: R = 4-tert-Butylbenzyl
d: R = -CH$_2$CH=C(CH$_3$)CH$_2$CH$_2$CH=C(CH$_3$)$_2$
(Geranyl)

Kurihara has recently prepared electroconductive LB films for sugar recognition. A polymerized electroconductive boronic acid LB film was transferred on to a gold surface; binding with an electroactive manoside (nitrobenzene) was monitored by cyclic voltammetry [103].

2.3.1.2 Transport and Extraction

The aim of saccharide extraction is to bind selectively with the saccharide in water and then to move the saccharide into a hydrophobic solvent (membrane). Saccharide transport is very similar to extraction; a saccharide must be bound in water then moved into a hydrophobic membrane. When in the membrane the saccharide must be transported across the membrane and then released into the water on the other side. Although the properties required for a good molecular extractor and transporter are similar, good transporters must balance extraction with release.

Efficient transport of saccharide-related water-soluble artificial drugs into individual cells via the cell membrane are critical to the future development of drug design and delivery. Many biomimetic systems capable of transporting neutral molecular species are known, although examples of systems that can transport such species actively are rare [104].

Boronic acid and its derivatives have been used as carriers for transport of saccharide through membranes [105]. Because formation of the anionic boronate is favored in alkaline pH and disfavored at low pH, this provides a means of actively transporting glucose by means of a pH gradient. In a study of saccharide transport by phenylboronic acid derivatives, the phenyl boronate ion has been accompanied by the lipophilic trioctylmethylammonium cation (TOMA) [105]. Not only saccharides but also saccharide-related biologically important molecular species such as uridine have been found to be effectively transported in this way [106, 107]. The negatively charged boronate complex accompanies the lipophilic cation (trioctylmethylammonium chloride, TOMAC) during the transport. Smith has used the assistance of F⁻ ions in saccharide transport [108]. Reaction of fluoride ions with boronic acid to form phenylfluoroborate has been proposed as the reason for this observation. This provides a means of active transport of saccharide related molecular species at neutral pH with a fluoride ion gradient. Interestingly, other halogens do not assist the transport. The incorporation of a cationic charge into the boronic acid could waive the requirement of an accompanying ammonium cation, for example pyridinium derivatives **99**. The strongly acidic pyridinium boronic acid **99c** transports saccharide-related species through membranes [109]. The transport of dopamine and related derivatives by the molecular receptor **100** has been reported [107]. Cooperative binding by two different receptor sites is possible in this instance. Transport of amino acid derivatives by boronic acids have also been reported [110].

Smith has investigated the capacity of twenty-one monoboronic acids to transport saccharides through lipid bilayers [111]. It was found that lipophilic boronic acids are capable of facilitating the transport of monosaccharides through lipid bilayers, but that disaccharides are not transported. Although the mechanism of transport requires complexation of the saccharide as the tetrahedral boronate, the species transported is the neutral conjugate acid. Smith has also investigated selective fructose

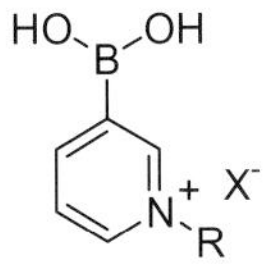

99 a. R = CH3   X = OMs
b. R = n-C18H37   X = OTs

c. R =

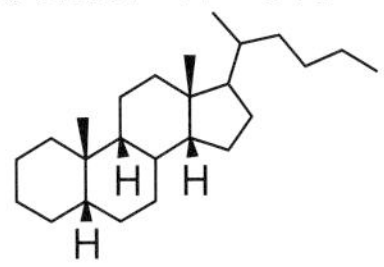

X = OMs

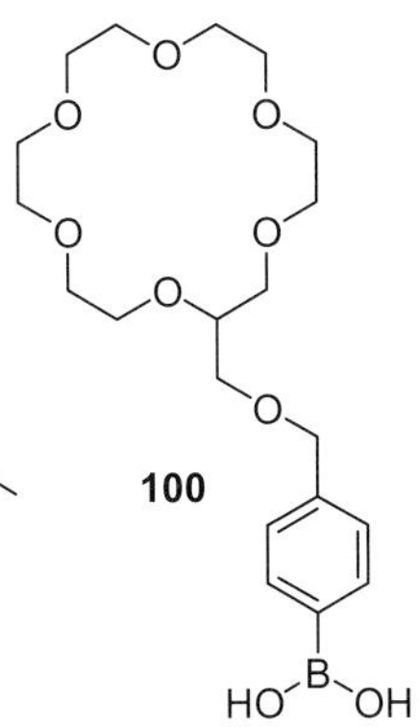

100

transport through supported liquid membranes using both mono- and diboronic acids **101**, **102a–c** [112, 113] and used a monoboronic acid **103** in combination with a diammonium cation **104** to facilitate the transport of ribonucleoside-5'-phosphates [114].

Boronic acids with linked ammonium ions have been used to aid saccharide extraction [115]. Using boronic acids in combination with crown ethers Smith has developed a sodium saccharide co-transporter **105** [116] and facilitated catecholamine transporters **106**, **107**, and **108** [117].

Smith has used microporous polypropylene impregnated with a boronic acid in 2-nitrophenyl octyl ether as a supported liquid membrane for the separation of fructose from fermentation broths [118].

Duggan has developed the highly lipophilic boronic acids **109** and **110** which transport fructose with very high selectivity [119, 120]. Duggan has achieved unprecedented fructose selectivity by using a rigid five-cavitand appended with boronic acid receptor groups [121].

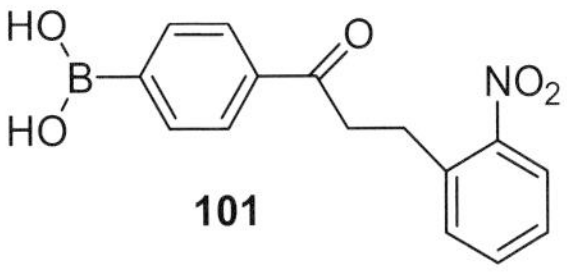

101

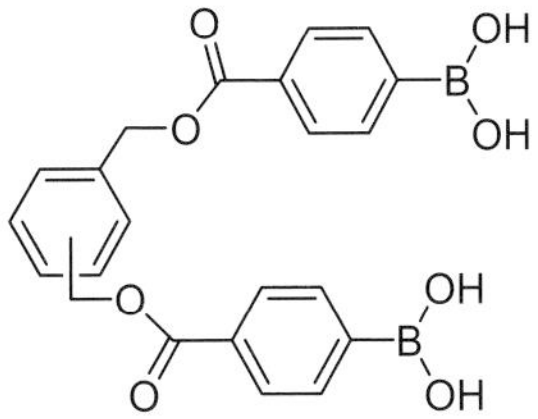

103

102 a: *ortho*
b: *meta*
c: *para*

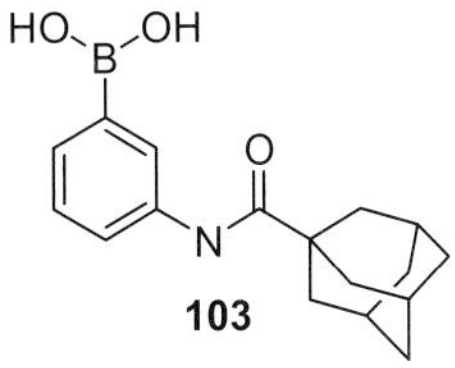

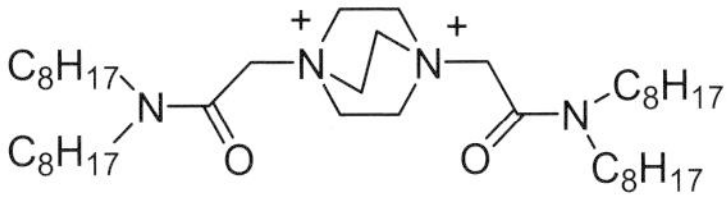

104

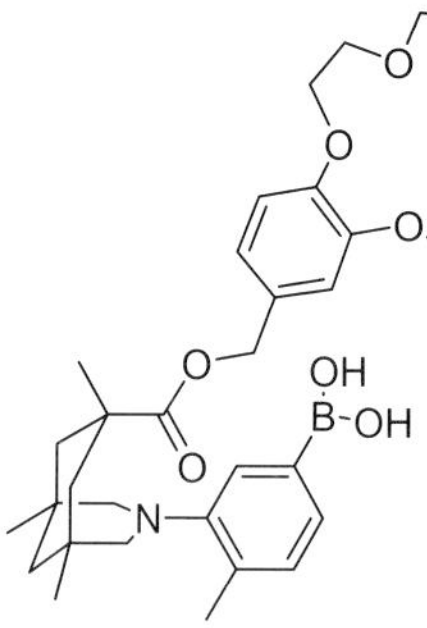

105

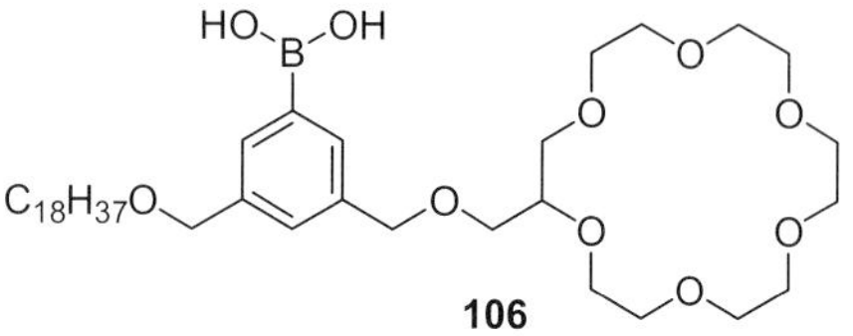

106

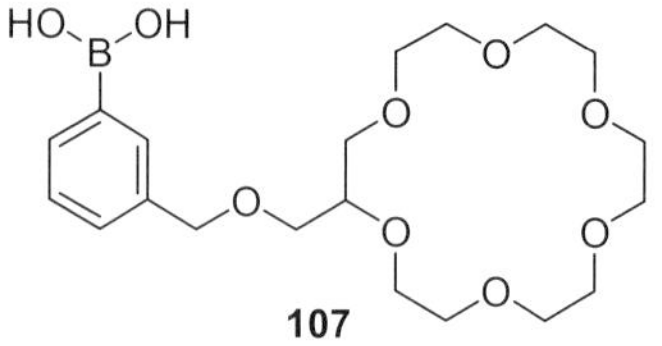

107

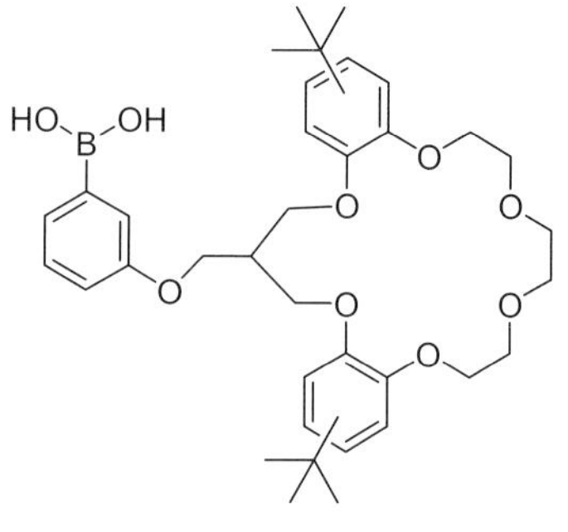

108

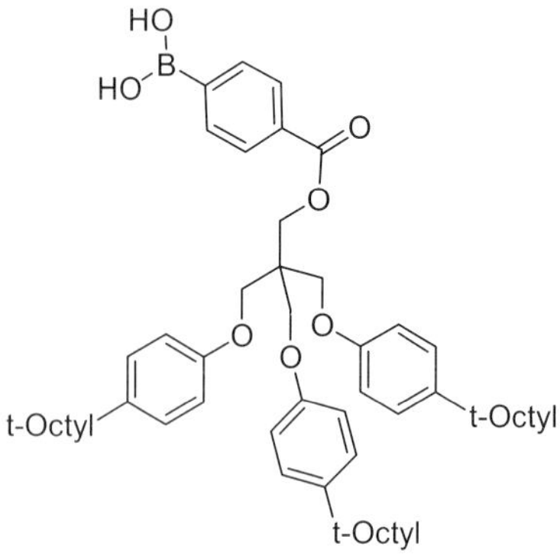

109

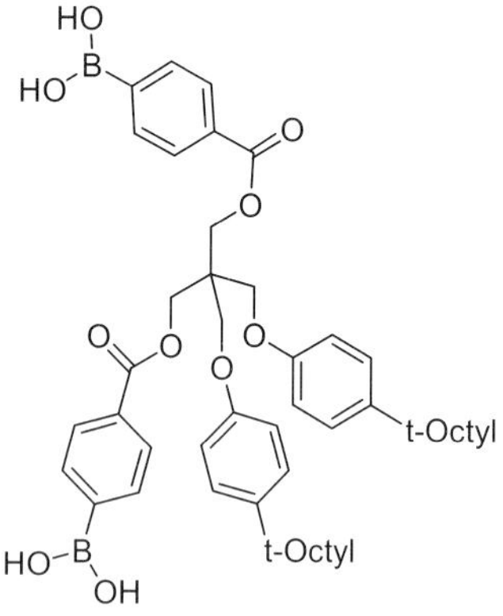

110

2.3.2
Carbohydrate Recognition in Water

2.3.2.1 Fluorescent Sensors

Fluorescent sensors for saccharides are of particular practical interest, partly because of the inherent sensitivity of the fluorescence technique. Only small amounts of a sensor are required (typically 10^{-6} M) offsetting the synthetic costs of such sensors. Fluorescent sensors have also found applications in continuous monitoring using fiber optics and intracellular mapping using confocal microscopy.

Internal charge transfer (ICT)

The first fluorescent sensors for saccharides were based on fluorophore-appended boronic acids. Czarnik showed that 2- and 9-anthrylboronic acids [122, 123] **111** and

112 could be used to detect saccharides. With these systems the negatively charged boronate has less fluorescence than the neutral boronic acid. The pK_a of a boronic acid is reduced on saccharide binding, therefore the fluorescence of these systems at a fixed pH decreases when saccharides are added.

The fluorescent stilbene boronic acid **113a** was originally investigated by Shinkai [124] but its importance went unnoticed until detailed investigation of this and a series of related systems **113a–f** by Lakowicz [125, 126]. In these systems, the neutral form of the boron group acts as an electron-withdrawing group whereas the anionic form acts as an electron-donating group. Changes in the electronic properties of the boron group are what cause the spectral changes of the fluorophore.

111

112

113a(n=1, X=H)   **113d** (n=1, X=N(CH₃)₂)
113b (n=1, X=CN)   **113e** (n=2, X=N(CH₃)₂)
113c (n=1, X=OCH₃) **113f** (n=3, X=N(CH₃)₂)

Lakowicz quickly recognized the importance of stilbene boronic acid **113d** and has since prepared several analogous ICT fluorophore systems, including oxazoline [127] **114**, chalcones [128] **115a,b**, and boron–dipyrromethene (BODIPY) [129] **116**. The oxazoline and chalcone systems produce large fluorescence changes whereas the BODIPY system results in only a small change in fluorescence. Although the BODIPY system **116** did not work particularly well, this class of fluorophore does require further exploration. The BODIPY chromophore has many advantages as a fluorescent probe – for example high extinction coefficients, high fluorescence quantum yields, good photostability, and a narrow emission band – and their building-block synthesis enables the development of many different analogs with emission maxima from 500 to 700 nm. Long-wavelength fluorescent probes for glucose are highly desirable for transdermal glucose monitoring and/or for measurements in whole blood. Narrow emission bands are desirable for high signal-to-noise ratio. Lakowicz and Geddes have recently demonstrated the usefulness of this type of fluorescent sensor by preparing contact lenses doped with **113b, d, e** or **115a, b** to prepare noninvasive monosaccharide sensors [130].

Wang has recently shown that a very simple naphthalene system **117** can produce very large fluorescence changes (41-fold on addition of 50 mм fructose) [131].

Saccharide binding with compound **118a**, a monoboronic fluorescent sensor, results in large shifts in emission wavelength [132]. The dual fluorescence of **118a** can be ascribed to locally excited (LE) and twisted internal charge transfer (TICT) states of the aniline fluorophore [133]. When saccharides interact with sensors **118a** in aqueous solution at pH 8.21 the emission maxima at 404 nm (TICT state) shifts to 362 nm (λ_{ex} 274 nm, LE state). The band at 404 nm is because of the TICT state of **118a** containing a B–N bond, i.e. the lone pair is coordinated with the boron and perpendicular to the π-system. The band at 274 nm (LE state) corresponds to the situation in which the B–N bond in **118a** has been broken (boronate).

When compounds **118b** and **118c** were excited at 240 nm and 244 nm, respectively, emission at 350 nm was observed (excitation at 274 nm resulted in no emission) [133]. Also, when compound **118a** was excited at 244 nm only emission at 360 nm was observed. In all these examples only the LE state is formed (no B–N bond is possible). The fluorescence enhancements (λ_{ex} 240 and 244 nm) obtained for **118a–c** on addition of D-fructose are 15, 18, and 25 fold respectively. These large fluorescence enhancements have been attributed to fluorescence recovery of the aniline fluorophore. With these systems in the absence of saccharides the normal fluorescence of the LE state of the aniline donor is quenched by energy transfer to the phenylboronic acid acceptor. When saccharides are added a negatively charged boronate anion is formed; under these conditions energy transfer from the aniline donor is unfavorable and fluorescence recovery of the LE state of the aniline donor is observed. Although these systems might be of limited practical use (aniline is not the best fluorophore) understanding this simple unit will aid in the development of improved ICT fluorescent systems [132].

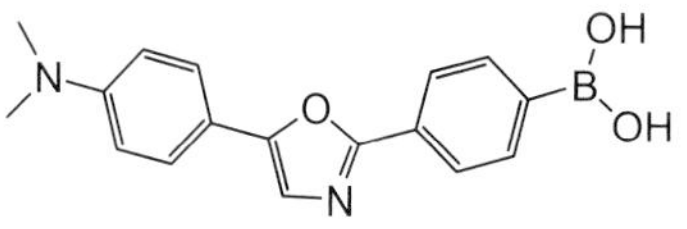

114

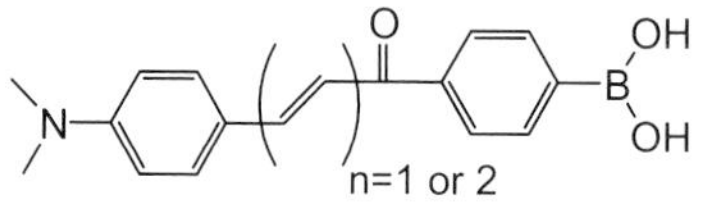

115a (n=1)
115b (n=2)

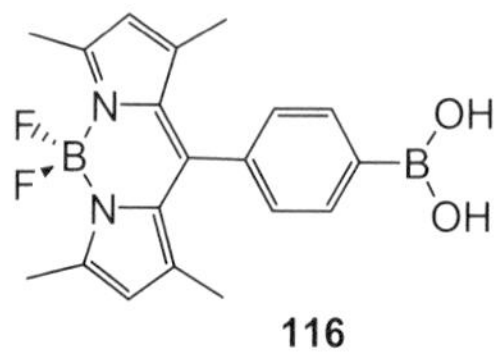

116

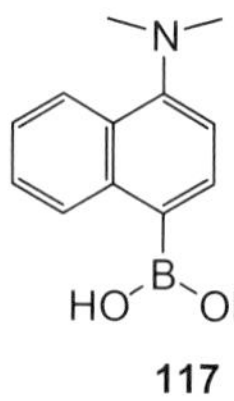

117

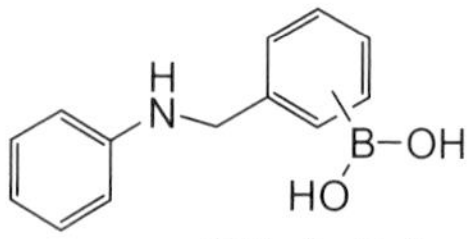

118a (*ortho*)
118b (*meta*)
118c (*para*)

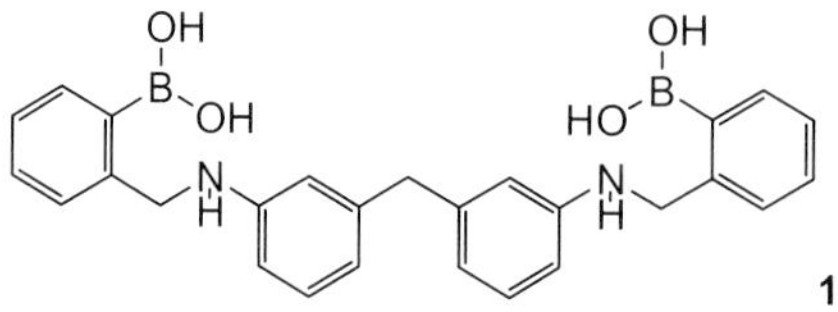

119

With diboronic acid sensor **119** the requirements of the signaling unit with those of a D-glucose-selective receptor have been combined [134]. Cooperative binding of the two boronic acid groups is clearly observed as illustrated by stability-constant differences between the mono- and diboronic acid compounds. The stability constant, *K*, of diboronic acid sensor **119** with D-glucose is 14 times greater than of monoboronic sensor **118a** whereas, the stability constant of diboronic acid sensor **119** with D-fructose is 0.6 times that of monoboronic acid sensor **118a**. This result can be explained since it is well known that D-glucose readily forms 1:1 cyclic complexes with diboronic acids whereas D-fructose tends to form 2:1 acyclic complexes with diboronic acids [95].

Photoinduced electron transfer (PET)
Photoinduced electron transfer (PET) has been widely used as the preferred tool in fluorescent sensor design for atomic and molecular species [135–138]. PET sensors usually consist of a fluorophore and a receptor linked by a short spacer. The changes in oxidation/reduction potential of the receptor on guest binding can affect the PET process creating changes in fluorescence.

The first rationally designed fluorescent PET saccharide sensor, compound **120**, was prepared in 1994 [139, 140]. This fluorescence sensor contains a boronic acid group and an amine group. The boronic acid group is required to bind with and capture sugar molecules in water. The amine group plays two roles in the system: (1) boronic acids with a neighboring amine facilitate the binding of saccharides at neutral pH; and (2) the fluorescence intensity is controlled by the amine.

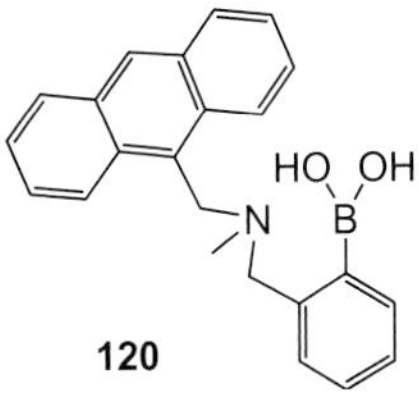

120

With **120** the "free" amine reduces the intensity of the fluorescence (quenching by PET). This is the "off" state of the fluorescent sensor. When saccharides are added, the amine becomes "bound" to the boron center. The boron-bound amine cannot quench the fluorescence and hence strong fluorescence is observed. This is the "on" state of the fluorescent sensor. The system described above illustrates the basic concept of an "off–on" fluorescent sensor for saccharides.

Cary and Satcher, while attempting to extend the fluorescence emission to longer wavelengths, have prepared the rhenium bipyridine system **121**. This system produces a good fluorescence response to glucose in methanol but fails to produce a response to glucose in 50 % (*v/v*) methanol–phosphate buffered saline [141]. The failure of this system can probably be traced back to competing equilibria, with the buffer overwhelming the system. Near IR fluorescence response to saccharides has been achieved by Akkaya using the diboronic acid squarene **122**. Sadly the system only produces an 8 % change in fluorescence intensity when D-glucose is added [142].

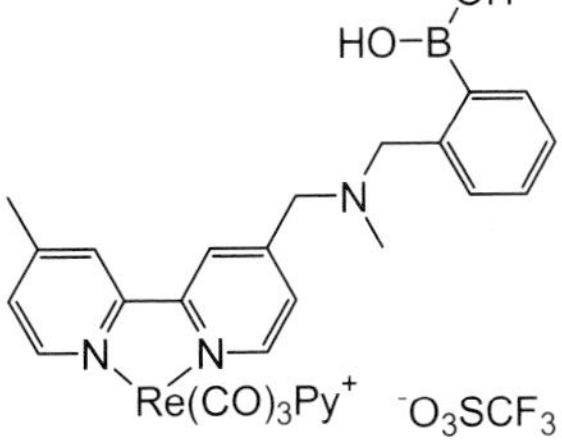

121 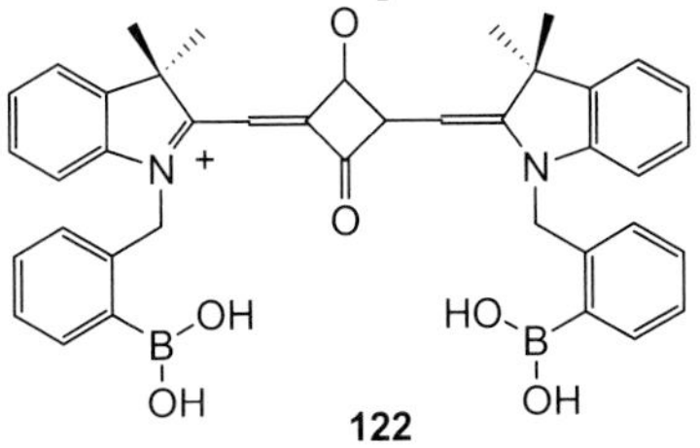   **122**

The simple "off–on" PET system **120** was improved by introduction of a second boronic acid group **123** [139, 143]. For compound **123** two possible saccharide-binding modes – a 2:1 complex and a 1:1 complex – can inhibit the electron-transfer process to give higher fluorescence. Because of fortuitous spacing of the boronic acid groups the diboronic acid was selective for D-glucose over other monosaccharides. Lakowicz has recently demonstrated that these two systems (**120** and **123**) could be used as fluorescence-lifetime sensors for saccharides [144]. These lifetime measurements could also be used to confirm unequivocally that the boronic acid–saccharide interaction is reversible.

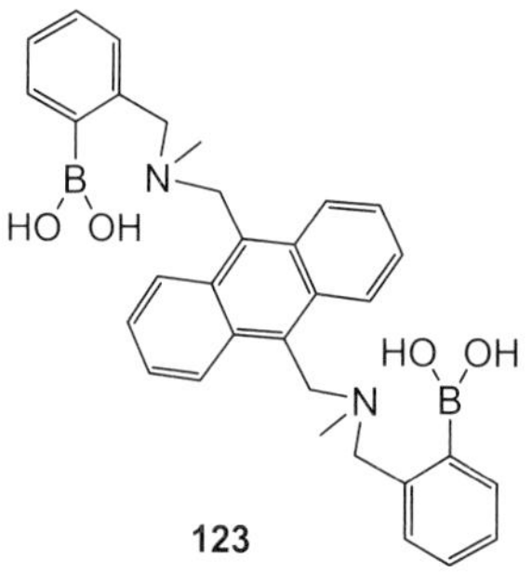

123

With sensor **124**, the amount of excimer emission can be directly correlated with the amount of non-cyclic saccharide complex formed [145]. The effect of linker length on this system has been investigated by Appleton, who found that as the linker length was increased selectivity for glucose was lost [146].

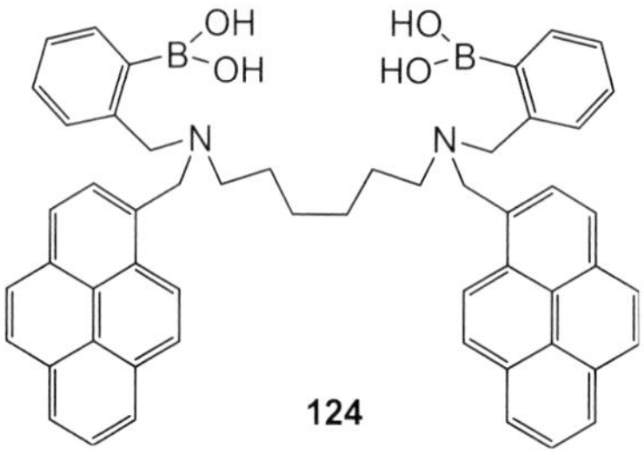

124

Sensors **125a–f** illustrate a modular approach to the design of saccharide selective fluorescence sensors [147, 148]. Compound **125d** with a hexamethylene linker and pyrene fluorophore has D-glucose selectivity whereas systems with longer linker units **125e–f** have enhanced selectivity for D-galactose [149].

Having determined the effect of the linker on saccharide selectivity other factors affecting saccharide selectivity were probed. The next logical component to vary was the fluorophore or "read-out" unit sensors **125d, g, h, i, j** [150]. Although not directly involved in saccharide binding, the nature of this unit directly influenced both the solvation and steric crowding of the binding site.

Sensors **125d, g** and **h** are more selective for D-glucose than for D-galactose [149] whereas with sensors **125i** and **j** the selectivity switches from D-glucose to D-galactose

[149]. These results show that in a PET saccharide sensor with two phenylboronic acid groups, a hexamethylene linker and a fluorophore, the choice of the fluorophore is crucial. Selectivity is fluorophore-dependent and careful choice of the fluorophore, such that it complements the polarity of the chosen guest species, is imperative.

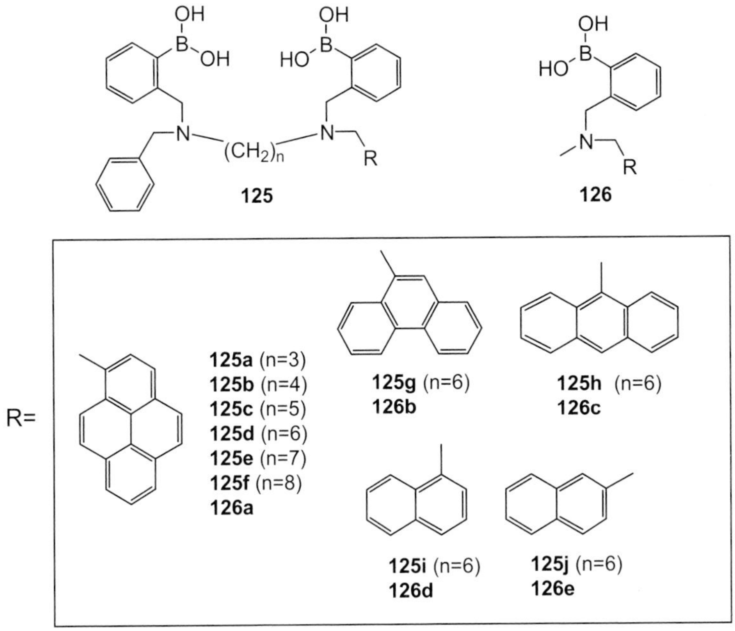

Wang has also prepared diboronic acid systems with two anthracene fluorophores with variable spacers **127a–j** [151, 152] and have demonstrated that **127e** is selective for sialyl Lewis X [151] and that **127f** is selective for D-glucose [152].

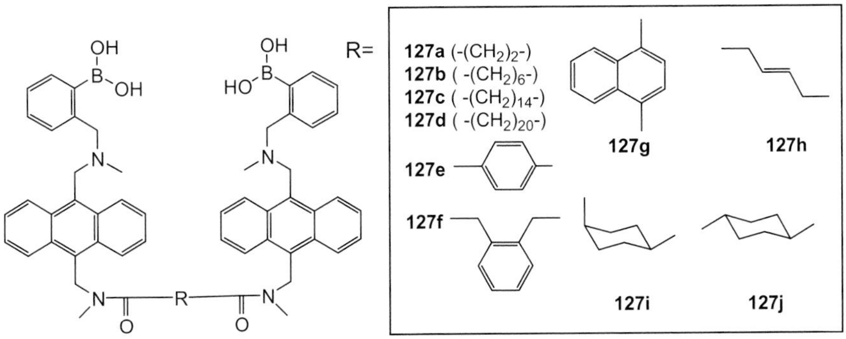

Hall took the modular approach a step further and employed a solid-phase synthetic approach to optimize structural and electronic properties of boronic acid sensors for oligosaccharides [153]. By this approach they were able to determine that para electron-withdrawing groups on the boronic acids were beneficial. The diboronic acid sensor **128** has a binding constant of 1870 M^{-1} with D-lactulose whereas the unsubstituted diboronic acid has a binding constant of 315 M^{-1} only. The research also indicated that saccharides with an isomerizable reducing end capable of forming a rigid furanose ring with *cis*-1,2-diols have higher binding constants. This phenomenon is well known and agrees with the recent evaluation of diboronic acids **125a–f** with disaccharides. Among these sensors **125d** binds particularly strongly with D-melibiose (binding constant 339 M^{-1}) whereas the monoboronic acid **126a** has a binding constant of 96 M^{-1}. The selectivity of **125a–f** for D-melibiose mirrors that for D-glucose, indicating that these diboronic acid sensors form stable 1:1 cyclic structures with the furanose segment of D-melibiose [154]. These observations and those of Hall with polysaccharides complement the observations made by Norrlid for monosaccharides [91, 92, 155].

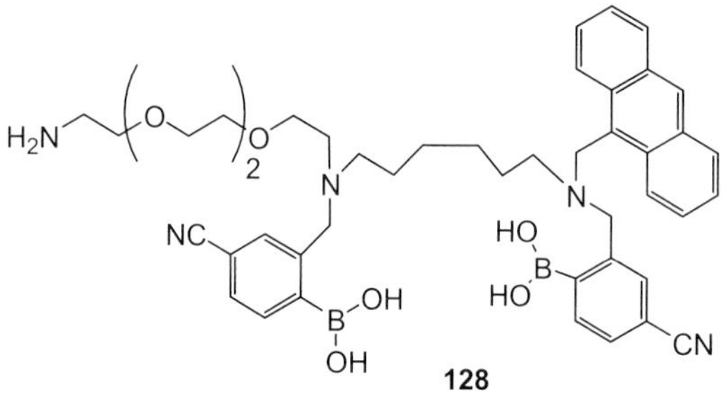

128

Compound **129** uses the efficiency of fluorescence energy transfer (ET) from phenanthrene to pyrene to monitor saccharide binding [156]. Sensor **129** is particularly interesting in that the differences between the observed fluorescence enhancements obtained when excited at phenanthrene (299 nm) and pyrene (342 nm) can be correlated with the molecular structure of the saccharide–sensor complex. The fluorescence enhancement of sensor **129** with D-glucose is 3.9 times greater when excited at 299 nm and 2.4 times greater when excited at 342 nm whereas with D-fructose the enhancement was 1.9 times greater when excited at 299 nm and 3.2 times greater when excited at 342 nm. These results indicate that energy transfer from phenanthrene (donor) to pyrene (acceptor) in a rigid 1:1 cyclic D-glucose complex is more efficient than in a flexible 2:1 acyclic D-fructose complex. This more efficient energy transfer leads to an enhanced fluorescence response to D-glucose.

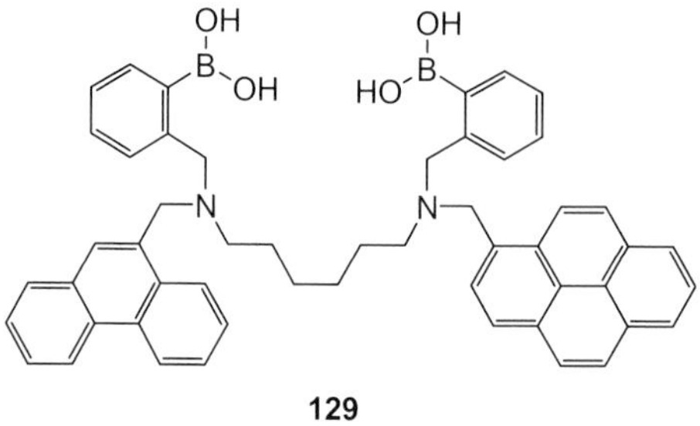

129

The boronic acid PET system has also been used in combination with other binding sites. The D-glucosamine-selective fluorescent systems **130a,b** based on a boronic acid and aza crown ether have been explored [157, 158]. Sensors **130a,b** consist of monoaza-18-crown-6 ether or monoaza-15-crown-5 as a binding site for the ammonium terminal of D-glucosamine hydrochloride whereas a boronic acid serves as binding site for the diol (carbohydrate) part of D-glucosamine hydrochloride. The nitrogen of the aza crown ether unit can participate in PET with the anthracene fluorophore; ammonium ion binding can then cause fluorescence recovery. This recovery is a result of hydrogen bonding from the ammonium ion to the nitrogen of the aza crown ether. The strength of this hydrogen-bonding interaction modulates PET from the amine to anthracene. As explained above, the boronic acid unit can also participate in PET with the anthracene fluorophore and diol binding can also cause fluorescence recovery. The anthracene unit serves as a rigid spacer between the two-receptor units, with the appropriate spacing for the glucose guest. This system behaves like an **AND** logic gate [138, 159], in that fluorescence recovery is only observed when two chemical inputs are supplied; for this system the two chemical inputs are an ammonium cation and a diol group.

A D-glucarate system **131** consisting of boronic acid and guanidinium receptor units has also been reported by Wang [160].

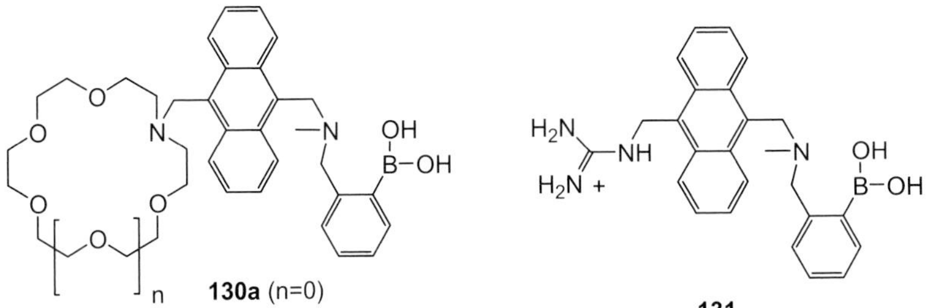

130a (n=0)
130b (n=1)

131

Uncategorized fluorescence sensors

Wang has recently reported that 8-quinoline boronic acid **132** responds to the binding of saccharides with over 40-fold increases in fluorescence intensity [161]. The authors ascribe the fluorescence changes to environmental factors.

Heagy and Lakowicz have investigated *N*-phenylboronic acid derivatives of 1,8-naphthalimide **133a,b**. With these systems the fluorescence is substantially quenched (by a factor of ca. five) on saccharide binding [162, 163]. The fluorescence change has been ascribed to PET from the boronate to the naphthalimide fluorophore. The nitro derivative **133c** was particularly interesting, because dual fluorescence was observed and was particularly sensitive for D-glucose [164].

Hayashita and Teramae prepared an interesting fluorescent ensemble comprising compound **134** and β-cyclodextrin [165]. With this system fluorescence enhancement is observed on saccharide binding and, as expected for a monoboronic acid, the highest binding was observed with D-fructose.

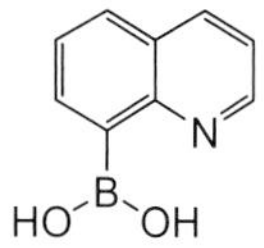

132

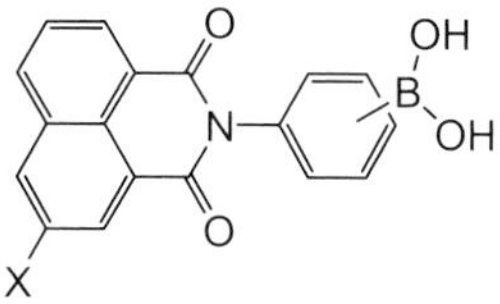

133a (X=H, meta)
133b (X=H, ortho)
133c (X=NO$_2$, meta)

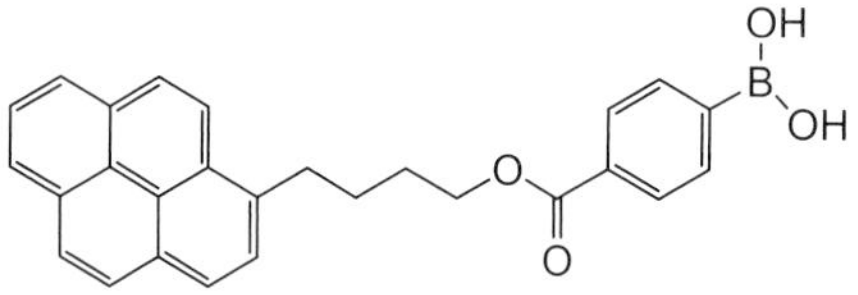

134

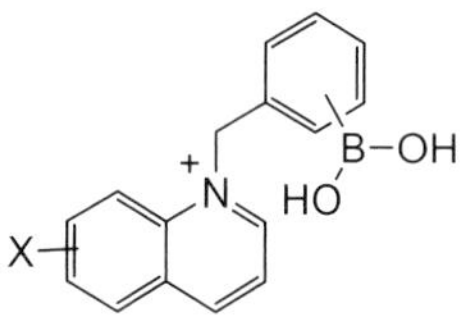

135a (X=OCH$_3$, meta)
135b (X=OCH$_3$, ortho)
135c (X=OCH$_3$, meta)
135d (X=CH$_3$, meta)
135e (X=CH$_3$, ortho)
135f (X=CH$_3$, meta)

Lakowicz and Geddes have explored the use of several quaternized quinolium boronic acids **135a–f** for glucose monitoring within contact lenses [166].

Norrild has developed an interesting diboronic acid system **136** [155]. The system works by reducing the quenching capacity of the pyridine groups of **136** on saccharide binding. The system selectively binds D-glucose with a log K of 3.4. The structure of the complex was determined to be a 1,2:3,5 bound α-D-glucofuranose. Evidence for the furanose structure was obtained from ^{1}H and ^{13}C NMR data with emphasis on information from $^1J_{C-C}$ coupling constants.

All of the systems described above for selective binding of D-glucose have been designed using the approximate positioning of two boronic acid units. Many of these systems bind D-glucose strongly and selectively. In systems in which the structure of the D-glucose complex has been determined the furanose rather than pyranose form of D-glucose is favored. Norrild has stated that "In our opinion binding of the more

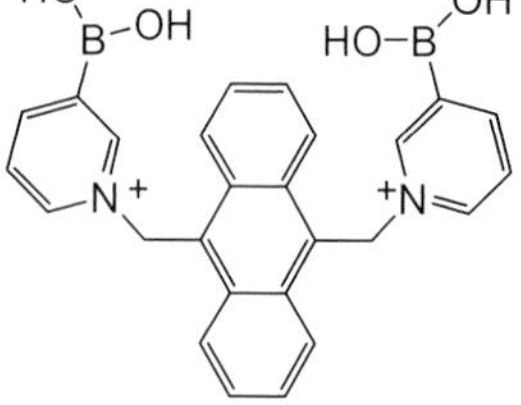

136

abundant α-pyranose form of glucose by boronic acids is not to be considered in the future design of boronic acid-based sensors for aqueous systems."[155].

However, using computational methods Drueckhammer has designed a system selective for the pyranose form of D-glucose. The approach resulted in the design and synthesis of compound **137** which binds very strongly with D-glucose [167]. The affinity of compound **137** for D-glucose was 400-fold that for the other saccharides (D-galactose, D-mannose, and D-fructose). ¹H NMR was used to confirm that the bound D-glucose was captured in the pyranose, not furanose, form.

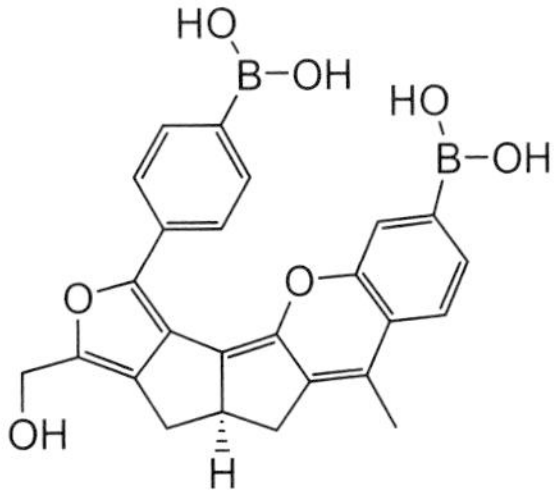

137

A novel approach has been taken by Hamachi, who coupled a natural receptor protein (Concanavalin A or ConA) with a fluorescent boronic acid PET system to prepare a semisynthetic biosensor **138** [168]. The system illustrates how both synthetic and natural systems can be combined to generate biosensors with enhanced selectivity toward specific saccharides.

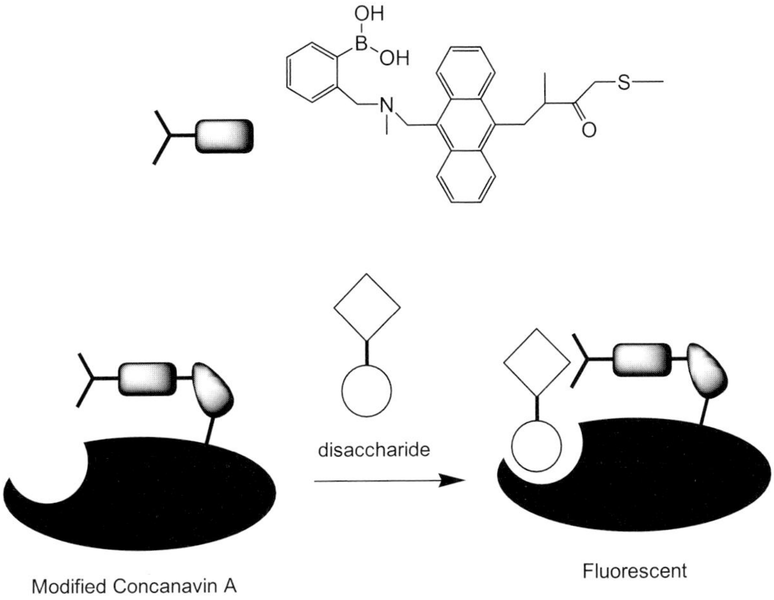

disaccharide

Modified Concanavin A

Fluorescent

138

2.3.2.2 **Colorimetric Sensors**

Colorimetric sensors for saccharides are of particular practical interest. If a system with a large color change could be developed it could be incorporated into a diagnostic test paper for D-glucose, similar to universal indicator paper for pH. Such a system would make it possible to measure D-glucose concentrations without the need of specialist instrumentation. This would be of particular benefit to diabetics in developing countries.

Mizuno has investigated the use of chiral salen cobalt(II) complexes **139** and **140** [169]. Spectroscopic changes in the metal complexes were used to monitor the formation of the saccharide complexes. Chiral discrimination was observed with **139**, which had twofold selectivity for L-allose over D-allose. Mizuno also used a prochiral salen cobalt(II) complex; the binding and chirality were monitored using circular dichroism (CD) spectroscopy [170].

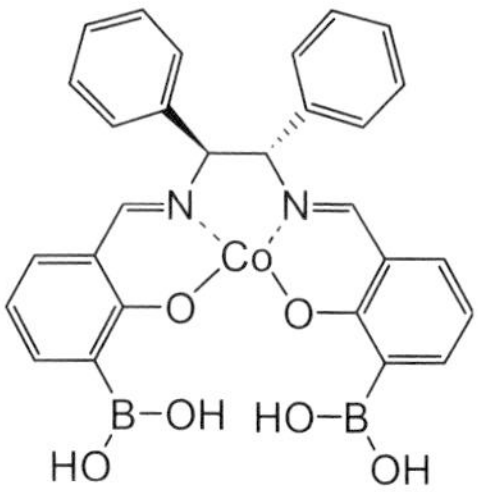

139 (*R*) 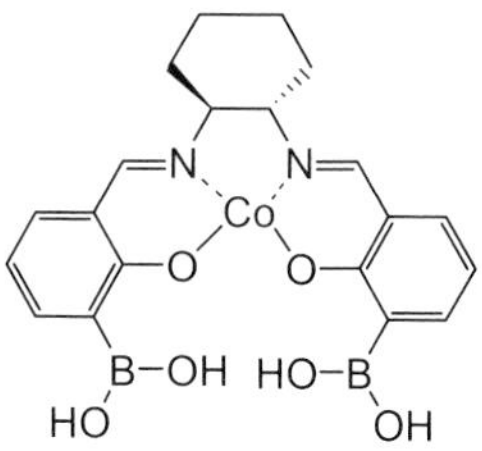   **140** (*R*)

Yam and Kai [171] and a detailed re-investigation by Mizuno [172] have explored the sensing properties of a boronic acid-appended rhenium(I) complex **141**. This system and **139** and **140** illustrate how metal chelation can be used to extend the working wavelength of a sensor.

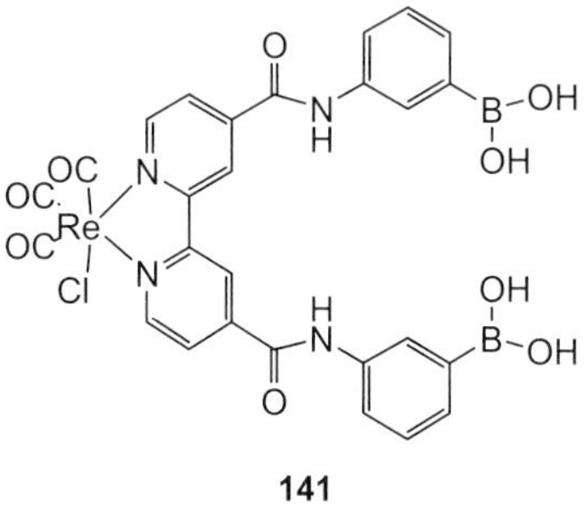

141

Strongin has prepared a tetraboronic acid–resorcinarene system **142** for visual sensing of saccharides [173]. Characteristic color changes were observed for specific carbohydrates, glucose phosphates, and amino sugars on gentle heating in DMSO. Further work by Strongin with another resorcinol derivative **143** has shown that oxygen promotes the color changes and that the resorcinol hydroxyl groups play a key role in the formation of color in the solutions [174]. The mechanism of color change points to xanthenes as in situ chromophores formed by heating resorcinols in DM-SO. Non-boronic acid receptors also produce colored solutions but to a lesser extent; in these systems the color is a result of hydrogen bonding between aldonic acids (heating sugars in DMSO produces aldonic acid derivatives) and the hydroxyl groups of the in-situ xanthene chromophore [175].

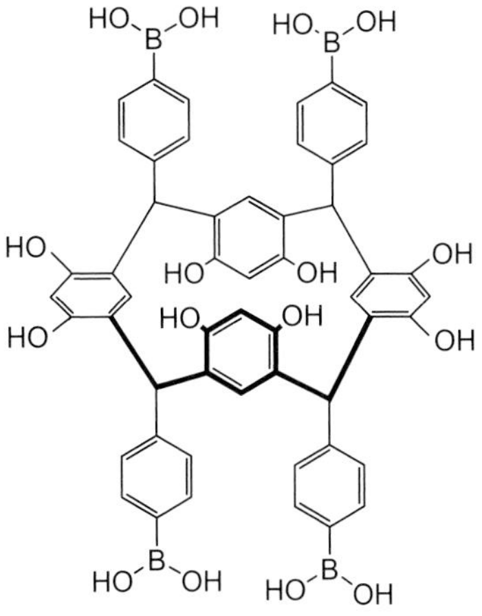

142

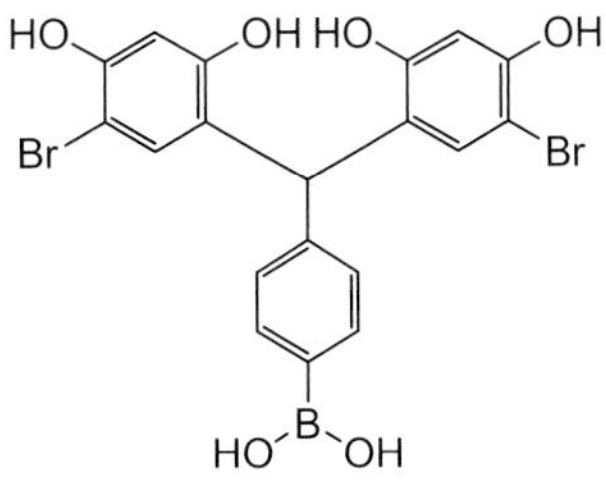

143

Koumoto demonstrated that azobenzene derivatives **144** and **145**, bearing one or two aminomethylphenylboronic acid groups, can be used for practical colorimetric saccharide sensing in "neutral" aqueous media [176].

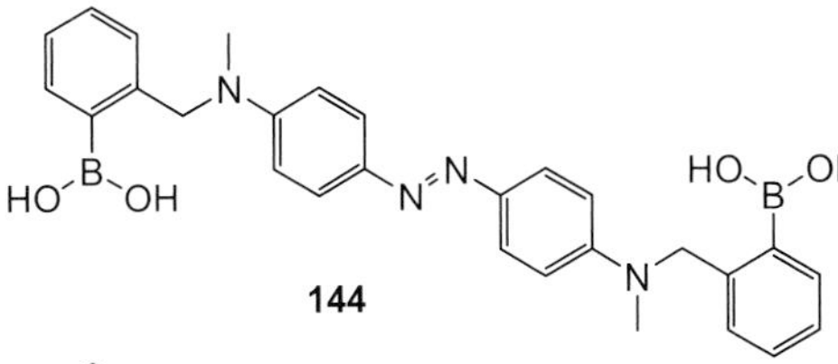

144

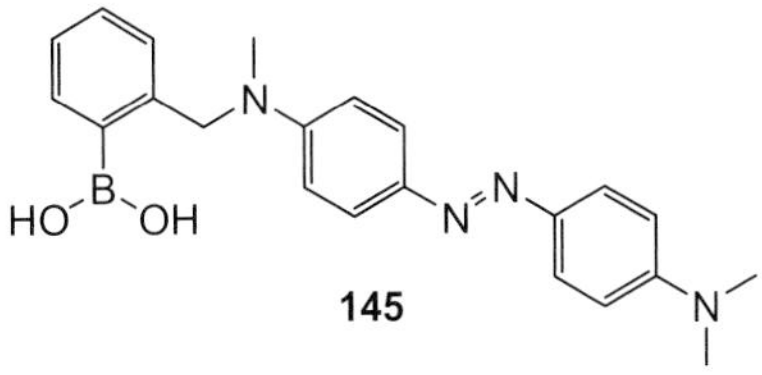

145

A large visible color change from purple to red is obtained with diazo dye system **146a** on saccharide binding [177]. With azo dye **146a** the wavelength maximum shifts by approximately 55 nm to a shorter wavelength on saccharide complexation. The concentration of the guest required to produce the change is different in each system, because of the different stability constants of the saccharides. The stability constants (log *K*) of the boronic acid dye–saccharide complexes are 3.75, 1.85, and 0.66 for D-fructose, D-glucose, and ethylene glycol, respectively.

What makes the equilibria of dye molecule **146a** interesting is that in the absence of saccharide at pH 11.32 the color of **146a** is purple whereas in the presence of saccharide the color is red. In the presence of saccharide the B–N interaction becomes stronger. The increased B–N interaction causes the N–H proton to become more acidic. At pH 11.32, therefore, the saccharide boronate dehydrates (loss of H⁺ from aniline and OH⁻ from boronate) to produce a red species with a strong B–N bond.

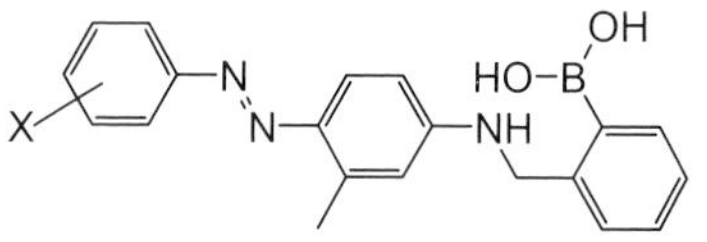

146a (*X* = *p*-NO₂)
146b (*X* = *p*-SO₃H)
146c (*X* = *p*-CO₂H)
146d (*X* = *p*-OMe)
146e (*X* = *m*-CO₂H)

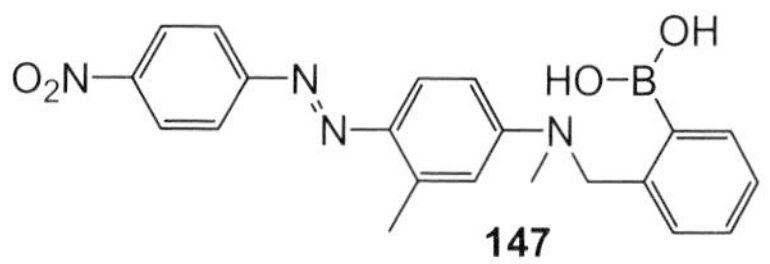

These equilibrium species explain why dye molecule **147** does not give a visible spectral shift on saccharide binding. With **147** the tertiary amine removes the possibility of dehydration, so a strong boron–nitrogen bond cannot be formed, hence no spectral shift is observed [178]. A detailed investigation of a series of azo dyes with both electron donating and withdrawing groups **146a-e** indicated that a strong electron-withdrawing group is required to produce a color change [178]. For the strongly electron withdrawing tricyanovinyl dye **148**, the p*K*ₐ (7.81) was much less than for **146a** (10.2), enabling a visible color change on addition of saccharides at much lower pH (8.21) [178].

Lakowicz has also prepared boronic acid azo dye molecules **149** and **150** in which direct conjugation with the boron center is possible. In particular the azo dye **150** produces a visible color change from yellow to orange at pH 7 [179].

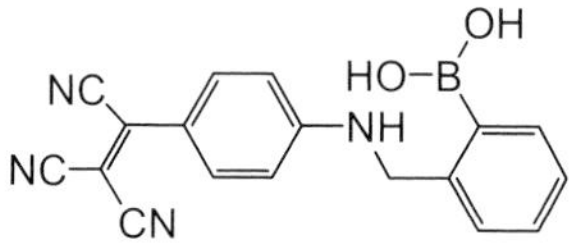

148

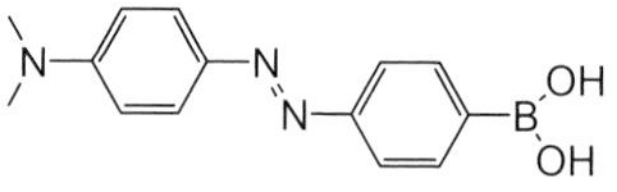

149

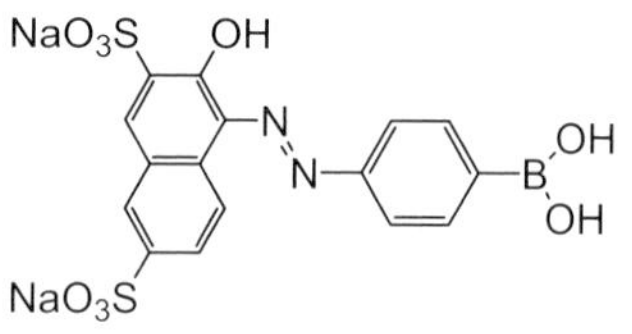

150

Sato has prepared stilbazolium boronic acids **151** and **152** and demonstrated the suitability of these units for optical sensing of saccharides [180].

Wang has prepared nitrophenol boronic acids **153** and **154** which give large shifts in the UV on addition of saccharides. The changes have been attributed to changes in the balance of the phenolate to boronate equilibria in the presence of saccharides [181].

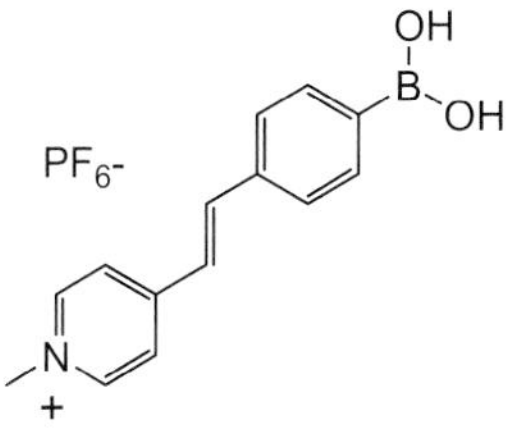

With the aid of better chromophores and better receptors it will be possible to develop selective and sensitive color sensors. Such sensory systems will have widespread applications in industry. It should, for example, be possible to provide cheap and stable "test papers" for the detection of blood glucose.

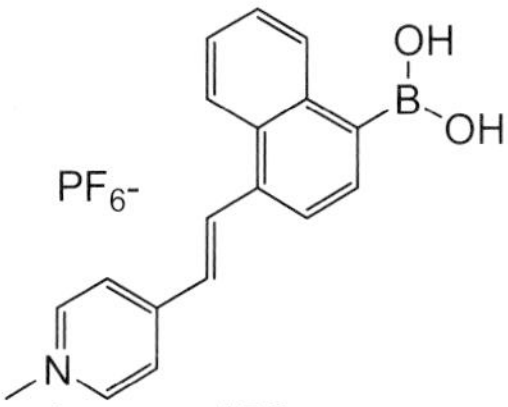

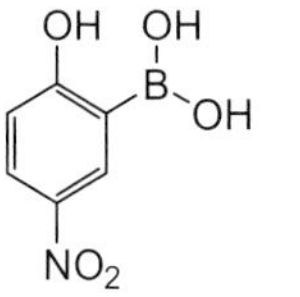

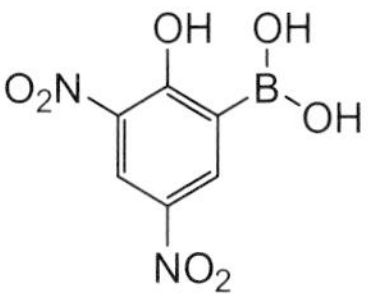

2.3.2.3 Electrochemical Sensors

Electrochemical detection of saccharides by enzymatic decomposition is well known [182]. The development of boronic acid-based electro-active saccharide receptors could provide selectivity for a range of saccharides.

Moore and Wayner have explored the redox switching of carbohydrate binding with commercial ferrocene boronic acid [183]. From their detailed investigations they determined that binding constants of saccharides with the ferrocenium form are approximately two orders of magnitude greater than with the ferrocene form. The increased stability is ascribed to the lower pK_a of the ferrocenium (5.8) than ferrocene (10.8) boronic acid.

Fabre has investigated the electrochemical sensing properties of a boronic acid-substituted bipyridine iron(II) complex **155** [184]. On addition of 10 mM D-fructose the oxidation peak was shifted by 50 mV towards more positive values.

Ferrocene monoboronic acid **156** and diboronic acid **157** have been prepared and evaluated [185]. The monoboronic acid system **156** has also been prepared and proposed as an electrochemical sensor for saccharides by Norrild [186]. The electro-

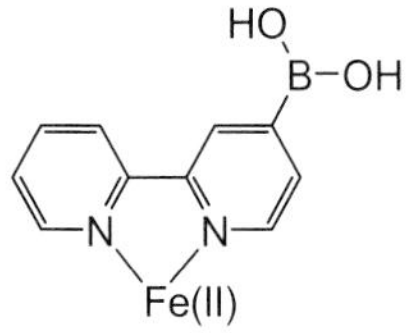

155

chemical saccharide sensor **157** contains two boronic acid units (saccharide selectiv-ity), one ferrocene unit (electrochemical read-out) and a hexamethylene linker unit (D-glucose selectivity). Electrochemical sensor **157** has greater selectivity for D-glu-cose (40-fold) and D-galactose (17-fold) than the monoboronic acid **156**.

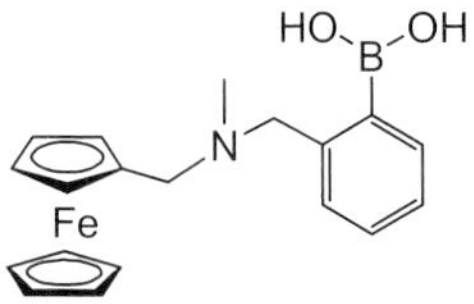

156

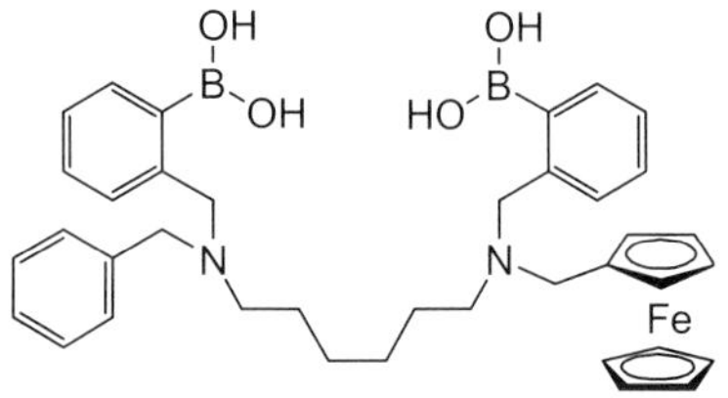

157

2.3.2.4 Sensor Assay

The systems described so far contain a receptor and a reporter (fluorophore or chro-mophore) as part of a discrete molecular unit. In another possible approach to boron-ic acid-based sensors, however, the receptor and a reporter unit are separate – a com-petitive assay. A competitive assay requires that the receptor and reporter (typically a commercial dye) associate under the measurement conditions. The receptor–re-porter complex is then selectively dissociated by addition of the appropriate guests. When the reporter dissociates from the receptor, a measurable response is produced.

The competitive assay approach to novel chemosensors was pioneered by Anslyn [187]. These competative systems are particularly interesting, because they reduce the synthetic complexity of the receptor. Wang has demonstrated that alizarin red S and phenyl boronic acid could be used in competitive assays for saccharides [188, 189]. The system is D-fructose-selective, the expected selectivity for a monoboronic acid system [90].

Anslyn has recently reported two very interesting systems based on boronic acid re-ceptors. Although the Anslyn systems involve competitive colorimetric assay, there is no reason the system cannot be extended to a fluorimetric assay by choice of an ap-propriate dye molecule. The first system is a receptor for glucose-6-phosphate **158** [190]. The binding of glucose-6-phosphate is measured via competitive displacement of 5-carboxyfluorescein. The second is a system in which the binding of heparin and **159** is monitored by displacement of pyrocatechol violet [191].

Lakowicz has also used competitive interactions between a ruthenium metal–ligand complex, a boronic acid derivative, and glucose [192]. The metal–ligand complex forms a reversible complex with 2-tolylboronic acid or 2-methoxyphenylboronic acid. Complexation is accompanied by a severalfold increase in the luminescent intensity of the ruthenium complex. Addition of glucose results in reduced luminescent intensity, which appears to be the result of reduced binding between the metal–ligand complex and the boronic acid. Ruthenium metal–ligand complexes are convenient for optical sensing, because their long luminescence decay times enable lifetime-based sensing with simple instrumentation.

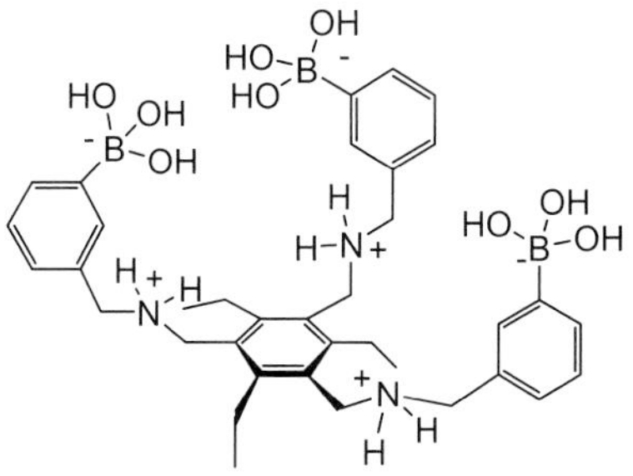

158

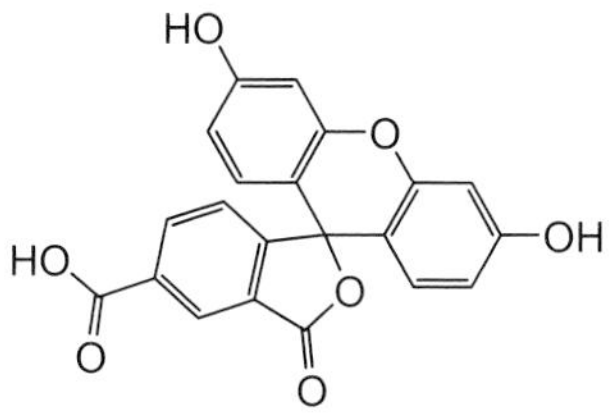

5-carboxyfluorescein

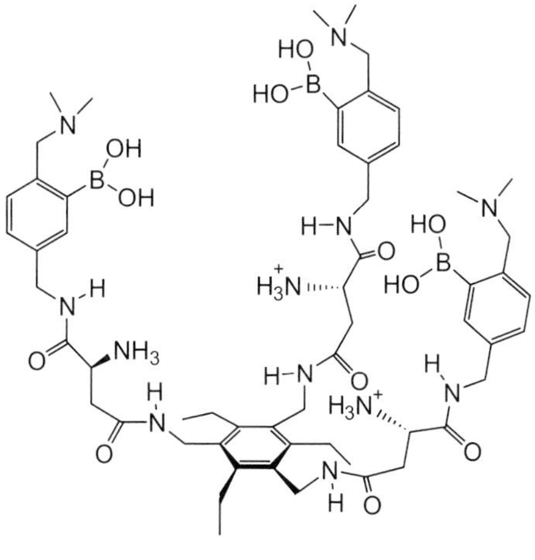

159

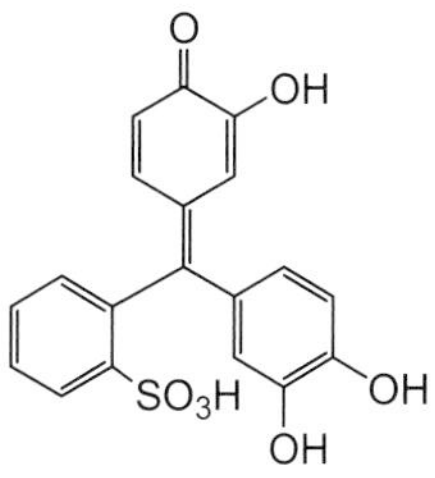

pyrocatechol violet

An interesting multicomponent system has also been devised by Singaram. In this quenching of a pyranine dye by bisboronic acid viologen units **160** and **161** is modulated by added saccharide [193, 194]. Although compound **160** binds well with D-fructose ($K = 2600$ M^{-1}) and weakly with D-glucose ($K = 43$ M^{-1}), the system only produces 4 % fluorescence recovery [194]. Compound **161** binds well with both D-fructose ($K = 3300$ M^{-1}) and D-glucose ($K = 1800$ M^{-1}); together with enhanced selectivity for D-glucose this system also produces a 45 % fluorescence recovery on addition of saccharides [193].

Lakowicz has also examined the quenching and recovery of a sulfonated poly(phenylene ethynylene) by a bisboronic acid viologen **162** on addition of saccharides [195]. The system is D-fructose selective and produces up to 70-fold fluorescence enhancement on addition of saccharides.

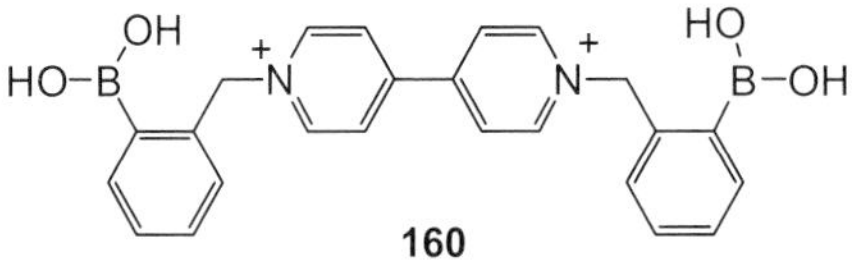

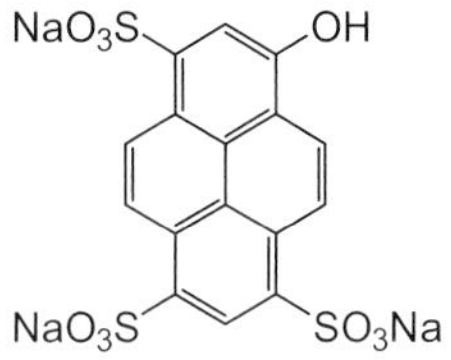

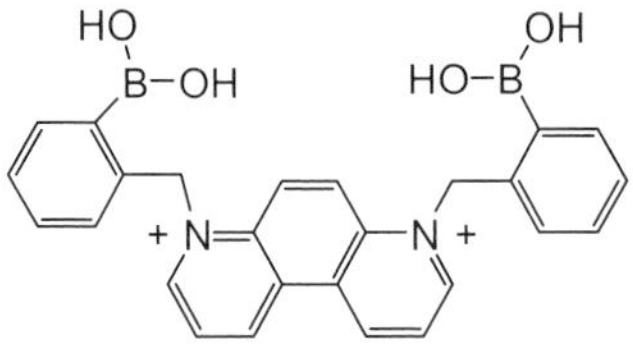

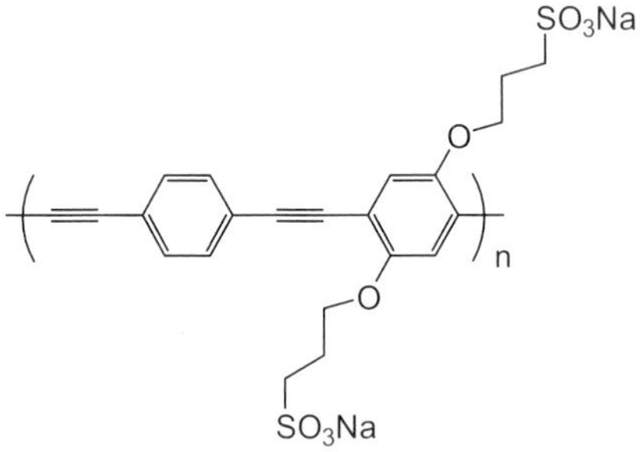

Based on Wang's general system [188, 189] the diboronic acid **163** and alizarin red S results in a very efficient D-glucose assay [196]. Sensor **163** and alizarin red S produces a sensor with a four-fold enhancement in binding over phenyl boronic acid (PBA) with D-glucose. Sensor **163** can also be used at a concentration one tenth that of PBA.

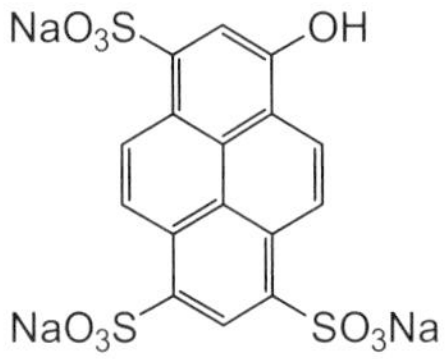

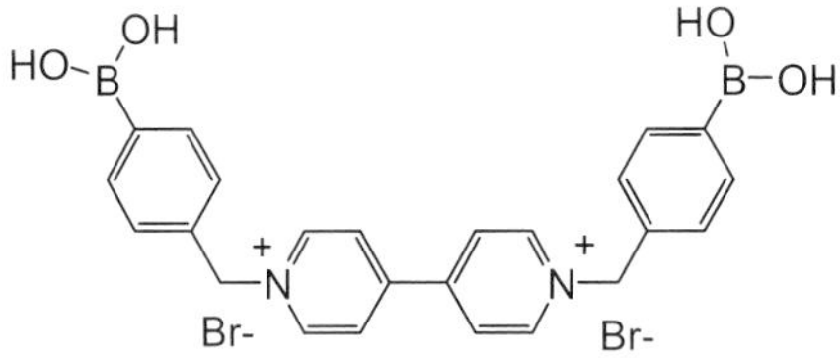

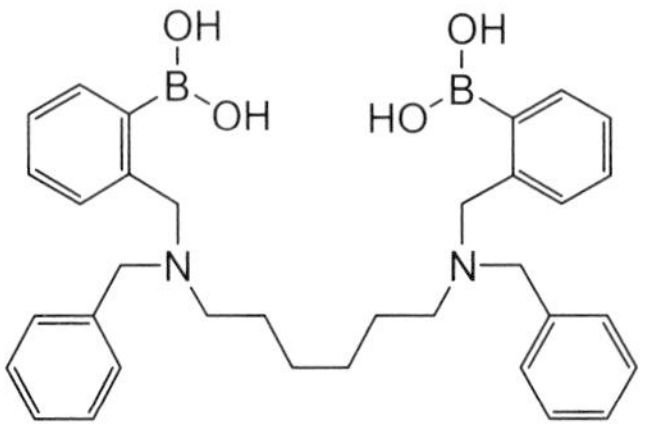

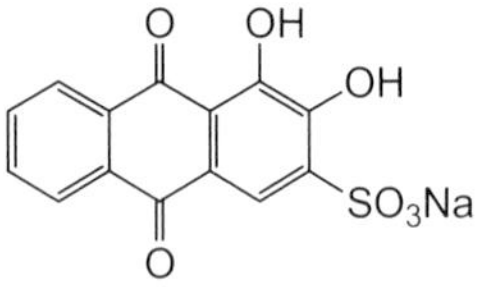

From the description above, the fluorescent assay method seems to be one of the best ways forward in the design of saccharide-selective sensors. In a competitive assay, however, all competition must be controlled so that the signal can be used to produce an analytical outcome. The occurrence of previously unrecognized interactions between boronic acids and buffer conjugate bases (phosphate, citrate, and imidazole) [96] to create ternary complexes (boronate–X–saccharide) will need to be considered in future assay design.

2.3.2.5 Polymer and Surface Bound Sensors

If practically useful sensors are to be developed from the boronic acid sensors described above, they will have to be integrated into a device. One way to help achieve this goal is to incorporate the saccharide-selective interface into a polymer support.

Wang has employed the template approach using monomer **164** to prepare a fluorescent polymer with enhanced selectivity toward D-fructose [197, 198]. Appleton has used a similar approach, using monomer **165** to prepare a D-glucose-selective polymer [146]. The Appleton polymer clearly shows the value of the imprinting technique, the monomer's selectivity for D-fructose over D-glucose has been reversed in the polymer.

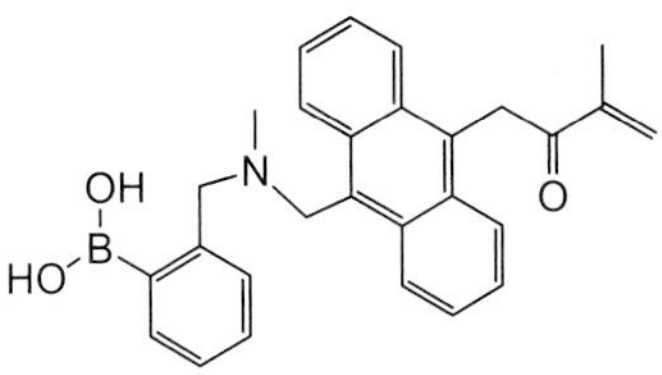

164 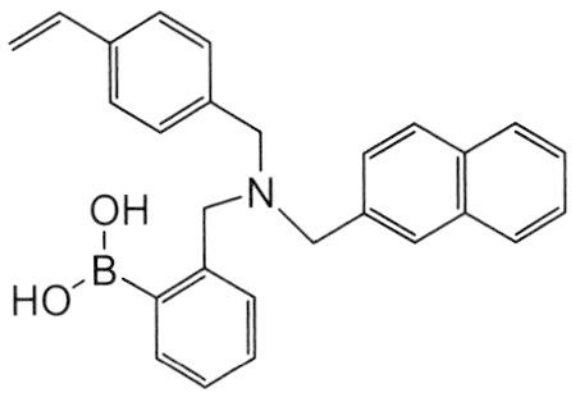   **165**

Another approach is to link a solution-based sensor to a solid support. Sensor **166** results when the glucose-selective receptor **125d** is attached to a polymer support [147]. The main difference between the polymer-bound system **166** and the solution-based system **125d** is the D-glucose selectivity. The D-glucose selectivity drops for

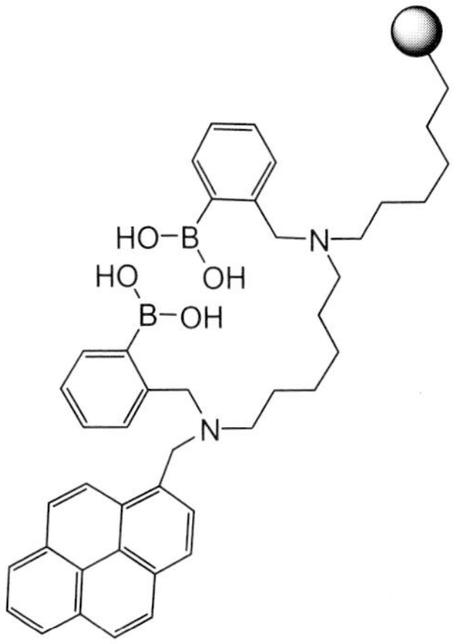

166

compound **166** whereas the selectivity for other saccharides is similar to those observed for compound **125d**; the polymeric system still has a ninefold enhancement for D-glucose over the monoboronic acid model compound, however. The differences between **125d** and **166** are believed to be due to the proximity of the receptor to the polymer backbone.

More recently, a membrane in which a PET-based glucose-sensing system is immobilized has been developed **167**. The amide group was introduced not only as a linker to the membrane but also to shift the excitation and emission maxima to longer wavelengths [199].

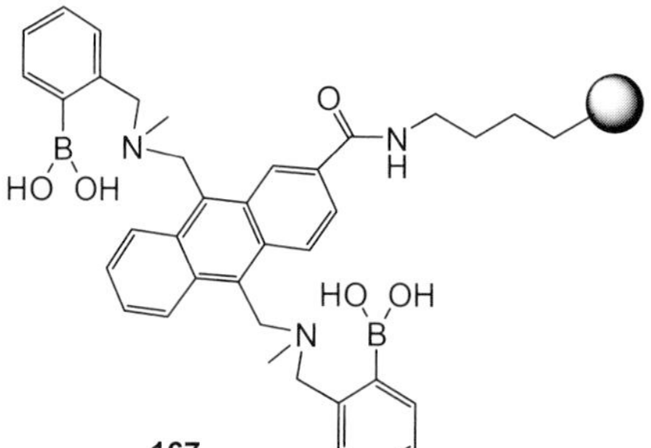

167

Singram has made a significant breakthrough in the development of a continuous glucose monitoring system by incorporating his assay system, comprising a pyranine dye and bisboronic acid viologen units, into a thin film hydrogel **168** [200]. The system can be used to detect glucose in the physiologicaly important range of 2.5–20 mM and operate reversibly under physiological conditions 37 °C, 0.1 µM ionic strength, and pH 7.4.

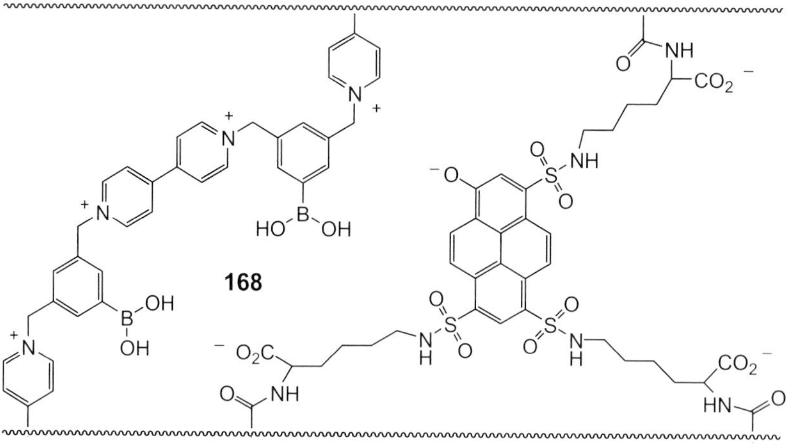

168

Asher has developed a very impressive system in which a crystalline colloidal array (CCA) is incorporated into a polyacrylamide hydrogel to create a polymerized crystalline colloidal array (PCCA). Two systems have been developed. One polyacrylamide has pendant boronic acid groups and works in low-ionic-strength solutions [201]. The other has pendant polyethylene glycol or 15-crown-5 and boronic acid groups and works in high-ionic-strength solutions [202]. The embedded CCA diffracts visible light and the PCCA diffraction wavelength is used to calculate the hydrogel volume. The PCCA photonic crystal sensing material responds to glucose by swelling and red shifting the diffraction as the glucose concentration increases.

Wolfbeis has prepared a polyaniline with a near-infrared optical response to saccharides. The film was prepared by copolymerization of aniline and 3-aminophenyl boronic acid. Addition of saccharides at pH 7.3 led to changes in absorption at 675 nm [203].

Nakashima has prepared a phenylboronic acid-terminated redox-active self-assembled monolayer on a gold electrode as an electrochemical sensor for saccharides. Self-

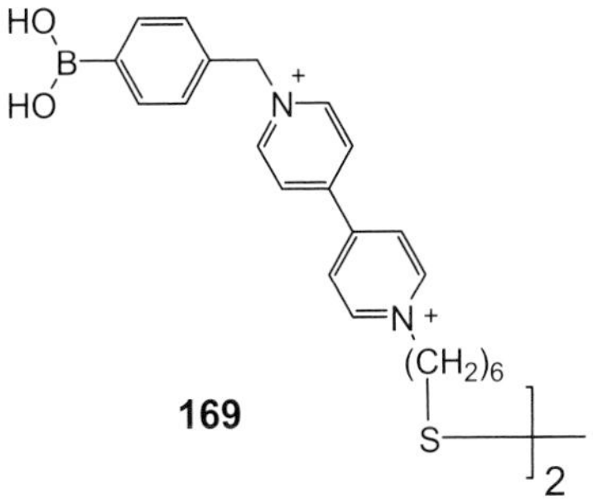

169

assembled monolayers of **169**, a phenylboronic acid terminated viologen alkyl disulfide, function as a sensitive saccharide sensor in aqueous solution [204].

Freund has prepared polyaniline boronic acids by electrochemical polymerization of 3-aminophenyl boronic acid [205, 206]. The electrochemical potential of the polymer is sensitive to the change in the pK_a of the polymer as a result of boronic acid–diol complex formation. Fabre has also used polyaniline boronic acids as conductiomeric sensors for dopamine [207].

Polymers containing vinylphenylboronic acid groups have been evaluated as electrochemical saccharide sensors [208].

References

1 (a) R. A. Laine, *Glycobiology* **1994**, *4*, 759; (b) B. G. Davis, *Chem. Rev.* **2002**, *102*, 579. The figure for hexasaccharides is based on just 6 different monomers; consideration of the full range of possibilities would yield astronomical numbers.

2 For leading references on carbohydrate recognition in biology ("glycobiology"), see: R. A. Dwek, T. D. Butters, *Chem. Rev.* **2002**, *102*, 283 (and succeeding articles); C. R. Bertozzi, L. L. Kiessling, *Science* **2001**, *291*, 2357; T. Feizi, B. Mulloy, *Curr. Opin. Struct. Biol.* **2001**, *11*, 585 (and succeeding articles); S. J. Williams, G. J. Davies, *Trends Biotechnol.* **2001**, *19*, 356; A. Imberty, S. Perez, *Chem. Rev.* **2000**, *100*, 4567; P. Sears, C. H. Wong, *Angew. Chem. Int. Ed.* **1999**, *38*, 2301; R. A. Dwek, *Chem. Rev.* **1996**, *96*, 683.

3 Leading references: K. Tohda, M. Gratzl, *ChemPhysChem* **2003**, *4*, 155; E. Katz, L. Sheeney-Haj-Ichia, A. F. Buckmann, I. Willner, *Angew. Chem. Int. Ed.* **2002**, *41*, 1343; Y. Wei, H. Dong, J. G. Xu, Q. W. Feng, *ChemPhysChem* **2002**, *3*, 802; K. E. Shafer-Peltier, C. L. Haynes, M. R. Glucksberg, R. P. Van Duyne, *J. Am. Chem. Soc.* **2003**, *125*, 588.

4 (a) T. K. Dam, C. F. Brewer, *Chem. Rev.* **2002**, *102*, 387; (b) H. Lis, N. Sharon, *Chem. Rev.* **1998**, *98*, 637; (c) W. I. Weis, K. Drickhamer, *Ann. Rev. Biochem.* **1996**, *65*, 441; (d) Y. C. Lee, R. T. Lee, *Acc. Chem. Res.* **1995**, *28*, 321; (e) S. D. Rosen, C. R. Bertozzi, *Curr. Opin. Cell Biol.* **1994**, *6*, 663.

5 J. J. Lundquist, E. J. Toone, *Chem. Rev.* **2002**, *102*, 555; C. O. Mellet, J. Defaye, J. M. G. Fernandez, *Chem. Eur. J.* **2002**, *8*, 1982; D. A. Fulton, J. F. Stoddart, *Bioconj. Chem.* **2001**, *12*, 655.

6 F. A. Quiocho, *Pure Appl. Chem.* **1989**, *61*, 1293.

7 (a) N. K. Vyas, M. N. Vyas, F. A. Quiocho, *Science* **1988**, *242*, 1290; (b) N. K. Vyas, M. N. Vyas, F. A. Quiocho, *Biochemistry* **1994**, *33*, 4762.

8 A. P. Davis, R. S. Wareham, *Angew. Chem. Int. Ed.* **1999**, *38*, 2978.

9 T. D. James, K. R. A. S. Sandanayake, S. Shinkai, *Angew. Chem. Int. Ed. Engl.* **1996**, *35*, 1911.

10 For a more detailed discussion of this issue, see Ref. [8].

11 Lemieux, R. U. *Acc. Chem. Res.* **1996**, *29*, 373; R. U. Lemieux in *Carbohydrate Antigens* (Eds.: P. J. Garegg, A. A. Lindberg), American Chemical Society, **1993**, pp. 5–18.

12 Toone, E. J. *Curr. Opin. Struct. Biol.* **1994**, *4*, 719.

13 Some weaker interactions do, however, show positive entropies, in accordance with entropy-enthalpy compensation.

14 W. Blokzijl, J. B. F. N. Engberts, *Angew. Chem. Int. Ed. Engl.* **1993**, *32*, 1545.

15 Meyer, E. A.; Castellano, R. K.; Diederich, F. *Angew. Chem., Int. Ed.* **2003**, *42*, 1210; Chervenak, M. C.; Toone, E. J. *J. Am. Chem. Soc.* **1994**, *116*, 10533.

16 Y. Kikuchi, Y. Tanaka, S. Sutaro, K. Kobayashi, H. Toi, Y. Aoyama, *J. Am. Chem. Soc.* **1992**, *114*, 10302.

17 R. P. Bonar-Law, J. K. M. Sanders, *J. Am. Chem. Soc.* **1995**, *117*, 259.

18 Y. Aoyama, Y. Tanaka, H. Toi, H. Ogoshi, *J. Am. Chem. Soc.* **1988**, *110*, 634.

19 Y. Kikuchi, K. Kobayashi, Y. Aoyama, *J. Am. Chem. Soc.* **1992**, *114*, 1351.

20 M. Segura, B. Bricoli, A. Casnati, E. M. Muñoz, F. Sansone, R. Ungaro, C. Vicent, *J. Org. Chem.* **2003**, *68*, 6296.

21 T. Ishi-i, M. A. Mateos-Timoneda, P. Timmerman, M. Crego-Calama, D. N. Reinhoudt, S. Shinkai, *Angew. Chem. Int. Ed.* **2003**, *42*, 2300.

22 A. P. Davis, R. P. Bonar-Law, J. K. M. Sanders, in *Comprehensive Supramolecular Chemistry, Vol. 4 (Supramolecular Reactivity and Transport: Bioorganic Systems)* (Ed.: Y. Murakami), Pergamon, Oxford, 1996, p. 257; P. Wallimann, T. Marti, A. Furer, F. Diederich, *Chem. Rev.* **1997**, *97*, 1567.

23 R. P. Bonar-Law, A. P. Davis, *J. Chem. Soc. Chem. Commun.* **1989**, 1050; R. P. Bonar-Law, A. P. Davis, B. A. Murray, *Angew. Chem. Int. Ed. Engl.* **1990**, *29*, 1407.

24 K. M. Bhattarai, R. P. Bonar-Law, A. P. Davis, B. A. Murray, *J. Chem. Soc. Chem. Commun.* **1992**, 752.

25 (a) K. M. Bhattarai, A. P. Davis, J. J. Perry, C. J. Walter, S. Menzer, D. J. Williams, *J. Org. Chem.* **1997**, *62*, 8463; (b) A. P. Davis,

J. J. Walsh, *Chem. Commun.* **1996**, 449; (c) S. Kohmoto, D. Fukui, T. Nagashima, K. Kishikawa, M. Yamamoto, K. Yamada, *Chem. Commun.* **1996**, 1869.

26 R. P. Bonar-Law, J. K. M. Sanders, *J. Am. Chem. Soc.* **1995**, *117*, 259.

27 (a) T. Mizutani, T. Murakami, N. Matsumi, T. Kurahashi, H. Ogoshi, *J. Chem. Soc. Chem. Commun.* **1995**, 1257; (b) T. Mizutani, T. Kurahashi, T. Murakami, N. Matsumi, H. Ogoshi, *J. Am. Chem. Soc.* **1997**, *119*, 8991; (c) K. Wada, T. Mizutani, S. Kitagawa, *J. Org. Chem.* **2003**, *68*, 5123.

28 K. Ladomenou, R. P. Bonar-Law, *Chem. Commun.* **2002**, 2108.

29 Y. H. Kim, J. I. Hong, *Angew. Chem. Int. Ed.* **2002**, *41*, 2947.

30 C.-Y. Huang, L. A. Cabell, E. V. Anslyn, *J. Am. Chem. Soc.* **1994**, *116*, 2778.

31 (a) S. Tamaru, S. Shinkai, A. B. Khasanov, T. W. Bell, *Proc. Natl. Acad. Sci. USA* **2002**, *99*, 4972; (b) S. Tamaru, M. Yamamoto, S. Shinkai, A. B. Khasanov, T. W. Bell, *Chem. Eur. J.* **2001**, *7*, 5270.

32 M. Inouye, T. Miyake, M. Furusyo, H. Nakazumi, *J. Am. Chem. Soc.* **1995**, *117*, 12416.

33 M. Inouye, J. Chiba, H. Nakazumi, *J. Org. Chem.* **1999**, *64*, 8170.

34 M. Inouye, K. Takahashi, H. Nakazumi, *J. Am. Chem. Soc.* **1999**, *121*, 341.

35 J. M. Fang, S. Selvi, J. H. Liao, Z. Slanina, C. T. Chen, P. T. Chou, *J. Am. Chem. Soc.* **2004**, *126*, 3559.

36 M. Inouye, M. Waki, H. Abe, *J. Am. Chem. Soc.* **2004**, *126*, 2022.

37 (a) J. Cuntze, L. Owens, V. Alcazar, P. Seiler, F. Diederich, *Helv. Chim. Acta* **1995**, *78*, 367; (b) D. K. Smith, F. Diederich, *Chem. Commun.* **1998**, 2501.

38 (a) M. Mazik, H. Bandmann, W. Sicking, *Angew. Chem. Int. Ed.* **2000**, *39*, 551; (b) M. Mazik, W. Sicking, *Chem. Eur. J.* **2001**, *7*, 664; (c) M. Mazik, W. Radunz, W. Sicking, *Org. Lett.* **2002**, *4*, 4579; (d) H. J. Kim, Y. H. Kim, J. I. Hong, *Tetrahedron Lett.* **2001**, *42*, 5049.

39 D. W. P. M. Löwik, C. R. Lowe, *Eur. J. Org. Chem.* **2001**, 2825.

40 (a) G. Das, A. D. Hamilton, *J. Am. Chem. Soc.* **1994**, *116*, 11139; (b) G. Das, A. D. Hamilton, *Tetrahedron Lett.* **1997**, *38*, 3675.

41 (a) S. Anderson, U. Neidlein, V. Gramlich, F. Diederich, *Angew. Chem. Int. Ed. Engl.* **1995**, *34*, 1596; (b) A. Bahr, B. Felber, K. Schneider, F. Diederich, *Helv. Chim. Acta* **2000**, *83*, 1346.

42 J. M. Coterón, F. Hacket, H. J. Schneider, *J. Org. Chem.* **1996**, *61*, 1429.

43 (a) U. Neidlein, F. Diederich, *Chem. Commun.* **1996**, 1493; (b) A. S. Droz, F. Diederich, *J. Chem. Soc. Perkin Trans. 1* **2000**, 4224; (c) A. S. Droz, U. Neidlein, S. Anderson, P. Seiler, F. Diederich, *Helv. Chim. Acta* **2001**, *84*, 2243.

44 (a) T. Schrader, *J. Org. Chem.* **1998**, *63*, 264; (b) T. Schrader, *J. Am. Chem. Soc.* **1998**, *120*, 11816.

45 J. Bitta, S. Kubik, *Org. Lett.* **2001**, *3*, 2637.

46 R. Liu, W. C. Still, *Tetrahedron Lett.* **1993**, *34*, 2573.

47 A. P. Davis, R. S. Wareham, *Angew. Chem. Int. Ed.* **1998**, *37*, 2270.

48 G. Lecollinet, A. P. Dominey, T. Velasco, A. P. Davis, *Angew. Chem. Int. Ed.* **2002**, *41*, 4093.

49 R. Welti, F. Diederich, *Helv. Chim. Acta* **2003**, *86*, 494.

50 P. B. Savage, S. H. Gellman, *J. Am. Chem. Soc.* **1993**, *115*, 10448.

51 A. Bahr, A. S. Droz, M. Puntener, U. Neidlein, S. Anderson, P. Seiler, F. Diederich, *Helv. Chim. Acta* **1998**, *81*, 1931.

52 Leading references: T. M. Altamore, E. S. Barrett, P. J. Duggan, M. S. Sherburn, M. L. Szydzik, *Org. Lett.* **2002**, *4*, 3489. M. F. Paugam, J. A. Riggs, B. D. Smith, *Chem. Commun.* **1996**, 2539.

53 N. Greenspoon, E. Wachtel, *J. Am. Chem. Soc.* **1991**, *113*, 7233.

54 Y. Aoyama, Y. Tanaka, S. Sugahara, *J. Am. Chem. Soc.* **1989**, *111*, 5397.

55 Y. Tanaka, C. Khare, M. Yonezawa, Y. Aoyama, *Tetrahedron Lett.* **1990**, *31*, 6193.

56 T. Rhlalou, M. Ferhat, M. A. Frouji, D. Langevin, M. Metayer, J. F. Verchere, *J. Membr. Sci.* **2000**, *168*, 63.

57 K. Kurihara, K. Ohto, Y. Tanaka, Y. Aoyama, T. Kunitake, *J. Am. Chem. Soc.* **1991**, *113*, 444.

58 T. J. Ryan, G. Lecollinet, T. Velasco, A. P. Davis, *Proc. Natl. Acad. Sci. USA* **2002**, *99*, 4863.

59 S. A. Galema, H. Hoiland, *J. Phys. Chem.* **1991**, *95*, 5321; S. A. Galema, M. J. Blandamer, J. Engberts, *J. Org. Chem.* **1992**, *57*, 1995.

60 T. Velasco, G. Lecollinet, T. Ryan, A. P. Davis, *Org. Biomol. Chem.* **2004**, *2*, 645.

61 K. Kobayashi, F. Ikeuchi, S. Inaba, Y. Aoyama, *J. Am. Chem. Soc.* **1992**, *114*, 1105.

62 K. Kasuga, T. Hirose, S. Aiba, T. Takahashi, K. Hiratani, *Tetrahedron Lett.* **1998**, *39*, 9699.

63 H. Ikeda, A. Matsuhisa, A. Ueno, *Chem. Eur. J.* **2003**, *9*, 4907.

64 I. V. Lyutikova, I. V. Pletnev, I. G. Matveeva, I. I. Torocheshnikova, *Russ. Chem. Bull.* **1998**, *47*, 177.

65 S. J. Choi, B. G. Choi, S. M. Park, *Anal. Chem.* **2002**, *74*, 1998.

66 For a discussion of this issue, see A. J. Ayling, S. Broderick, J. P. Clare, A. P. Davis, M. N. Pérez-Payán, M. Lahtinen, M. J. Nissinen, K. Rissanen, *Chem. Eur. J.* **2002**, *8*, 2197.

67 K. Kobayashi, Y. Asakawa, Y. Kato, Y. Aoyama, *J. Am. Chem. Soc.* **1992**, *114*, 10307.

68 M. Nishio, Y. Umezawa, M. Hirota, Y. Takeuchi, *Tetrahedron* **1995**, *51*, 8665.

69 R. Yanagihara, Y. Aoyama, *Tetrahedron Lett.* **1994**, *35*, 9725.

70 (a) O. Rusin, V. Král, *Tetrahedron Lett.* **2001**, *42*, 4235; (b) O. Rusin, K. Lang, V. Král, *Chem. Eur. J.* **2002**, *8*, 655.

71 (a) B.-L. Poh, C. M. Tan, *Tetrahedron* **1993**, *49*, 9581; (b) B.-L. Poh, C. M. Tan, *Tetrahedron Lett.* **1994**, *35*, 6387.

72 Y. Aoyama, Y. Nagai, J. Otsuki, K. Kobayashi, H. Toi, *Angew. Chem. Int. Ed. Engl.* **1992**, *31*, 745.

73 F. Hacket, J. M. Coterón, H. J. Schneider, V. P. Kazachenko, *Can. J. Chem.* **1997**, *75*, 52.

74 A. F. D. de Namor, P. M. Blackett, M. C. Cabaleiro, J. M. A. Alrawi, *J. Chem. Soc. Faraday Trans.* **1994**, *90*, 845.

75 See commentary in Ref. [73].

76 A. V. Eliseev, H.-J. Schneider, *J. Am. Chem. Soc.* **1994**, *116*, 6081.

77 J. Otsuki, K. Kobayashi, H. Toi, Y. Aoyama, *Tetrahedron Lett.* **1993**, *34*, 1945.

78 N. Sugimoto, D. Miyoshi, J. Zou, *Chem. Commun.* **2000**, 2295.

79 E. Klein, M. P. Crump, A. P. Davis, *Angew. Chem. Int. Ed.* **2005**, *44*, 298.

80 (a) O. Rusin, V. Král, *Chem. Commun.* **1999**, 2367; (b) V. Král, O. Rusin, J. Charvatova, P. Anzenbacher, J. Fogl, *Tetrahedron Lett.* **2000**, *41*, 10147; (c) J. Charvatova, O. Rusin, V. Král, K. Volka, P. Matejka,

Sens. Actuators, B **2001**, *76*, 366; (d) V. Král, O. Rusin, F. P. Schmidtchen, *Org. Lett.* **2001**, *3*, 873.

81 J. Burger, C. Gack, P. Klufers, *Angew. Chem. Int. Ed. Engl.* **1995**, *34*, 2647. S. D. Kinrade, J. W. Del Nin, A. S. Schach, T. A. Sloan, K. L. Wilson, C. T. G. Knight, *Science* **1999**, *285*, 1542.

82 (a) S. Striegler, E. Tewes, *Eur. J. Inorg. Chem.* **2002**, 487; (b) S. Striegler, M. Dittel, *J. Am. Chem. Soc.* **2003**, *125*, 11518; (c) S. Striegler, M. Dittel, *Anal. Chim. Acta* **2003**, *484*, 53.

83 G. H. Chen, Z. B. Guan, C. T. Chen, L. T. Fu, V. Sundaresan, F. H. Arnold, *Nature Biotechnology* **1997**, *15*, 354; S. Striegler, *Bioseparation* **2001**, *10*, 307; S. Striegler, *Tetrahedron* **2001**, *57*, 2349; S. Striegler, *Macromolecules* **2003**, *36*, 1310.

84 J. M. Coterón, C. Vicent, C. Bosso, S. Penadés, *J. Am. Chem. Soc.* **1993**, *115*, 10066; J. Jimenez-Barbero, E. Junquera, M. Martinpastor, S. Sharma, C. Vicent, S. Penadés, *J. Am. Chem. Soc.* **1995**, *117*, 11198; J. C. Morales, D. Zurita, S. Penadés, *J. Org. Chem.* **1998**, *63*, 9212.

85 S. A. Staley, B. D. Smith, *Tetrahedron Lett.* **1996**, *37*, 283; M. A. Lipton, *Tetrahedron Lett.* **1996**, *37*, 287.

86 J. Billing, H. Grundberg, U. J. Nilsson, *Supramolecular Chemistry* **2002**, *14*, 367.

87 R. D. Hubbard, S. R. Horner, B. L. Miller, *J. Am. Chem. Soc.* **2001**, *123*, 5810.

88 H. G. Kuivila, A. H. Keough, E. J. Soboczenski, *J. Org. Chem.* **1954**, *19*, 780.

89 G. L. Roy, A. L. Laferriere, J. O. Edwards, *J. Inorg. Nucl. Chem.* **1957**, *114*, 106.

90 J. P. Lorand, J. O. Edwards, *J. Org. Chem.* **1959**, *24*, 769.

91 J. C. Norrild, H. Eggert, *J. Am. Chem. Soc.* **1995**, *117*, 1479.

92 J. C. Norrild, H. Eggert, *J. Chem. Soc., Perkin Trans. 2* **1996**, 2583.

93 J. Rohovec, T. Maschmeyer, S. Aime, J. A. Peters, *Chem. Eur. J.* **2003**, *9*, 2193.

94 A. E. Martell, R. M. Smith, *Critical Stability Constants, Vol. 4*, Plenum Press, New York, **1976**.

95 T. D. James, S. Shinkai, *Top. Curr. Chem.* **2002**, *218*, 159.

96 L. I. Bosch, T. M. Fyles, T. D. James, *Tetrahedron* **2004**, *60*, 11175.

97 A. Laschewsky, *Angew. Chem., Int. Ed. Engl.* **1989**, *28*, 1574.

98 S. Shinkai, K. Tsukagoshi, Y. Ishikawa, T. Kunitake, *J. Chem. Soc., Chem. Commun.* **1991**, 1039.

99 R. Ludwig, T. Harada, K. Ueda, T. D. James, S. Shinkai, *J. Chem. Soc., Perkin Trans. 2* **1994**, 697.

100 R. Ludwig, K. Ariga, S. Shinkai, *Chem. Lett.* **1993**, 1413.

101 R. Ludwig, Y. Shiomi, S. Shinkai, *Langmuir* **1994**, *10*, 3195.

102 C. Dusemund, M. Mikami, S. Shinkai, *Chem. Lett.* **1995**, 157.

103 T. Miyahara, K. Kurihara, *J. Am. Chem. Soc.* **2004**, *126*, 5684.

104 T. Araki, H. Tsukube, *Liquid Membranes: Chemical Applications*, CRC Press, Boca Raton, FL, USA, **1990**.

105 T. Shinbo, K. Nishimura, T. Yamaguchi, M. Sugiura, *J. Chem. Soc., Chem. Commun.* **1986**, 349.

106 B. F. Grotjohn, A. W. Czarnik, *Tetrahedron Lett.* **1989**, *30*, 2325.

107 M. F. Paugam, L. S. Valencia, B. Boggess, B. D. Smith, *J. Am. Chem. Soc.* **1994**, *116*, 11203.

108 M. F. Paugam, B. D. Smith, *Tetrahedron Lett.* **1993**, *34*, 3723.

109 L. K. Mohler, A. W. Czarnik, *J. Am. Chem. Soc.* **1993**, *115*, 2998.

110 L. K. Mohler, A. W. Czarnik, *J. Am. Chem. Soc.* **1993**, *115*, 7037.

111 P. R. Westmark, S. J. Gardiner, B. D. Smith, *J. Am. Chem. Soc.* **1996**, *118*, 11093.

112 M. F. Paugam, J. A. Riggs, B. D. Smith, *Chem. Commun.* **1996**, 2539.

113 S. J. Gardiner, B. D. Smith, P. J. Duggan, M. J. Karpa, G. J. Griffin, *Tetrahedron* **1999**, *55*, 2857.

114 J. A. Riggs, K. A. Hossler, B. D. Smith, M. J. Karpa, G. Griffin, P. J. Duggan, *Tetrahedron Lett.* **1996**, *37*, 6303.

115 M. Takeuchi, K. Koumoto, M. Goto, S. Shinkai, *Tetrahedron* **1996**, *52*, 12931.

116 J. T. Bien, M. Y. Shang, B. D. Smith, *J. Org. Chem.* **1995**, *60*, 2147.

117 M. F. Paugam, J. T. Bien, B. D. Smith, L. A. J. Chrisstoffels, F. de Jong, D. N. Reinhoudt, *J. Am. Chem. Soc.* **1996**, *118*, 9820.

118 M. Di Luccio, B. D. Smith, T. Kida, C. P. Borges, T. L. M. Alves, *J. Membr. Sci.* **2000**, *174*, 217.

119 S. P. Draffin, P. J. Duggan, S. A. M. Duggan, *Org. Lett.* **2001**, *3*, 917.

120 S. P. Draffin, P. J. Duggan, S. A. M. Duggan, J. C. Norrild, *Tetrahedron* **2003**, *59*, 9075.

121 T. M. Altamore, E. S. Barrett, P. J. Duggan, M. S. Sherburn, M. L. Szydzik, *Org. Lett.* **2002**, *4*, 3489.

122 J. Yoon, A. W. Czarnik, *J. Am. Chem. Soc.* **1992**, *114*, 5874.

123 J. Yoon, A. W. Czarnik, *Bioorg. Med. Chem.* **1993**, *1*, 267.

124 H. Shinmori, M. Takeuchi, S. Shinkai, *Tetrahedron* **1995**, *51*, 1893.

125 N. Di Cesare, J. R. Lakowicz, *J. Photochem. Photobiol., A* **2001**, *143*, 39.

126 N. DiCesare, J. R. Lakowicz, *J. Phys. Chem. A* **2001**, *105*, 6834.

127 N. DiCesare, J. R. Lakowicz, *Chem. Commun.* **2001**, 2022.

128 N. DiCesare, J. R. Lakowicz, *Tetrahedron Lett.* **2002**, *43*, 2615.

129 N. DiCesare, J. R. Lakowicz, *Tetrahedron Lett.* **2001**, *42*, 9105.

130 R. Badugu, J. R. Lakowicz, C. D. Geddes, *Anal. Chem.* **2004**, *76*, 610.

131 X. Gao, Y. Zhang, B. Wang, *Org. Lett.* **2003**, *5*, 4615.

132 S. Arimori, L. I. Bosch, C. J. Ward, T. D. James, *Tetrahedron Lett.* **2001**, *42*, 4553.

133 L. I. Bosch, M. F. Mahon, T. D. James, *Tetrahedron Lett.* **2004**, *45*, 2859.

134 S. Arimori, L. I. Bosch, C. J. Ward, T. D. James, *Tetrahedron Lett.* **2002**, *43*, 911.

135 A. P. de Silva, H. Q. N. Gunaratne, T. Gunnlaugsson, A. J. M. Huxley, C. P. McCoy, J. T. Rademacher, T. E. Rice, *Chem. Rev.* **1997**, *97*, 1515.

136 A. P. de Silva, T. Gunnlaugsson, C. P. McCoy, *J. Chem. Educ.* **1997**, *74*, 53.

137 A. P. de Silva, D. B. Fox, T. S. Moody, S. M. Weir, *Trends Biotechnol.* **2001**, *19*, 29.

138 G. J. Brown, A. P. de Silva, S. Pagliari, *Chem. Commun.* **2002**, 2461.

139 T. D. James, K. R. A. S. Sandanayake, R. Iguchi, S. Shinkai, *J. Am. Chem. Soc.* **1995**, *117*, 8982.

140 T. D. James, K. R. A. S. Sandanayake, S. Shinkai, *J. Chem. Soc., Chem. Commun.* **1994**, 477.

141 D. R. Cary, N. P. Zaitseva, K. Gray, K. E. O'Day, C. B. Darrow, S. M. Lane, T. A. Peyser, J. H. Satcher, Jr., W. P. Van

Antwerp, A. J. Nelson, J. G. Reynolds, *Inorg. Chem.* **2002**, *41*, 1662.

142 B. Kukrer, E. U. Akkaya, *Tetrahedron Lett.* **1999**, *40*, 9125.

143 T. D. James, K. R. A. S. Sandanayake, S. Shinkai, *Angew. Chem., Int. Ed. Engl.* **1994**, *33*, 2207.

144 N. DiCesare, J. R. Lakowicz, *Anal. Biochem.* **2001**, *294*, 154.

145 K. R. A. S. Sandanayake, T. D. James, S. Shinkai, *Chem. Lett.* **1995**, 503.

146 B. Appleton, T. D. Gibson, *Sens. Actuators, B* **2000**, *65*, 302.

147 S. Arimori, M. L. Bell, C. S. Oh, K. A. Frimat, T. D. James, *Chem. Commun.* **2001**, 1836.

148 S. Arimori, M. L. Bell, C. S. Oh, K. A. Frimat, T. D. James, *J. Chem. Soc., Perkin Trans. 1* **2002**, 803.

149 To help visualize the trends in the observed stability constants, the stability constants of the diboronic acid sensors **125** were compared with the stability constants of the equivalent monoboronic acid analogues **126**. The relative stability illustrates that an increase in selectivity is obtained by cooperative binding by formation of 1:1 cyclic systems. The large enhancement of the relative stability observed for the 1:1 cyclic systems (D-glucose, D-galactose) clearly contrast with the small twofold enhancement observed for the 2:1 acyclic systems (D-fructose, D-mannose).

150 S. Arimori, G. A. Consiglio, M. D. Phillips, T. D. James, *Tetrahedron Lett.* **2003**, *44*, 4789.

151 W. Yang, S. Gao, X. Gao, V. V. R. Karnati, W. Ni, B. Wang, W. B. Hooks, J. Carson, B. Weston, *Bioorg. Med. Chem. Lett.* **2002**, *12*, 2175.

152 V. V. Karnati, X. Gao, S. Gao, W. Yang, W. Ni, S. Sankar, B. Wang, *Bioorg. Med. Chem. Lett.* **2002**, *12*, 3373.

153 D. Stones, S. Manku, X. Lu, D. G. Hall, *Chem. Eur. J.* **2004**, *10*, 92.

154 S. Arimori, M. D. Phillips, T. D. James, *Tetrahedron Lett.* **2004**, *45*, 1539.

155 H. Eggert, J. Frederiksen, C. Morin, J. C. Norrild, *J. Org. Chem.* **1999**, *64*, 3846.

156 S. Arimori, M. L. Bell, C. S. Oh, T. D. James, *Org. Lett.* **2002**, *4*, 4249.

157 C. R. Cooper, T. D. James, *Chem. Commun.* **1997**, 1419.

158 C. R. Cooper, T. D. James, *J. Chem. Soc., Perkin Trans. 1* **2000**, 963.

159 A. P. de Silva, H. Q. N. Gunaratne, C. P. McCoy, *Nature* **1993**, *364*, 42.

160 W. Yang, J. Yan, H. Fang, B. Wang, *Chem. Commun.* **2003**, 792.

161 W. Yang, J. Yan, G. Springsteen, S. Deeter, B. Wang, *Bioorg. Med. Chem. Lett.* **2003**, *13*, 1019.

162 N. DiCesare, D. P. Adhikari, J. J. Heynekamp, M. D. Heagy, J. R. Lakowicz, *J. Fluorescence* **2002**, *12*, 147.

163 D. P. Adhikiri, M. D. Heagy, *Tetrahedron Lett.* **1999**, *40*, 7893.

164 H. Cao, D. I. Diaz, N. DiCesare, J. R. Lakowicz, M. D. Heagy, *Org. Lett.* **2002**, *4*, 1503.

165 A. J. Tong, A. Yamauchi, T. Hayashita, Z. Y. Zhang, B. D. Smith, N. Teramae, *Anal. Chem.* **2001**, *73*, 1530.

166 R. Badugu, J. R. Lakowicz, C. D. Geddes, *J. Fluorescence* **2003**, *13*, 371.

167 W. Yang, H. He, D. G. Drueckhammer, *Angew. Chem., Int. Ed. Engl.* **2001**, *40*, 1714.

168 E. Nakata, T. Nagase, S. Shinkai, I. Hamachi, *J. Am. Chem. Soc.* **2004**, *126*, 490.

169 T. Mizuno, M. Takeuchi, S. Shinkai, *Tetrahedron* **1999**, *55*, 9455.

170 T. Mizuno, M. Yamamoto, M. Takeuchi, S. Shinkai, *Tetrahedron* **2000**, *56*, 6193.

171 V. W. W. Yam, A. S. F. Kai, *Chem. Commun.* **1998**, 109.

172 T. Mizuno, T. Fukumatsu, M. Takeuchi, S. Shinkai, *J. Chem. Soc., Perkin Trans. 1* **2000**, 407.

173 C. J. Davis, P. T. Lewis, M. E. McCarroll, M. W. Read, R. Cueto, R. M. Strongin, *Org. Lett.* **1999**, *1*, 331.

174 P. T. Lewis, C. J. Davis, L. A. Cabell, M. He, M. W. Read, M. E. McCarroll, R. M. Strongin, *Org. Lett.* **2000**, *2*, 589.

175 M. He, J. R. Johnson, J. O. Escobedo, P. A. Beck, K. K. Kim, N. N. St. Luce, C. J. Davis, P. T. Lewis, F. R. Fronczeck, B. J. Melancon, A. A. Mrse, W. D. Treleaven, R. M. Strongin, *J. Am. Chem. Soc.* **2002**, *124*, 5000.

176 K. Koumoto, S. Shinkai, *Chem. Lett.* **2000**, 856.

177 C. J. Ward, P. Patel, P. R. Ashton, T. D. James, *Chem. Commun.* **2000**, 229.

178 C. J. Ward, P. Patel, T. D. James, *J. Chem. Soc., Perkin Trans. 1* **2002**, 462.

179 N. DiCesare, J. R. Lakowicz, *Org. Lett.* **2001**, *3*, 3891.

180 K. Sato, A. Sone, S. Arai, T. Yamagishi, *Heterocycles* **2003**, *61*, 31.

181 W. Ni, H. Fang, G. Springsteen, B. Wang, *J. Org. Chem.*, **2004**, *69*, 1999.

182 G. S. Wilson, H. Yibai, *Chem. Rev.* **2000**, *100*, 2693.

183 A. N. J, Moore, D. D. M. Wayner, *Can. J. Chem.* **1999**, *77*, 681.

184 M. Nicolas, B. Fabre, J. Simonet, *Electrochim. Acta* **2001**, *46*, 1179.

185 S. Arimori, S. Ushiroda, L. M. Peter, A. T. A. Jenkins, T. D. James, *Chem. Commun.* **2002**, 2368.

186 J. C. Norrild, I. Sotofte, *J. Chem. Soc., Perkin Trans. 2* **2002**, 303.

187 S. L. Wiskur, H. Ait-Haddou, J. J. Lavigne, E. V. Anslyn, *Acc. Chem. Res.* **2001**, *34*, 963.

188 G. Springsteen, B. Wang, *Chem. Commun.* **2001**, 1608.

189 G. Springsteen, B. Wang, *Tetrahedron* **2002**, *58*, 5291.

190 L. A. Cabell, M. K. Monahan, E. V. Anslyn, *Tetrahedron Lett.* **1999**, *40*, 7753.

191 Z. Zhong, E. V. Anslyn, *J. Am. Chem. Soc.* **2002**, *124*, 9014.

192 Z. Murtaza, L. Tolosa, P. Harms, J. R. Lakowicz, *J. Fluorescence* **2002**, *12*, 187.

193 J. T. Suri, D. B. Cordes, F. E. Cappuccio, R. A. Wessling, B. Singaram, *Langmuir* **2003**, *19*, 5145.

194 J. N. Camara, J. T. Suri, F. E. Cappuccio, R. A. Wessling, B. Singaram, *Tetrahedron Lett.* **2002**, *43*, 1139.

195 N. DiCesare, M. R. Pinto, K. S. Schanze, J. R. Lakowicz, *Langmuir* **2002**, *18*, 7785.

196 S. Arimori, C. J. Ward, T. D. James, *Tetrahedron Lett.* **2002**, *43*, 303.

197 W. Wang, S. Gao, B. Wang, *Org. Lett.* **1999**, *1*, 1209.

198 S. Gao, W. Wang, B. Wang, *Bioorg. Chem.* **2001**, *29*, 308.

199 T. Kawanishi, M. Holody, M. Romey, S. Shinkai, *Fluorescence* **2004**, *14*, 499.

200 J. T. Suri, D. B. Cordes, F. E. Cappuccio, R. A. Wessling, B. Singaram, *Angew. Chem., Int. Ed. Engl.* **2003**, *42*, 5857.

201 S. A. Asher, V. L. Alexeev, A. V. Goponenko, A. C. Sharma, I. K. Lednev, C. S. Wilcox, D. N. Finegold, *J. Am. Chem. Soc.* **2003**, *125*, 3322.

202 V. L. Alexeev, A. C. Sharma, A. V. Goponenko, S. Das, I. K. Lednev, C. S. Wilcox, D. N. Finegold, S. A. Asher, *Anal. Chem.* **2003**, *75*, 2316.

203 E. Pringsheim, E. Terpetschnig, S. A. Piletsky, O. S. Wolfbeis, *Adv. Mater.* **1999**, *11*, 865.

204 H. Murakami, H. Akiyoshi, T. Wakamatsu, T. Sagara, N. Nakashima, *Chem. Lett.* **2000**, 940.

205 E. Shoji, M. S. Freund, *J. Am. Chem. Soc.* **2002**, *124*, 12486.

206 E. Shoji, M. S. Freund, *J. Am. Chem. Soc.* **2001**, *123*, 3383.

207 B. Fabre, L. Taillebois, *Chem. Commun.* **2003**, 2982.

208 F. H. Arnold, W. Zheng, A. S. Michaels, *J. Membr. Sci.* **2000**, *167*, 227.

3
Ammonium, Amidinium, Guanidinium, and Pyridinium Cations

Thomas Schrader and Michael Maue

3.1
Introduction

The beginning of supramolecular chemistry coincided with the discovery of crown ethers for metal cation and ammonium recognition [1]. The triple Nobel prize went to Lehn, Cram, and Pederson for their pioneering work and the necessary development of analytical tools for the exact measurement of free binding energies [2]. Much energy was initially devoted, e.g. by the Lehn group, to the development of ever more efficient structures for ammonium complexation, ultimately resulting in very complicated cryptand types, capable of encapsulating free ammonium ions and primary mono- and diammonium alkanes [3]. Thus, complete desolvation was achieved, although most often in organic solvents of low polarity, for example halogenated hydrocarbons.

An application which was explored relatively early was the transport of charged analytes across a membrane or membrane model. Thus, lipophilic hosts (often podands) were able to transport ammonium guests through a chloroform membrane separating two water reservoirs, one containing the ammonium substrate and the other "empty" [4]. Although passive downhill transport is usually observed, in an elegant combination with simultaneous proton-backtransfer the active transport of cations against a concentration gradient has been realized [5].

Cram and Lehn introduced chirality into the crown ether hosts and achieved chiroselective complexation of adrenaline derivatives or amino acids [6]. To this end, the 18-crown-6 moiety was synthesized from fragments containing the C_2-symmetric unit of enantiomerically pure tartaric acid or binaphthol. The binaphthyl moieties form a chiral cavity above the crown ether which distinguish with great selectivity between appropriately substituted alkylammonium substrates [7]. Cram later invented the sensational so-called chiral resolution machine, a continuously working apparatus which could be used to separate the enantiomers of racemic primary ammonium ions, e.g. aromatic amino acid esters [8]. In a continuous extraction experiment, chiral resolution of a large quantity of substrate was achieved. Thus, ammonium ion complexation reflected the development of supramolecular chemistry in general, as

Functional Synthetic Receptors. Edited by T. Schrader, A. D. Hamilton
Copyright © 2005 WILEY-VCH Verlag GmbH & Co. KGaA, Weinheim
ISBN: 3-527-30655-2

it moved from initial exploration of a new playground toward higher efficiency and selectivity and competitive solvents of higher polarity, ultimately, hopefully, reaching aqueous physiological conditions.

The early work on ammonium ion complexation has been reviewed in several excellent articles and monographs and in the nobel lectures [9]. In this chapter recent developments in the past decade are summarized, with new theoretical and fundamental physical organic work and further optimizations of existing sensor systems. Finally, much emphasis is placed on peptide and protein receptors, which are still in their infancy but will become increasingly important in the near future. We will also include, for the first time, an overview of artificial receptors developed for the other biologically important types of organic cation, i.e., amidinium, guanidinium, and pyridinium cations. Each subdivision will include biological examples of the molecular recognition of the respective guest species, introduce new synthetic receptor systems, their binding characteristics and selectivity profile, and (potential) applications. These topics will be outlined in the order ammonium cations, amidinium cations, guanidinium cations and pyridinium cations.

3.2
Ammonium Cations

3.2.1
New Receptor Structures

Few new host structures were found in the past decade. Saalfrank recently presented tetrahemispheraplexes with ammonium and alkylammonium ions as exohedral guests. This is one of the rare examples of ammonium recognition at the surface of coordination polygons; the adamantane-like complexes based on Mg^{2+} cations and doubly bidentate bridging ligands are generated by self-assembly or by guest ex-

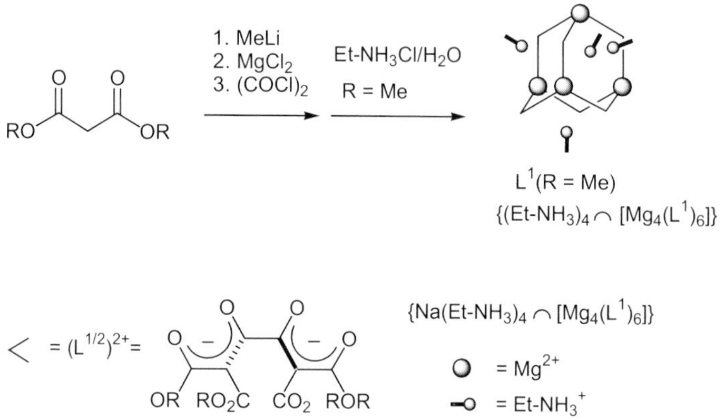

Figure 3.1 Alkylammonium ions as exohedral guests at the surface of adamantane-like tetrahemispheraplexes.

change [10]. Four alkylammonium ions are responsible for charge compensation; they are hydrogen-bonded to the oxygen donors of the triangular faces of the tetraanionic core. It was observed that the space available on the surface of the tetraanionic complex depends largely on the steric demand of the ester groups on the bridging ligands; thus tetrahemispheraplexes can be fine-tuned at will to receive smaller or larger alkylammonium guests (Fig. 3.1).

In a series of papers Kim and Ahn developed neutral tripodal receptor species for the NH_3^+-group present in all primary ammonium cations [11]. Three five-membered heterocycles are covalently attached to a 2,4,6-trimethylbenzene framework. The basic tripodal oxazoline system (Fig. 3.2) enables remarkably selective recognition of NH_4^+ compared with K^+, and is suited to biomedical applications. High sensitivity to biogenic amines is valuable in the development of sensors for GABA, dopamine, and related clinically relevant species [12]. It was achieved when additional phenyl substituents were introduced on the oxazoline ring. X-ray crystallographic analyses and NMR spectroscopic studies indicate the formation of a hydrophobic wall surrounding the "encapsulated" alkylammonium guest, leading to additional π–cation and hydrophobic interactions. Exquisite size selectivity was observed, enabling distinction of *n*-butylammonium from *t*-butylammonium by a factor of 700. Finally, preorganization of the trimethylbenzene framework was markedly improved by transition to a triethylbenzene skeleton. The best receptor molecule approaches a K_a value of 10^8 M^{-1} in chloroform, determined by UV titrations. Kim and Ahn claim that these tripodal benzene trisoxazolines are the strongest and most selective receptors yet reported for linear alkylammonium ions. Because oxazolines adorned with phenyl substituents are chiral, these new receptor molecules, which are often readily available from simple building blocks, are also suitable for enantioselective recognition of chiral alkylammonium species. Isomeric butyl ammonium ions have also been selectively recognized by calixarene-based host molecules, described by Chang and Pappalardo [13]. Preorganization is also crucial for the efficiency of this system, as demonstrated by the dramatically improved selectivity of the cone-shaped calix[5]arenes over their highly flexible calix[6]arene counterparts.

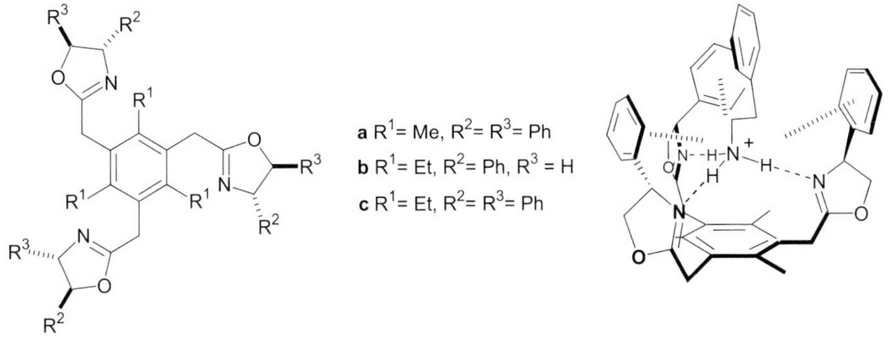

a R^1= Me, R^2= R^3= Ph
b R^1= Et, R^2= Ph, R^3 = H
c R^1= Et, R^2= R^3= Ph

Figure 3.2 Left: the strongest alkylammonium receptors reported to date are tripodal benzene trisoxazolines. Right: proposed structure of the alkylammonium complex; note the three linear hydrogen bonds reinforced by π–cation interactions with the aromatic sidewalls.

Bell and Shinkai recently found that a host originally designed for the recognition of flat guanidinium cations is also a good molecular receptor for cationic amino sugars in polar protic solvents such as methanol. The term "arginine cork" was coined for this halfmoon-shaped array of a central naphthyridine with two annulated adjacent pyridine carboxylates, because of its high affinity toward the amino acid arginine even in an aqueous environment [14]. However, closer inspection of its molecular structure revealed a twisted overall conformation and a cleft composed of four basic nitrogen atoms with a size comparable with that of monosaccharides. It was thus reasoned that it might also receive cationic monosaccharides such as glucosamine by virtue of a salt bridge and both ion-pair-reinforced and ionic hydrogen bonds. Bell and Shinkai found that this is indeed true – glucosamine, mannosamine, and galactosamine are all bound with K_a values between 8×10^4 and 2×10^5 M^{-1} in pure methanol [15]. When 20 % water was present, however, the 1:1 complexes were completely dissociated, confirming the absence of significant hydrophobic contributions. Deletion of the ammonium functionality or, to a lesser extent, of some sugar hydroxyls, led to a marked decrease in binding energy, indicative of a multipoint recognition event. Interestingly, the complexes between the host and its amino sugars became strongly CD-active. This is only possible if the saccharide is bound by the artificial receptor at several points, forming a pseudo-cyclic structure [16]. On the basis of these results, supporting chemical shift changes in the NMR, and force-field calculations the authors developed a model which explains the ICD (induced circular dichroism). Selective stabilization of one enantiomeric conformation of the planar chiral host molecule occurs as soon as it pairs its own hydrogen bond acceptor groups with the guest's hydrogen bond donor groups. The "amino sugar cork" (Fig. 3.3) is a rare example of sugar recognition by purely noncovalent interactions in protic solvents. Previously, only Diederich et al. achieved binding of mono- and oligosaccharides in 12 % methanolic acetonitrile with a macrocyclic cyclophane featuring anionic phosphodiester binding sites for ionic hydrogen bonds to the sugar hydroxyls [17]. More recently, Kral et al. claimed that porphyrins with polycationic groups are capable of binding trisaccharides even in water [18]. By systematic variation of the annulated oligopyridine skeleton Bell later identified a selective lysine binder. With exchange of one of the carboxylate-bearing pyridines for a simple benzene ring high and selective affinity for lysine derivatives in methanol was found ($K_a > 10^5$ M^{-1}) [19].

Another new powerful binding motif for the recognition of ammonium ions is a chelate arrangement of benzylic bisphosphonates; the ammonium ion is placed exactly above the host's benzene ring and seized by both bisphosphonate anions [20]. The free binding energy comes mainly from a double salt bridge, complemented by a network of corresponding ion-pair-reinforced hydrogen bonds. In contrast with crown ethers, secondary alkylammonium ions found in many β-blockers are bound as well as primary ions, because there is enough space left between both phosphonate arms to accommodate the second alkyl group. Schrader et al. introduced this basic recognition element for efficient complexation of adrenaline derivatives and found that in polar solvents such as DMSO extra hydroxyl groups present in the guest molecule add incrementally to the overall binding energy by means of ionic hydrogen bonding with the phosphonate oxygen anions [21]. Thus the new host unit is very

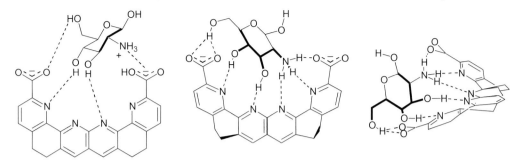

Figure 3.3 The multipoint interactions necessary for amino sugar recognition in polar solvents – the "amino sugar cork". Left: schematic diagram. Middle and right: minimized structure (SYBYL).

suitable for amino alcohols and, especially, amino sugars. Glucosamine and related compounds are bound in DMSO with K_a values >10^5 M^{-1}. On formation of the maximum number of ionic hydrogen bonds a well-defined complex geometry evolves indicated by large diastereotopic splittings of otherwise shiftisochronic methylene protons. Thus, even the simple *m*-xylylene bisphosphonate is able to distinguish with high sensitivity between anomeric amino sugars; de values up to 90 % are obtained [22]. However, transition from DMSO to water results in a complete loss of affinity for adrenaline. In an attempt to introduce addditional hydrophobic interactions a variety of aromatic ester alcohols were introduced into the phosphonate binding sites, but these open-chain hosts had too many rotatory degrees of freedom and no significant improvement was observed. Much more successful was the switch to a macrocyclic hydrophobic cavity adorned with benzylic bisphosphonates [23]. Strong binding was now observed in methanol, and variable temperature experiments proved the whole guest molecule was included inside the cavity, because, depending on the size of the guest, an increasing number of rotatable benzene rings in the macrocyclic core was shown to become decelerated and, eventually, frozen in their dynamic motion. To achieve selectivity for catecholamines over amino acid derivatives the extent of host preorganization was significantly increased. A coat of complementary shape was designed which carried the benzylic bisphosphonates on one end and an aromatic cleft on the other end, with an isophthalamide head group capable of grasping the catechol OH groups by two inward-projecting amide NH functionalities [24]. Now, adrenaline derivatives were not only bound in aqueous solution, but also with a high preference over similar structures lacking the amino alcohol or the catecholamine. Thus, amino acid derivatives were completely rejected [25]. This high selectivity was also achieved in methanol by a much simpler approach – covalent attachment of two rigid biphenylic sidewalls with terminal phosphonate binding sites to a bisphenylpiperazine template produced molecular tweezer **1**, which binds adrenaline derivatives with an efficiency at least one order of magnitude higher than for any related structure or any other neurotransmitter species [26] (Fig. 3.4). Finally, combination of rigid tolane side-walls with a complementary coat around the noraa)

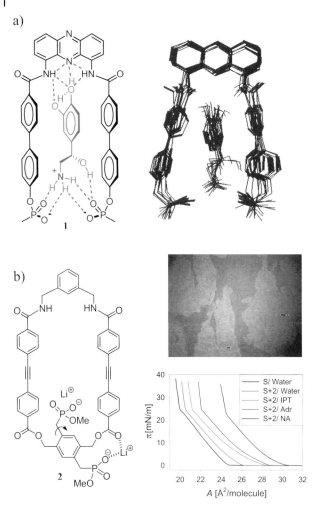

b)

Figure 3.4 New generations of bisphosphonate adrenaline receptors exploiting the powerful combination of chelate salt bridges and the hydrophobic effect. (a) Complex between highly shape-selective tweezer **1** and noradrenaline according to Monte-Carlo simulations in water with subsequent molecular dynamics. (b) Macrocyclic preorganized adrenaline host **2**.

Top right: self-aggregated receptor domains observed in the Brewster-angle microscope (BAM) after embedding **2** in a stearic acid monolayer. Bottom right: pressure–area isotherms of stearic acid (S) and receptor **2** in a monolayer over water, with isoproterenol (IPT), adrenaline (Adr), and noradrenaline (NA).

drenaline molecule led to new macrocyclic host systems (e.g. **2**). These were moderately good binders in free methanol solution but became highly efficient and selective in the new microenvironment of a stearic acid monolayer at the air/water interface [27]. Because of its high tendency to self-associate, large areas of aggregated host molecules were formed inside the monolayer, which could be visualized as gray patches in Brewster angle microscopy (BAM). Subsequent subinjection of different agonists and antagonists of the membrane-bound adrenaline receptor resulted in characteristic changes in the pressure/area (π–A) diagrams (Fig. 3.4). Even 10^{-6} M aqueous solutions of noradrenaline produced a marked increase in the corresponding binding isotherm. High structure-sensitivity was found which makes the new macrocycle an ideal candidate for new sensors of adrenaline receptor substrates. Thus, agonists and antagonists differing only in a methyl or allyl group can be readily distinguished by a film-balance experiment.

It should be noted, that other groups have also created ditopic adrenaline receptors by combining known binding motifs for ammonium recognition with others for the catechol ring. Thus, Smith et al. have covently fused a benzocrown with a boronic acid and found that, because of the reduced pK_a of the catechol, a neutral zwitterionic species with tetrahedral boronate ester moiety could be transported across a liquid membrane [28]. Nolte et al. adorned glycoluril-based clefts with carboxylate moieties and produced efficient hosts for amino acids and amino alcohols in water [29].

The new concept of dynamic combinatorial chemistry (DCC, Fig. 3.5) [30] was exploited by Sanders et al. resulting in the identification and isolation of a receptor for *N*-methyl alkylammonium salts [31].

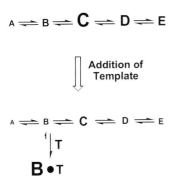

Figure 3.5 Dynamic combinatorial chemistry – the principle of receptor amplification in an equilibrating mixture containing a guest (= template).

Building on Kubik's cyclopeptide receptor system for the inclusion of tetraalkylammonium ions [32], a proline-derived building block was developed featuring an acylhydrazine and a dimethylacetal at the respective α,ω-termini (Fig. 3.6). This was subjected to acidic equilibrium conditions and a dynamic combinatorial library with a predominant dimeric cyclopeptide resulted from intermolecular hydrazone formation. In the presence of acetylcholine as guest template, however, the equilibrium was dramatically shifted to a 86:13 ratio between trimer and dimer. This ratio corresponds to a 50-fold amplification of the best binder, the highest factor ever observed. By simple silica gel chromatography the cyclic trimer could be isolated on a 100-mg scale. Binding between the cations and the amplified cyclic trimer was later unequivocally proven by mass spectrometry (M^+ ion peaks for the 1:1 adduct) and ^{1}H NMR spectroscopy (NOESY of the complex with intermolecular crosspeaks). The dynamic combinatorial approach combines rational design of the building blocks with self-sorting

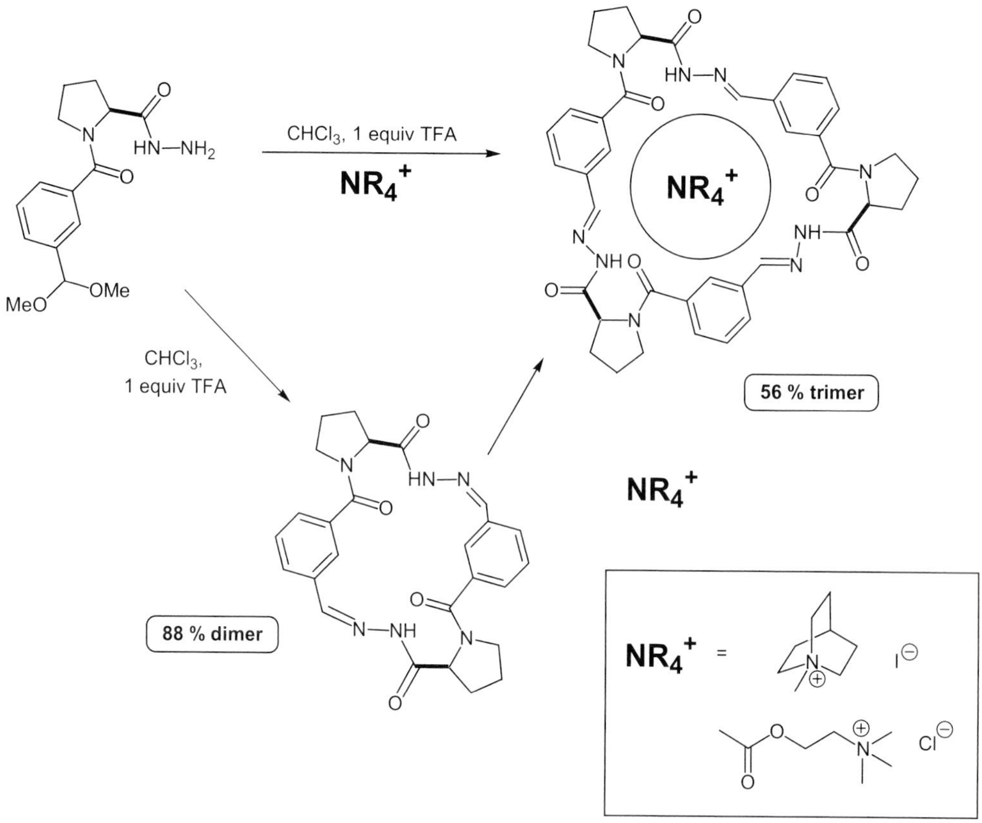

Figure 3.6 Application of DCC to the dynamic generation and selection of an artificial host for tetraalkylammonium ions.

mechanisms; it offers a much improved means of finding receptor systems of high selectivity and will definitely be used to identify new enzyme-mimicks, catalysts, etc.

In an imitation of the size, shape, and charge-recognition elements found in the natural acetalcholinesterases, Rebek et al. reported new resorcinearene-based deep cavitands **3** for the efficient inclusion of trimethylalkylammonium guests, preferably acetylcholine ($K_a \approx 3 \cdot 10^4$ M^{-1}) [33]. Four negative charges surround the entrance of a deep pocket which is lined with aromatic residues. Dramatic upfield shifts of >4 ppm are observed for the guest's trimethylammonium group, showing they are deeply buried within the aromatic binding pocket of the cavitand (Fig. 3.7). Thus, hydrophobic interactions, i.e. release of water molecules from the pocket plus cation–π interactions between its aromatic faces and the quaternary ammonium ion, are likely to be the forces driving complex formation. Favorable entropy and enthalpy terms found with ITC measurements support this assumption. Later Rebek et al. also showed that unbranched surfactants were inserted into the binding pocket the other way round, with their long alkyl chain deeply buried in the cavitand. Large upfield shifts of all CH$_2$ groups within the guest's alkyl chain prove it adopts a helical con-

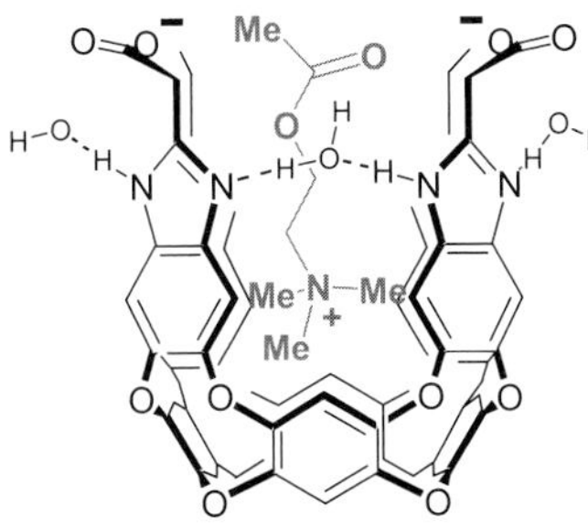

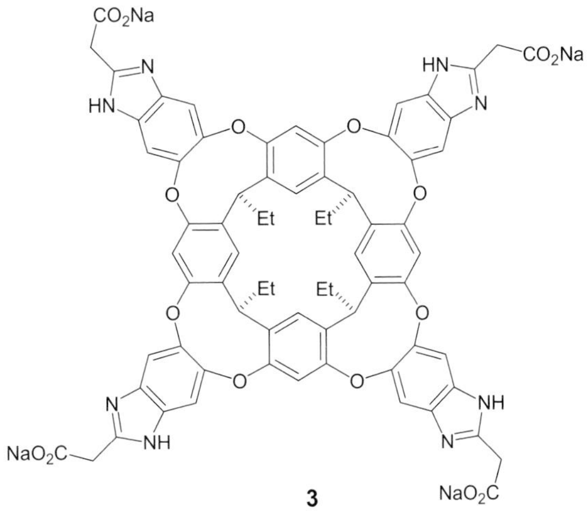

Figure 3.7 Deep cavitand with acetylcholine guest inside.

formation to maximize the hydrophobic effect and van-der-Waals interactions within the cavity [34].

Dougherty's well-known macrocyclic platform for organic cations has also been used for recognition of a variety of tetraalkylammonium ions, mainly by π–cation interactions. Additional carboxylates at the host's periphery, which significantly enhance binding of arginine derivatives, have virtually no effect on $RNMe_3^+$ compounds, however [35]. It is argued that only guests with focused regions of large, positive electrostatic potential, for example guanidinium moieties, experience this novel induced-dipole mechanism. Dougherty's investigations on the π–cation interaction and new applications in synthetic receptors will be explained in more detail in the next section, on theoretical investigations [36].

3.2.2
Theoretical Investigations

Schneider et al. made important fundamental contributions to our understanding of the well-known crown ether–ammonium complexes. To elucidate the binding mechanism and solvent effects, they conducted numerous thermodynamic investigations by ITC [37]. Strong contributions from entropy terms counterbalance the dominant enthalpy of complexation, and a linear increase in ΔG is observed from R_3NH^+ to NH_4^+ with the number of polarized $NH^{\delta+}$ bonds in the guest. Tetraalkylammonium ions have no appreciable affinity for crown ethers. Association constants determined in a wide range of solvents varied by factors of up to 1000 (e.g. methanol lg K = 4.2, DMSO lg K = 1.3). These solvent effects can largely be attributed to the hydrogen bond-accepting power of the solvent molecules. Hence, the best correlations ($R \approx 0.9$) for lg K are obtained with terms proportional to the electron-donor capacitiy of the solvent (C_a, β^*, DN). In summary, gas-phase calculations and solvent effects point to a dominant contribution of hydrogen bonds over any other noncovalent attractive force in these complexes (Fig. 3.8).

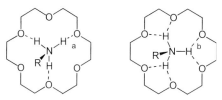

Figure 3.8 Complex between 18-crown-6 and alkylammonium ions with either linear or bifurcated hydrogen bonds. O⋯H distances: a ~1.9 Å; b ~2.1 Å.

Another fundamental noncovalent interaction has been revisited and intensively examined by the Dougherty group. Systematic screening of crystal structures for arginine and lysine-containing proteins led to the rediscovery of π–cation interactions between aromatic and basic amino acid sidechains as an important factor controlling protein folding and stability [38]. During the following years, single π–cation interactions between various metal and organic cations and different aromatic moieties have been characterized in the gas phase and in solution with sophisticated model systems [39]. Further optimization of these models eventually led to the design of highly potent macrocyclic arginine, tetraalkylammonium, and quinolinium receptors [40]; theoretical calculations building on these experimental results deepened our understanding of electrostatic and dispersive contributions to this important noncovalent force [41]. Recently, unnatural amino acid mutagenesis and heterologous expression were used to successfully introduce tryptophan analogs into the binding site of two ion channels 542]. In the ligand-gated nicotinic acetylcholine receptor (Fig. 3.9) the energetic contribution of the observed π–cation interaction between a tryptophan and the quaternary ammonium ion of ACh was thus estimated at 2 kcal mol^{-1}. In contrast, the respective interaction in the serotonin 5-HT$_{3A}$ receptor between a tryptophan and the primary ammonium ion of serotonin amounts to 4 kcal mol^{-1}. By using the same techniques as mentioned above Dougherty et al. also incorporated tyrosine derivatives with tethered quaternary, tertiary, and secondary ammonium groups into the nicotinic acetylcholine receptor (nAChR) at the agonist binding site [43]. Constitutively active receptors, which open the ion channel even in the absence of acetylcholine, were permanently formed only with quaternary ammonium tethers. The other tethers worked only at low pH, when they were fully protonated. It was found, that the phenomenological pK_a values of the tethered amines are much smaller than their values in free solution. In contrast, no such pK_a shift was observed for the tightly bound nicotine agonist, and it was argued that a new nicotine-like binding mode must be added to the pharmacophore model describing the muscle-type nAChR.

Another important concept which is found throughout natural recognition events between macromolecules such as DNA and proteins, was coined "multivalency". This was extensively investigated by Whitesides et al., and natural examples and applications in synthetic receptor molecules have recently been reviewed [44]. Self-as-

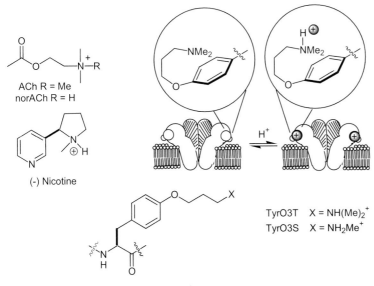

Figure 3.9 Left: Acetylcholine, norACh and nicotine. Right: Incorporation of TyrO3T at α149 in the nAChR for a pH-sensitive receptor. Bottom: Tethered agonists TyrO3S and TyrO3T.

sembly of complex artificial structures from up to 36 parts was reported by Whitesides [45] and, among others, by the Reinhoudt [46] and Lehn [47] groups. In this concept, the intelligent design of strategically preorganized building blocks is combined with a favorable entropy balance obtained by simultaneous interaction at multiple binding sites. Similar to nature, high specificity and efficiency are achieved by cooperative combination of many weak noncovalent interactions. Thus, the first host–guest interaction facilitates the second by restricting the degrees of freedom of both complexation partners. This is true for duplex formation between complementary DNA strands and also for the cyanuric acid–melamine rosettes described by Whitesides et al. An impressive example is the dramatic enhancement of the affinity of the naturally occurring antibiotic vancomycin toward the D-Ala-D-Ala fragment found in bacterial cell walls. Dimeric vancomycin **4** (Fig. 3.10) linked at the C-terminus by a *p*-xylylene diamine spacer, and the dimeric succinic amide derivative of L-Lys-D-Ala-D-Ala form a divalent complex with a K_d of ~1 nM, determined by a competitive assay using affinity capillary electrophoresis (ACE) [48]. Here, the divalency improves the entropy balance dramatically, and the affinity toward the peptidic guest is enhanced by a factor of 1000. A trimeric vancomycin was shown to be even more effective – association constants achieved with trimeric D-Ala-D-Ala motifs were superior to any other found in nature [49]. The biological relevance lies at hand – not only can the biological action of vancomycin-type glycopeptides be better understood, but the tight binders can also be developed into potent antibiotics. Multivalency was later explored in many oligomeric receptor structures. Whereas polymers used for molecular recognition purposes are often heterogeneous, linear, flexible, and not well defined, dendrimers tend to be monodisperse, globular, with well-defined size and shape [50]. Thus, dendrimers have recently entered the field, e.g. with rigid calixarene scaffolds and peptidic dendrimeric core units (Rudkevich) and many others [51]. A dendrimer can, e.g., build a nonpolar environment around a central binding site, and thus mimic the active sites in enzymes [52]. More elaborate systems using biomimetic multipoint recognition of lysines and arginines will be discussed in Sect. 3.2.4 on peptide and protein recogniton (vide infra).

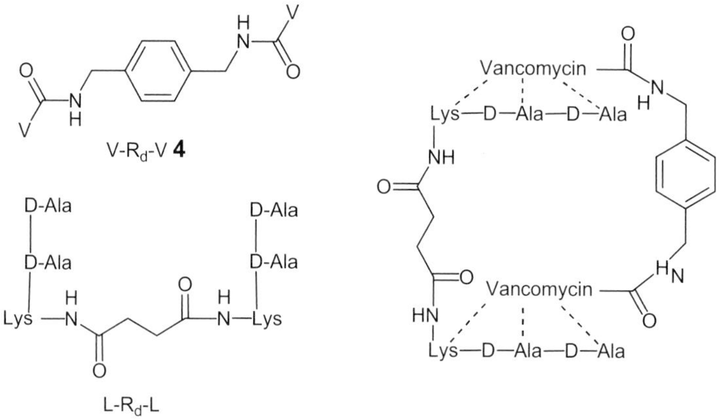

Figure 3.10 Dimeric vancomycin V-R$_d$-V **4** and its corresponding dimeric ligand L-R$_d$-L.

3.2.3
New Functions

Many applications of crown ethers and related compounds have been suggested and in fact also realized in sensor systems for a wide range of ammonium cations. In an elegant approach, Bergbreiter and Crooks [53] reported optional selectivity enhancement by introducing pH-sensitive filter layers on top of immobilized receptor molecules; thus, electrochemical sensors were prepared from Au electrodes coated with β-cyclodextrin-functionalized, hyperbranched polyacrylic acid (PAA) films capped with a chemically grafted ultrathin polyamine layer (Fig. 3.11). Whereas the PAA film contains the covalently bound receptor molecules, the thin polyamine overlayer acts as a pH-sensitive molecular filter that selectively passes suitably charged analytes to the underlying β-cyclodextrin. At low pH, when the polyamines are fully protonated, positively charged redox probe molecules, for example benzylviologen dications, do not permeate the filter layer. At higher pH, however, when the polyamines become uncharged the guest molecules pass on to the cation-receptor sites where they are reduced, as demonstrated by cyclovoltammetry measurements.

In a related approach, alternating layers of anionic gold nanoparticles and cationic biscycloparaquat catecholamine receptor molecules immobilized on ITO electrodes have been successfully developed by Willner et al. for highly sensitive electrochemical detection of catecholamines (detection limit ~1 μM) [54].

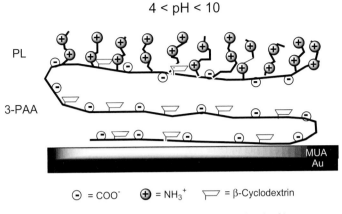

$4 < pH < 10$

PL

3-PAA

MUA
Au

$\ominus$ = COO⁻ $\oplus$ = NH₃⁺   ▽ = β-Cyclodextrin

Figure 3.11 A polyamine overlayer as a pH-sensitive molecular filter in a β-cyclodextrin-functionalized hyperbranched PAA film.

Membrane transport has always challenged chemists to build artificial models which imitate the passive, gated, and active transport of ions across lipid bilayer membranes, and many crown ether derivatives and other cation receptors have been used for this purpose. A recent review focused on such artificial transporters which model mechanisms of natural membrane transport [55]. Last year's nobel-prize winner MacKinnon demonstrated that prokayotic and eukaryotic K^+ channels have the same pore structures [56]. Detailed understanding of the structure and mechanism of naturally occurring, peptide-based ionophores stimulated many chemists to build artificial ionophores as models for biological cation transport across membranes. These efforts resulted in new sensors, gated carriers and self-assembled capsules; they have recently been reviewed [57]. Enantioselective binding and transport of amino acids is a particularly challenging problem. Sessler et al. presented a lipophilic complexing agent for zwitterionic amino acids by covalently connecting a natural ammonium binding ionophore (lasalocid) with a cationic carboxylate receptor (sapphyrin) [58]. U- and W-tube experiments revealed an intrinsic preference for Phe over Trp and Tyr. Finally, photoisomerizable azobis(crowns) **5** (Fig. 3.12) were among the

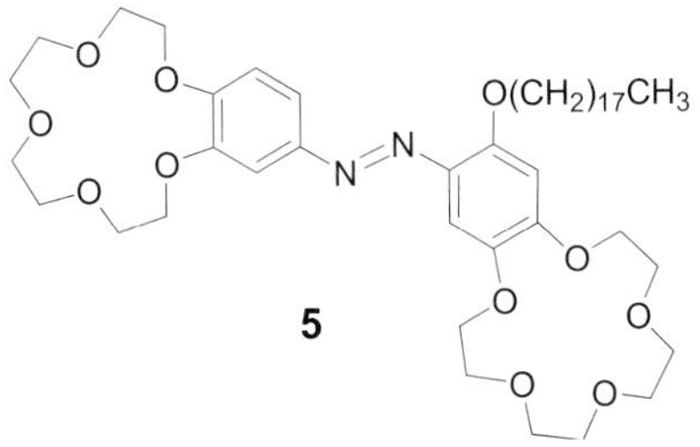

$O(CH_2)_{17}CH_3$

5

Figure 3.12 Photoswitchable lipophilic crown ether receptor for gated cation transport.

first molecules that can be switched between high- and low-affinity states. They have been used as molecular probes for phase boundary potentials in ion-selective electrode membranes. These systems pave the way toward artificial systems which can mimic gated transport across biomembranes by means of an external signal [59].

Organic cations have also been encapsulated inside neutral molecular capsules formed in a self-assembly process from two identical self-complementary building blocks (Rebek, Böhmer) [60]. Crystal structures and ESI–MS experiments reveal that π–cation interactions and dispersive forces govern the encapsulation process [61]. Occasionally hydrogen-bonded capsules are formed only in the presence of a template. More recently, Schrader [62] and Timmermans [63] extended the design of calixarene-based molecular capsules to ionic complementary structures with four positive or four negative charges located at the upper rim of each half-sphere. These cap-

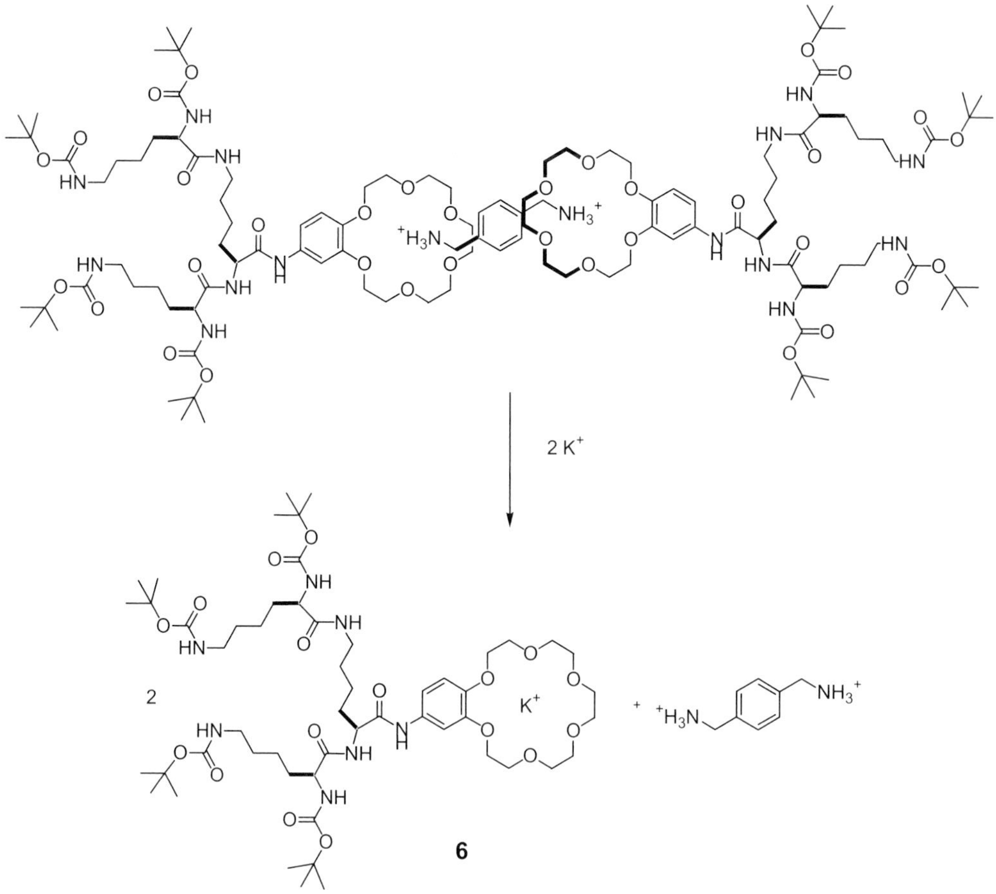

Figure 3.13 Disassembly of the supramolecular dendrimer induced by the addition of KPF$_6$.

sules are stable even in highly competitive solvents such as alcohols and water. The specific inclusion of tetralkylammonium salts such as ACh was shown by Timmermanns in a variety of experiments. The inner voids of supramolecular metal/ligand cages are also possible sites for encapsulated cationic species, as was beautifully explored by the Raymond group [64]. Crown ethers with dendritic branches based on L-lysine repeat units **6** have been synthesized by Smith et al. for reversible encapsulation of ammonium ions in the altered microenvironment inside the dendron [65]. By addition of excess K^+ cations to the complex, complete controlled release of the template back into solution could be affected (Fig. 3.13).

3.2.4
Peptide and Protein Recognition

In cooperation with the Finocchiaro group Schrader et al. developed chiral sensors for basic amino acids in a peptidic environment; the macrocyclic receptor features a spirobisindane skeleton with planar chirality which recognizes two ammonium groups of the guest by means of its two phosphonate anions [66]. If the chiral bridge comes into van-der-Waals distance of the alkyl chains of arginine or lysine their enantiomers are distinguished. The new host molecule can therefore be used as a chiral shift reagent to determine the enantiomeric excess in dicationic amino acid derivatives.

The combination of a porphyrin or other moiety with carboxylate affinity and a crown ether was found to be a very powerful means of access to molecular receptors for free small peptides. This concept was beautifully developed by Schneider and Hossain, who synthesized hybrid hosts **7** (Fig. 3.14) from benzocrowns attached to peralkylammonium ions via hydrophobic spacers, and achieved sequence-selective peptide binding in water [67]. With a pendant dansyl group in the spacer this new receptor system became selective for tripeptides with central tryptophane units and complex formation could be monitored by fluorescence spectroscopy. A porphyrin unit could serve the same purpose [68].

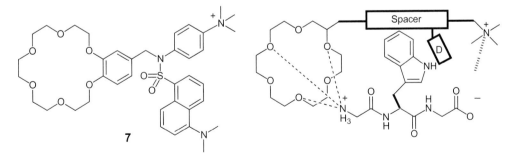

Figure 3.14 A host (**7**) for the tripeptide Gly–Trp–Gly with D = Dansyl as fluorescence-label capable of forming π-stacking interactions and its 3-point-interaction with the free peptide.

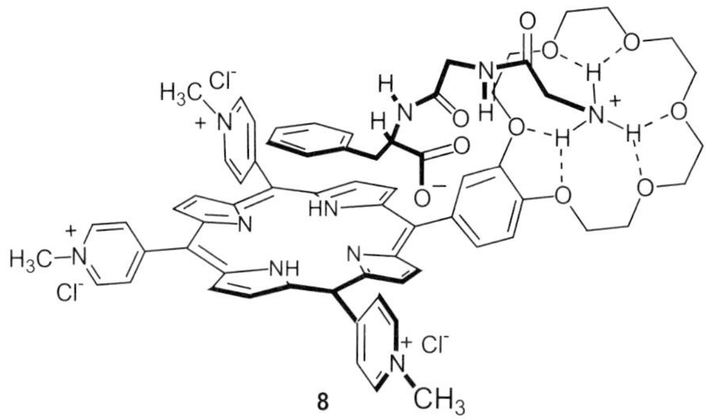

Figure 3.15 A projection of the three-dimensional structure of the complex of the porphyrin-based tripeptide receptor **8** with Gly–Gly–Phe.

In a logic optimization, Schneider et al. recently reported that zwitterionic peptides are especially tightly bound if their N-terminus is inserted into a pendant crown ether while their C-terminus is drawn into the center of the positive charges of a triply pyridinium-substituted porphyrin [69] (Fig. 3.15). Complexation constants can be conveniently measured in water by UV spectroscopy, because complex formation with the free peptide is always accompanied by changes of the Soret band extinctions and red-shifts of the absorption maxima. The highest K_a values exceed 10^5 M^{-1} and are markedly dependent on the length of the peptide and the number of aromatic amino acids, especially Phe. COSY, HSQC, HMBC, and NOESY NMR experiments on the Gly–Gly–Phe 1:1-adduct provide ample evidence for the complex geometry with extensive π-stacking of the Phe residues on the porphyrin ring. A corresponding structural model obtained by a CHARMm simulation explains both the high efficiency of **8** and its selectivity for aromatic amino acids.

Porphyrins have also been used by Mizutani et al. for peptide recognition in water (Fig. 3.16) [70]. In this work the porphyrin carried four meso (p-carboxyphenyl) units with additional carboxyalkyloxy groups for ion pairing with the peptidic ammonium group. Simultaneously, the alkyl spacers form a hydrophobic cavity above the porphyrin plane and lead to substantial hydrophobic interactions. A coordinative bond with the central Zn^{2+} ion complements the binding pattern and leads to superior affinity for histidine-containing peptides (K_a up to 10^5 M^{-1} at pH 8, $I = 0.1$ M KCl). The large dependence of the binding constants on the ionic strength indicated that the electrostatic interaction between the ammonium group of histamine and the carboxylate groups of the receptor contributes significantly to tight binding in water. This was also true for complexes with arginine derivatives, and it was argued that a water-exposed salt bridge involving hydrogen-bonding can indeed be used as a significant recognition force even in this highly competitive solvent.

The interaction of glycosylaminoglycans (GAG) with polypeptides plays a key regulatory role in many physiological processes [71]. Because p-sulfonatocalix[n]arenes can act as GAG mimics [72] and the respective recognition sites are rich in arginines

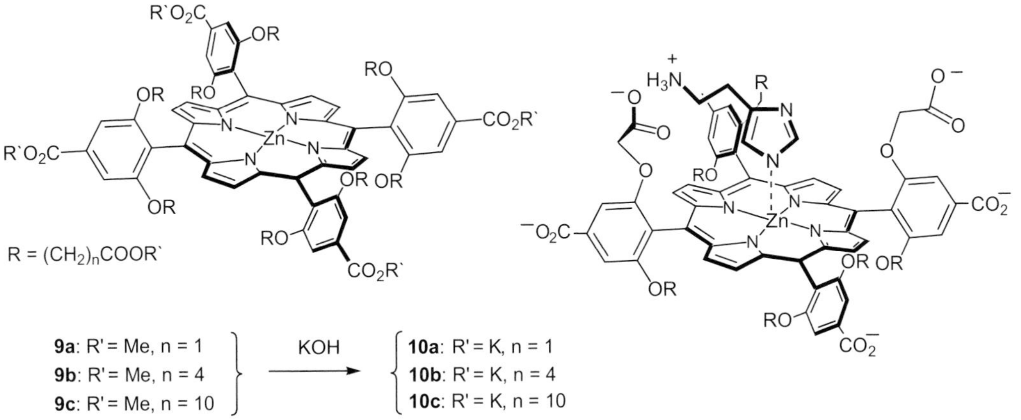

9a: R' = Me, n = 1
9b: R' = Me, n = 4
9c: R' = Me, n = 10

KOH
⟶

10a: R' = K, n = 1
10b: R' = K, n = 4
10c: R' = K, n = 10

Figure 3.16 Water-soluble porphyrin receptors **10** for free histidine-containing peptides.

and lysines [73], a systematic binding study has been conducted with di- and tripeptides composed of these amino acids. Earlier work had already revealed the high affinity of the cyclotetramer and the cyclohexamer for acetylcholine and other quaternary ammonium ions, involving π–cation interactions with their electron rich concave inner aromatic surface [74]. Coleman et al. investigated their mechanism of binding to di- and trilysine by titration microcalorimetry and NMR spectroscopy in aqueous buffer at pH 8 and ambient temperature [75]. They observed especially tight binding for lysine-containing di- and tripeptides with the calix[4]arenes (K_a >3 × 10^4 M^{-1}), resulting from van-der-Waals interactions between the apolar part of the guest and the host cavity (favorable enthalpy term) and desolvation of their charged groups on ionic interaction with the sulfonates (favorable entropy term). These water-soluble anionic calixarenes are especially promising for potential development into drugs, because of their low toxicity [76]. Rigid functionalization of the upper rim of the calix[4]arene skeleton with phosphonate monoanions also leads to arginine and lysine-specific receptor molecules (**11**) (Fig. 3.17) [77]. In this case, however, π–cation interactions with the calixarene are minimized and a network of ionic hydrogen-bonds between the cationic sidechain and all negatively charged phosphonate groups is preferentially formed. The alkoxy groups on their lower rim render these host molecules highly amphiphilic; hence, as the Schrader group recently showed, they can be readily incorporated into stearic acid monolayers at the air/water interface [78]. The BAM picture shows even distrubution in the membrane model, even in the pseudocrystalline state. Subinjection of basic peptides leads to large shifts in the π–A diagram (>5 Å^2 for triarginine at 10^{-7} M), owing to partial neutralization of the tetraanionic receptors in solution and their reincorporation into the monolayer. These positive shifts are also observed for aqueous solutions of basic peptides at nanomolar concentrations (~5 Å^2 for lysine-rich histone H1 at 10^{-9} M). A close correlation is observed between the size of the π–A shift and the respective pI of the corresponding proteins. Thus, acidic proteins such as acyl carrier proteins produce minute shifts ($\Delta A_{rec} \leq 1$ Å^2; $pI = 4.2$), whereas neutral proteins such as thrombin furnish medium size shifts ($\Delta A_{rec} = 2$ Å^2; $pI = 7.5$). Even in the solid state of a monolayer the embedded receptor molecules can move freely around in all directions, facilitating automatic self-assembly over polytopic guests in the subphase [79]. The above-described simple model system has thus successfully mimicked the efficient and selective binding of certain proteins to cell surfaces by multipoint recognition. Because only one basic domain occurs in many of the proteins examined, they must be oriented parallel to each other to dock on to the anionic binding sites on the monolayer "ceiling". This might in the future facilitate their crystallization [80].

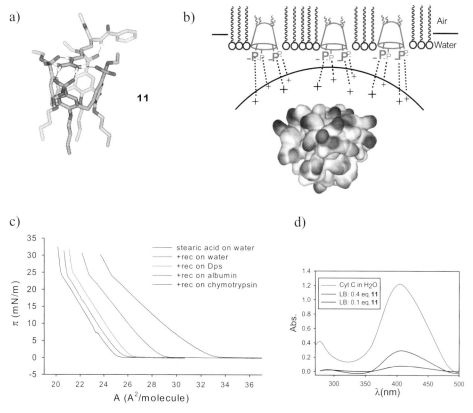

Figure 3.17 Lysine-rich protein recognition at the air/water interface: (a) Optimized complex structure formed between *N/C*-protected arginine and calixarene tetraphosphonate **11** (Monte-Carlo simulation in water). (b) Schematic representation of the proposed multipoint binding process – the receptor molecules self-organize in the fluid monolayer over the positive domains on the protein surface; the Connolly surface of cytochrome C is patterned with the electrostatic surface potential (ESP), showing basic (blue) and acidic (red) domains. (c) Pressure–area isotherms for acidic, neutral, and basic proteins at 10^{-8} M on the Langmuir film balance. (d) UV/Vis spectra of cytochrome C in water and in LB films drawn from the film balance experiments.

A very successful strategy has been pursued throughout the past six years by the Hamilton group [81]: They discovered that in many proteins playing key roles in biochemical processes a hydrophobic patch at the binding site is surrounded by an array of basic amino acids, i.e. arginine and lysine units. This combination affords a unique opportunity to achieve strong binding with an artifical host molecule which is essentially complementary both in charge and topology [82]. A nonpolar extended core unit for hydrophobic interactions should be surrounded by anionically charged functional groups for multiple ion pairing with the ammonium and guanidinium sidechains. Their first approach featured a calix[4]arene skeleton, preorganized in the cone conformation and adorned with four identical cyclodecapeptides carrying Asp residues. It was argued that this arrangement mimics the hypervariable loops in antibody-combining regions. With affinity and gel filtration chromatography strong binding of the best receptor towards the surface of cytochrome c was established, even at high phosphate buffer concentrations ($K_a = 3 \times 10^6$ M^{-1}). Molecular modeling and inhibition experiments strongly suggest that the receptor adopts a flattened conformation and binds to the heme edge region, which is surrounded by several lysine residues. This is exactly the binding site for the natural protein partners such as cytochrome (per)oxidases. The receptor indeed potently inhibits reduction of Fe(III) cytochrome c to its Fe(II) form [83]. Interestingly, the nature of the substituent on the lower rim was critical for efficient protein recognition, indicating the high importance of this receptor region's geometry. In their optimized conformation, these receptor molecules can cover a surface area on the proteins >600 Å^2. As a related example, the approach of small molecule substrates to α-chymotrypsin were also inhibited at submicromolar concentrations [84]. In extension of this powerful concept, Hamilton et al. combined the hydrophobic core of a tetraphenylporphyrin with multiple anionic binding sites composed of terminal Asp residues and obtained the strongest receptor molecules (**12**) for cytochrome c reported to date (Figs. 3.18 and 3.19), with K_a values approaching 10^8 M^{-1} [85]. It should be mentioned, however, that

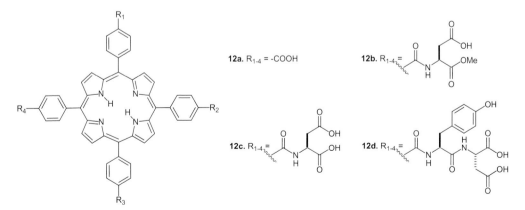

Figure 3.18 Tetraphenylporphyrin receptors **12a–d** adorned with aspartates for cytochrome c recognition at the heme edge region.

Fisher et al. first discovered this geometrical complementarity between tetraphenyl-porphyrins and cytochrome c and thus had already pioneered this work 15 years earlier [86]. The strategic cooperation between the large central hydrophobic patch and the ionic binding sites demonstrates the validity of their general concept for efficient molecular recognition of water-exposed protein surfaces under near physiological conditions. The porphyrin offers the additional advantage of highly sensitive fluorescence quenching upon binding of the protein. In recent investigations, the affinity of porphyrin-based synthetic receptors toward basic proteins was further opimized

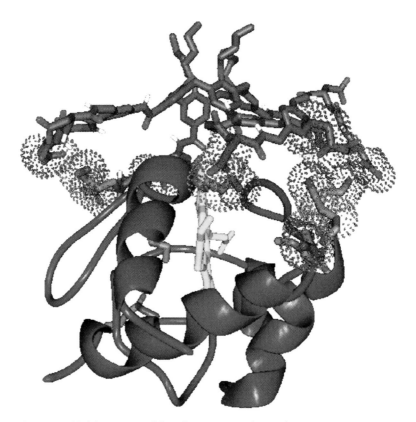

Figure 3.19 Modeling picture of the calixarene tetracyclopeptide receptor docking on to cytochrome c's basic amino acids surrounding its heme group.

by means of small combinatorial libraries and reached the sub-nanomolar range [87]. In addition, these hosts were shown to be strong, but reversibly acting denaturants at stoichiometric ratios. Clearly, many potential applications of selective disruption of important protein–protein interactions can be envisaged, with many medicinal implications.

Some linear anionic polymers e. g., aurintricarboxylic acid and oligophenoxyacetic acid derivatives **13** (Fig. 3.20) also bind to basic peptide regions. Earlier reports from the Chang group mentioned heparin-like properties, e.g. reversal of basic fibroplast growth factor (bFGF)-mediated autocrine cell transformation [88]. This makes sense, because the overall structure of these artificial protein binders bears some similarity to that of the polysulfated polysaccharide heparan sulfate. More recently, potent inhibitors of human leukocyte elastase with K_i values of 17–200 nM were identified as diphenylmethane-based 6mer oligomers [89]. They presumably interact with basic residues adjacent to the active site, surrounded by hydrophobic regions.

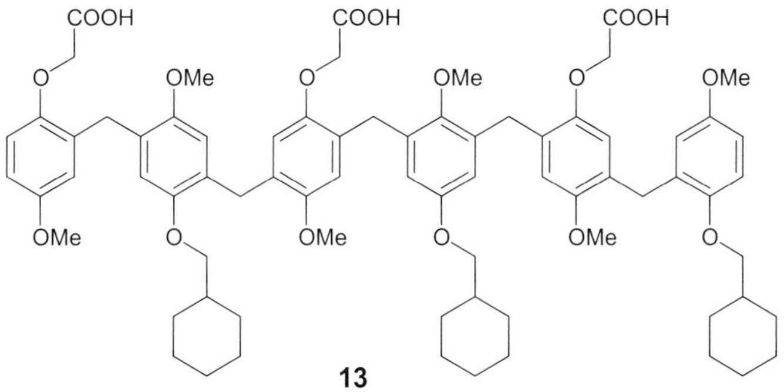

13

Figure 3.20   Oligophenoxyacetic acids such as **13** bind to basic peptide regions.

3.2.5
Conclusion and Outlook

Several promising new alkylammonium receptors include tripodal oxazolines, benzylic bisphosphonates, and resorcinearene-based deep cavitands; these could, in principle, all be amplified in an equilibrating mixture containing a guest. For crown ether complexes with alkylammonium ions gas-phase calculations and solvent effects point to a dominant contribution of hydrogen bonds over any other noncovalent attractive force. The rediscovery of π–cation interactions for artificial receptors and the invention of unnatural amino acid mutagenesis and heterologous expression led to deepened understanding of natural ion channels. Natural concepts such as multivalency or the altered microenvironment inside active sites are being mimicked by chemists with polytopic hosts or dendrimer encapsulation. pH-sensitive filters or nanoparticle-based receptors lead to highly efficient electrochemical sensor devices. Membrane transport, photoswitchable receptors and size and shape-selective encapsulation characterize other innovative areas of intense research. Finally, the first sequence-selective artificial receptors for unprotected small peptides have been presented, many of these based on porphyrin cores. Occasionally, enantiomers can be discriminated. Protein recognition is still in its infancy. Small calixarenes are capable of immobilizing charged proteins at the air/water interface, with high sensitivity; in free buffered aqueous solution, proteins with a hydrophobic patch located at the binding site, surrounded by an array of basic amino acids, can be efficiently inhibited by a complementary array of binding sites in synthetic hosts. The Hamilton group has pioneered the development of some highly efficient new receptor systems based on a calixarene or porphyrin core surrounded by acidic peptide units. Subnanomolar K_D values and reversible denaturation effects accompany these exciting new functional synthetic receptors in action. High affinity molecular recognition of protein surfaces combined with high specificity remains a premier challenge, however, especially in the light of their solvent-exposed nature.

3.3
Amidinium Cations

3.3.1
Introduction

There are only very few receptor molecules for amidinium ions compared with other *N*-cations such as ammonium or guanidinium ions; this is mainly because of the lack of biologically relevant targets. The few examples of artificial amidinium receptors are motivated by the important role of amidines in antithrombotic, [90–92] antibacterial and antiprotozoal [93, 94] agents. Amidines (or, better, amidinium ions) play a crucial role in the development of thrombin inhibitor drugs. Thrombotic vascular diseases are a major cause of mortality in the industrialized world. Thus the development of selective artificial thrombin inhibitors is a key goal for many research groups and pharmaceutical companies. A prominent representative of this class of inhibitor is the so-called [R]-NAPAP **14** synthesized by Stürzebecher (Fig. 3.21). The benzamidine moiety is essential for its inhibition activity.

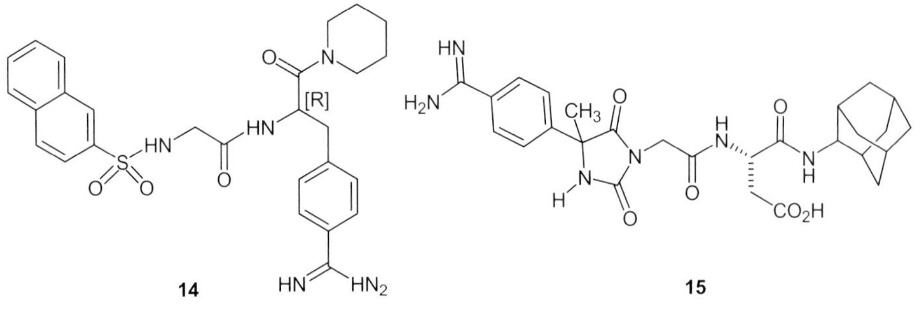

Figure 3.21 Structure of [R]-NAPAP **14** and the RGDS-mimeticum **15**.

Compound **15**, an RGDS (Arg–Gly–Asp–Ser) mimetic, antagonistically inhibits the activation and thus the complex formation of the GIIb/IIIa protein with fibrinogen. GIIb/IIIa belongs to the integrin superfamily and is jointly responsible for the development of platelet blood aggregation, in the worst case leading to thrombi [95, 96]. The benzamidinium group is here used for complexation of a specific Asp residue of the GIIb/IIIa protein. This Asp residue binds originally to the Arg of the RGDS sequence.

3.3.2
Artificial Receptors

A very simple host–guest system for amidinium ions was investigated in detail by Kraft [97, 98]. The *N,N'*-diethyl-substituted benzamidinium-ion is complexed by a benzoic acid derivative which binds by forming two hydrogen bonds and even more importantly by coulombic interactions between opposite charges. The substituted benzamidine occurs normally as a mixture of (E,Z) and (E,E) isomers, the (E,Z) configuration is thermodynamically favored. Kraft showed that in organic solvents, a strongly binding ligand such as benzoic acid can force the amidine into the thermodynamically less favored (E,E) configuration (Fig. 3.22). This result was confirmed by NMR experiments and X-ray crystal structure determination. Although binding constants of 10^6 M^{-1} in chloroform were observed for related complexes, addition of more competitive solvents such as DMSO drastically reduced complexation free energy.

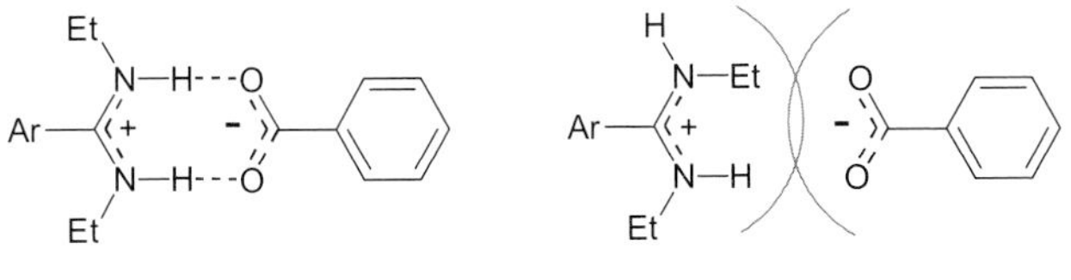

(E,E) *(E,Z)*

Figure 3.22 Left: interaction between the preferred (E,E) isomer and benzoic acid; right: sterically hindered interaction with the (E,Z)-isomer.

Bell et al. have reported two highly preorganized synthetic receptors which bind benzamidinium ions via a network of convergent hydrogen bonds and/or π-stacking interactions (Fig. 3.23) [99]. Complexation of benzamidinium ions by **16** and **17** was detected by NMR and UV/Vis spectroscopy. UV/Vis titrations revealed minimum complex stability of 5×10^5 M^{-1} and 2×10^7 M^{-1}, respectively, in dichloromethane–ethanol, 95:5. The smaller K_a values of **16** might be because of steric shielding of the binding cleft by the aromatic moiety, indicating that the binding process requires more conformational reorganization. This aromatic moiety was originally introduced for a potentially useful π contact between guest and host.

Initially these receptors were synthesized against the background of developing a host for pentamidine isothionate, an important DNA-binding bis(benzamidine) drug used for treatment of *Pneumocystis carinii* in AIDS patients. According to the results

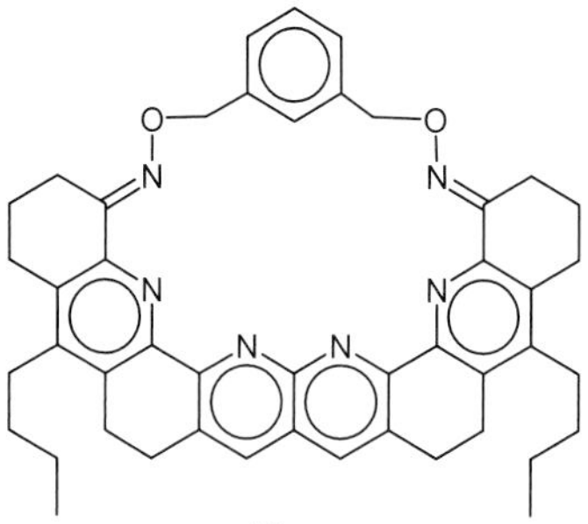

16

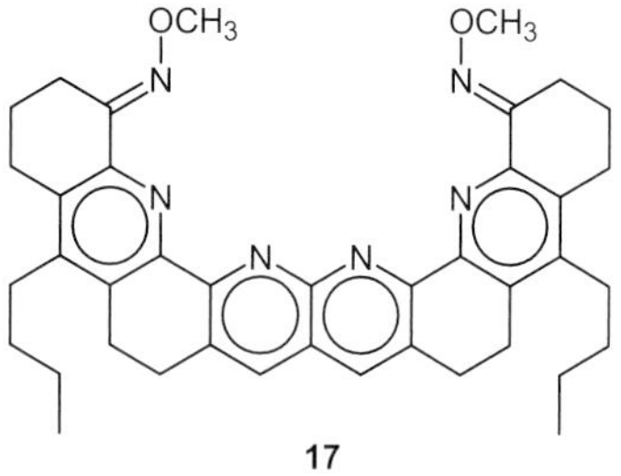

17

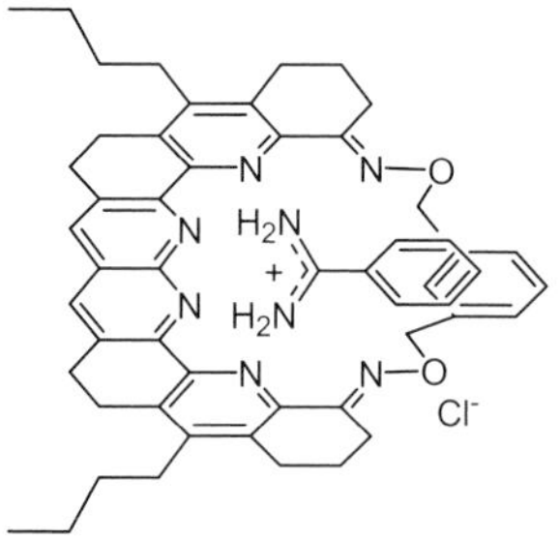

18

Figure 3.23 Bell's benzamidinium receptors **16** and **17** and the proposed complex **18** between **17** and benzamidinium hydrochloride.

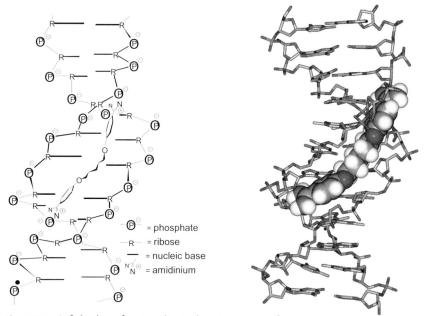

Ⓟ = phosphate
R = ribose
— = nucleic base
N⁺N = amidinium

Figure 3.24 Left: binding of pentamidine in the minor groove of DNA; right: modeling structure (MacroModel 7.0, Amber*, water).

of numerous research groups the bis(benzamidine) moiety binds to the minor groove of the DNA along its phosphodiester backbone (Fig. 3.24) [100–101].

Recently Schrader et al. reported artificial receptors for exactly this kind of bis(benzamidine) drug. (Fig. 3.25) [102]. Starting point for their receptor design was the sim-

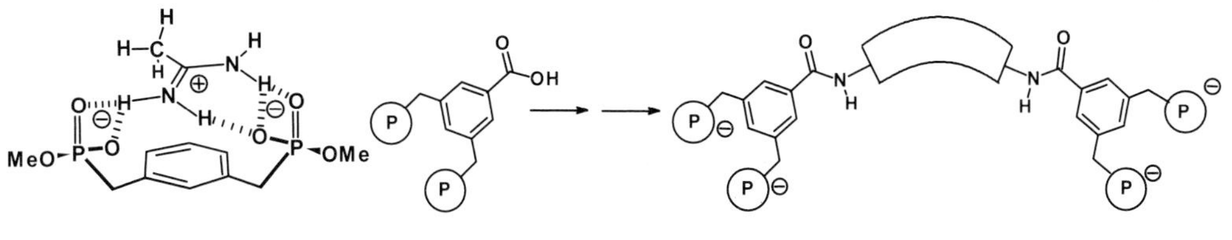

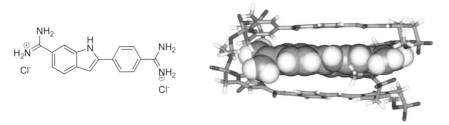

19

Figure 3.25 Left: energy-minimized structure of the complex between the acetamidinium ion and bisphosphonate **19**; right: schematic illustration of the new ditopic receptor molecules.

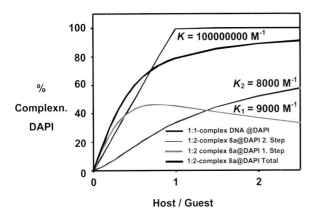

Figure 3.26 Left: bisamidine DAPI; right: lowest energy structure for the 2:1-complex between **22** and DAPI according to Monte-Carlo simulations in water (MacroModel 7.0, Amber*, 2000 steps).

$K = 100000000 \text{ M}^{-1}$

$K_2 = 8000 \text{ M}^{-1}$

$K_1 = 9000 \text{ M}^{-1}$

% Complexn. DAPI

- 1:1-complex DNA @DAPI
- 1:2-complex 8a@DAPI 2. Step
- 1:2 complex 8a@DAPI 1. Step
- 1:2-complex 8a@DAPI Total

Host / Guest

Figure 3.27 Calculation of the amount of bound DAPI in its 1:2 complex with **24** and its 1:1 complex with DNA at 10^{-4} M.

ple bisphosphonate **19**, which was found to bind different benzamidinium chlorides in DMSO with association constants of approximately 10^5 M^{-1}. The assumed complex geometry is shown in Fig. 3.25. The complex is formed by means of four ion pair-reinforced hydrogen bonds and extensive chelate type electrostatic interactions between both phosphonate anions and the benzamidinium cation. π–cation assistence is highly probable. The observed high K_a values for the monomeric complex led to the design of receptors containing two bisphosphonate moieties covalently linked with different spacers.

The α,ω-diamine spacers were varied from rigid through semirigid to flexible to cover a wide range from highly preorganized to highly flexible artificial receptors (complexation via induced fit). Binding constants between selected receptor molecules and pentamidine and DAPI are shown in Tab. 3.1. High free energies of binding, independent of spacer size or flexibility, are found for the 1:1 complexes with K_a values always surpassing 10^5 M^{-1} in methanol and 10^4 M^{-1} in water.

The small decrease in K_a from methanolic to aqueous solutions suggests the contribution of substantial solvophobic forces, complemented by van der Waals attractions from the close contact between host and guest. Job-plot experiments revealed stoichiometry (host:guest) was 1:1 or 1:2 for most of the complexes. In contrast, complex formation between DAPI and the receptor molecule **22** led to a 2:1 complex (host:guest). Figure 3.26 depicts the proposed calculated complex geometry; it is strikingly similar to that of bis(benzamidinium) cations in DNA's minor grove.

With such high association constants for **24** and DAPI (1:2 complex) in water, it was obvious to compare DNA–bisamidine affinity with this obtained. Figure 3.27 shows total and single-step complexation curves separately. These reveal that when bisamidine is in excess (host/guest < 1)host **24** binds at least the same amount of guest as does DNA. This host system for bisamidinium cations might, therefore, be used for transport, detection, or controlled release of these drugs even in the presence of DNA.

Another interesting approach for the complexation of bisamidines by artficial receptors was realized by Sellergren. He used an imprinted polymer capable of recognizing pentamidine in a solid-phase extraction procedure to selectively retain pen-

Table 3.1 Binding constants between tetrakis-phosphonate receptor molecules and pentamidine and DAPI.

Tetrakis-phosphonate	Bisamidinium salt Pentamidine			DAPI		
	K_a	ΔG (kcal mol^{-1})	Stoichio-metry	K_a	ΔG (kcal mol^{-1})	Stoichio-metry[b]
m-Phenylene **20**	2×10^5 M^{-1}	−7.1	1:1	2×10^5 M^{-1}	−7.1	1:1
m-Xylylene **21**	4×10^{10} M^{-2}	−14.2	2:1	1×10^5 M^{-1}	−6.7	1:1
p-Phenylene **22**	4×10^{11} M^{-2}	−15.6	2:1	1×10^{10} M^{-2}	−13.4	2:1
Fluorenylene **23**	2×10^4	−5.8	1:1	4×10^4 M^{-1}	−6.2	1:1
p-Xylylene **24**	2×10^4	−5.8	1:1	7×10^7 M^{-2}	−10.5	1:2
n-Decylene **25**	3×10^4	−6.0	1:1	5×10^8 M^{-2}	−11.7	1:2

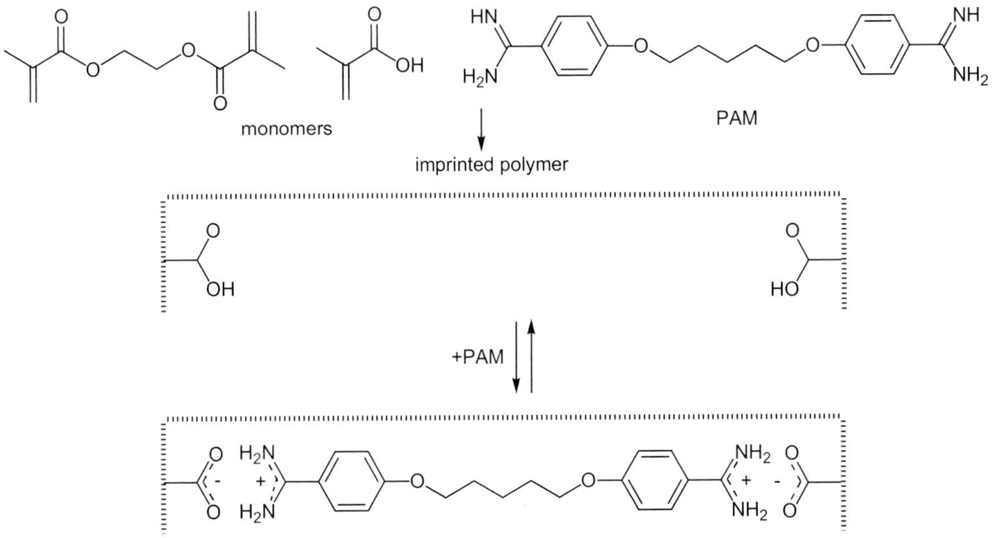

Figure 3.28 Synthesis and binding experiment of PAM-imprinted polymer from monomers ethylene glycol dimethacrylate, methacrylic acid and pentamidine.

tamidine from dilute analyte solutions [103]. The imprinted polymers were synthesized from a mixture of two types of monomer, one containing a free carboxylic acid moiety for amidinium recognition, by following elaborated imprinting methods. Pentamidine and benzamidine were used as reference compounds (Fig. 3.28) [104–108].

The pentamidine-imprinted polymer achieved an enrichment factor of 54 for the offered template at physiological concentrations (30 nM), whereas the benzamidine-imprinted polymer only achieved 14, so the selectivity factor is an impressive 4.0. With this method one can enrich and directly analyze pentamidine even at low concentration in urine samples. Especially in view of the extensive use of such drugs with AIDS patients, their simple and cheap detetction in urine by use of imprinted polymers is of great interest.

Zimmerman used the complexation of benzamidium ions by naphthyridine units to determine the nanoenvironment of different dendrimers all containing naphthyridine in their cores [109]. The hydrogen bonds formed by the complexation of 26 with 27 (Fig. 3.29) are highly dependent on solvent polarity. Therefore the guest molecule serves as a sensitive probe of dendrimer porosity and core polarity. The experiments revealed that even for the most sterically demanding dendrimers complexation strength is relatively insensitive to dendrimer size; the similar responses to changes in solvent polarity indicate that these dendrimers are highly porous.

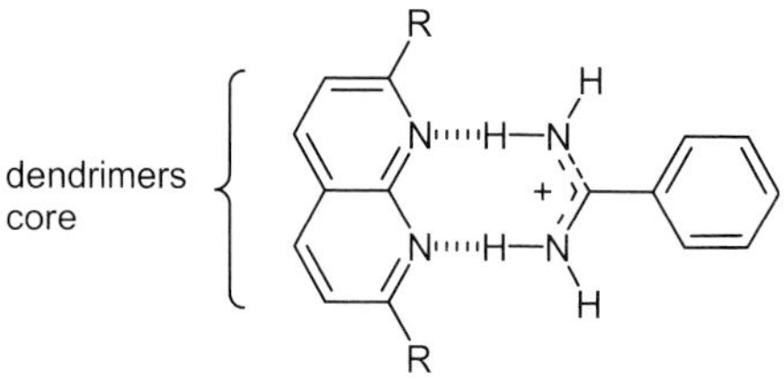

Figure 3.29 Complexation of benzamidinium cations with a naphthyridine moiety containing dendrimers as sensitive probes of dendrimer porosity and core polarity.

3.3.3
Conclusion

The development of new receptor systems for amidinium ions has been almost completely disregarded by supramolecular scientists in recent years. The groups involved in this area focused their attention mainly on molecular recognition of bisamidines, which have been developed as drugs against *Pneumocystis carinii* in AIDS patients, as mentioned above. Special attention was, and will be, directed toward transport, controlled release, and detection of these kinds of drugs. Further investigation of these three topics will aid discovery of the correct dosage to minimize side effects of the benzamidine drugs.

Until now the synthesis of amidinium receptors has been based on rational design and the advantages of combinatorial chemistry have been mainly neglected. Both approaches should be used synergetically and thus will help to find systems with enhanced binding properties.

3.4
Guanidinium Cations

3.4.1
Introduction

The most important biomolecule containing a guanidinium moiety is the basic amino acid arginine. Individual arginine residues are found to be critical for the function of many RNA-binding proteins that mediate a wide range of biological processes [110, 111]. Whereas the "basic-domain" class of RNA-binding proteins contain arginine-rich sequences of 10–15 amino acids, highly conserved arginine residues are often involved in specific electrostatic contacts with phosphate groups of the polyribonucleotide backbone. Two prominent examples of such proteins regulate important steps in the replication of the HIV-1 – the transcriptional activator Tat and the Rev proteins [114].

Other biologically important arginine-dependent proteins are the integrins. These are heterodimeric transmembrane proteins necessary for many biological processes, for example migration regulation, cell communication, cell proliferation, and apoptosis [115]. The occurrence of an extraordinarily high number of integrin subtypes in cancer cells has aroused the interest of many scientists and pharmaceutical companies and stimulated investigations of integrin–protein interactions [116, 117].

Study of the crystal structures of the extracellular domain revealed that the key recognition element in peptidic ligands is their RGD sequence [118, 119]. The arginine (i.e. its guanidinium moiety) is complexed by two aspartates of the integrin α subunit whereas the aspartate is bound by a metal ion of the β subunit. Interestingly enough, additional van-der-Waals contacts between the integrin and the RGD protein, the main driving force for most protein–protein interactions, were not observed. Thus, the specific development of RGD mimetics or RGD binders is of special interest.

3.4.2
Artificial Receptors

The above-described chelate arrangement of two aspartates binding the guanidinium ion of the RGD sequence has inspired chemists to attempt biomimetic design of artificial guanidium receptors. Thus, Schrader et al. have introduced two covalently linked phosphonate moieties mimicking the carboxylate groups of the two aspartates as a potent binding site for the guanidinium moiety in peptidic arginines in polar solution. By varying the spacer between the phosphonates a series of different host molecules for guanidinium derivatives was synthesized [120–124].

A very simple but still effective host molecule for alkylguanidinium cations was reported by Schrader et al. in the late nineties. The molecular tweezer contains two phosphonate groups forming a strong chelate complex with alkylguanidinium cations by primarily electrostatic interactions reinforced by a network of hydrogen bonds. This binding motif is similar to the so-called "arginine-fork" postulated by

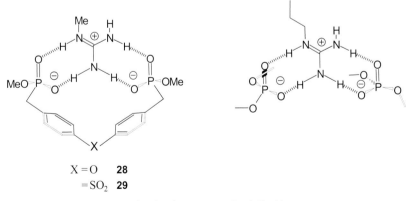

X = O   **28**
= SO₂   **29**

Figure 3.30 Left: complexes of molecular tweezers **28**and **29** with the methylguanidinium ion; right: the so-called "arginine-fork".

Frankel for a simple sequence-selective binding mode used by Nature – the Tat-regulator protein recognizes an RNA conformation that enables single arginine residue to bind simultaneously to two phosphates. The evolving structure was coined the "arginine-fork".

NMR spectroscopic studies of binding in complexes between host **29** and different kinds of organic cations in DMSO revealed remarkable selectivity for arginine derivatives over ammonium ions (Tab. 3.2). Even other guanidinium derivatives such as creatinine or methylguanidine are bound much more weakly.

Table 3.2

Guest	$K_{1:1}$ 28 ($\times 10^3$ M^{-1})	$K_{1:1}$ 29 ($\times 10^3$ M^{-1})
Methylguanidine	5.6	9.3
α-*N*-Tosylarginine methyl ester	21.8	62.8
Benzylamine	3.7	1.9
1,1-Dimethylguanidine	6.6	2.9

Two phosphonate moieties were also used by Schrader and Finocchiaro in the design of a chiral sensor for lysine and arginine. The spirobisindane-based hosts **30** and **31** have a cleft-like form, high rigidity caused by the spirobisindane skeleton, and, in the form of the phosphonate moieties, two binding sites for α,ω-dicationic substrates (Fig. 3.31). Host **31** also contains a mesitylene bridge, acting as a hydrophobic binding site. This leads to strong and chiral recognition of longer α,ω-diamines in DMSO.

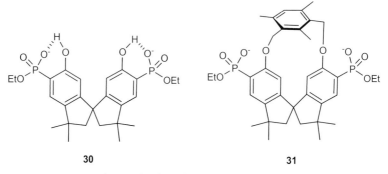

30   **31**

Figure 3.31 Two new host-molecules – the open-chain **30** has C_2-symmetry whereas macrocycle **31** has C_1-symmetry.

Binding constants derived from NMR titrations were in the range 10^3 to 10^4 M^{-1} for both arginine and lysine derivatives, but only host **31** is capable of dicriminating between their enantiomers. Obviously, only the sterically more demanding host **31** can distinguish between enantiomeric guests with an N$^+$···N$^+$ distance of more than five bonds. Although the ee are not very high (**31**: L-Arg 17 %; **31**: L-Lys 33 %), well resolved NMR signals are produced during the NMR titration, so optically pure **31** can be used as a shift reagent for the quantitative determination of the enantiomeric purity of arginine and lysine derivatives.

In a further modification, another benzylic phosphonate group was introduced into the existing bisphosphonate moiety, to enforce electrostatic attraction of a centrally bound guanidinium ion. NMR titrations in DMSO for trisphosphonate **32** with different organic cations revealed that the guanidinium cation is complexed even more strongly than organic trications, although it carries only one positive charge. The authors attribute this to additional π–cationic interactions that become effective when the guanidinium moiety is located directly on the top of the central phenyl ring at a calculated, almost ideal, distance of 3.2 Å. Interestingly enough, the mesitylene-based all-benzylic trisphosphonate-triammonium complexes form capsule-like assemblies even in water, although they are still too small for guest encapsulation (Fig. 3.32).

Bell recently described a series of interesting host molecules for guanidinium cations, obtained by means of a slightly different approach [126–129]. One of the most intriguing examples of guanidinium–arginine recognition is the so-called "arginine-cork" **33**. The host **33** has nearly perfect preorganization for a guest-induced network of hydrogen bonds. In addition, the carboxylate groups not only increase the water solubility of the receptor but also play an important role in the complexation of guanidinium cations by Coulomb interactions.

Conjugation between the carboxylate groups in **33** and the neighboring pyridine rings stabilizes their mutual coplanar conformation leading to almost perfect preorganization. This rational design was mainly confirmed by X-ray crystallography of the host–guest complexes. The association constants between host **33** and guanidinium derivatives were measured by NMR spectroscopy and microcalorimetry. The respective nonlinear regression analysis reveals that the primary interaction between the cationic guest and the dicationic host is electrostatic, but molecular recognition based on hydrogen bond complementarity distinguishes between guests of the same charge (Fig. 3.33). Thus, the dipeptide ArgArg is more much tightly bound than the dipeptide LysLys with its two ammonium functionalities.

The importance of hydrogen bonds in the complexation of guanidinium ions by Bell's hosts was proven by systematic comparison of the association constants for hosts **34** and **35**. Exchange of one or two nitrogen atoms for carbon in hosts **34** and **35** leads to markedly decreased stability constants for the investigated complexes (Fig. 3.34).

Interestingly enough, the selectivity of **33** for guanidinium cations over ammonium cations is inverted in **34**, because of its smaller binding site. Because of the sterically demanding C–H group, **34** cannot comfortably accommodate the larger guanidinium cation, but it accepts instead the smaller ammonium sidechain of lysine.

With receptor **35**, binding of almost all guests is weak. The "cavity" size is apparently too small to accommodate either ammonium or guanidinium ions. In general, host **35** does not discriminate between the offered cationic guests, thus providing additional experimental evidence that a hydrogen-bond network is indeed essential for high selectivity.

Host **36** belongs to the class of torand with a large, highly preorganized cavity (Fig. 3.35). It has been confirmed by mass spectrometric and NMR studies that the

32

Figure 3.32 Energy-minimized structure of the complex between 1,3,5-trisphosphonate **32** and its analogous triammonium salt: (a) Lewis structure; (b) front view of the corresponding CPK model; (c) calculated van-der-Waals surface shown as Connolly surface [125] (dotted solvent-accessible area around the complex) with internal cavity.

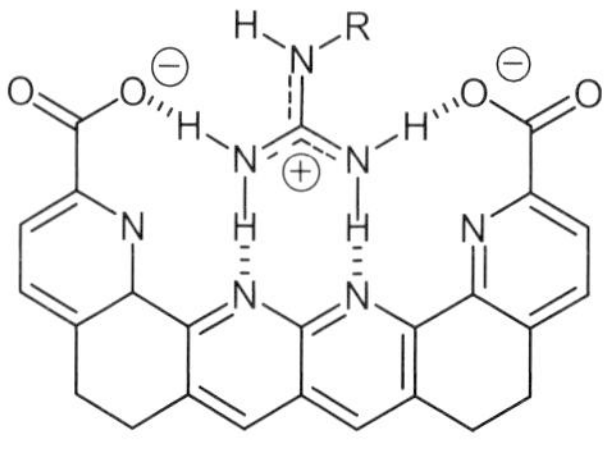

Guest	K_d [µm] (H_2O)	$\Delta H°$ [kcal mol^{-1}] (H_2O)
Methyl-guanidinium	4300 ± 400	$+1.3 \pm 0.3$
Arginine (Arg)	1100 ± 100	$+0.49 \pm 0.05$
Lysine (Lys)	7000 ± 2000	-0.14 ± 0.02
ArgArgNH$_2$	60 ± 6	-5.6 ± 0.6
ArgArgOH	50 ± 20	-2.1 ± 0.2
LysLysOH	250 ± 30	-1.5 ± 0.2

33

Figure 3.33 Left: the so-called "arginine-cork" in a complex with a guanidinium cation. Right: thermodynamic data for 1:1 complexes with **33** as determined by ^{1}H NMR spectroscopic and microcalorimetric titrations.

free guanidinium cation fits perfectly into this cavity. Even the non-macrocyclic host **37** has relatively high affinity for guanidinium hydrochloride, although it must switch from its thermodynamically preferred helical conformation into a flat conformation to form the maximum number of hydrogen bonds. Thus, the increased conjugation of the π-system of **37** in the complex causes a bathochromic shift of the absorption maximum in the UV/Vis spectrum.

Bell also designed a chromogenic sensor for creatinine, a guanidinium derivative which is used as a key indicator of renal function (Fig. 3.36). Host **38** entraps creatinine by formation of three hydrogen bonds. The complexation process involves proton transfer from the phenolic hydrogen to the imino nitrogen in **39**. Creatinine is extracted from water into dichloromethane solutions of **38** with a concomitant color change from yellow to brown–orange. This chromogenic response was found to be selective for creatinine with no color change occurring for histidine, proline, uric acid, urea, or creatine as guests.

As already mentioned, the RGD sequence is a worthwhile target for supramolecular chemists, because of its fundamental biological importance. On the basis of the previously developed host system for arginine recognition, Schrader et al. designed the first synthetic receptor for the RGD sequence. Starting from the trisphosphonate

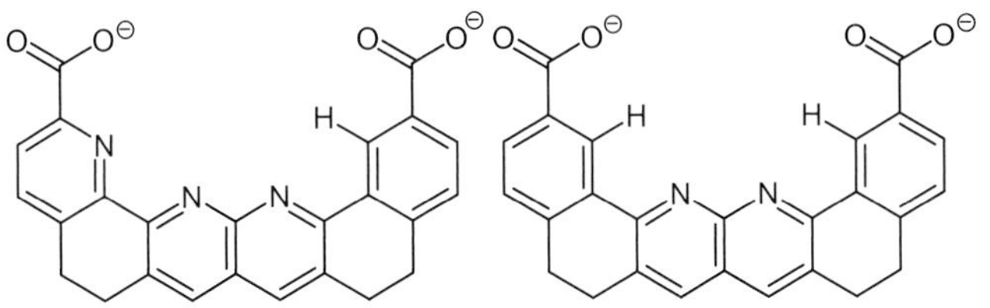

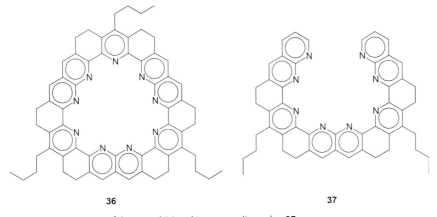

34   **35**

Guest	receptor		
	33	*34*	*35*
Methylguanidine HCl	> 100	3.9 ± 0.4	1.0 ± 0.1
Creatine	14.5 ± 1.0	< 1.0	< 1.0
N-α-acetyl-L-arginine	> 100	1.2 ± 0.2	3.9 ± 2
L-Arginine	> 100	> 100	1.1 ± 0.2
L-Lysine	29 ± 3	> 100	1.1 ± 0.1
N-α-acetyl-L-lysine	> 100	> 100	< 1.0
6-Aminocaproic acid	27 ± 10	> 100	< 1.0
1-Propylamine HCl	> 100	> 100	1.6 ± 0.2
N-ε-acetyl-L-lysine	1.2 ± 0.1	1.4 ± 0.3	1.4 ± 0.3

Figure 3. 34 Structure of the "arginine-cork" analogs **34** and **35** and binding constants K_a ($\times 10^3$ M^{-1} in methanol) of **33–35** with selected organic cations.

32, already shown to have pronounced arginine selectivity, the new host **40** was synthesized (Fig. 3.37).

The carboxylate sidechain of the Asp amino acid was intended to be recognized by the ammonium group of **40** and, indeed, NMR titration revealed that the additional recognition site increases association constants with free RGD. Even Kessler's cyclo(RGDfV), a model compound for the RGD loop in fibronectin [130], was bound

36   **37**

Figure 3.35 Structure of the torand **36** and its non-cyclic analog **37**.

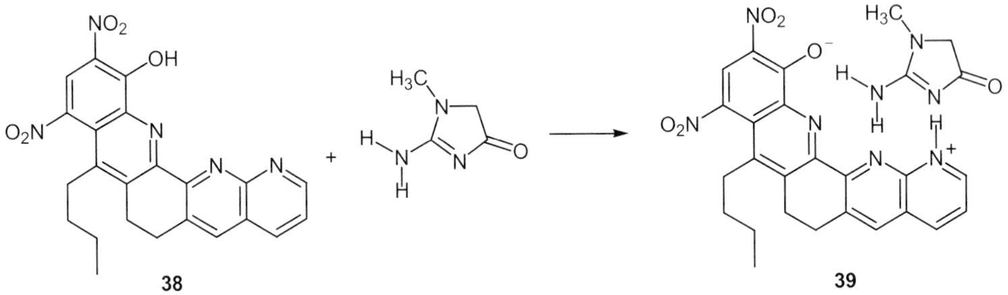

Figure 3.36 Chromogenic sensor for creatinine according to Bell et al.

RGD peptide or mimetic	40 [M^{-1}]	self-association (dilution)
free RGD	1300 ±5 %	no shifts
cyclo(RGDfV)	700 ±14 %	no shifts
benzamidine	no shifts	40 alone: no shifts

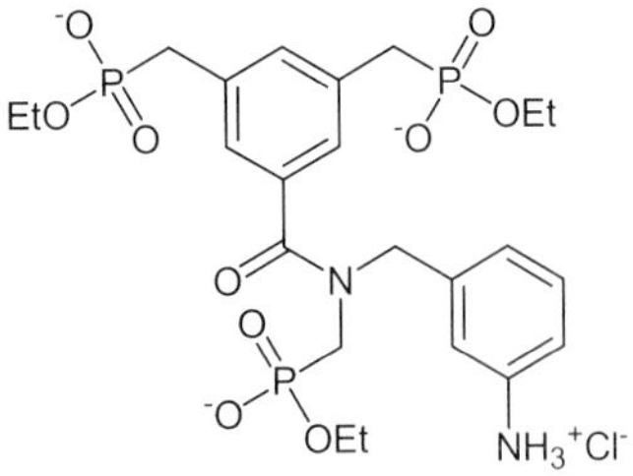

Figure 3.37 Left: binding constants resulting from NMR titration experiments between RGD-containing peptides or mimetics and trisphosphonate receptor molecule **40**, in water at 25 °C. Right: artificial RGD receptor **40**.

by **40** in water. Gratifyingly, the binding study between host **40** and the RGD-mimetic **41** (Fig. 3.38), a fibrinogen receptor antagonist [131], revealed no association between host and potential guest.

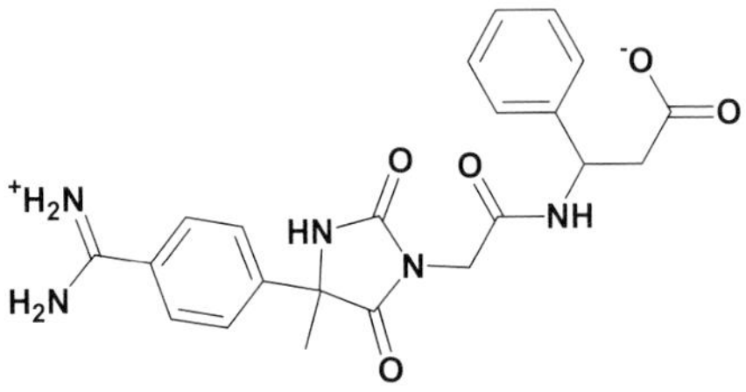

Figure 3.38 RGD-mimetic

The cyclophane derivatives introduced by Dougherty [132] are very potent but also very unselective receptors. Positively charged or even merely electron-poor guests are complexed in water inside the host cavity by π–cation interactions and hydrophobic effects. Moderate free energies of complexation were observed for guanidinium

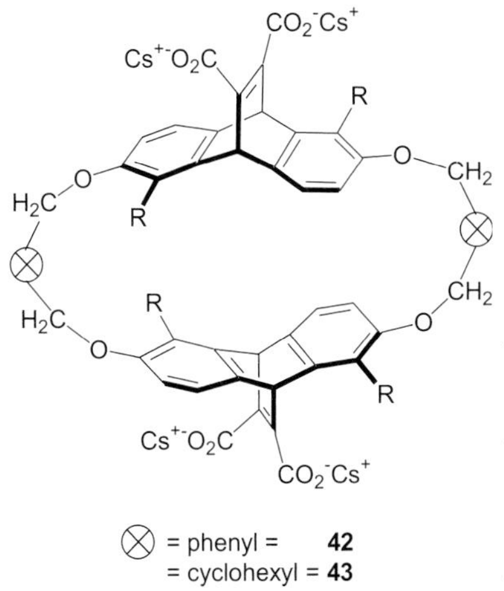

Guest	$-\Delta G°$ [kcal mol^{-1}] with 42
N,N,N',N'-Tetramethylguanidinium	4.7
hexamethylguanidinium	4.6
N-Methylquinoline	8.4

Figure 3.39 Left: structures of the cyclophane molecules **42** and **43**. Right: binding constants of **42** with selected organic cations determined by NMR in H$_2$O–borate buffer.

cations with host **42** (Fig. 3.39) whereas for the pyridinium cation *N*-methylquinoline the value was remarkably high – up to 8.4 kcal mol^{-1}.

Interesting results were obtained with the hexamethylguanidinium–cyclophane complex. The methyl signal of the propeller-like D_3-symmetric hexamethylguanidinium in the ^{1}H NMR spectrum is split into two signals of *unequal* intensity. This indicates that the host acts as a chiral shift reagent (splitting of the methyl signal) providing not only the chiral environment but also binding one enantiomer more tightly than the other (unequal intensity). A more accurate presentation of this class of receptor will be given in Sect. 3.5.2 on pyridinium cations.

In recent years some very interesting applications have been found for guanidinium complexation agents. The most extraordinary examples are reviewed in Chap. 4. Rheinhoudt et al. recently presented an electrochemical sensor for guanidinium cations [133]. The sensor was synthesized by incorporation of calix[4]- and calix[6]arenes in PVC membranes containing a cation exchanger. It works via complexation of guanidinium cations by the calix[*n*]arenes inside the membrane, while the cation exchanger supports their absorption by the membrane. The increasing voltage between the two opposite faces of the membrane was found to depend on the concentration of the guanidinium cations. The sensors also have very good selectivity for guanidinium ions in competition with potassium, sodium, or ammonium ions.

Molecular recognition of methylguanidinium hydrochloride was used by Elizeev to control the composition of an equilibrating mixture of simple arginine receptors [134]. This experiment was set up as a model for mutation and selection, two major forces driving evolutionary processes. The simple receptor **44** occurs as three different isomers, the (trans,trans), (trans,cis), and (cis,cis) isomers. According to molecular modeling, only the (cis,cis) isomer of **44** has geometry suited for complexation with the guanidinium moiety of an arginine by formation of two salt bridges and

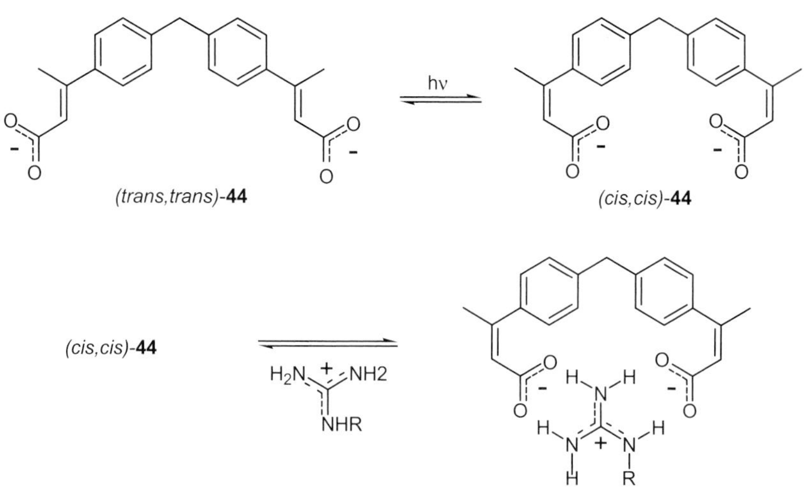

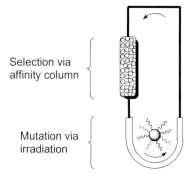

Figure 3.40 Photoisomerization of **44** and complex formation between guanidinium and the resulting (cis,cis) isomer of **44**.

four hydrogen bonds (Fig. 3.40). Indeed, in binding experiments the (cis,trans) and, in particular, (trans,trans) isomers had much lower affinity for guanidinium, because of the longer distance between their carboxylate tips.

The critical experiment was performed in the apparatus shown in Fig. 3.41. After the isomerization step exposing all isomers to broadband UV-light the mixture (cis,cis/cis,trans/trans/trans = 3:28:69) is pumped through an affinity column containing arginine immobilized on a silica gel support. While passing through the column most of the high-affinity cis,cis isomer was retained. After 30 circulations the concentration of **44** in solution dropped to 11 % of the total original value and an isomer ratio of cis,cis/cis,trans/*trans*/trans = 48:29:23 was detected for **44**. HPLC analysis of the mixture obtained from the affinity column yielded a distribution of isomers cis,cis/cis,trans/trans/trans = 85:13:2.

Selection via
affinity column

Mutation via
irradiation

Figure 3.41 Apparatus used by Elizeev in the photoisomerization experiment.

This method leads to considerable amplification of the total amount of the effective binder in the system and thus provides the option of isolating only those compounds with the desired binding affinity, and to separate them from byproducts or weak binders. Future applications might be envisaged in driving dynamic combinatorial pools of compounds toward particular subsets of structures.

Hunter recently designed a similar photosensitive receptor–guest system in which he used two photoisomerizable azobenzene moieties each bearing a carboxylate as the recognition motif [134a]. As Fig. 3.42 shows, only the cis,cis isomer of **45** has the correct conformation for chelate complexation of the guanidinium moiety. The thermodynamically favored trans,trans isomer was isomerized by irridiation to produce a 1:1:1 mixture of all three possible isomers. Of these the cis,cis isomer has tenfold higher affinity toward guanidinium ions than both other isomers, as proven experimentally by NMR titrations with guanidine hydrochloride.

A very interesting example of the use of carboxylate–guanidinium interactions was recently demonstrated by Yano et al. [135]. They found that melamine derivatives bearing a guanidinium ion can act as templates for self-assembly of a thymine-linked thiazolium ion next to a pyruvate via hydrogen bonding. In this ternary complex the rate of oxidative decarboxylation is markedly enhanced, probably because the guanidinium ion not only binds pyruvate but also activates it for initial nucleophilic attack

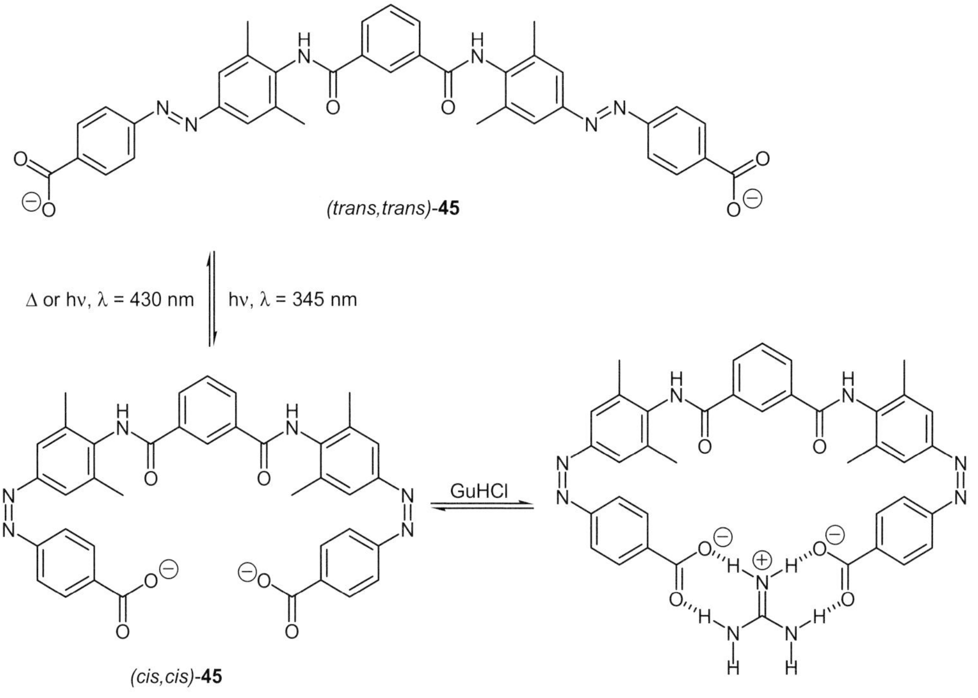

Figure 3.42 Photoisomerization of the (trans,trans) isomer of **45** into the corresponding (cis,cis) isomer, and subsequent formation of strong complexes with guanidine hydrochloride.

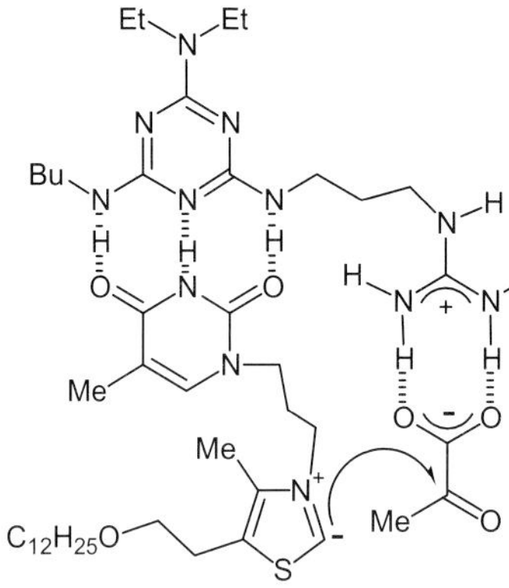

Figure 3.43 Noncovalently assembled system enabling acceleration of the rate of oxidative decarboxylation of pyruvate.

by the carbene (Fig. 3.43). The work presented is a first, very promising, step toward an artificial enzyme system mimicking pyruvate-decarboxylase activity.

3.4.3
Conclusion

The most important natural compounds bearing a guanidinium moiety are certainly the amino acid arginine and peptides derived therefrom. Consequently, development of new receptor systems has focused on this particular amino acid. An outstanding new representative is the highly selective so-called "arginine-cork" combining strong Coulombic interactions with an extended hydrogen-bond network for efficient guest differentiation. Even more important was the development of receptor systems for the RGD sequence, because of its biological relevance in integrin proteins. Because guanidinium phosphate or carboxylate ion pairs are involved in central recognition motifs for biological processes, the design of potent and specific receptor molecules will be of enormous pharmaceutical relevance. A plethora of diseases is connected with pathologic interaction of RGD-containing proteins and cell surfaces or arginine-rich regulatory proteins and viral RNA.

Very promising results where obtained with photoisomerizable recognition systems for guanidinium ions. Incorporation into switches or even more sophisticated artificial enzyme systems seems only a question of time.

3.5
Pyridinium Cations

3.5.1
Introduction

Flat aromatic pyridinium ions have been used as guests in many different artificial host molecules. The most important attractive force between these hosts and the pyridinium guest is often the π–cation interaction [136]. Although its importance has been overlooked for quite a long time, this interaction has now been rediscovered as a broadly applicable and efficient noncovalent binding force primarily from extensive study of protein crystal structure [137] and binding studies involving artificial, cyclophane receptors in aqueous media [138–140].

The interest in artificial pyridinium hosts is again because of their biochemical relevance. NADH and NADPH are two of the most important cofactors in enzymatic reductions involving formal hydride transfer [141]. The reverse process uses the oxidized form NAD$^+$ or NADP. In dehydrogenases pyridinium ions are noncovalently buried in a cleft called the Rossman fold [142–144]. The adenine and nicotinamide moieties are both enclosed in cavities with at least one hydrophobic face, facilitated by strong hydrophobic interactions and dispersive forces. Additional electrostatic interactions with metal ions and hydrogen bonds with basic amino acids are used to keep the phosphate anions in place.

Other important pyridinium cations are the viologens, which are used in such diverse applications as crop protection, electrochromism, solar energy conversion, molecular electronics, and supramolecular chemistry [145–147]. These are all appropriate guests for artificial pyridinium receptor molecules.

3.5.2
Artificial Receptors

In the late eighties Dougherty reported his above-mentioned cyclophane receptors which had extraordinarily high affinity for pyridinium derivatives in water (Fig. 3.39) [148]. The guests were bound by hydrophobic forces and electrostatic attraction inside the host cavity with remarkable association constants of up to 4×10^5 M^{-1}. Dougherty also demonstrated that less water-soluble guests are bound even more strongly if the more hydrophobic cyclohexyl receptor was used instead of the phenyl receptor. On the other hand the phenyl receptor produces higher association constants with cationic pyridinium derivatives, because of π–cation interactions. They also postulated an intriguing benzene-transmitted long-range electrostatic interaction between the carboxylate groups of the cyclophane host and its cationic guest. These assumptions were supported by computational studies on a sodium ion–benzene–chloride ion system [149].

The receptors both have another extraordinary property – they catalyze the alkylation and dealkylation of quinoline derivatives ($k_{cat}/k_{uncat} = 80$ with host **42** (phenyl)) [150]. The catalytic activity of these cyclophanes can be explained by preferential binding of the charged transition state. Because close proximity of the reaction partners or hydrophobic effects can be ruled out, the cation–π interaction is indeed a prime candidate as the source for such an effect. Both host molecules **42** and **43** have a high affinity for positively charged guests and, of course, the alkylation reaction develops a partial positive charge in the transition state. The π–cation effect is normally larger in host **42** than in host **43** and, moreover, host **42** is also the better catalyst.

When the authors compared their artificial alkylation catalysts with biologically important methylation enzymes involving SAM **46** (Fig. 3.44) as methylation agent [151, 152] they found transition states were similar for both systems. It was argued that this supports the assumption of π–cation and polarization effects at the active sites.

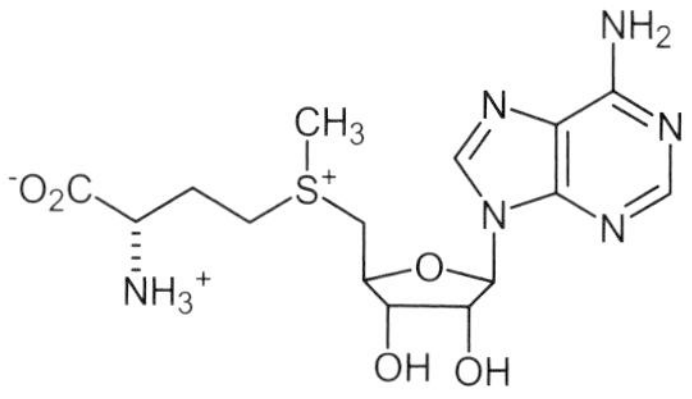

46

Figure 3.44 Structure of the methylation agent SAM.

A type of bicyclophane with a large spheroidal cavity was introduced by Lehn et al. (Fig. 3.45) [153]. Initially compound **47** was synthesized for complexation of quaternary ammonium ions, because of its potential complexation of the neurotransmitter acetylcholine [154, 155]. Although 1:1 complex formation was observed with several

Guest	logK$_{ass}$ with 47
Tetramethylammonium	3.0
N-Methylpyridinium	3.5
N-Methylquinoline	5.5
Methylviologen	6.6

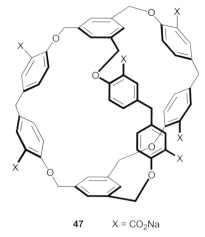

47   X = CO$_2$Na

Figure 3.45 Structure of a bicyclophane with a spheroidal cavity described by Lehn, and its binding constants in complexes with selected organic cations in water.

ammonium and pyridinium ions, extraordinarily high association constants compared with simple ammonium ions were measured for the viologen derivatives in water (Fig. 3.45). This increased affinity for flat aromatic guests can probably be explained by their higher hydrophobicity. Other effects, for example the π–cation interaction's preference for aromatic guests or strong π-stacking donor–acceptor interactions can be excluded if the binding constants for *N*-methylpyridinium (log $K_{ass} = 3.5$) and *N*-methylquinoline (log $K_{ass} = 5.5$) are considered. Nevertheless, host **47** has remarkably high affinity for other organic cations in water.

A different binding motif for the complexation of electron-poor aromatic compounds was introduced independently by Klärner and Nolte, who used molecular clips or molecular tweezers which bind the guest between their extended tips. Clip-shaped hosts based on glycoluril such as **48** were presented by Nolte [156, 157]. This kind of receptor molecule binds viologen and viologen polymers between their parallel sidewalls (Fig. 3.46). The force driving complexation is again the π–cation interaction between the positively charged, electron-deficient aromatic moiety of the guest and the electron-rich tips of the host.

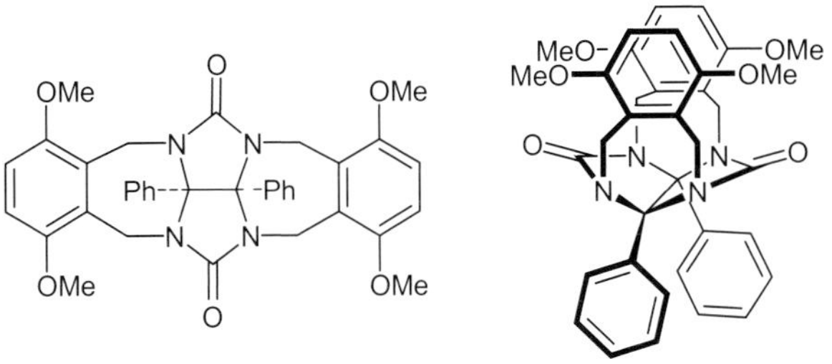

48

Figure 3.46 Lewis-structure of **48** and its side view.

The complexation process was monitored by mass spectrometry (FAB-MS), NMR spectroscopy, and viscosity measurements (for the paraquat polymers), and binding constants up to 5×10^4 M^{-1} were observed in acetonitrile. The electrochemical properties of free and complexed paraquat derivatives were also investigated. A remarkable shift of –100 mV for the first electron transfer of the complexed compared with uncomplexed paraquat was observed, indicating the increased stability of the guest after complexation. The first electron transfer destroys the complex because the guest loses its positive charge and thus its ability to exert π–cation attraction. Therefore, the second electron transfer of the "complexed" paraquat is of the same size as for uncomplexed paraquat. The fact that such a guest is released after single-electron oxidation has already been observed [158, 159]. For the electrochemically tuned complexed polymers in particular a potential application as optical information storage material or as a molecular switch is proposed.

Another interesting derivative of these diphenylglycoluril-derived clips forms liquid crystalline complexes on addition of an aromatic guest. Nolte even adorned glycoluril clips with porphyrins and showed that they accommodate *N*-alkylpyridinium

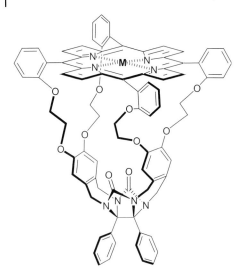

Figure 3.47 Diphenylglycouril-derived clips modified with porphyrins, synthesized by the Nolte group.

cations between their tips (Fig. 3.47) [160]. Unfortunately, these second generation host molecules also bind more electron-rich aromatic guests such as pyridine, hydroxypyridine, and phenols.

Klärner et al. synthesized molecular tweezers with an aromatic backbone (Fig. 3.48) which bind neutral and positively charged electron-poor aromatic guests via π–π or multiple π–cation interactions [161]. The guests (pyridinium and viologen derivatives) are complexed between both sidewalls of the host, leading to a drastic upfield shift of specific guest protons (up to 6.0 ppm) in the ^{1}H NMR spectra. The bind-

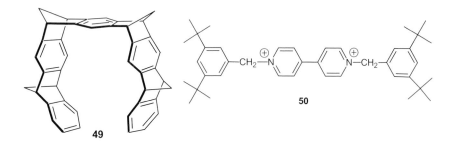

Figure 3.48 Molecular tweezers **49** and its guests **50** and **51**.

ing constants (up to 130 M^{-1}) were measured in organic solvents of low polarity; however, in these, hydrophobic interactions remain small.

Detailed studies were conducted on the mechanism of complexation of the viologen derivatives **50** and **51** with host **49**. The end groups of these guests are too bulky to enable complexation by threading the host **49** over the side chains of **50** and **51**. The observed complexation of these guests must therefore occur by an alternative mechanism, most probably a clipping process through the tips of the tweezers over the alkane chains or the paraquat units of the substrates. Indeed, computational studies have shown that the tips can widen the cleft by approximately 2.81 Å, requiring only 4.2 kcal mol^{-1} extra strain energy. The attempt to synthesize the clipped-rotaxane-type supermolecule therefore failed, because after attaching the stoppers to the complexed paraquat the supermolecule did not have thermodynamic stability, resulting in a temperature-dependent mixture of complexed and uncomplexed paraquat derivative.

In a joint effort Klärner and Schrader modified related clips by introducing two phosphonate moieties into their aromatic spacer (**52**, Fig. 3.49) [162].These new water-soluble hosts were tested with many different guests, among which ammonium ions, including guanidinium and imidazolium ions, were bound weakly in DMSO and not at all in water. The complexation-induced shifts for the guest protons remain very small, indicating that the guests are not bound inside the host's cleft between the tips but most likely above the aromatic bridge grasped by both phosphonate arms. In contrast, alkylated pyridinium ions are excellent guests for the new clip. Very large chemically induced upfield shifts are observed reaching maximum values of up to +3.5 ppm at saturation. In methanol, strong binding occurred between the clip and a variety of pyridinium species with association constants in the range 10^3–10^4 M^{-1}. Even stronger binding was observed in water, pointing clearly to the importance of hydrophobic interactions. The formation of 1:1 complexes was proven by Job plots and ESI-MS spectra.

Substrate	K_a [M^{-1}]^{a,b} in CD$_3$OD	K_a [M^{-1}]^{a,b} in D$_2$O
Kosower salt	3200 ± 350	4800 ± 1300
Pyrazinium salt	600 ± 30	9400 ± 880
Nicotinamide	6900 ± 540	12700 ± 5800
Adenosine	–	5300 ± 320
NAD$^+$	precipitation	6500 ± 710
NADP	–	1400 ± 200

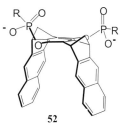

52

Figure 3.49 Phosphonate-modified tweezer **52** and its binding constants with selected organic cations.

The proposed complexation mode of the NAD$^+$ guest inside the clip **52**, found by Monte-Carlo simulations, is shown in Fig. 3.50. Interestingly enough, the nicotinamide moiety is preferentially included inside the host's cavity whereas the adenosine ring can simultaneously be placed on top of one of the host sidewalls.

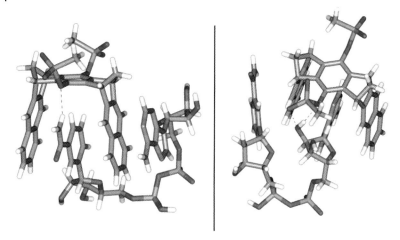

Figure 3.50 Two Monte-Carlo simulations for the complex between NAD⁺ and clip **52**. Note the overall similarity of both inclusion complexes; according to NMR, a) is formed predominantly.

One potential application of this new receptor species with NAD⁺ affinity involves its electrochemistry. By embracing the catalytically active nicotinamide site the redox properties of NAD⁺ will be modified. In addition, it is planned to create purely artificial enzyme models from the clips with noncovalently bound zinc ions and complexed NAD⁺, in the complete absence of any protein.

Tweezers and clips are not the only binding motif for cations relying on π–π and π–cation interactions. The very well investigated calix[*n*]arenes use the same interactions for complexation of organic and inorganic cations. These are an important class of artificial pyridinium receptor, because π-basic cavities of varying size and shape can be created [163–168]. Shinkai investigated calix[*n*]arenes and compared the magnitudes of the relative upfield shifts in the NMR spectrum and the relative peak intensity in mass spectrometry to determine which cavity size was ideal for pyridinium binding. Interestingly, conflicting results were produced by these two methods. In solution (NMR) K_{ass} were in the order $n = 8 > 4 > 6$ (guest = *N*-methylpyridinium) whereas in the gas phase (MS) the corrected relative peak intensities were in the order $n = 8 > 6 \gg 4$. These differences can be explained by the effect of solvent molecules on the binding process. In solution the guest inclusion into the host cavity is accompanied by a release of solvent molecules, resulting in increased entropy but also consuming energy for desolvation. In the example discussed above the latter effect seems to play the dominant role leading to a reduced binding ability for calix[6]arenes in solution. In contrast, solvent effects certainly do not affect host–guest interactions in the gas phase.

Self-assembled hydrogen-bonded molecular cages of calix[6]arene carboxylic acid derivatives were reported by Secchi et al. Two calix[6]arenes, both with three carboxylic acid moieties on their upper rims, form stable but, of course, reversible capsules in chloroform (Fig. 3.51). During binding experiments with *N*-methyl-4-picol-

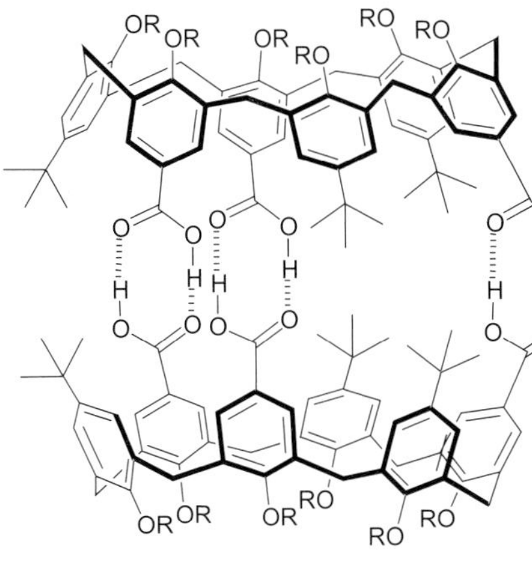

R = Me, n-C$_{18}$H$_{17}$

Figure 3.51 Dimeric hydrogen-bonded capsule formed between two calix(6)arenes.

inium iodide the authors observed a large upfield shift for all the guest protons caused by the anisotropy of the concave aromatic host moieties; this is strongly indicative of endo-cavity complexation and furnished a binding constant of 230 M^{-1}. That N-methyl-2-picolinium iodide is not bound indicates the high sensitivity of the complexation process to the size and shape of the guest.

Encouraged by this success, Secchi et al. synthesized dimeric calix[6]arenes covalently linked at their upper rims. These were designed as carrier systems which should be able to transport cations through membranes. Two of the three capsules

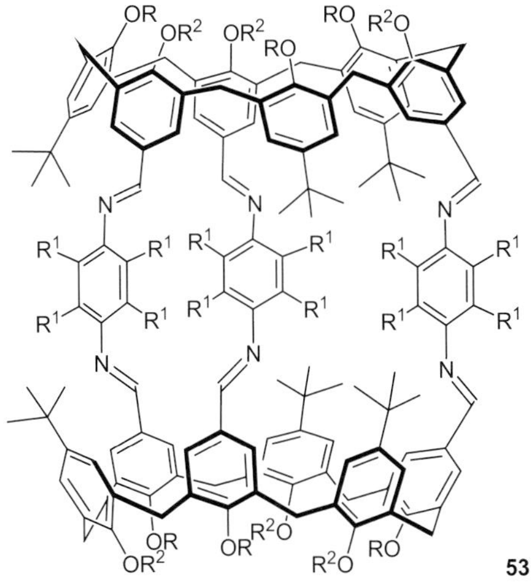

53

R = Me; R^1= H; R^2= C$_8$H$_{17}$

Figure 3.52 Structure of the covalently-linked dimeric capsule synthesized by Secchi.

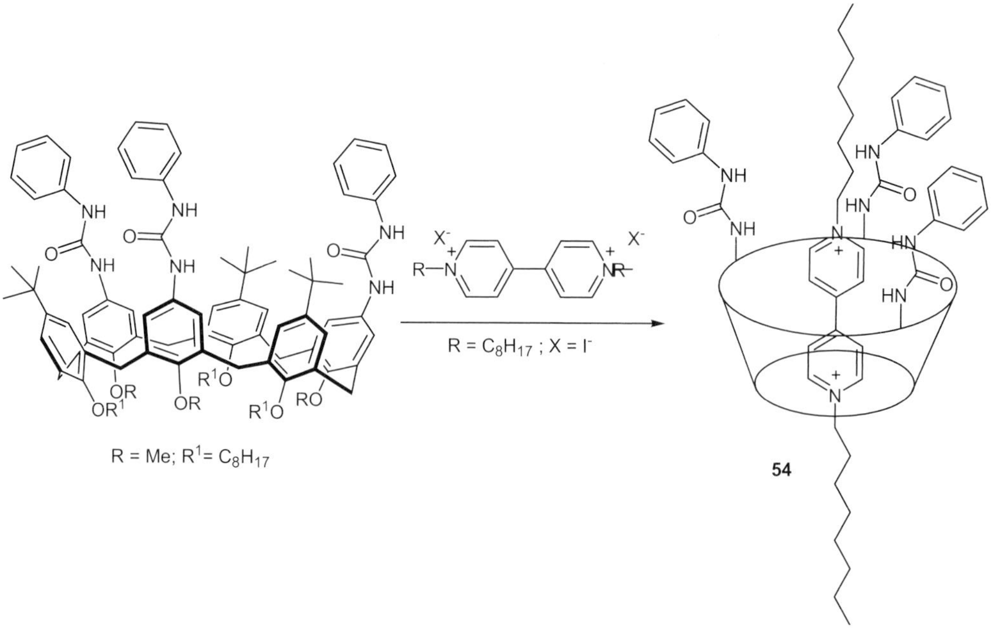

Figure 3.53 Synthesis of the pseudorotaxane **54**.

had no recognition properties for *N*-methylpicolinium iodide derivatives, because of steric hinderance, but capsule **53** binds *N*-methyl-4-picolinium iodide with an association constant of ~890 M^{-1} in chloroform (Fig. 3.52). Again all guest protons undergo a strong upfield shift in the ^1H NMR spectrum, indicative of complexation inside the cavity.

In another attempt Secchi et al. prepared a rotaxane from a calix[6]arene and a paraquat derivative (Fig. 3.53). Efficient complexation of the paraquat cation by the calix[6]arene core was a key prerequisite. The high stability of this complex is proven by the fact that the pseudorotaxane **54** is stable even during thin layer chromatography with aprotic solvents. Evidence for the proposed structure **54** came from NMR, MS, and elemental analysis.

Kim recently reported the synthesis and binding studies of multi-calix[4]arenes. In particular the double calix[4]arenes **55–59** (Fig. 3.54) were tested in binding experiments (NMR titrations) with paraquat derivatives as guests. Interestingly, only the double calix[4]arene **55** with a thiophene linker includes the guests; association constants were up to 7×10^2 M^{-1} in a 2:1 chloroform–methanol mixture. The fact that uncharged viologen derivatives were not bound by **55** indicates that the driving force for complexation is again the π–cation interaction and not simply a π–π interaction. It is argued that the extremely electron-rich thiophene linker makes the best host for electron-deficient guests. This is quite surprising in the light of Dougherty's observations on related sytems. He found that thiophene linkers in his cyclophanes did not lead to higher binding constants than simple aromatic linkers. Dougherty claimed that it is dangerous to extrapolate conventional criteria for electron-richness, which

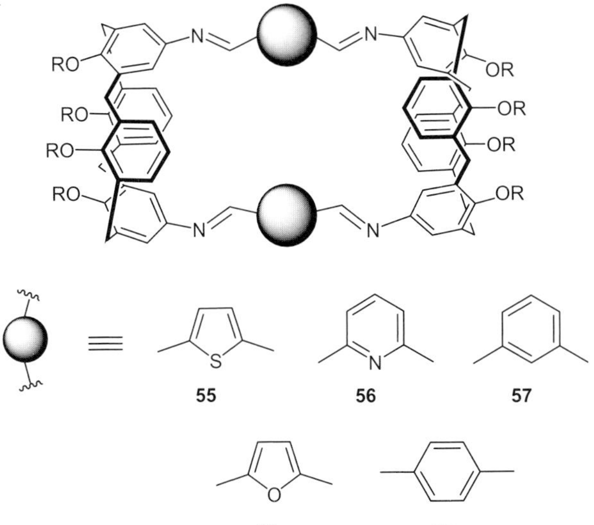

Figure 3.54 Covalently-linked capsules **55-59** synthesized from dimeric calix[4]arenes.

are usually based on reactivity patterns, to binding events. For supramolecular interactions the ground-state wave functions of the aromatic groups are most important.

3.5.3
Conclusion

During the last two decades a fairly large number of new receptor molecules, with many different recognition motifs, have been synthesized for molecular recognition of pyridinium cations. The best studied system with the widest range of application is probably the cyclophane system introduced by Dougherty; this can be used not only as an efficient pyridinium receptor molecule but also as a catalyst for the alkylation or dealkylation of quinolines. Even more interesting will be future approaches toward artificial enzymes in which pyridinium receptors are not used as catalysts but rather to direct a cofactor such as NAD$^+$ into the catalytic pocket. In summary, the trend is leading from pyridinium receptor systems for guests of pure academic interest to those with potential biomedical applications.

3.6
Conclusions and Outlook

There is a clear trend toward nonrational synthetic pathways for the development of new artificial receptors. On the one hand, evolutive methods have been adopted from molecular biology and will be applied to many areas of molecular recognition, whenever high sensitivity and selectivity are desired. On the other hand, combinatorial libraries have just begun to combine the task of receptor synthesis and selection in dynamic equilibria. Large emphasis will be placed on new host species capable of binding biologically relevant cationic guests under physiological conditions. Carriers, switches, and gates will be developed for drugs, sensors, and markers, preferably in water or across lipid bilayers. Artificial signal transduction across membranes triggered by messengers is overdue [169]! The same is true for molecular machines and motors, using the ubiquitous *alkylammonium* cations. Synthetic ammonium receptors, which are among the oldest known artificial host systems, and are hence well studied both experimentally and theoretically, will lead the way toward new applications and functions of supramolecular entities.

Although numerous drugs are based on the *amidinium* moiety, this area has been largely neglected in the past. However, analytical and medicinal applications of bisamidine receptors have been studied in recent years. Because these are important drugs for AIDS patients, further improvements in transport, controlled release, and detection of bisamidines will be challenges for functional synthetic amidinium receptors in the future.

Much effort has been devoted to the development of *guanidinium* receptors, because of the paramount importance of arginine in biological processes. Although the best artificial receptor seems to be the "arginine cork" developed by Bell et al., immobilized calixarenes have been presented as sensor arrays, photoswitchable hosts can be used in continuously driven molecular evolution devices, and the first RGD-selective molecular receptors promise future medicinal applications by preventing pathologic integrin-mdiated cell–cell recognition.

Pyridinium guests are found in NAD$^+$ and paraquats (viologens). Many systems have been developed for their efficient binding, often using π–cation interactions. Selectivity is best with Klärner's clips, which bind pyridinium derivatives such as NAD$^+$ selectively in water. Interference with enzymatic processes is envisaged. The broad guest spectrum of Dougherty's cyclophanes has spurred intense theroetical investigations and deepened our understanding of the π–cation interaction. Transport of pyridinium derivatives across membranes and redox-tunable pyridinium receptors pave the way toward practical applications of these new receptor systems.

Acknowledgment. We thank Mrs. A Bamberger for the structural drawings in this chapter.

References

1 Early monographs: (a) M. Hiraoka, *Crown Compounds. Their Characteristics and Applications*, Kodansha, Tokyo, **1978**; (b) *Crown Ethers and Analogs* (Eds.: S. Patai and Z. Rappoport), J. Wiley, New York, **1989**; (c) *Cation Binding by Macrocycles* (Eds.: Y. Inoue, G. Gokel), Marcel Dekker, New York, **1990**; (d) G. Gokel, *Crown Ethers and Cryptands*, RSC, Cambridge, **1991**.

2 Nobel lectures in Chemistry **1987** by C. J. Pedersen, D. J. Cram, J.-M. Lehn: (a) *Nobel Lectures*, Chemistry 1981–1990, World Scientific Publishing Co., Singapore; (b) *Angew. Chem.* **1988**, *27*, 89, 1009, 1053.

3 (a) F. Kotzyba-Hibert, J.-M. Lehn, K. Saigo, *J. Am. Chem. Soc.* **1981**, *103*, 4266; (b) A. D. Hamilton, J.-M. Lehn, J. Sessler, *J. Chem. Soc. Chem. Commun.*, **1984**, 311.

4 Separation of dopamine from alkali metal ions: H. Tsukube, *Tetrahedron Lett.* **1982**, *23*, 2109.

5 T. M. Fyles, V. A. Malik-Diemer, C. A. McGavin, D. M. Whitfield, *Can. J. Chem.* **1982**, *60*, 2259.

6 (a) J.-P. Behr, J.-M. Lehn, P. Vierling, *Helv. Chim. Acta* **1982**, *65*, 1853–1867; (b) J.-M. Lehn, *Pure Appl. Chem.* **1978**, *50*, 871; (c) D. J. Cram in: *Application of Biochemical Synthesis in Organic Systems*, part II (Eds.: J. B. Jones, C. J. Sih, D. Perlman), Techniques of Chemistry, Vol. X, Wiley, New York, **1976**, p. 815.

7 D. J. Cram, J. M. Cram, *Acc. Chem. Res.* **1978**, *11*, 8.

8 M. Newcomb, J. L. Toner, R. C. Helgeson, D. J. Cram, *J. Am. Chem. Soc.* **1979**, *101*, 4941.

9 Inter alia: D. J. Cram, J. M. Cram, *Container Molecules and their Guests*, RSC, Cambridge, **1994**; J.-M. Lehn, Supramolecular Chemistry, Verlag Chemie, Weinheim, **1995**.

10 R. W. Saalfrank, B. Demleitner, H. Glaser, H. Maid, S. Reihs, W. Bauer, M. Maluenga, F. Hampel, M. Teichert, H. Krautscheid, *Eur. J. Inorg. Chem.* **2003**, 822–829.

11 (a) K.-H. Ahn, S.-G. Kim, J. Jung, J. Kim, J. Chin, K. Kim, *Chem. Lett.* **2000**, 170–171; (b) tripodal pyrazole receptors: J. Chin, C. Walsdorff, B. Stranix, J. Oh, H. J. Chung, S.-M. Park, K. Kim, *Angew. Chem.*

1999, *111*, 2923–2926; *Angew. Chem. Int. Ed. Engl.* **1999**, *38*, 2756–2759.

12 (a) P. Bühlmann, E. Pretsch, E. Bakker, *Chem. Rev.* **1998**, *98*, 1593–1687; (b) P. Kroogsgaard-Larsen, in *Comprehensive Medicinal Chemistry, Vol. 3* (Eds.: P. G. Sammes, J. B. Taylor, J. C. Emmett), Pergamon, Oxford, **1990**, pp. 493–537.

13 (a) S.-K. Chang, M. J. Jang, S.-Y. Han, *Chem. Lett.* **1992**, 1937–1940; (b) S.-Y. Han, M.-H. Kan, Y.-E. Jung, S.-K. Chang, *J. Chem. Soc. Perkin 2* **1994**, 835–839; (c) F. Arnaud-Neu, S. Fuangswasdi, A. Notti, S. Pappalardo, M. F. Parisi, *Angew. Chem.* **1998**, *110*, 120–122; *Angew. Chem. Int. Ed. Engl.* **1998**, *37*, 112–114.

14 T. W. Bell, A. B. Khasanov, M. G. B. Drew, A. Filikov, T. James, *Angew. Chem. Int. Ed.* **1999**, *38*, 2543–2547.

15 S. Tamaru, M. Yamamoto, S. Shinkai, A. B. Khasanov, T. W. Bell, *Chem. Eur. J.* **2001**, *7*, 5270–5276.

16 S. Topiol, *Chirality*, **1989**, *1*, 69–79.

17 (a) U. Neidlein, F. Diederich, *J. Chem. Soc. Chem. Commun.* **1996**, 1493–1494; (b) A. Bähr, B. Felber, K. Schneider, F. Diederich, *Helv. Chim. Acta* **2000**, *83*, 1346–1376.

18 V. Král, O. Rusin, F. P. Schmidtchen, *Org. Lett.* **2001**, *3*, 873–876.

19 T. W. Bell, A. B. Khasanov, M. G. B. Drew, *J. Am. Chem. Soc.* **2002**, *214*, 14092–14103.

20 T. Schrader, *Angew. Chem.* **1996**, *108*, 2816.

21 T. Schrader, *J. Org. Chem* **1998**, *63*, 264–272.

22 T. Schrader, *J. Am. Chem. Soc.* **1998**, *120*, 11816–11817.

23 T. Schrader, M. Herm, *Chem. Eur. J.* **2000**, *6*, 47–53.

24 T. Schrader, M. Herm, O. Molt, *Angew. Chem.* **2001**, *113*, 3244–3248.

25 T. Schrader, M. Herm, O. Molt, Full Paper, *Chem. Eur. J.* **2002**, *8*, 1485–1499.

26 O. Molt, D. Rübeling, T. Schrader, *J. Am. Chem. Soc.* **2003**, *125*, 12086–12087.

27 O. Molt, T. Schrader, *Angewandte Chem.* **2003**, *115*, 5667–5671.

28 (a) M.-F. Paugam, L. S. Valencia, B. Boggess, B. D. Smith, *J. Am. Chem. Soc.* **1994**, *116*, 11203; (b) M.-F. Paugam, J. T. Biens, B. D. Smith, A. J. Christoffels, F. de Jong,

D. N. Reinhoudt, *J. Am. Chem. Soc.* **1996**, *118*, 9820–9825.

29 B. Escuder, A. E. Rowan, M. C. Feiters, R. J. M. Nolte, *Tetrahedron* **2004**, *60*, 291–300.

30 (a) J.-M. Lehn, *Chem. Eur. J.* **1999**, *5*, 2455–2463; (b) G. R. L. Cousins, S.-A. Poulsen, J. K. M. Sanders, *Curr. Opin. Chem. Biol.* **2000**, *4*, 270–279.

31 G. R. L. Cousins, R. L. E. Furlan, Y.-F. Ng, J. E., Redman, J. K. M. Sanders, *Angew. Chem. Int. Ed.* **2001**, *40*, 423–428.

32 (a) S. Kubik, R. Goddard, *J. Org. Chem.* **1999**, *64*, 9475–9486; (b) S. Kubik, *J. Am. Chem. Soc.* **1999**, *121*, 5846–5855.

33 F. Hof, L. Trembleau, E. C. Ullrich, J. Rebek, Jr., *Angew. Chem. Int. Ed.* **2003**, *42*, 3150–3153.

34 (a) A. Scarso, L. Trembleau, J. Rebek, Jr., *Angew. Chem. Int. Ed.* **2003**, *42*, 5499–5502; (b) L. Trembleau, J. Rebek, Jr., *Science* **2003**, *301*, 1219–1220.

35 S. M. Ngola, P. C. Kearney, S. Mecozzi, K. Russell, D. A. Dougherty, *J. Am. Chem. Soc.* **1999**, *121*, 1192–1201.

36 The cation–p interaction: (a) D. A. Dougherty, *Science* **1996**, *271*, 163–168; (b) J. C. Ma, D. A. Dougherty, *Chem. Rev.* **1997**, *5*, 1303–1324 (Special issue: Molecular Recognition).

37 V. Rüdiger, H.-J. Schneider, V. P. Solov'ev, V. P. Kazachenko, O. A. Raevsky, *Eur. J. Org. Chem.* **1999**, 1847–1856.

38 S. K. Burley, G. A. Petsko, *FEBS Lett.* **1986**, *203*, 139–143; J. B. O. Mitchell, C. L. Nandi, I. K. McDonald, J. Thornton, S. L. Price, *J. Mol. Biol.* **1994**, *239*, 315–331; S. Karlin, M. Zuker, L. Brocchieri, *J. Mol. Biol.* **1994**, *239*, 227–248; A. M. de Vos, M. Ultshc, A. A. Kossiakoff, *Science* **1992**, *255*, 306–312.

39 (a) S. Mecozzi, A. P. West, Jr., D. A. Dougherty, *J. Am. Chem. Soc.* **1996**, *118*, 2307–2308; (b) S. Mecozzi, A. P. West, Jr., D. A. Dougherty, *Proc. Natl. Acad. Sci. USA* **1996**, *93*, 10566–10571; (c) H.-J. Schneider, T. Schiestel, P. Zimmerman, *J. Am. Chem. Soc.* **1992**, *114*, 7698–7703; H.-J. Schneider, *Chem. Soc. Rev.* **1994**, 227–234.

40 (a) T. J. Shepodd, M. A. Petti, D. A. Dougherty, *J. Am. Chem. Soc.* **1986**, *108*, 6085–6087; (b) D. A. Dougherty, D. A. Stauffer, *Science* **1990**, *250*, 1558–1560.

41 J. P. Gallivan, D. A. Dougherty, *J. Am. Chem. Soc.* **2000**, *122*, 870–874.

42 D. L. Beene, G. S. Brandt, W. Zhong, N. M. Zacharias, H. A. Lester, D. A. Dougherty, *Biochemistry* **2002**, *41*, 10262–10269.

43 E. J. Petersson, A. Choi, D. S. Dahan, H. A. Lester, D. A. Dougherty, *J. Am. Chem. Soc.* **2002**, *124*, 12662–12663.

44 M. Mammen, S.-K. Choi, G. M. Whitesides, *Angew. Chem.* **1998**, *37*, 2754–2794.

45 Tripledecker complexes: J. P. Mathias, E. E. Simanek, C. T. Seto, G. M. Whitesides, *Angew. Chem. Int. Ed.* **1993**, *32*, 1766.

46 Rosette assemblies: V. Paraschiv, M. Crego-Calama, P. Timmerman, D. N. Reinhoudt, *J. Org. Chem.* **2001**, *66*, 8297–8301.

47 Self-instructed helices: U. Koert, M. M. Harding, J.-M. Lehn, *Nature* **1990**, *346*, 339.

48 J. Rao, G. M. Whitesides, *J. Am. Chem. Soc.* **1997**, *119*, 10286–10290.

49 J. Rao, J. Lahiri, L. Isaacs, R. M. Weis, G. M. Whitesides, *Science* **1998**, *280*, 708–711.

50 (a) J. M. J. Frechet, *Proc. Natl. Acad. Sci. USA* **2002**, *99*, 4782–4787; (b) A. W. Bosmann, H. M. Janssen, E. W. Meijer, *Chem. Rev.* **1999**, *99*, 1665–1688; (c) F. Zeng, S. C. Zimmerman, *Chem. Rev.* **1997**, *97*, 1681–1712.

51 H. Xu, G. R. Kinsel, J. Zhang, M. Li, D. M. Rudkevich, *Tetrahedron* **2003**, *59*, 5837–5848.

52 Review: F. Diederich, B. Felber, *Proc. Natl. Acad. Sci. U S A* **2002**, *99*, 4778–4781.

53 D. L. Dermody, R. F. Peez, D. E. Bergbreiter, R. M. Crooks, *Langmuir* **1999**, *15*, 885–890.

54 A.N. Shipway, E. Katz, I. Willner, *ChemPhysChem* **2000**, *1*, 18.

55 T. M. Fyles, *Curr. Opin. Chem. Biol.* **1997**, *1*, 497–505.

56 (a) D. A. Doyle, J. M. Cabral, R. A. Pfuetzner, A. Kuo, J. M. Gulbis, S. L. Cohen, B. T. Chait, R. MacKinnon, *Science* **1998**, *280*, 69–77; (b) R. MacKinnon, S. L. Cohen, A. Kuo, A. Lee, B. T. Chait, Science **1998**, *280*, 106–109.

57 T. W. Bell, *Curr. Opin. Chem. Biol.* **1998**, *2*, 711–716.

58 J. L. Sessler, A. Andrievsky, *Chem. Eur. J.* **1998**, *4*, 158–167.

59 K. Tohda, S. Yoshiyagawa, M. Kataoka, K. Odashima, Y. Umezawa, *Anal. Chem.* **1997**, *69*, 3360–3369.

60 (a) A. Shivanyuk, E. F. Paulus, V. Böhmer, *Angew. Chem. Int. Ed.* **1999**, *38*, 2906–2909; (b) C. A. Schalley, J. M. Rivera, T. Martin, J. Santamaria, G. Siuzdak, J. Rebek, Jr., *Eur. J. Org. Chem.* **1999**, 1325–1331.

61 (a) C. A. Schalley, R. K. Castellano, M. S. Brody, D. M. Rudkevich, G. Siuzdak, J. Rebek, Jr., *J. Am. Chem. Soc.* **1999**, *121*, 4568–4579; (b) C. A. Schalley, T. Martin, U. Obst, J. Rebek, Jr, *J. Am. Chem. Soc.* **1999**, *121*, 2133–2138.

62 (a) R. Zadmard, T. Schrader, T. Grawe, A. Kraft, *Org. Lett.* **2002**, *4*, 1687–1690; (b) R. Zadmard, M. Junkers, T. Schrader, T. Grawe, A. Kraft, *J. Org. Chem.* **2003**, *68*, 6511–6521.

63 F. Corbellini, R. Fiammengo, P. Timmerman, M. Crego-Calama, K. Versluis, A. J. R. Heck, I. Luyten, D. N. Reinhoudt, *J. Am. Chem. Soc.* **2002**, *124*, 6569–6575.

64 D. L. Caulder, R. E. Powers, T. N. Parac, K. N. Raymond, *Angew. Chem. Int. Ed.* **1998**, *37*, 1840–1843.

65 G. M. Dykes, D. K. Smith, *Tetrahedron* **2003**, *59*, 3999–4009.

66 T. Grawe, T. Schrader, P. Finocchiaro, G. Consiglio, S. Failla, *Org. Lett.* **2001**, *3*, 1597–1600.

67 M. A. Hossain, H.-J. Schneider, *J. Am. Chem. Soc.* **1998**, *120*, 11208–11209.

68 M. Sirish, H.-J. Schneider, *Chem. Commun.* **1999**, 907–908.

69 M. Sirish, V. A. Chertkov, H.-J. Schneider, *Chem. Eur. J.* **2002**, *8*, 1181–1188.

70 T. Mizutani, K. Wada, S. Kitagawa, *J. Am. Chem. Soc.* **1999**, *121*, 11425–11431.

71 L. Kjellen, U. Lindahl, *Annu. Rev. Biochem.* **1991**, *60*, 443.

72 (a) Genelabs Techologies Inc., *Patent US* 93-07441, WO 94-03165 ; (b) D. J. S. Hulmes, E. Aubert-Foucher, A. W. Coleman, *Patent FR* 98-10074.

73 H. E. Conrad, *Heparin-binding proteins*, Academic Press, San Diego, **1997**.

74 J.-M. Lehn, R. Meric, J. P. Vigneron, M. Cesario, J. Guilhem, C. Pascard, Z. Asfari, J. Vicens, *Supramol. Chem.* **1995**, *5*, 97.

75 N. Douteau-Guével, F. Perret, A. Coleman, J.-P, Morel, N. Morel-Desrosiers, *J. Chem. Soc. Perkin Trans. 2*, **2002**, 524–532.

76 (a) M. B. H. Grote Gansey, A. S. de Haan, E. S. Bos, W. Verboom, D. N. Reinhoudt, *Bioconjugate Chem.* **1999**, *10*, 613; (b) M. Mazzorana, G. Esposito: private communication.

77 R. Zadmard, Ph. D. thesis, Marburg University, Germany, **2003**.

78 R. Zadmard, T. Schrader, submitted to *J. Am. Chem. Soc.*

79 A. Ulman, *An Introduction to Ultrathin Organic Films from Langmuir–Blodgett to Self-assembly*, Academic Press, New York, **1991**.

80 E. Pechkova, C. Nicolini, *J. Cell. Biochem.* **2002**, *85*, 243–251.

81 For a general overview about peptide and protein receptors, including those of the Hamilton group, see: M. W. Peczuh, A. D, Hamilton, *Chem. Rev.* **2000**, *100*, 2479–2494.

82 Y. Hamuro, M. C. Calama, H. S. Park, A. D. Hamilton, *Angew. Chem. Int. Ed. Engl.* **1997**, *36*, 2680–2683.

83 Q. Lin, H. S. Park, Y. Hamuro, C. S. Lee, A. D. Hamilton, *Biopolymers* **1998**, *47*, 285–297.

84 H. S. Park, Q. Lin, A. D. Hamilton, *J. Am. Chem. Soc.* **1999**, *121*, 8–13.

85 R. K. Jain, A. D. Hamilton, *Org. Lett.* **2000**, *2*, 1721–1723.

86 K. K. Clark-Ferris, J. Fisher, *J. Am. Chem. Soc.* **1985**, *107*, 5007–5008.

87 Plenary lecture at „Synthetic Receptors 2003", Lisboa, Portugal, Oct 14–18, 2003.

88 M. Benezra, I. Vlodavsky, A. Yayon, R. Bar-Shavit, J. Regan, M. Chang, S. Ben-Sasson, *Cancer Res.* **1992**, *52*, 5656–5662.

89 J. Regan, D. McGarry, J. Bruno, D. Green, J. Newman, C.-Y. Hsu, J. Kline, J. Barton, J. Travis, Y. M. Choi, F. Volz, H. Pauls, R. Harrison, A. Zilberstein, A. A. Ben-Sasson, M. Chang, *J. Med. Chem.* **1997**, *40*, 3408–3422.

90 W. Bode, D. Turk, J. Stürzebecher, *Eur. J. Biochem.* **1990**, *193*, 175–182.

91 K. Hilpert, J. Ackermann, D. W. Banner, A. Gast, K. Gubernator, P. Hadváry, L. Labler, K. Müller, G. Schmid, T. B. Tschopp, H. v. d. Waterbeemd, *J. Med. Chem.* **1994**, *37*, 3889–3901.

92 H. U. Stilz, W. Guba, B. Jablonka, M. Just, O. Klingler, W. König, V. Wehner, G. Zoller, *J. Med. Chem.* **2001**, *44*, 1158–1176.

93 W. P. Gluth, G. Kaliwoda, O. Dann, *J. Chromatogr.* **1986**, *378*, 183–193.

94 R. J. Grout, In *The Chemistry of Amidines and Imidates*, S. Patai, Ed.: John Wiley, New York, 1975, Chapter 6, pp 263–264.

95 D. R. Phillips, I. F. Charo, L. V. Parise, L. A. Fitzgerald, *Blood* **1988**, *71*, 831–843.

96 J. Hawiger, *J. Hum. Pathol.* **1987**, *18*, 111.

97 A. Kraft, A. Reichert, *Tetrahedron* **1999**, *55*, 3923–3930.

98 A. Kraft, L. Peters, H. R. Powell, *Tetrahedron* **2002**, *58*, 3499–3505.

99 T. W. Bell, V. J. Santora, *J. Am. Chem. Soc.* **1992**, *114*, 8300–8302.

100 D. G. Brown, M. R. Sanderson, E. Garman, S. Neidle, *J. Mol. Biol.* **1992**, *226*, 481–490.

101 C. A. Laughton, F. Tanious, C. M. Nunn, D. W. Boykin, W. D. Wildon, S. Neidle, *Biochemistry* **1996**, *35*, 5655–5661.

102 T. Grawe, G. Schäfer, T. Schrader, *Org. Lett.* **2003**, *5*, 10, 1641–1644.

103 B. Sellergren, *Anal. Chem.* **1994**, *66*, 1578–1582.

104 G. Wulff, In *Polymeric Reagents and Catalysts*, W. T. Ford, Ed.: ACS Symposium Series 308, American Chemical Society, Washington DC, 1986.

105 G. Wulff, W. Vesper, R. Grobe-Einsler, A. Sarhan, *Makromol. Chem.* **1977**, *178*, 2799–2816.

106 K. J. Shea, T. K. J. Dougherty, *J. Am. Chem. Soc.* **1986**, *108*, 1091.

107 G. Wulff, B. Heide, G. J. Helfmeier, *J. Am. Chem. Soc.* **1986**, *108*, 1089.

108 L. Andersson, B. Sellergren, K. Mosbach, *Tetrahedron Lett.* **1984**, *25*, 5211.

109 S. C. Zimmermann, Y. Wang, P. Bharathi, J. S. Moore, *J. Am. Chem. Soc.* **1998**, *120*, 2172–2173.

110 A. D. Ellington, *Curr. Biol.* **1993**, *3*, 375–377.

111 G. Viriani, *Acc. Chem. Res.* **1997**, *30*, 189–195.

112 B. J. Calnan, B. Tidor, s. Biancalana, D. Hudosn, A. D. Frankel, *Science* **1991**, *252*, 1167–1171.

113 A. Mujeeb, T. G. Parslow, Y.-C. Yuan, T. L. James, *J. Biomol. Struct. Dyn* **1996**, *13*, 649–659.

114 J. L. Battiste, H. Mao, N. S. Rao, R. Tan, D. R. Muhandiram, L. E. Kay, A. D. Frankel, J. R. Williamson, *Science* **1996**, *273*, 1547–1551.

115 A. E. Aplin, A. K. Howe, R. L. Juliano, *Curr. Opin. Cell. Biol.* **1999**, *11*, 737–744.

116 C. C. Kumar, L. Armstrong, Z. Yin, M. Malkowski, E. Maxwell, H. Ling, B. Yaremko, M. Liu, J. Varner, E. M. Smith, B. Neustadt, T. Nechuta, *Adv. Exp. Med. Biol.* **2000**, *476*, 169–180.

117 G. P. Curley, H. Blum, M. J. Humphries, *Cell Mol. Sci.* **1999**, *56*, 427–441.

118 J.-P. Xiong, T. Stehle, B. Diefenbach, R. Zhang, R. Dunker, D. L. Scott, A. Joachimiak, S. L. Goodman, M. A. Arnaout, *Science* **2001**, *294*, 339–345.

119 J.-P. Xiong, T. Stehle, R. Zhang, A. Joachimiak, M. Frech, S. L. Goodman, M. A. Arnaout, *Science* **2002**, *296*, 151–155.

120 T. Schrader, *Chem. Eur. J.* **1997**, *3*, 1537–1541.

121 M. Wehner, T. Schrader, P. Finocchiaro, S. Failla, G. Consiglio, *Org. Lett.* **2000**, *2*, 605–608.

122 T. Grawe, T. Schrader, P. Finocchiaro, G. Consiglio, S. Failla, *Org. Lett.* **2001**, *3*, 1597–1600.

123 T. Grawe, T. Schrader, R. Zadmard, A. Kraft, *J. Org. Chem.* **2002**, *67*, 3755–3763.

124 S. Rensing, T. Schrader, *Org. Lett.* **2002**, *4*, 2161–2164.

125 (a) Connolly, M. L. *Science* **1983**, *221*, 709–713. (b) Connolly, M. L. *J. Appl. Cryst.* **1983**, *16*, 548.

126 T. W. Bell, J. Liu, *Angew. Chem.* **1990**, *102*, 931–933.

127 T. W. Bell, Z. Hou, Y. Luo, M. G. B. Drew, E. Chapoteau, B. P. Czech, A. Kumar, *Science* **1995**, *269*, 671–674.

128 T. W. Bell, A. B. Khasanov, M. G. B. Drew, A. Filikov, T. L. James, *Angew. Chem.* **1999**, *38*, 2543–2547.

129 T. W. Bell, A. B. Khasanov, M. G. B. Drew, *J. Am. Chem. Soc.* **2002**, *124*, 14092–14103.

130 M. A. Dechantsreiter, E. Planker, B. Mathä, E. Lohof, G. Hölzemann, A. Jonczyk, S. L. Goodmann, H. Kessler, *J. Med. Chem.* **1999**, *42*, 3033–3040.

131 H. U. Stilz, W. Guba, B. Jablonka, M. Just, O. Klingler, W. König, V. Wehner, G. Zoller, *J. Med. Chem.* **2001**, *44*, 1158–1176.

132 P. C. Kearney, L. S. Mizoue, R. A. Kumpf, J. E. Forman, A. McCurdy, D. A. Dougherty, *J. Am. Chem. Soc.* **1993**, *115*, 9907–9919.

133 F. J. B. Kremer, G. Chiosis, J. F. J. Engbersen, D. N. Reinhoudt, *Chem. Soc. Perk. Trans. 2* **1994**, 677–681.

134 A. V. Eliseev, M. I. Nelen, *J. Am. Chem. Soc.* **1997**, *119*, 1147–1148.

134a C. A. Hunter, M. Togrul, S. Tomas, *Chem. Commun.* **2004**, 108–109.

135 A. Takaki, K. Utsumi, T. Kajiki, T. Kuroi, T. Nabeshima, Y. Yano, *Chem. Lett.* **1997**, 75–76.

136 J. C. Ma, D. A. Dougherty, *Chem. Rev.* **1997**, *97*, 1303–1324.

137 S. K. Rurley, G. A. Petsko, *FEBS Lett.* **1986**, *203*, 139–143.

138 T. J. Shepodd, M. A. Petti, D. A. Dougherty, *J. Am Chem. Soc.* **1986**, *108*, 6085–6087.

139 T. J. Shepodd, M. A. Petti, D. A. Dougherty, *J. Am Chem. Soc.* **1988**, *110*, 1983–1985.

140 M. A. Petti, T. J. Shepodd, J. Barrans, D. A. Dougherty, *J. Am Chem. Soc.* **1988**, *110*, 6825–6840.

141 L. Stryer, *Biochemistry*, 3rd ed., W. H. Freeman, New York, 1988, p. 320.

142 M. G. Rossmann, D. Moras, K. W. Olsen, *Nature* **1974**, *250*, 194–199.

143 A. M. Lesk, *Curr. Opin. Struct. Biol.* **1995**, *5*, 775–783.

144 C. A. Bottoms, P. E. Smith, J. J. Tanner, *Protein Sci.* **2002**, *11*, 2125–2137.

145 P. M. S. Monk, in *The Viologens*, Wiley, Chichester, 1998..

146 M. Asakawa, C. L. Brown, S. Menzer, F. M. Raymo, J. F. Stoddart, D. J. Williams, *J. Am. Chem. Soc.* **1997**, *119*, 2614–2627.

147 D. B. Amabilino, P. R. Ashton, S. E. Boyd, J. Y. Lee, S. Menzer, J. F. Stoddart, D. J. Williams, *Angew. Chem. Int. Edt. Engl.* **1997**, *36*, 2070–2072.

148 T. J. Shepodd, M. A. Petti, D. A. Dougherty, *J. Am Chem. Soc.* **1988**, *110*, 1983–1985.

149 S. M. Ngola, P. C. Kearney, S. Mecozzi, K. Russel, D. A. Dougherty, *J. Am . Chem. Soc.* **1999**, *121*, 1192–1201.

150 A. McCurdy, L. Jimenez, D. A. Stauffer, D. N. Dougherty, *J. Am. Chem. Soc.* **1992**, *114*, 10314–1321.

151 G. L. Cantoni, *Ann. Rev. Biochem.* **1975**, *49*, 433–451.

152 G. A. Maw, In *The Chemistry of the Sulphonium Group*, C. J. M., S. Patai, Edt.:

John Wiley and Sons, New York, 1981, Chapter 17; E. Q. Lederer, *Rev. Chem. Soc.* **1969**, *23*, 453–481.

153 R. Meric, J.-P. Vigneron, J.-M. Lehn, *J. Chem. Soc. Chem. Commun.* **1993**, 129–131.

154 M. Dhaenens, L. Lacombe, J.-M. Lehn, J.-P. Vigneron, *J. Chem. Soc. Chem. Commun.* **1984**, 1097–1099.

155 M. Dhaenens, M.-J. Fernandez, J.-M. Lehn, J.-P. Vigneron, *New. J. Chem.* **1991**, *15*, 873–877.

156 A. P. H. J. Schenning, B. de Bruin, A. E. Rowan, H. Kooijman, A. L Spek, R. J. M. Nolte, *Angew. Chem.* **1995**, *107*, 2288–2289.

157 A. E. Rowan, J. A. A. W. Elemans, R. J. M. Nolte, *Acc. Chem. Res.* **1999**, *32*, 995–1006.

158 B. L. Allwood, N. Spencer, H. Shahriari-Zavareh, J. F. Stoddart, D. J. Williams, *J. Chem Soc. Chem. Commun.* **1987**, 1064–1066.

159 P. L. Anelli, P. R. Ashton, R. Ballardini, V. Balzani, M. Delgado, M. T. Gandolfi, T. T. Goodnow, A. E. Kaifer, D. Philip, M. Pietraszkiewicz, L. Prodi, M. V. M. V. Reddington, A. M. Z. Slawin, N. Spencer, J. F. Stoddart, C. Vicent, D. J. Williams, *J. Am. Chem. Soc.* **1992**, *114*, 193–218.

160 J. A. A. W. Elemans, M. B. Claase, P. P. M. Aarts, A. E. Rowan, A. P. H. J. Schenning, R. J. M. Roeland, *J. Org. Chem.* **1999**, *64*, 7009–7016.

161 M. Kamieth, F.-G. Klärner, *J. Prakt. Chem.* **1999**, *341*, 245–251.

162 C. Jasper, T. Schrader, J. Panitzky, F.-G. Klärner, *Angew. Chem.* **2002**, *114*, 1411–1414, *Angew. Chem. Int. Ed.* **2002**, *41*, 1355–1358.

163 K. Araki, H. Shimizu, S. Shinkai, *Chem. Lett.* **1993**, 205–208.

164 F. Inokuchi, K. Araki, S. Shinkai, *Chem. Lett.* **1994**, 1383–1386.

165 A. Arduini, L. Domiano, L. Ogliosi, A. Pochini, A. Secchi, R. Ungaro, *J. Org. Chem.* **1997**, *62*, 7866–7868.

166 A. Arduini, R. Ferdani, A. Pochini, A. Secchi, F. Ugozzoli, *Angew. Chem. Int. Ed.* **2000**, *39*, 3453–3456.

167 A. Arduini, R. Ferdani, A. Pochini, A. Secchi, *Tetrahedron* **2000**, *56*, 8573–8577.

168 G. T. Hwang, B. H. Kim, *Tetrahedron* **2002**, *58*, 9019–9028.

169 The first artificial signal transduction system without a messenger: C. A. Hunter, *Angew. Chem.* **2002**, *20*, 114, 4034–4037.

4
Artificial Pyrrole-based Anion Receptors

Won-Seob Cho and Jonathan L. Sessler

4.1
Introduction

Anions are among the most important species in the natural world and are critically involved in almost the full range of known chemical processes. For example, not only are the vast majority of enzyme substrates and cofactors anionic in nature, anions are also among the most pervasive of all known pollutants (e.g. nitrate, phosphate). They are at the center of energy transduction (e.g. ATP) and biological information processing (e.g. DNA and RNA). The synthesis and study of synthetic receptors for the purpose of anion recognition continues, therefore, to attract increasing attention within the supramolecular community [1–3]. The field of anion recognition, the origins of which can be traced to Simmons and Park's seminal report in 1968 [4], recently marked its 35th anniversary. In honor of this milestone, a special issue of Coordination Chemistry Reviews [5–16] has just appeared. It, in conjunction with an earlier Chemical Review [17] and a monograph on the subject [1], provides a comprehensive overview of the field, including a full discussion of approaches to construction of anion receptors, for example those based on high-charge, alternative hydrogen-bond-donor schemes, the use of metal centers, etc., that must necessarily lie outside the scope of a book chapter such as this that must, by its very nature, be limited in its length and focus. In this chapter the anion-recognition chemistry of artificial pyrrole-based receptors will be reviewed. The first section will set the stage for the rest of the chapter by describing a few examples of anion-binding systems found in nature, with those involving pyrrole-based recognition receiving particular attention. The next theme, introduced in the second section, will center around a discussion of positively charged, pyrrole-based synthetic receptors that rely on a combination of hydrogen bonds and electrostatic interactions to effect anion binding. The third portion of this chapter will cover the chemistry of neutral pyrrole-based anion receptors for which the basis of anion recognition is almost exclusively hydrogen-bonding interactions. The fourth part will introduce the still-evolving theme of using pyrrole-containing systems as anion carriers to enhance the efficiency of through-membrane anion transport. This focus on potential practical applications will be developed more

Functional Synthetic Receptors. Edited by T. Schrader, A. D. Hamilton
Copyright © 2005 WILEY-VCH Verlag GmbH & Co. KGaA, Weinheim
ISBN: 3-527-30655-2

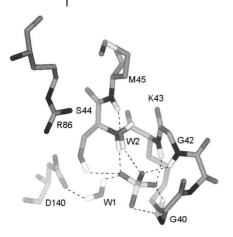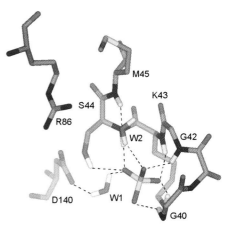

Figure 4.1 Stereo view of the ATPase active site showing the interaction of the bound sulfate anion with various P-loop residues. This structure was originally reported in Ref. [26] and was redrawn using data downloaded from the Protein Data Bank.

fully in a section that details the use of pyrrolic receptors as the basis for new optical and electrochemical anion-sensing systems. Finally, to put the use of pyrrole-based anion receptors into a more general context, the chapter will conclude with a brief review of other anion-binding systems.

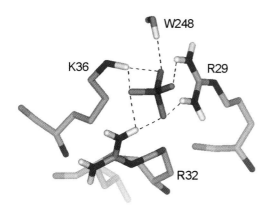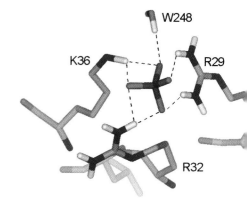

Figure 4.2 Stereo view of the amino acid residues that interact with the phosphate anion in a structurally characterized histone octamer–phosphate complex. Only one of five bound phosphates is shown. This structure was originally reported in Ref. [29] and was redrawn using data downloaded from the Protein Data Bank.

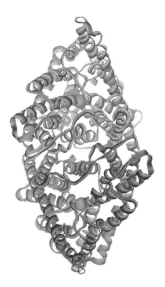
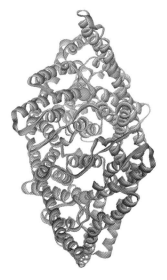

Figure 4.3 Stereo view of a ribbon representation of the StClC dimer visualized from the extracellular side. The two subunits are cyan and brown. A chloride anion in the selectivity filter is represented as a green sphere. This structure was originally reported in Ref. [45] and has been redrawn using data downloaded from the Protein Data Bank.

4.2
Anions in Biological Systems

Anions are ubiquitous in biology. They are present in approximately 70 % of all enzyme sites, play essential structural roles in many proteins, and are critical in the manipulation and storage of genetic information (DNA and RNA are polyanions). Anions are also involved in regulating osmotic pressure, activating signal transduction

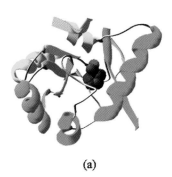

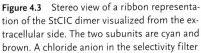

(a)   (b)

Figure 4.4 (a) Ribbon diagram of *Methanosarcina barketi* momomethylamine methyltransterase subunit. The L-pyrrolysine is shown using a space-filling model. (b) Stick-diagram of proposed L-pyrrolysine amino acid. This X-ray structure was originally reported in Ref. [49] and has been redrawn using data downloaded from the Protein Data Bank.

pathways, maintaining cell volume, and in the production of electrical signals. Not surprisingly, therefore, the disruption of anion flux across cell membranes (especially chloride, present in cells at 5–15 mM concentration levels [18]) is increasingly recognized as being the primary determinant of many diseases, including cystic fibrosis [19], Bartter's syndrome [20], Dent's disease [21, 22], Pendred's syndrome [22, 23], and osteopetrosis [24]. In fact, the transport of anions through cell phospholipid bilayers is known to be mediated by a variety of channels and anion transport proteins with at least fourteen mitochondrial anion-transport systems having been identified [25]. These include (among others) systems responsible for the trafficking of ADP, ATP, phosphate, citrate, maleate, oxaloacetate, sulfate, glutamate, fumarate, and halide anions.

Recently, several X-ray crystal structures have been solved that have enabled the direct visualization of enzyme–anionic substrate complexes stabilized by multiple hydrogen-bonding interactions. In particular, the structure of the DNA helicase RepA sulfate complex, solved to 1.95 Å resolution, has six hydrogen-bonding interactions between the sulfate anion and the RepA protein scaffold. The sulfate anion is also hydrogen bonded to Asp140 via an intervening water molecule (Fig. 4.1) [26]. Another example of a biological sulfate anion complex is the *Salmonella typhumurium* (St)–sulfate anion complex, a structure that is stabilized by seven hydrogen-bonding interactions [27, 28]. For the phosphate anion, another tetrahedral oxoanion of prime physiological significance, an X-ray crystal structure of the histone octamer–phosphate complex reveals that five separate phosphate anions interact with lysine and arginine residues at five different sites. Fig. 4.2 shows one of the phosphate ions and emphasizes how it is bound by several basic amino acid residues [29].

Acyclic polyamines such as spermine, $^+NH_3(CH_2)_3NH(CH_2)_3NH(CH_2)_3NH_3^+$, are involved in promoting cell growth, inducing the biosynthesis of DNA, RNA, and proteins [30], and regulating a variety of enzymatic activities [31]. It is also well known that polyamines have high affinity for phosphate anion and bind well, for example, to phenylalanine transfer RNA [32–35]. Ohishi and coworkers recently reported the X-ray crystal structure of a spermine–phosphate anion complex [36, 37]. This structure revealed that the protonated spermine interacts with the bound phosphate anion via hydrogen bonds involving intervening water molecules.

The guanidinium group is widely distributed in biological systems as a sidechain of arginine and, not surprisingly given its charge, this moiety plays a major role in the binding of anionic substrates. Synthetic guanidinium-based receptors are also well known and have played an important role in the development of anion-recognition chemistry. As a general rule, these systems have strong affinity for carboxylate, phosphate, sulfate, and nitrate anions, binding these substrates by a combination of hydrogen bonding and electrostatic interactions [17, 38, 39]. New guanidinium-based hosts capable of catalyzing biological reactions and effecting phosphate anion transport have recently been designed and synthesized by Hamilton [40, 41] and by Schmidtchen [42–44].

In 2002, the 3-Å-resolution X-ray structure of the StClC chloride ion channel was solved [45]. This long-awaited crystal structure revealed a homodimer-derived channel that is notable for its hour-glass shape and the complete absence of positively charged amino-acid sidechains anywhere near where the chloride anions would pass (Fig. 4.3). On the other hand, the structure does indicate the presence of a negatively charged glutamate sidechain just above the channel entrance. This residue is thought to act as an anion-regulating gate. By swinging out to open the channel, it allows Cl⁻ ions to enter the channel pore from whence they are pulled (presumably) toward the constricted, neutral center of the channel by surfaces rich in positively polarized (but not charged) residues. Despite the elegance of this structure, and a considerable body of work devoted to understanding biological ion transport in general, our understanding of through-membrane anion transport remains extremely limited. One way of addressing this deficiency is through the synthesis and study of synthetic anion receptors and artificial anion channels; this, in turn, is providing motivation to produce such systems [46, 47].

The year 2002 was also marked by the exciting discovery of a new, genetically encoded amino acid, L-pyrrolysine, whose proposed biological function relies in part on pyrrole-like NH-anion recognition (Fig. 4.4) [48, 49]. Pyrrole NH-chloride anion recognition and transport have also been proposed recently as an explanation of the anticancer and immunosuppressive activity of prodigiosin, a naturally occurring tripyrrolic pigment known since the 1930s (*vide infra*), although an alternative explanation of this activity based on copper complex formation and DNA cleavage has also been proposed (Sect. 4.5). A desire to distinguish between these two limiting mechanisms and to test whether synthetic oligopyrrole species can act as anion carriers provides a further incentive to make and study pyrrole-based anion receptors.

Prodigiosins **1** are a family of naturally occurring tripyrrolic red pigments characterized by a common pyrrolylpyrromethene skeleton that were first isolated in the 1930s [50–55] from microorganisms including *Serratia* and *Streptomyces* [56, 57]. These molecules, especially prodigiosin 25-C, **1b**, have been studied extensively for their promising immunosuppressive [58] and anticancer activity [59]. A variety of prodigiosins have also been found to induce apoptosis in dozens of human cancer cell lines, including liver cancer [60], human breast cancer [61], human colon cancer [59], gastric cancer [62], and hematopoietic cancer cell lines [63]. Taken in concert, this combination of properties has made the prodigiosins attractive targets for use in a variety of combination drug therapies. It is also inspiring the synthesis of new pyrrole-based anion carriers, as discussed further in Sect. 4.5.

1a

1b

4.3
Cationic Pyrrole-based Receptors

Over the past two decades substantial effort has been devoted to the preparation of synthetic anion receptors. It has been known from the earliest days of the field that protonated polyamines and other highly charged species (e.g. guanidiniums) can bind anions well, even in protic media, as the result of strong electrostatic and hydrogen-bonding interactions. Such an appreciation makes polypyrrolic compounds of interest as potential anion receptors. These species often contain sp²-hybridized "imine-like" pyrrolic nitrogen atoms that are partially or completely protonated in neutral media. They also usually contain normal pyrrolic NH subunits and this unique combination of characteristics makes them attractive as potential anion-binding or transporting agents. In this section we review the chemistry of a variety of oligopyrrolic systems whose documented anion-binding ability is predicated on the prior protonation of one or more basic imine-like nitrogen atoms. In point of fact, this means the primary focus of this section will be on so-called "expanded porphyrins", a class of compounds that has recently been reviewed from a completely different perspective [64].

4.3.1
Cyclic Receptors

4.3.1.1 Pentapyrrolic Systems
The fact that polypyrrolic macrocycles, rich in NH hydrogen-bond-donor groups, can act as strong anion-complexing agents in their protonated forms is now well established [65–71]. To date, anion binding in the solid state has been observed for three types of pentapyrrolic macrocycle, sapphyrin **2**, isosmaragdyrin **3**, and pentaphyrin-(2.1.0.0.1) **4**, although anion binding has only been studied extensively for the first of these. Indeed, as detailed below, it was with this system (sapphyrin) that pyrrole-based anion binding was first noted and appreciated.

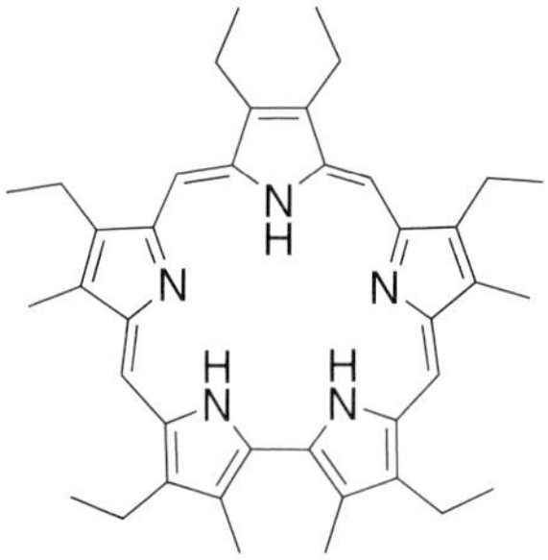

| 2 | 3 | 4 |

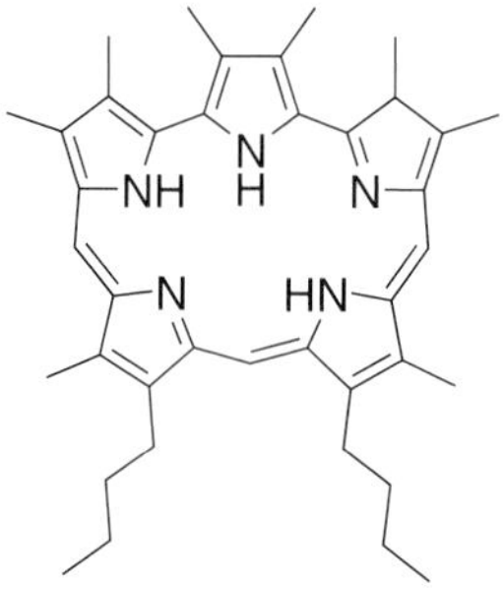

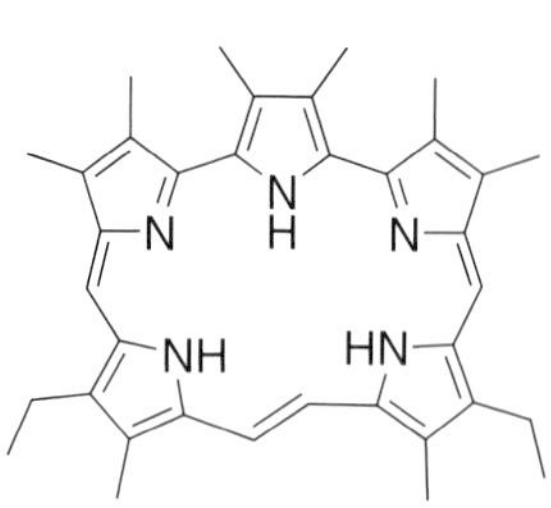

Sapphyrin, a class of pentapyrrolic macrocycle discovered by Woodward and his group [72], contains five pyrroles and four meso carbon atoms. This macrocycle also has an aromatic 22 π-electron framework. The free-base form of sapphyrin has three pyrrolic NH and two sp²-hydridized imine-like nitrogen atoms. These latter nitrogen atoms are relatively basic, with pK_a values of ca 4.8 and 8.8 for the corresponding conjugate acids in water [73, 74]. This basicity has the consequence that sapphyrins occur in one of two protonated forms, mono- or di-, depending on the conditions; it is these species that are important as anion-chelating agents.

In 1990, Sessler, Ibers, and coworkers reported the first X-ray crystal structure of a sapphyrin–anion complex [65]. Here, analysis of a crystal, initially thought to be the bis-PF_6^- salt of the doubly protonated form of sapphyrin **2**, revealed the presence of only one PF_6^- counteranion. The same analysis, however, indicated that a fluoride atom was bound inside the sapphyrin core, being held there via a regular array of five N–H···F hydrogen bonds with a N–H distance of ca. 2.7 Å (Fig. 4.5). Presumably, the bound fluoride anion was either extracted from the PF_6^- counteranion or trapped from adventitious fluoride anion present in solution. In any event, the key fact that F⁻ was bound by sapphyrin provided *prima facie* evidence that this and other protonated expanded porphyrins could act as anion receptors, at least in the solid state.

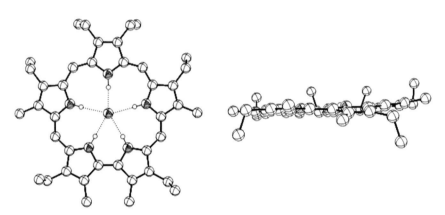

Figure 4.5 Single-crystal X-ray structure (top and side views) of the fluoride anion complex of diprotonated sapphyrin **2**. This figure was generated using information downloaded from the Cambridge Crystallographic Data Centre (CCDC) and corresponds to a structure originally reported in Ref. [65].

Further support for the notion that protonated sapphyrin could act as an anion-binding agent came when the X-ray structure of the bis-hydrochloride salt of sapphyrin **2** was elucidated in 1992 [66]. In contrast with the fluoride anion complex, neither of the chloride counteranions was located in the plane of the diprotonated sapphyrin (Fig. 4.6). Rather, they were found bound to, and above, opposite faces of the sapphyrin ring via a combination of hydrogen bonding and electrostatic interactions. This out-of-plane binding mode occurs because anionic chloride is larger than anionic fluoride and only the latter is of an appropriate size to fit within the relatively

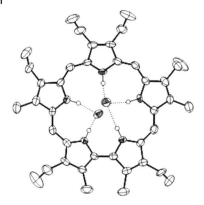 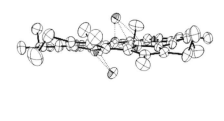

Figure 4.6 Single-crystal X-ray structure (top and side views) of the bis-hydrochloride salt of sapphyrin **2**. This figure was generated using information downloaded from the CCDC and corresponds to a structure originally reported in Ref. [66].

small sapphyrin core. On a related note, the observation of out-of-plane binding provided a strong "hint" that chloride anion would be bound less well than fluoride anion in solution.

The desire to study the halide-binding properties of sapphyrin in solution was largely inspired by the solid-state anion complex structures described above (cf. Figs. 4.5 and 4.6). This solution-phase anion-binding behavior was studied using fluorescence and UV–visible spectroscopic methods. In dichloromethane solution the most intense absorbance feature seen in the UV–visible spectrum of **2**•2HF is a Soret-like band at $\lambda_{max} = 446$ nm. This spectral feature was found to be shifted to the red for **2**•2HCl and **2**•2HBr ($\lambda_{max} = 456$ and 458 nm, respectively) [75]. This difference was magnified in the corresponding fluorescence emission spectra. From associated spectral titrations the fluoride anion binding constant of **2** in dichloromethane was calculated to be $>10^8$ M^{-1}. In contrast, those for chloride and bromide were found to be lower (1.8×10^7 M^{-1} and 1.5×10^6 M^{-1}, respectively) [66]. Because the association between sapphyrin **2** and fluoride was found to be too large to measure accurately in dichloromethane, the association constants in methanol for sapphyrin **2** with the fluoride, chloride, and bromide anions were determined and found to be 2.8×10^5, ca. 10^2, and $<10^2$ M^{-1}, respectively [66].

Although, as noted above, most anion-binding studies performed to date have focused on the use of protonated sapphyrins as the complexing ligand, it is important to appreciate that other pentapyrrolic systems also bind anions. Consistent with this conclusion is the finding that both isosmaragdyrin **3** [76] and pentaphyrin-(2.1.0.0.1) **4** [77] bind chloride anion in the solid state (Fig. 4.7).

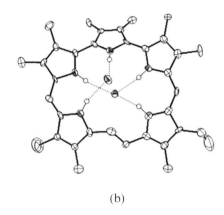

Figure 4.7 (a) Single-crystal X-ray structure of the hydrochloride salt of isosmaragdyrin **2**. (b) Single-crystal X-ray structure of the bis-hydrochloride salt of pentaphyrin-(2.1.0.0.1) **3**. These figures were generated using information downloaded from the CCDC and correspond to structures originally reported in Refs. [76, 77].

4.3.1.2 Hexapyrrolic Systems

The discovery that the diprotonated form of sapphyrin could act as an anion-binding agent provided a stimulus for Sessler and coworkers to make and study new expanded porphyrins. Among the numerous systems obtained are the hexapyrrolic macrocycles rubyrin **5**, rosarin **6**, and cyclo[6]pyrrole **7**.

5	**6**	**7**

Rubyrin **5** was prepared early on by the Sessler group and was the first hexapyrrolic system to be characterized by X-ray diffraction analysis. By analogy with sapphyrin, the diprotonated form of rubyrin was found to stabilize a 1:2 chloride anion complex in the solid state containing two bound chloride anions, as illustrated in Fig. 4.8. Again, the two chloride anions were found to be held above and below the plane of the relatively planar macrocycle. The anion-binding capacity of rubyrin in solution was inferred from anion-transport experiments. Specifically, in the presence of C-

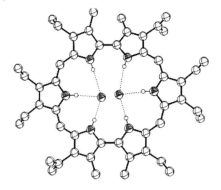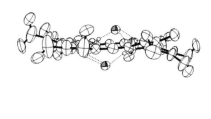

Figure 4.8 Single-crystal X-ray structure of the bis-hydrochloride salt of rubyrin **5**. This structure was generated using information downloaded from the CCDC and corresponds to a structure originally reported in Ref. [71].

Tips (triisopropylsilyl cytidine), added as a cofactor, the rate of rubyrin-mediated GMP (guanosine monophosphate) transport was found to be roughly 30 times faster than that mediated by the smaller sapphyrin system [78]. Complementing these studies is recent work by Umezawa and coworkers. As detailed in the sensor-related section of this chapter, these researchers found that rubyrin-based membranes have an EMF response toward the 3,5-dinitrobenzoate anion [79].

The fact that rubyrin is capable of forming anion complexes in solution and in the solid state inspired the synthesis of additional hexapyrrolic systems. Rosarin **6** and cyclo[6]pyrrole **7** are examples of such hexapyrrolic macrocycles. Rosarin is a nonaromatic 24 π-electron system that contains three pyrrolic NH and three basic sp²-hy-

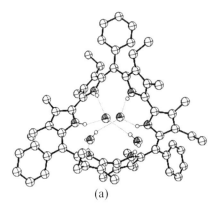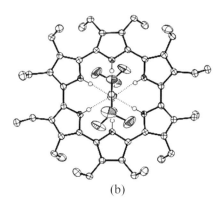

(a)   (b)

Figure 4.9   (a) Single-crystal X-ray structure of the tris-hydrochloride salt of rosarin **6**. Only two chloride anions (with two water molecules) are bound in the core of macrocycle. A third, nonchelated chloride anion (not shown) is found within the crystal lattice. (b) Single-crystal X-ray structure of the bis-TFA salt of cyclo[6]pyrrole **7**. These two structures were generated using information downloaded from the CCDC and correspond to structures originally reported in Refs. [80, 81].

dridized nitrogen atoms. The ability of the fully protonated form to bind chloride anions in the solid state has been confirmed by X-ray diffraction studies. In this structure two of the three chloride anions were found to be bound directly to the triprotonated core with the remaining Cl⁻ appearing elsewhere in the crystal lattice (Fig. 4.9). The presence of the three chloride counteranions supports the fact that rosarin has an overall 3⁺ positive charge.

Cyclo[6]pyrrole **7**, recently synthesized, is another example of an hexapyrrolic system capable of binding anions. Not only were interactions between two trifluoroacetate counteranions apparent from the solid-state structure of the protonated form, cyclo[6]pyrrole itself was only isolated as the result of what seemed to be a specific anion template effect.

4.3.1.3 Oligopyrrolic Systems

Underscoring the idea that expanded porphyrins could define a general new approach to anion binding is the observation that the protonated forms of many so-called "higher order" systems, for example cyclo[n]pyrroles **8** (n = 7 and 8), heptaphyrin **9**, octaphyrins **10** and **11**, and the decapyrrolic macrocycle turcasarin **12**, form anion complexes in the solid state (Figs. 4.10–4.12) [81–87]. It was also observed that the neutral form of cycle[8]pyrrole **8b** binds hydrogen sulfate anion with an affinity constant of 5.8×10^3 M⁻¹ in DMSO, as judged from isothermal titration calorimetry (ITC) analyses [88]. This result, which might reflect both partial proton transfer and H-bonding, is consistent with X-ray crystallographic results obtained in the solid state (Fig. 4.11a).

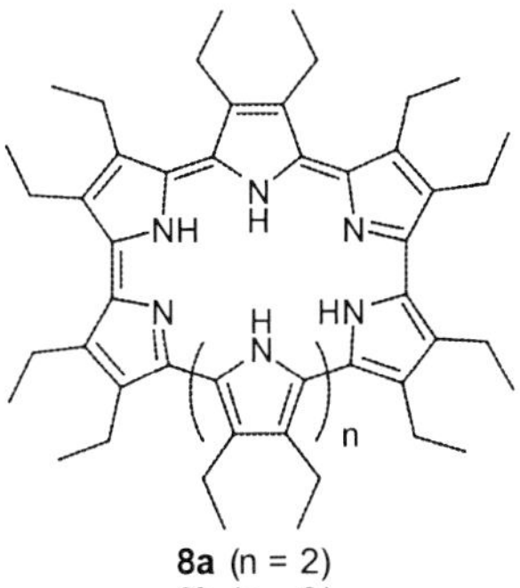

8a (n = 2)
8b (n = 3)

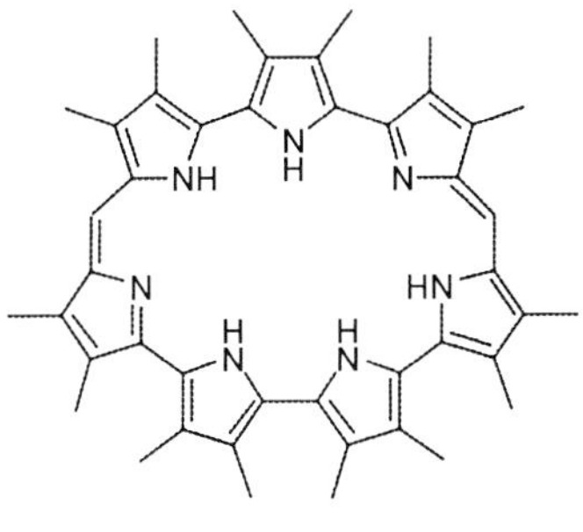

9

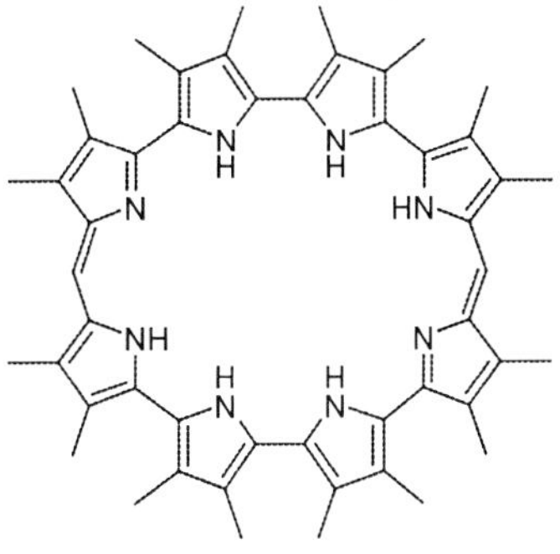

10

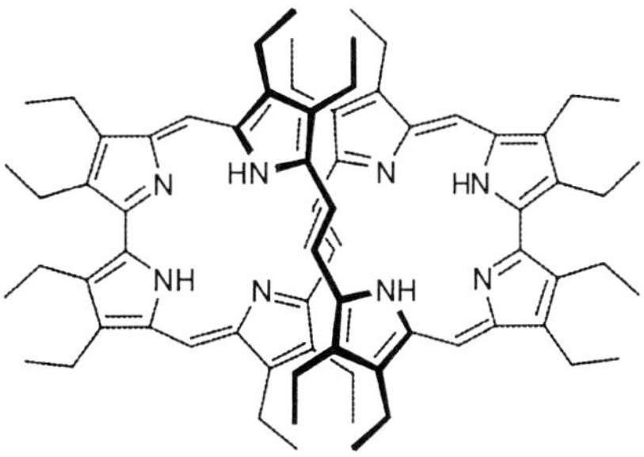

11

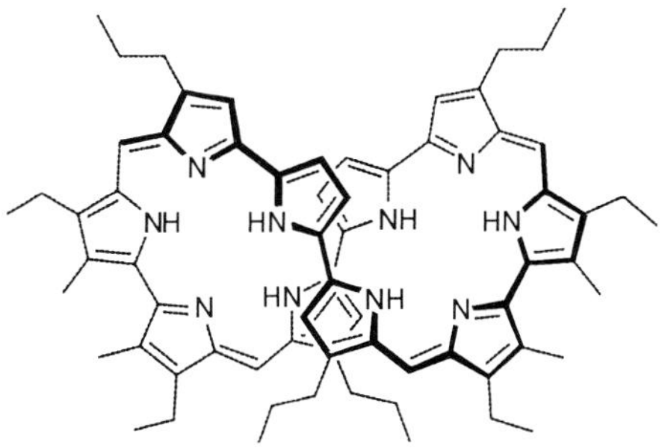

12

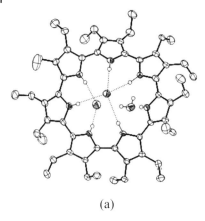

(a)

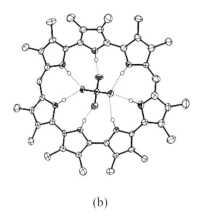

(b)

Figure 4.10 (a) Single-crystal X-ray structure of the bis-hydrochloride salt of cyclo[7]pyrrole **8a**. (b) Single-crystal X-ray structure of the sulfuric acid salt of heptaphyrin **9**. These two structures were generated using information downloaded from the CCDC and correspond to structures originally reported in Refs. [81, 86].

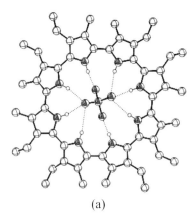

(a)

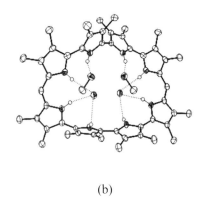

(b)

Figure 4.11 (a) Single-crystal X-ray structure of the dihydrogen sulfate salt of cyclo[8]pyrrole **8b**. (b) Single-crystal X-ray structure of the bis-hydrochloride salt of octaphyrin **10**. These two structures were generated using information downloaded from the CCDC and correspond to structures originally reported in Refs. [86, 87].

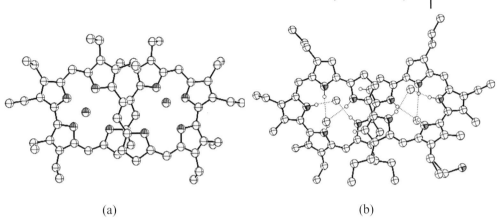

(a)   (b)

Figure 4.12 (a) Single-crystal X-ray structure of the bis-hydrochloride salt of octaphyrin **11**. (b) Single-crystal X-ray structure of the tetrahydrochloride salt of turcasarin **12**. These two structures were generated using information downloaded from the CCDC and correspond to structures originally reported in Refs. [84, 85].

4.3.1.4 Imine-linked Receptors and Other Related Systems

The observation that the basicity of pyrrolic imine groups enables protonated expanded porphyrins to function as effective anion-complexing agents, led to the consideration that other imine-based polypyrrolic systems could act as anion receptors. In accord with such thinking, the Schiff base macrocycles **13** and **14** were prepared by Sessler and coworkers and explored as anion-binding agents [89, 90].

13   **14**

The first of these products, anthraphyrin **13**, is an historically important system. This is because it is the second expanded porphyrin, after sapphyrin, to be recognized as being an anion-binding agent and thus played an important role in terms of helping generalize the concepts of pyrrole-based anion recognition. Its anion binding behavior differs from that of sapphyrin, however. In particular, only one chloride anion, not two, is bound in the solid state. This anion is found near the center of the protonated anthraphyrin ring core, being held there via four hydrogen bonds and by associated electrostatic interactions (Fig. 4.13) [89]. This Schiff-base system also has much higher affinity for chloride anion ($K_a = 2 \times 10^5$ M^{-1}) than for fluoride anion ($K_a = 1.4 \times 10^4$ M^{-1}) in CH$_2$Cl$_2$ solution, as judged from standard Benesi–Hildebrand and Scatchard plots of the anion-induced UV–visible spectral shifts (both species studied in the form of their respective tetrabutylammonium salts). This relatively higher affinity for chloride anion is in contrast with the behavior of diprotonated sapphyrin. Perhaps as a consequence, anthraphyrin acts as a much more effective mediator of the through-membrane transport of fluoride anion than sapphyrin, as judged from model Pressman-type U-tube model experiments [89]. On the other hand, the fluoride-anion-transport ability of anthraphyrin is significantly inhibited by chloride anion. These findings were considered consistent with the proposal that it is the rate of anion release, rather than the overall binding affinity, which determines the efficacy of anion transport [89].

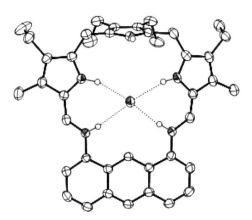 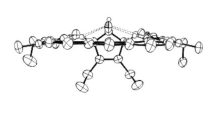

Figure 4.13 Single-crystal X-ray structure of the mixed hydrochloride and hydrotetrafluoroborate salt of anthraphyrin **13**. The chloride anion is bound in the diprotonated core (as illustrated). The BF$_4^-$ counteranion, however, is not proximate to the core and is not shown. This structure was generated using information downloaded from the CCDC and corresponds to a structure originally reported in Ref. [89].

More recently, another expanded Schiff base macrocycle **14**, which is actually a cal-ix[4]pyrrole–texaphyrin hybrid, was synthesized [90]. This macrocycle has the four pyrrole NH donor elements characteristic of the calix[4]pyrroles (cf. Sect. 4.4.1) with the larger cavity size characteristic of texaphyrin. According to X-ray diffraction analysis one chloride anion is bound to the cleft of what is an overall V-shaped Schiff base macrocyclic structure by four pyrrolic NH groups and two protonated imine hydrogen atoms (Fig. 4.14). This result provided support for the notion that the protonated form would bind chloride anion in solution. Efforts to determine the association constants of the neutral, nonprotonated form with chloride and bromide revealed that these species (studied as the corresponding tetrabutylammonium salts) were not bound, at least within the limits of detection, in CH$_3$CN solution. On the other hand, protonation of this Schiff base macrocycle (**14**) serves to increase dramatically its chloride anion-binding affinity ($K_a = 4.6 \times 10^5$ M^{-1} for the monoprotonated form;

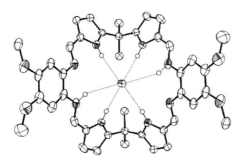 

Figure 4.14 Two views of the single-crystal X-ray structure of the bis-hydrochloride salt of the Schiff base macrocycle **14**. Only one chloride anion is found in the core of macrocycle; a second nonchelated anion (not shown) is found within the crystal lattice. This structure was generated using information downloaded from the CCDC and corresponds to a structure originally reported in Ref. [90].

K_{a1} = 4.46 × 10^6 M^{-1} and K_{a2} = 2.2 × 10^5 M^{-1} for the bis-protonated form in acetonitrile) [90]. This finding is consistent with the increased positive charge present in the protonated forms serving to strengthen the electrostatic interaction between the nitrogen-rich host and the bound chloride anion.

Calixphyrin is a term recently coined by Sessler to denote hybrid systems that are chemical "blends" of calixpyrrole (containing sp^3-hybridized pyrrole–pyrrole bridges) and porphyrins (characterized by sp^2-hybridized meso carbons) [91, 92]. In practice, this definition has been extended to include all porphyrin-like materials containing at least one sp^3 meso-carbon bridging atom. In 2001, calix[6]phyrin **15** was synthesized and it is observed that the neutral form of the macrocycle does not act as an effective anion receptor, at least for the anions tested (chloride, bromide, iodide, nitrate, and hydrogen sulfate). An accompanying X-ray diffraction study revealed, however, that crystals produced in the presence of HCl were monoprotonated, with one of the pyrrolic imine units being protonated. As a consequence, one chloride anion is bound in the cavity; it is held there by five hydrogen-bonding interactions (Fig. 4.15) [93]. The fact that the monoprotonated form of **15** binds chloride in the sol-

15

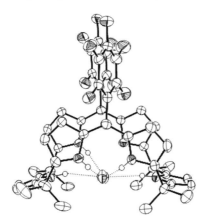

Figure 4.15 Single-crystal X-ray structure of the hydrochloride salt of cal-ix[6]phyrin **15**. The structure was generated using information downloaded from the CCDC and corresponds to a structure originally reported in Ref. [93].

id state led to the proposal that the protonated forms of calix[6]phyrin could function as anion receptors in solution. In acetone solution containing a high concentration of fuming sulfuric acid to ensure full protonation UV–visible spectroscopic studies revealed that the diprotonated form of calix[6]phyrin, which predominates under such conditions, bound chloride, bromide, and iodide anions strongly. The affinity constants for I^- and HSO_4^-, which were determined quantitatively, were found to be 2.55×10^4 and 3.5×10^3 M^{-1}, respectively, in this solvent mixture.

4.3.2
Linear Receptors

Many linear oligopyrroles have recently been prepared and their roles as building blocks for polypyrrolic macrocycles have been extensively studied. Smaller di- and, particularly, tripyrrolic species have also been made in the context of studies involving prodigiosins (e.g. **1**). As mentioned in the introduction, one current proposal is that prodigiosins mediate their physiological effect as a result of effecting the into-cell transport of HCl. As an part of a general effort to provide support for (or against) this proposal, anion-binding studies of linear receptors such as the prodigiosin analogs tetrapyrrin **16**, and hexapyrrin **17** were undertaken. These studies are particular important, because the anion-binding behavior of prodigiosin has yet to be analyzed either kinetically or thermodynamically. Further, no solid-state structural proof has been proposed to substantiate the claim that protonated prodigiosins can bind chloride anions. The structure of the HCl complex of a dipyrromethene derivative, **1c**, was recently elucidated by Melvin and Manderville; the structure involves single-point chloride anion binding, as illustrated in Fig. 4.16 [94]. It has also been shown that protonated prodigiosins can bind to DNA and can complex copper ions. This has led to the suggestion by these workers and by Fürstner [95] that the biological activity of prodigiosins will probably prove to be correlated with these chemical features, rather than with the capacity to transport H^+ + Cl^-.

1c

1d R = H
1e R = CH$_3$

16a R$_1$ = H, R$_2$ = R$_3$ = C$_2$H$_5$
16b R$_1$ = CN, R$_2$ = C$_2$H$_5$, R$_3$ = CH$_3$

17a R$_1$ = C$_3$H$_7$, R$_2$ = H
17b R$_1$ = R$_2$ = C$_2$H$_5$

Structure–function studies of prodigiosin analogs are still in their early days [94, 96–98]. Most of these studies have only involved systems containing the key tripyr-role moiety or its dipyrromethene subunits. Nonetheless, these analyses have been extremely valuable; they have served to emphasize the apparent necessity of the 5-

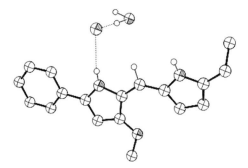

Figure 4.16 Single-crystal X-ray structure of the hydrochloride salt of the dipyrrolic prodi-giosin analog **1c**. This structure was generated using information downloaded from the CCDC and corresponds to a structure original-ly reported in Ref. [94].

methoxy group, the benefit of alkyl substituents on the C ring, and an apparent lack of correlation between pyrrole nitrogen basicity and biological activity [94].

Sessler and coworker recently completed the synthesis of two simple prodigiosins **1d** and **1e** and determined their chloride anion-binding affinity by performing ITC measurements in anhydrous acetonitrile. The chloride anion affinity of the protonated forms ([H•**1d**]$^+$ and [H•**1e**]$^+$) were found to be quite high ($K_a = 1.1 \times 10^5$ M^{-1} for both) [99]. The fact that the protonated forms [H•**1d**]$^+$ and [H•**1e**]$^+$ bind chloride anion with essentially equal affinity, coupled with the high chloride anion affinity previously recorded for sapphyrin (a species that is also monoprotonated at neutral pH [68, 73]) [74], led to the consideration that other oligomeric pyrrole systems, for example the tetrapyrrole **16** and the hexapyrrole **17**, might function as HCl receptors or as interesting prodigiosin analogs. In the tetrapyrrin **16** the presence of one more pyrrole unit than in the tripyrrolic prodigiosins was expected to lead to increased binding affinity. This proved true for the protonated form of the electron-poor dicyano derivative [H•**16a**]$^+$, for which a K_a value of 4.9×10^5 M^{-1} was recorded. This affinity for the chloride anion is greater than that of its α-free analog [H•**16b**]$^+$ by a factor of approximately 1.7. For both species binding was better than for [H•**1d**]$^+$ and [H•**1e**]$^+$.

Hexapyrrin **17** which is formally an α,α′-linked prodigiosin dimer, also has promise as an anion receptor. This system contains two potential chloride anion-binding sites and was expected to enable the formation of a 2:1 anion-to-receptor complex under appropriate conditions. Consistent with this expectation is the structure of the bis-HCl salt of **17a** determined by X-ray crystal structural analysis (Fig. 4.17) [100]. It reveals two chloride anions tethered to the flat diprotonated hexapyrrin via two sets of three hydrogen bonds each. In a solution-phase study involving a slightly different hexapyrrin (**17b**), the diprotonated system was found to bind two chloride anions strongly in acetonitrile ($K_{a1} = 1.2 \times 10^6$, $K_{a2} = 3.2 \times 10^4$ M^{-1}) [99].

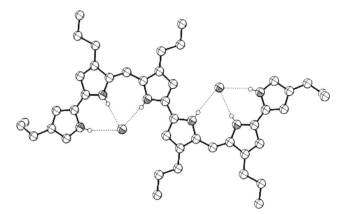

Figure 4.17 Single-crystal X-ray structure of the bis-hydrochloride salt of hexapyrrin **17a**. This structure was generated using information downloaded from the CCDC and corresponds to a structure originally reported in Ref. [100].

4.4
Neutral Pyrrole-based Anion Receptors

As a general rule, neutral, organic-based anion receptors rely primarily on hydrogen-bonding interactions to stabilize their complexes with bound anions. This is in contrast with cationic receptors which exploit both electrostatic and hydrogen-bonding interactions. Potential advantages of neutral receptors include the possibility of more selective binding. This is because, in contrast with binding effects resulting from positive charges, hydrogen-bonding interactions are usually anisotropic. Such interactions thus enable the construction, at least in principle, of receptors selective for a particular anion. Another advantage associated with the use of a neutral system is that there is no inherent competition with a counteranion, an effect that often tends to complicate analysis of the systems in question. In pyrrolic systems the use of neutral receptors obviates the need for protonation, either before or concurrent with anion recognition. Not surprisingly, considerable effort has recently been devoted to the construction of neutral pyrrole-based anion receptors. Most interestingly, however, the history of such systems dates back only to 1996, when the use of calix[4]pyrrole as an anion receptor was first introduced.

4.4.1
Cyclic Receptors

As implied above, the calix[4]pyrroles (e.g. **18a**) were the first neutral pyrrole-based anion-binding systems to be reported. They were regarded as attractive as receptors, because they are not only easy to prepare but also potentially "tunable" in terms of their inherent anion-binding characteristics. This promise, which is still being realized in several laboratories worldwide, has made the generalized class of calix[n]pyrroles (in which n can be ≥ 4) one of the more attractive approaches to anion receptor development currently being pursued.

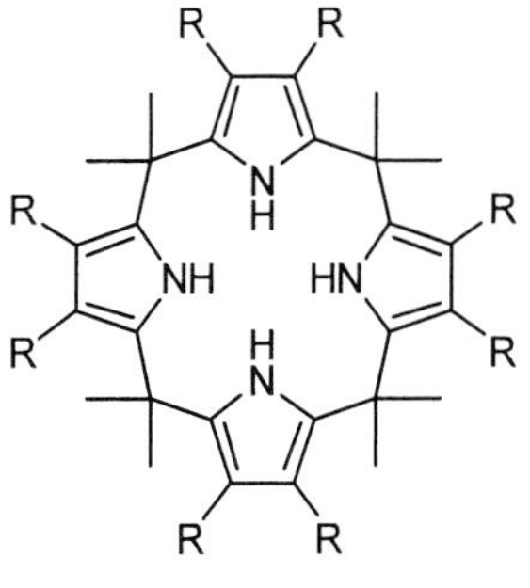

18a R = H
18b R = Br

The calix[4]pyrroles, in contrast with their higher homologs (*vide infra*) are a venerable class of materials. They were first prepared by Baeyer [101] in 1886 but only proposed as possible anion-binding agents in 1996 [102]. It has been found that both fluoride and chloride anions were bound to octamethylcalix[4]pyrrole **18a** in the solid state (Fig. 4.18) [102, 103]. Although these two structures are similar, for the fluoride anion complex, the average of N···F distance is 2.767 Å whereas for the corresponding chloride complex the N···Cl distance is 3.303 Å. Thus, fluoride anion seems to be more tightly bound in the solid state. In both complexes it is important to appreciate that the cone conformation, seen in the presence of anions, is very different from the 1,3-alternate form observed in their absence.

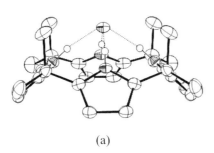

(a)   (b)

Figure 4.18 (a) Single-crystal X-ray structure of the fluoride anion complex of calix[4]pyrrole **18a**. (b) Single-crystal X-ray structure of the chloride anion complex of calix[4]pyrrole **18a**. These two structures were generated using in- formation downloaded from "Acta Crystallographica Section E" and the CCDC, respectively, and correspond to structures originally reported in Refs. [102, 103].

In CD_2Cl_2 solution, calix[4]pyrrole **18a** was found to bind fluoride and chloride anions (studied in the form of their tetrabutylammonium salts) with affinities, K_a, of 1.7×10^4 and 3.5×10^2 M^{-1}, respectively, as judged from EQNMR experiments [102, 103]. As part of this generalized study, several derivatives, including the β-octabromo-*meso*-octamethylcalix[4]pyrrole **18b**, were also prepared. The synthesis of **18b** was straightforward and simply involved reaction of calix[4]pyrrole **18a** with NBS. The anion-binding ability of this compound was also measured by EQNMR under conditions analogous to those used for **18a**. On this basis it was found that **18b** was an improved fluoride and chloride anion receptor ($K_a = 2.7 \times 10^4$ and 4.3×10^3 M^{-1} for fluoride and chloride, respectively) [104]. This higher anion-binding affinity was rationalized in terms of the eight bromine substituents; these electron-withdrawing atoms serve to increase the acidity of the pyrrolic NH, thereby enhancing the anion-binding affinity.

Schmidtchen recently applied isothermal titration calorimetry (ITC) to the study of calix[4]pyrrole anion-recognition and used his report to emphasize further the advantages of this method for studying problems in molecular recognition [105]. These latter include an increased dynamic range, higher detection limits, and greater reproducibility compared with NMR spectroscopic methods. Table 4.1 summarizes the anion affinity constants of calix[4]pyrrole **18a** measured by ITC that have proved reproducible in the authors' hands. Inspection of this table reveals that for systems with high affinity, the recorded K_a values are usually greater than those obtained by NMR spectroscopic methods, presumably as the result of increased precision. Nonetheless, at least among the set of anions studied, the same general selectivity trends are revealed by both ITC and NMR titrations. (Despite considerable effort, the authors have been unable to obtain reliable K_a values for fluoride anion-binding by ITC. For this reason, the claim made by Schmidtchen [105] that fluoride and chloride are bound with approximately the same affinity in CH_3CN has yet to be independently substantiated, at least to the best of the author's knowledge.)

Table 4.1 Association constants, K_a, for compound **18a** (M^{-1}) with a variety of anionic substrates, as measured by ITC at 30 °C using the corresponding tetrabutylammonium salts. Data taken from Refs. [106–109].

Anion	Cl⁻	Br⁻	CN⁻	NO$_2^-$	CH$_3$CO$_2^-$	C$_6$H$_5$CO$_2^-$	H$_2$PO$_4^-$
Solvent	CH$_3$CN	CH$_3$CN	CH$_3$CN	CH$_3$CN	CH$_3$CN	CH$_3$CN	DMSO
K_a	140,000	3400	9400	1700	290,000	120,000	5100

4.4.1.1 Extended-cavity Systems

To improve the anion binding selectivity of calix[4]pyrrole, two kinds of so-called "deep"- or "extended"-cavity calix[4]pyrrole (**19**) have been prepared. In both the size of the cavity was enhanced by "appending" bulky groups, substituted aryl [110–113] subunits [114] on to the four meso positions. The resulting sterically congested environment, enforced around the anion-binding site, was found to improve the anion-binding selectivity. In the aryl systems electronic effects were also observed [115].

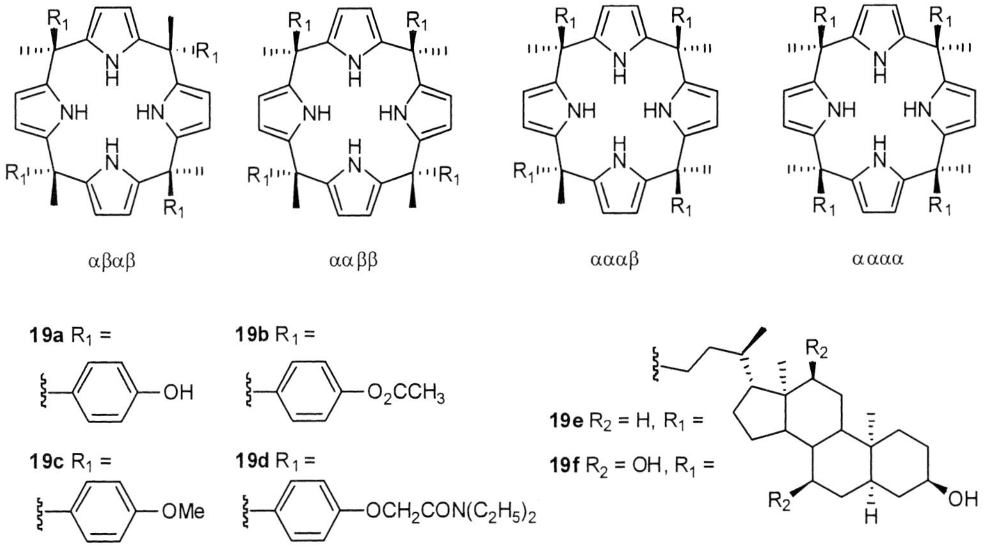

$\alpha\beta\alpha\beta$ $\alpha\alpha\beta\beta$ $\alpha\alpha\alpha\beta$ $\alpha\alpha\alpha\alpha$

19a R$_1$ =

19b R$_1$ =

19c R$_1$ =

19d R$_1$ =

19e R$_2$ = H, R$_1$ =

19f R$_2$ = OH, R$_1$ =

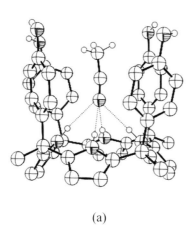

(a)   (b)

Figure 4.19 (a) Single-crystal X-ray structure of the acetonitrile complex of calix[4]pyrrole **19a**. (b) Single-crystal X-ray structure of the chloride anion complex of calix[4]pyrrole **19b**. These two structure were generated using information downloaded from the CCDC and correspond to structures originally reported in Refs. [112, 113].

Both the *meso*-tetraaryl and tetrasteroidyl deep-cavity systems were prepared by acid-catalyzed condensations involving pyrrole and the requisite ketone. Such reactions provide a mixture of four configurational isomers which are termed αβαβ, ααββ, αααβ, and αααα, respectively, by analogy with the porphyrin literature. X-ray structural analysis of the two aryl systems **19a** and **19b** revealed receptor–substrate complexes with high walls and well-defined binding cavities (Fig. 4.19), at least for the αααα isomers [112, 113].

Not surprisingly, the presence of a well-defined binding cavity resulted in greatly enhanced selectivity for fluoride anion relative to chloride anion and dihydrogen phosphate anion. This enhancement is presumably because small guests such as fluoride anion can fit into the size-limited binding cavities. Consistent with this assumption, significantly different anion-binding affinity was observed in acetonitrile solution, as determined by standard ^{1}H NMR titration methods (Table 4.2). For example, the ααββ isomer of **19a**, the isomer with the most open binding cavity, was found to bind chloride anion roughly four times as well as the other isomers. Data from this same solution-phase binding study also revealed obvious electronic effects, despite the substantial distance between the pyrrolic NH donor atoms and the aryl substituents. For instance, each isomer of **19a** has considerably greater anion affinity than do the corresponding isomers in the methoxy-substituted series **19c**.

In DMSO-d$_6$, a more polar solvent system than acetonitrile, the αααα-isomers of the two extended cavity receptors **19b** and **19c** were found to bind only fluoride anion within the limits of detection [110, 112]. Even for this anion, however, the binding affinity was low. For example, in DMSO-d$_6$ the fluoride anion-binding affinity of **19b** is ca. 74 M^{-1}.

Table 4.2 Association constants for compounds **18a**, **19a**, and **19c** with anionic substrates[a] in acetonitrile-d$_3$ (0.5 % v/v D$_2$O) at 22 °C. Data taken from Ref. [113].

	Compound						
	18a	**19a**			**19c**		
Isomer	ααββ	αααβ	αααα	ααββ	αααβ	αααα
F⁻	>10,000	>10,000	5000[b]	>10,000[c]	460	1100[b]	>10,000
Cl⁻	>5000	1400[d]	260	320	<100	220	300
H₂PO₄⁻	1 300	520[d]	230	230	<100	< 80	< 100

[a] Acetonitrile-d$_3$ (0.5 % v/v D$_2$O) solutions of receptors **18a**, **19a**, and **19c** were titrated by adding increasing amounts of concentrated acetonitrile-d$_3$ (0.5 % v/v D$_2$O) solutions of the anions in question (in the form of their tetrabutylammonium salts). To account for dilution effects and to simplify analysis these latter solutions also contained **18a**, **19a**, and **19c** at their initial concentrations. Estimated errors were <15 %. Binding stoichiometries, determined by means of Job plots, were 1:1 unless otherwise indicated
[b] Fit by following the change of two different β-pyrrolic CH resonances
[c] At high [F⁻]/[calixpyrrole] ratios a second binding process involving, presumably, interactions between the fluoride and the phenolic OH residues, is observed
[d] Fit by following the change in both the *meso*-aryl CH and β-pyrrole CH resonances

The steroid-based calix[4]pyrrole systems **19e** and **19f** were introduced by Král and Sessler in 2002. FAB-MS screening was used to determine the ability of these systems to effect the enantioselective recognition of tartaric acid and mandelic acid (Fig. 4.20)

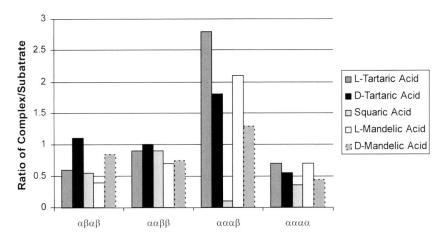

Figure 4.20 Results of FAB-MS screening experiments designed to probe the interactions of **19f** with selected anions. In these studies a large excess (>100-fold) of the carboxylic acid in question was added to a MeOH solution of the receptor and subjected to FAB-MS analysis. The values given are the ratios in percentages relative to the parent-ion peak. It should be noted, of course, that these results give only a semiquantitative approximation of the relative concentrations of the species in question under more normal solution-phase conditions. Data taken from Ref. [114].

[114]. Evidence for enantioselective binding was found for the polyhydroxylated $\alpha\alpha\alpha\beta$ configurational isomer of **19f**. This finding is rationalized in terms of multiple substrate–receptor hydrogen-bonding interactions that involve not only specific anion pyrrolic NH contacts but also less well-defined steroid–substrate interactions. The importance of these latter ancillary contacts is emphasized by the observation that **19e** proved far less effective than **19f**, as judged by both mass spectrometric screening and more direct extraction methods.

4.4.1.2 Higher-order System

Although not isolable from the normal acid-catalyzed reaction of pyrrole and simple ketones, ever since Sessler's 1996 report there has been interest in preparing so-called higher-order calix[*n*]pyrrole (*n* > 4) systems. Such systems, it could be expected, might have different anion-binding affinities or display selectivity for larger anionic substrates. In recent years several groups have, in fact, succeeded in preparing higher-order calix[*n*]pyrroles (e.g. **21–24**) and bipyrrole-derived analogs (e.g. **25** and **26**).

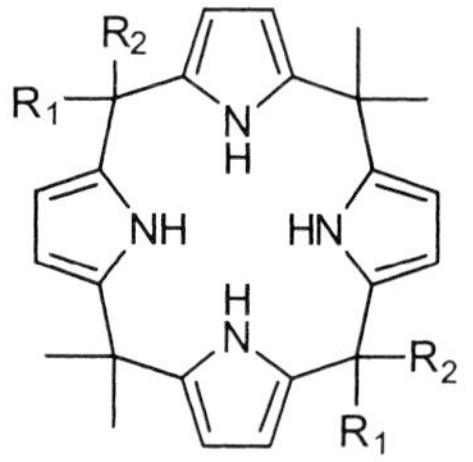

20a R$_1$ = R$_2$ = phenyl
20b R$_1$ = R$_2$ = 2-pyridyl
20c R$_1$-R$_2$ = 9,9-fluorenediyl

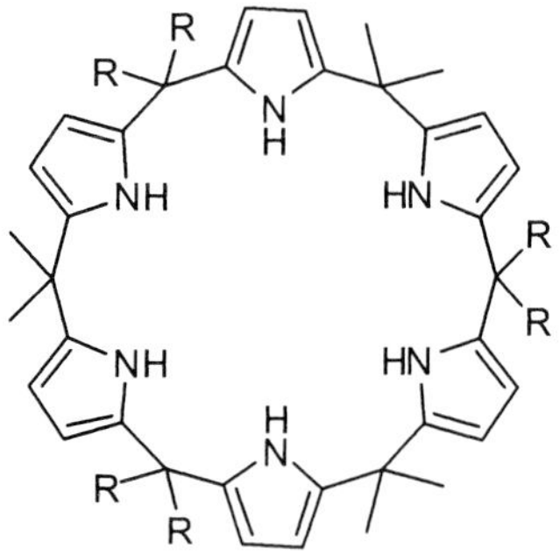

21a R = phenyl
21b R = 2-pyridyl

The first synthesis of a free-standing higher-order calix[6]pyrrole was achieved by Eichen and coworkers [116, 117]. The resulting receptors, compounds **21a** and **21b**, have two more pyrrole units and an expanded core relative to calix[4]pyrrole. Table 4.3 lists the association constants of calix[4]pyrrole and calix[6]pyrrole, as derived from ^{1}H NMR titration studies performed in acetone-d$_3$–CDCl$_3$ (1:9). For the *meso*-octamethylcalix[4]pyrrole **18a** and its phenyl-substituted congener **20a** clearly greater affinity for fluoride anion is observed. In contrast, the extended calix[6]pyrrole system **21a** has greater affinity for iodide anion (the binding order is I$^-$ > F$^-$ >> Cl$^-$ > Br$^-$ rather than F$^-$ > Cl$^-$ > Br$^-$ > I$^-$ as seen for **20a**). The greater affinity of these systems for larger anions emphasizes the importance of a proper geometric fit between the host and guest. Whereas the core size of the expanded system enables six pyrrolic NH to interact with the bound iodide anion, the smaller anion, fluoride, is presumably unable to benefit from the full complement of such multiple interactions. Similar effects have been invoked to rationalize the cation selectivity of crown ethers [118, 119].

Table 4.3 Anion affinity constants for the binding of anions to calix[6]pyrrole **21a** and calix[4]pyrrole **20a**, as determined from ^{1}H NMR spectroscopic titrations conducted in acetonitrile-d$_3$–CDCl$_3$ (1:9) at 298 K. The anions were studied in the form of their tetrabutylammonium salts. Data taken from Ref. [120].

	F$^-$	Cl$^-$	Br$^-$	I$^-$	BF$_4^-$	CF$_3$CO$_2^-$
20a	23,800	6800	270	<10	<10	70
21a	1080	650	150	6600	2350	1150

Using a very different synthetic approach, but one that is also characterized by extreme elegance, Kohnke and coworkers reported the synthesis of the *meso*-decamethylcalix[5]pyrrole and *meso*-dodecamethylcalix[6]pyrrole systems **22** and **23a**. These compounds were prepared from calix[n]furan (n = 5, 6) by first using *m*-CPBA to open the furan heterocycles and then using ammonium acetate to form the requisite pyrrole rings [121, 122]. Single crystals of both the chloride and bromide complexes of **23a** were obtained and elucidation of the corresponding X-ray diffraction structures revealed that the anions were coordinated at the center of the macrocycle via six hydrogen bonds (Fig. 4.21). Although these two structures are quite similar, for the chloride anion complex the N$\cdots$Cl distances range from 3.265 to 3.305 Å (compared with a range of 3.264–3.331 Å for the corresponding calix[4]pyrrole–chloride structure [102]), whereas the N$\cdots$Br distances range from 3.344 to 3.404 Å [123].

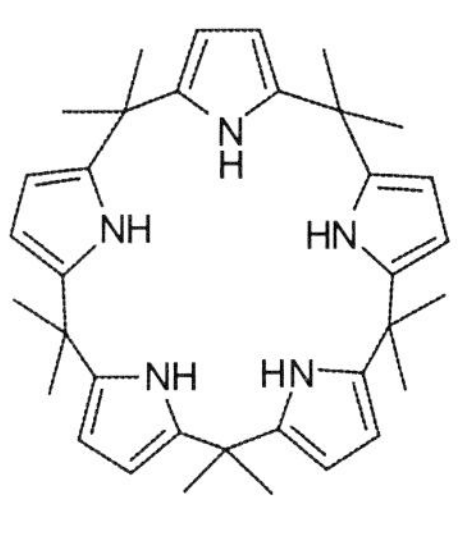

22

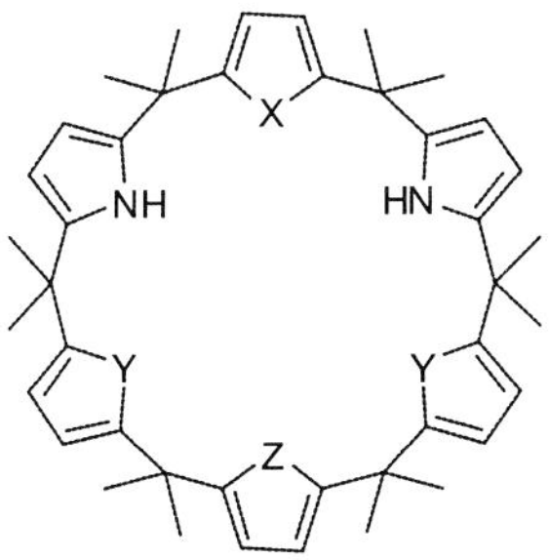

23a X = Y = Z = NH
23b X = Y = O, Z = NH
23c X = Z = O, Y = NH
23d X = O, Y = Z = NH

Results from anion-binding studies, performed with **22** and **23**, are shown in Table 4.4. It is difficult to make comparisons within this data set, because not all the K_a values were obtained under identical conditions or using the same method. Nonetheless, to the extent that such comparisons may be made, it can be seen from Table 4.4 that for simple halide anions the association constants increase with the number of pyrrolic NH units present in the macrocycles. Further, the inherent selectivity changes. In the calix[4]pyrrole **18a**, for instance, the fluoride-to-chloride K_a ratio is approximately 59. In contrast, the corresponding ratio for calix[6]pyrrole **23a**

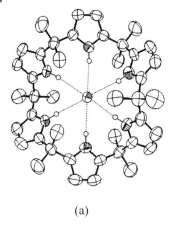
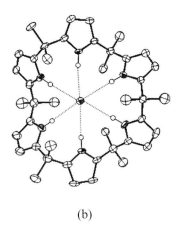

(a)   (b)

Figure 4.21 (a) Single-crystal X-ray structure of the chloride anion complex of calix[6]pyrrole **23a**. (b) Single-crystal X-ray structure of the corresponding bromide anion complex. These two structures were generated using information downloaded from the CCDC and correspond to structures originally reported in Ref. [123].

is ca 27, indicating fluoride selectivity is lower than for the parent calix[4]pyrrole. Receptors **22** and **23c**, in contrast, have greater selectivity for fluoride than for chloride, with the relevant K_a ratios being ca 400 and 880, respectively.

Table 4.4 Association constants (K_a, M^{-1}) for a variety of 1:1 anion complexes. The counter cation was always n-Bu$_4$N$^+$; detailed experimental conditions are indicated in the footnotes. Data taken from Refs. [122, 124, 125].

Anion	18a	22	23a	23c	23d
F$^-$	2700[a]	14,000	ca. 320,000[g]	57,000[g]	_[f]
Cl$^-$	46[a]	35[b]	12,000[e]	65[a]	5500[e]
Br$^-$	10[d]	–	710[c]	<10[a]	69[c]
I$^-$	<10[d]	–	_[I]	_[i]	_[i]
H$_2$PO$_4$$^-$	97[d]	–	_[h]	_[i]	_[i]
HSO$_4$$^-$	<10[d]	–	ca 10[c]	_[i]	_[i]
NO$_3$$^-$	<10[d]	–	16[c]	_[i]	<10[c]
CN$^-$	<10[a]	–	ca 100[a]	<100[a]	_[h]

[a] Measured by NMR spectroscopic titration in D$_2$O-saturated CD$_2$Cl$_2$ at 20 °C
[b] Measured by NMR spectroscopic titration in D$_2$O-saturated CD$_2$Cl$_2$ at 22 °C
[c] Measured by NMR spectroscopic titration in "dry" CD$_2$Cl$_2$ at 20 °C
[d] Measured by NMR spectroscopic titration in "dry" CD$_2$Cl$_2$ at 25 °C
[e] Determined using the Cram extraction method at 16 °C
[f] K_a proved too high to be determined by NMR spectroscopic titration methods. Moreover, the compound also proved ineffective as a phase-transfer agent between D$_2$O and CD$_2$Cl$_2$, meaning the Cram extraction method could likewise not be used
[g] Measured by competition experiment using calix[4]pyrrole **18a** as the competing species
[h] An interaction is observed, but the NH resonances disappeared on addition of the salt; thus, a K_a value could not be determined
[i] No detectable binding observed in either dry or wet CD$_2$Cl$_2$

In 2000 a direct synthesis of higher-order calix[*n*]pyrroles (**24b**, *n* = 5; **24d**, *n* = 8) was discovered. It relied on the use of 3,4-difluoropyrrole rather than simple pyrrole [126]. Because 3,4-difluoropyrrole is less reactive than pyrrole, the standard acid-catalyzed macrocyclization reaction proceeds under kinetic, rather than thermodynamic, control at ambient temperature. For presumably related electronic reasons the higher-order systems, which are always obtained in lower yields than the calix[4]pyrrole species **24a**, are also quite stable. Subsequent to the initial report it also proved possible to isolate the calix[6]pyrrole species (**24c**) from an analogous condensation reaction performed under modified conditions [127].

X-ray diffraction analysis revealed that the fluoride anion complex of the *meso*-octamethyloctafluorocalix[4]pyrrole **24a** adopts a normal cone-conformation in the solid state (Fig. 4.22a). In this conformation the average N···F distance is 2.766 Å, nearly the same as that in the fluoride anion complex of calix[4]pyrrole **18a**. In contrast, considerable distortion is seen in the crystal structure of *meso*-decamethyldecafluorocalix[5]pyrrole **24b**, although it should be noted that this structure reflects the

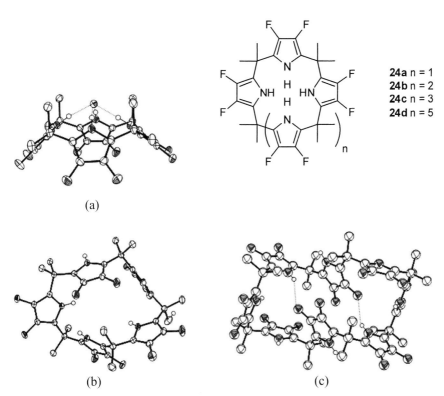

24a n = 1
24b n = 2
24c n = 3
24d n = 5

(a)

(b)   (c)

Figure 4.22 (a) Single-crystal X-ray structure of the fluoride anion complex of octafluorocalix[4]pyrrole **24a**. (b) Single-crystal X-ray structure of the decafluorocalix[5]pyrrole **24b**. (c) Single-crystal X-ray structure of the hexade-cafluorocalix[8]pyrrole **24d**. These three structures were generated using information downloaded from the CCDC and correspond to structures originally reported in Refs. [126, 128].

anion-free form (Fig. 4.22b). A high level of distortion, coupled with the presence of interannular $CF \cdots H$ hydrogen bonds, also characterizes the structure of **24d** (Fig. 4.22c).

Most recently, the comparable anion-binding affinities of the fluorinated calix[n]pyrroles were measured by ITC in solution under well-defined conditions. They were also studied by means of extraction experiments. In the extraction experiments, which involved partitioning between organic and aqueous layers, octafluoro-calix[4]pyrrole **24a** discriminated little among chloride, bromide, and iodide anions. These halide anions were extracted more effectively than nitrate and fluoride anions, however. The decafluorocalix[5]pyrrole **24b**, in contrast, was found to have greater affinity for nitrate and fluoride anions [129].

ITC titration studies, performed under very different conditions (e.g. dry acetonitrile and DMSO) revealed slightly different trends. In these it was found that the relative association constants ($K_{rel} = K_{a(Cl)}/K_{a(Br)}$) decrease with increasing macrocycle size (Table 4.5). This behavior is the opposite of that expected on the basis of anion electronegativity. Specifically, chloride anion, being more electronegative and having a higher charge density, was expected to be bound more strongly than bromide anion. Whereas this is true in an absolute sense, for **24c** the K_{rel} value is rather small, presumably because the relatively large cavity present in calix[6]pyrrole is better able to accommodate bromide anion than chloride anion. Interestingly, for $H_2PO_2^-$, an inherently less spherical anion, the association constants were seen to increase as the number of β-fluorinated pyrrolic subunits increased.

Table 4.5 Association constants (K_a, M^{-1}) determined in CH_3CN or DMSO solvents by ITC at 30 °C using the n-Bu_4N^+ salts of the indicated anions.[a] Data taken from Refs. [106, 109].

Anion	Solvent	24a	24b	24c
Cl^-	CH_3CN	530,000	41,000	280,000
Br^-	CH_3CN	8500	4500	110,000
I^-	CH_3CN	_[b]	_[b]	610
$H_2PO_4^-$	DMSO	17,000	9600	15,000
$H_2PO_2^-$	DMSO	3300	13,000	35,000
$C_6H_5CO_2^{-[c]}$	CH_3CN	1,390,000	83,000	580,000
$CH_3CO_2^{-[c]}$	CH_3CN	2,360,000	520,000	1,020,000
$K_{rel} = K_{a(Cl)}/K_{a(Br)}$		62.4	9.1	2.5

[a] The host (macrocycle) solution was titrated with the guest (anion) solution unless otherwise indicated
[b] No evidence of anion binding was observed
[c] The guest solution was titrated with the host solution (reverse titration)

To make more efficient receptors for larger anions, several groups have recently begun working on the generation of new extended calixpyrrole-type systems based on different combinations of bipyrrole, furan, thiophene, and pyrrole. Not all of these systems have been shown to be useful anion receptors. One set of compounds for which this utility has been established is the bipyrrole macrocycles **25a** and **25b**. For the calix[3]bipyrrole **25a**, X-ray structural analysis of the chloride anion complex re-

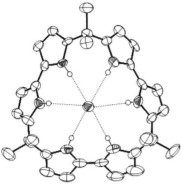

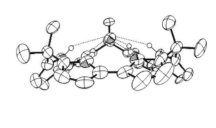

Figure 4.23 Single-crystal X-ray structure of the chloride anion complex of calix[3]bipyrrole **25a**. The structure was generated using infor- mation downloaded from the CCDC and cor- responds to a structure originally reported in Ref. [108].

vealed that the macrocycle adopts a cone conformation and that all six pyrrolic NH groups are involved in hydrogen-bonding interactions involving the chloride anion (Fig. 4.23). The N···Cl distances are in the range 3.338–3.382 Å [108], which is longer than that observed for calix[4]pyrrole (3.264–3.331 Å [102]).

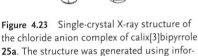

25a n = 1
25b n = 2

The X-ray structure of the chloride anion complex of calix[4]bipyrrole **25b** was also solved recently. It reveals that a simple chloride anion is bound to the V-shaped cleft of this receptor and is held in place in the resulting "pocket" by interactions involving all eight pyrrolic NH groups (Fig. 4.24). The N···Cl distances range from 3.422 to 3.571 Å.

The observation of bound anions in the solid-state structures of the various cal- ixbipyrrole macrocycles systems led to suggestions that these systems might be ca- pable of acting as anion receptors in organic solution. It was also expected that these receptors, to the extent they bound anions, would favor larger species (e.g. bromide anion in preference to chloride anion). Table 4.6 provides support for this latter no-

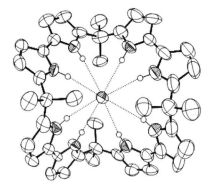
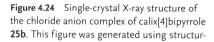
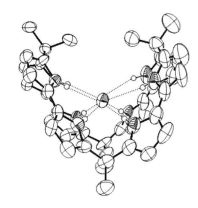

Figure 4.24 Single-crystal X-ray structure of the chloride anion complex of calix[4]bipyrrole **25b**. This figure was generated using structur- al data provided by Dr Vincent M. Lynch of the Department of Chemistry and Biochemistry, The University of Texas at Austin.

tion. Specifically, it contains data which emphasize that receptor **25a**, more effective- ly binds bromide and iodide anions than calix[4]pyrrole. For the larger calix[4]bipyr- role homolog **25b**, a receptor containing eight potential pyrrolic NH donor groups, the association constant for chloride was observed to be much higher than that for bromide. It was also found that the absolute anion affinity of this system was usual- ly greater than that of calix[4]pyrrole. Presumably, this set of combined observations reflects the fact that both size- and geometry-based matching between the macro- cyclic receptor and anionic substrates, and the total number of hydrogen-bond donors available for complexation (pyrrolic-NH in this instance), are important in regulating anion-binding affinity.

Table 4.6 Association constants (K_a, M^{-1}) for interaction of **25** and **26** with anions at 30 °C.[a] Data taken from Refs. [88, 107, 108].

Anion	Solvent	25a	25b	26a	26b
Cl^-	CH_3CN	110,000	2,940,000	1100	940
Br^-	CH_3CN	100,000	120,000	37[b]	103[b]
I^-	CD_3CN	9300[b]	56[b]	–	–
HSO_4^-	$CDCN_3$	–[c]	– 1	25[b]	28[b]
$C_6H_5CO_2^-$	CH_3CN	940,000	–	63,000	100,000
$CH_3CO_2^-$	CH_3CN	–[c]	–	78,000	140,000

[a] K_a values derived from ITC titrations at 30 °C unless otherwise indicated
[b] Value obtained from ^1H NMR titrations at 25 °C
[c] No evidence of anion binding is observed

A new set of anion receptors, **26a** and **26b**, based on bipyrrole "combined" with ei- ther furan or thiophene was recently prepared. These systems have good affinity for Y-shaped anions, for example benzoate and acetate, yet bind such classic spherical anions as chloride and bromide less well. As can be seen from an inspection of

Table 4.6, association constants for binding of benzoate to **26a** and **26b** are approximately 60 times higher than the corresponding chloride anion-binding affinities [107]. In contrast, calix[4]pyrrole **18a** (cf. Table 4.1) has no benzoate–chloride anion selectivity, whereas calix[3]bipyrrole **25a** has intermediate selectivity behavior (approximately ninefold). Taken together, these observations are consistent with the notion that controlling the internal cavity size and overall shape of pyrrole-derived receptors can lead to systems with selectivity optimized for certain classes of anion.

26a X = O
26b X = S

4.4.1.3 Strapped Systems and Other Related Receptors

One of the guiding themes in supramolecular chemistry is that increasing the extent of preorganization can lead to receptors with improved selectivity, enhanced affinity, or both. For the calixpyrroles this possibility has been explored by creating deep cavities as detailed in Sect. 4.4.1.1 or by expanding the core size as discussed in Sect. 4.4.1.2. It can also be achieved by increasing the dimensionality of the receptors. This last approach, which actually defines one of the current frontiers in pyrrole-based anion receptor design, is leading to the construction of strapped, capped, and cryptand-like systems. The first example of a "strapped" or "capped" calix[4]pyrrole was reported by Lee and coworkers in 2002 [130]. This system, **27**, is the first of a growing class of calix[4]pyrrole receptors bearing trans-substituted straps on one face of the molecule. Anion-binding affinity was assessed qualitatively by means of ^{1}H NMR spectroscopy, following the chemical shift changes or integrated intensity changes of a variety of receptor-derived signals as a function of anion concentration.

Anion-binding affinity was assessed quantitatively for receptor **27** using ITC methods. From these studies it was found that the chloride anion-binding affinity is ca 1.0 × 10^5 M^{-1} in DMSO. The corresponding fluoride-binding affinity, obtained from an ^{1}H NMR spectroscopic competition experiment, was found to be approximately 3.9 × 10^6 M^{-1} [130]. On the other hand, no appreciable binding interactions were observed in the presence of bromide, iodide, sulfate, or dihydrogen phosphate anions.

One year after Lee's original report, the X-ray crystal structure of the chloride anion complex of **27** was obtained. In the structure, shown in Fig. 4.25, the calix[4]pyrrole core assumes a cone-like conformation and the chloride anion is encapsulated within the three-dimensional binding cavity.

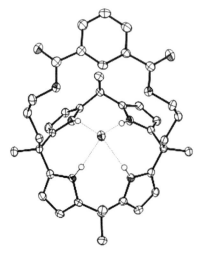

27

28a n = 1
28b n = 2
28c n = 3

29a n = 1
29b n = 2

On a different level, it was recognized that by varying the nature of the strap, fundamental insights into the relationship between the structure of the receptor, the size and shape of the anion, and the binding affinity might be obtained. In 2002 Sessler and Lee reported the synthesis of a new set of strapped calix[4]pyrroles, **28**, bridged by ether-containing straps of different lengths [109]. They have recently extended this approach to include the amide-strapped analogs **29** [131]. Table 4.7 summarizes affinity constants for binding of chloride, bromide, and iodide anion to a variety of strapped calix[4]pyrroles as determined by ITC in dry acetonitrile. Inspection of this table emphasizes the advantages that accrue as the result of "strapping" one face of a calix[4]pyr-

Figure 4.25 Single-crystal X-ray structure of the chloride anion complex of strapped calix[4]pyrrole **27**. This figure was generated using information downloaded from the CCDC and corresponds to a structure originally reported in Ref. [109].

role. For instance, use of a tight C-4 ether strap (system **28a**) gives rise to an extremely high chloride anion-binding affinity. In contrast, the use of a C-6 ether strap (**28c**) produces a system with enhanced bromide anion binding ability, at least relative to the parent calix[4]pyrrole system **18a**. The chloride anion-binding affinity of the C-4 and C-5 amide strapped calix[4]pyrroles **29a** and **29b** was found to be almost identical but still generally higher than those of the other strapped systems. Greater selectivity was observed for bromide anion. In particular, the C4-amide **29a** has the highest bromide anion association constant yet recorded for a calix[4]pyrrole-type receptor whereas the C5-amide **29b**, which has a larger cavity than the C4-amide system **29a**, has very high affinity for iodide anion. Presumably, these findings reflect the intuitively appealing expectation that the use of a longer strap provides a receptor whose size better complements that of the larger halide anions, bromide and iodide.

Table 4.7 Halide anion affinity constants for receptors **27**, **28**, and **29**. Determinations were made by ITC in acetonitrile at 30 °C using the corresponding n-Bu$_4$N$^+$ salts. Data taken from Refs. [109, 131].

	18a	27	28a	28b	28c	29a	29b
Cl$^-$	140,000	1,380,000	3,630,000	1,370,000	1,370,000	3,350,000	3,240,000
Br$^-$	3400	~0	30,000	31,000	120,000	1,250,000	700,000
I$^-$	N.D.[a]	–	–	–	–	2300	3100

[a] No evidence of anion binding was observed

In 1996 Sessler and coworkers reported the first structurally characterized three-dimensional oligopyrrole system, the "tripod" **30**. It was prepared by acid-catalyzed condensation of mono-formylbipyrrole and mono-α-free bipyrrole [132]. X-ray crystal structure analysis revealed that tripod **30a** occurs as a diastereogenic dimer in the solid state, wherein each monomer has a right or left-handed twist. This dimerized structure is apparently stabilized by twelve hydrogen-bond interactions involving the

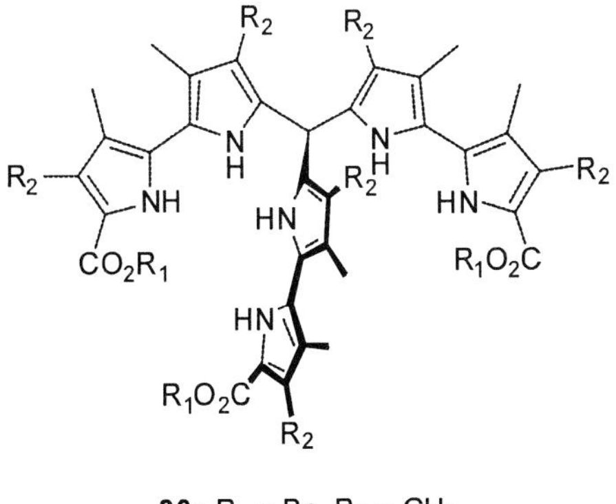

30a R$_1$ = Bn, R$_2$ = CH$_3$
30b R$_1$ = R$_2$ = C$_2$H$_5$

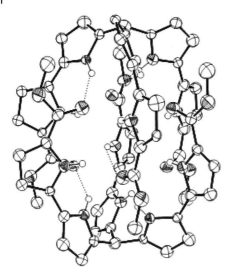

Figure 4.26 Single-crystal X-ray structure of tripod **30a** showing the dimer formed between two molecules of **30a** in the solid state. In this view the benzyl and methyl groups have been removed for clarity. This figure was generated using information downloaded from the CCDC and corresponds to a structure originally reported in Ref. [132].

pyrrolic NH groups and the benzoate oxygen atoms (Fig. 4.26). In the gas phase a dimer peak was observed by FAB mass spectrometric analysis. In protic solvents ^{1}H NMR spectroscopic studies provided evidence that tripod **30a** also exists as a dimer in solution. The fact that such a self-assembled dimer was found to exist over such a range of conditions, provided an incentive to try building corresponding three-dimensional cryptand-like species wherein the top and bottom "halves", analogous to **30**, would be linked via covalent bonds.

The first synthesis of an oligopyrrole-based cryptand was achieved by Beer and coworkers in 2001 [133]. An α-formylated tripyrrolylmethane was condensed with either ethylenediamine or butylenediamine to produce the imine-linked products **31**. X-ray crystal structural analysis revealed that the neutral diamine, used as a reactant, was trapped inside the three-dimensional cavity of cryptand **31a**, being held there by two different sets of hydrogen-bonding interactions involving the pyrrolic NH groups, the Schiff base nitrogen atoms and the diamine hydrogen atoms (Fig. 4.27). The inclusion of the diamine suggests that it acts not only as a reactant but also as a template for formation of the cryptand. Although this hypothesis has yet to be tested, quantitative determinations of the stability constants for binding of the diamine were made using ^{1}H NMR spectroscopic titration methods. Specifically, it was found that the neutral ethane-1,2-diamine and ethane-1,2-diol were bound with affinities of ca 1500 and 1060 M^{-1}, respectively, in CDCl$_3$.

Nearly contemporaneous with the Beer report, Sessler and coworkers published the first carbon–carbon-linked polypyrrole cryptand, the "three dimensional calix-pyrrole" **32a** [134]. Although receptor **32a** is too small to enable internal binding of

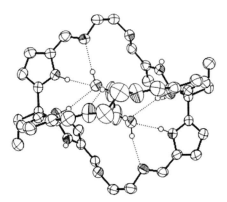

31a n = 1
31b n = 2

substrates (it contains essentially no cavity-like void), it has three identical binding surfaces and was found to bind solvents in the solid state, as shown in Fig. 4.28. This led to the suggestion that it could stabilize anion–receptor complexes of 1:1, 1:2, or 1:3 binding stoichiometry in solution. In fact, ^{1}H NMR spectroscopic experiments, conducted in CD_2Cl_2 and THF-d_8, revealed host–guest stoichiometries that were even more complicated and depended directly on the choice of anion. For example, fluoride, when added to **32a** as an anionic guest, was found to be bound to six of the nine pyrrolic subunits, in an apparent 1:1 fashion. In contrast, analysis of the binding data revealed that a single chloride anion interacts with two molecules of receptor **32a** in solution (1:2 anion:receptor stoichiometry), being bound with a stability constant of 3.1×10^6 M^{-2} in CD_2Cl_2. Finally, 2:1 anion:receptor stoichiometry was observed on addition of tetrabutylammonium nitrate. In this case, the stability constants, K_{a1} and K_{a2}, in CD_2Cl_2 solution were estimated from ^{1}H NMR spectroscopic analyses to be ca 1700 and 420 M^{-1}, respectively.

The oxidized form of **32b** was also reported, quite recently, by Sessler and coworkers [135]. This interesting structure contains sp^2 bridging carbon atoms at the bridge-

Figure 4.27 Single-crystal X-ray structure of the ethane-1,2-diamine complex of bis(tripyrroyl)cryptand **31a**. This figure was generated using information downloaded from the CCDC and corresponds to a structure originally reported in Ref. [133].

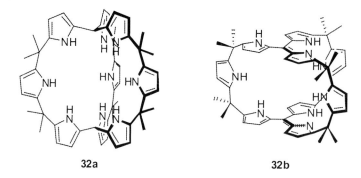

32a　　　　　　　　**32b**

head positions. It can thus be viewed as being a "3D calixphyrin". The capacity of this new system to act as a possible anion receptor has not yet been probed, however.

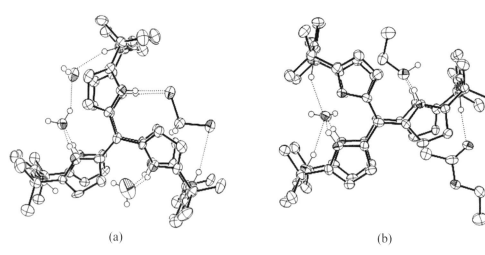

(a)　　　　　　　　　　　　　　(b)

Figure 4.28 (a) Single-crystal X-ray structure of the cryptand-like calixpyrrole **32a** showing the three bound water molecules and one bound molecule of dichloromethane. (b) Single-crystal X-ray structure of the 3D calixphyrin **32b** showing the one bound water molecule, one bound ethanol molecule, and one bound ethyl acetate molecule. These two structures were generated using information downloaded from the CCDC and correspond to the structures originally reported in Refs. [134, 135].

4.4.2
Linear Receptors

Whereas the anion-binding affinity of cyclic oligopyrrole receptors has been extensively studied over the last decade, until recently little attention had been paid to linear (acyclic) pyrrole-based anion receptors, because of a perception they would have low anion-binding affinity. Belying this misconception, Gale and his group, building

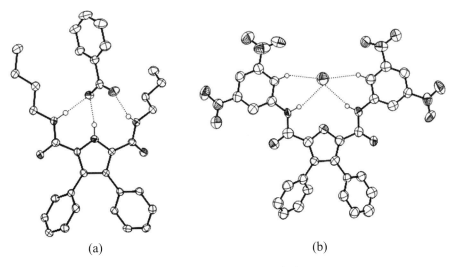

33a R = n-Bu
33b R = Ph
33c R = 4-nitrophenyl
33d R = 3,5-dinitrophenyl

34a R = H
34b R = COCH₃

35a R = n-Bu
35b R = Ph

on precedent from Schmuck and Crabtree [136–147], reported a series of highly effective acyclic mono- and dipyrrole-based anion receptors.

This series of receptors, represented by structures **33**, **34**, and **35**, were synthesized by reaction of pyrrolic diacid chlorides and a variety of amine derivatives. The crystal structure of the benzoate complex of the linear receptor **33a** was published in 2002. It revealed that all three hydrogen-bond donors are indeed involved in hydrogen bonding to this particular anionic guest (Fig. 4.29a) [148]. More recently, the single-crystal X-ray diffraction structure of the chloride anion complex of the deprotonated receptor **33d** was reported. In this structure the deprotonated receptor binds the

(a)   (b)

Figure 4.29 (a) Single-crystal X-ray structure of the benzoate anion complex of 2,5-diamide-pyrrole **33a**. (b) Single-crystal X-ray structure of the chloride anion complex of the deprotonated form of the linear pyrrole-based anion receptor **33d**. These two structures were generated using information downloaded from the CCDC and correspond to structures originally reported in Refs. [148, 149].

bound chloride anion via two amide NH hydrogen bonds and two phenyl CH$\cdots$Cl interactions (Fig. 4.29b) [149].

Table 4.8 summarizes the different anion-binding affinity of receptors **33** and **34**, as inferred from ^{1}H NMR spectroscopic analysis. These data reveal that **33** and **34** act as selective receptors for oxo-anions in polar organic solvents (e.g. CD$_3$CN or DMSO-d$_6$). Within each receptor series, however, some differences between the individual binding characteristics were observed. For instance, **33a** had a higher benzoate anion association constant than its congeners [150]. For **33c** and **33d**, receptors bearing attached electron-withdrawing groups on the amide phenyl subunit, anion-binding affinities were recorded that are generally higher than those of **33b** (for **33c** $K_a = 1250$, 40, and 4150 M^{-1} for fluoride, chloride, and benzoate, respectively, and for **33d** $K_a = 53$ and 4200 M^{-1} for chloride and benzoate; DMSO-d$_6$–0.5 % water) [149].

Table 4.8 Anion affinity constants of linear receptors (M^{-1}) determined by ^{1}H NMR spectroscopic titrations performed at 25 °C.[a] Data taken from Refs. [150–153].

Anion	33a	33b	34a	34b	35a	35b
	CD$_3$CN	DMSO-d$_6$	DMSO-d$_6$	DMSO-d$_6$	DMSO-d$_6$[d]	DMSO-d$_6$[d]
F$^-$	85	74	–	–	7560	8990
Cl$^-$	138	11	<20	<20	23	43
Br$^-$	<10	<10	_[b]	_[b]	13	10
H$_2$PO$_4^-$	357	1450	2 050[c]	525	_[e]	e
C$_6$H$_5$CO$_2^-$	2500	560	47.6	152	354	424
HSO$_4^-$	–	–	>10^4	_[b]	44	128

[a] Anions were added in the form of the corresponding tetrabutylammonium salts. Acetonitrile water content = 0.03 % and DMSO water content = 0.5 %
[b] No binding was observed
[c] Large errors were obtained when fitting this data to a 1:1 binding model for titrations performed in DMSO-d$_6$–0.5 % water. This titration was therefore repeated in DMSO-d$_6$–5 % water solution.
Although a better fit was obtained, the error (29 %) remained large
[d] Measured in DMSO-d$_6$–5 % water
[e] An adequate fit could not be obtained

Receptors **35a** and **35b** contain two pyrrole units and, not surprisingly, their anion binding behavior differs from that of **33** and **34**. For example, the halide anion association constants are higher and the oxyanion association constants are lower for both **35a** and **35b** than for receptors **33** and **34**. Further, these receptors were found to have high selectivity for fluoride anion in partially mixed organic–aqueous solutions [151].

4.5
Anion Carriers in Transport Applications

That cationic polypyrrole-based receptors bind anions in solution provides an incentive to test their utility as anion carriers. Such carriers, capable of transporting a variety of anionic substrates across normally impermeable biological membranes, could prove extremely useful. They could be used, for example, in remediation applications, being applied to the extraction- or membrane-based removal of common, eutrophication-inducing pollutants such as phosphate and nitrate anions from ground waters, or used to remove pertechnetate anion from radioactive wastes. They also could find application in dialysis applications, helping to remove, for example, excess phosphate from patients suffering from hyperphosphatemia and related disorders. Further, they could be considered as transport-enhancing devices for the intra-cell delivery of phosphorylated drugs, thus potentially facilitating the treatment of viral disorders such as AIDS.

As mentioned in the Introduction, it has been suggested that protonated prodigiosin acts as a naturally occurring pyrrole-based anion carrier. Specifically, it has been proposed that the prodigiosins mediate their observed immunosuppressive and anticancer activity by effecting the concurrent transport ("symport") of H^+ and Cl^- across cellular membranes [154, 155]. First advanced in 1998 by Ohkuma and Wasserman [154, 155], the Cl^-/H^+ symport mechanism is supported by a considerable body of experimental data. For example, before its initial suggestion, it was recognized that when baby hamster kidney cells were treated with acridine orange dye, as an indicator of pH, orange granules were observed in the cytoplasm of the cells [156]. On treatment with prodigiosin 25-C **1b** at a concentration of 2.5 μM, these yellow granules, presumably acidic vacuolar organelles, were found to disappear. It was found that prodigiosin 25-C did not affect cellular ATP levels, implying that the deacidification process is decoupled from ATP hydrolysis by V-ATPase [156]. Ohkuma et al. also demonstrated that both prodigiosin activity and prodigiosin-based lysosomal acidification depended on chloride anion concentration and that chloride anion channels were not responsible for these effects [154]. On the other hand, other proposed mechanisms have been proposed [55, 94, 95, 97, 98, 157–161]. This makes resolution of the mechanistic issues challenging, especially because no detailed anion-transport studies have yet been performed with prodigiosin or with any other open-chain oligopyrrole. However, the need to do so is clear, making the study of prodigiosin and analogs obvious targets for future research.

Hyperphosphatemia is an ailment caused by a build-up of excess phosphate in the body. It is prevalent among hemodialysis patients and is characterized by heightened calcium phosphate levels that, over time, lead to increased incidence of cardiovascular mortality [162, 163]. Removing the excess phosphate, perhaps by use of a carrier-mediated, out-of-cell transport-based approach, could serve to reduce this risk. The "other side" of this problem would involve development of selective carriers for phosphorylated species, including the active forms of known antiviral agents and antisense nucleotides. In this case, through-membrane into cell anion transport would enable potentially improved therapeutic treatments. In both cases what is needed are agents capable of binding to a phosphate (or phosphate-like) anion, neutralizing the charge, and producing what would be net hydrophobic species that can pass through the targeted biological membrane. Effective release of the phosphorylated entity after transport is also important.

By virtue of being the first pyrrole-based anion-binding systems to be discovered, sapphyrin derivatives have been extensively studied as possible phosphate anion carriers. In the context of such studies, sapphyrins **2** and **36** were considered very attractive, because they are hydrophobic and consequently would be expected to have high affinity for the relatively nonpolar environments of biological membranes. They have also been used to effect both nucleoside [164] and nucleotide [165] recognition under well defined conditions, and a cytosine-sapphyrin analog has proved useful as a solid support for the HPLC-based separation of monomeric and short oligomeric nucleotides at pH 7 (*vide infra*) [166].

36

Both spectroscopic and single-crystal X-ray diffraction methods were used to probe the phosphate anion behavior of sapphyrins **2** and **36**. The latter analyses proved particularly revealing and served to show that a variety of "pure" and "mixed" anion complexes could be formed in the presence of phosphates, including 1:1 and 1:2 (sapphyrin:phosphate) species. In the structurally characterized 1:1 complex a dihydrogen phosphate counteranion was found bound to the [**36**•H$_2$]$^{2+}$ core via five hydrogen-bonding interactions that served to link the phosphate oxyanion to the pyrrolic hydrogen atoms of the diprotonated sapphyrin core (Fig. 4.30a). A second kind of interaction is illustrated by the structure of the 1:2 complex formed between [**2**•H$_2$]$^{2+}$ and monobasic phenylphosphate (Fig. 4.30b). Here, two and three hydrogen bonds serve to link the anionic substrates to the top and bottom faces of the sapphyrin, respectively. The mixed chloride and cyclic-AMP complex of [**36**•H$_2$]$^{2+}$ was also solved and is shown in Fig. 4.30c. Here again, the two anions are bound on opposite faces of the sapphyrin macrocycle, with the phosphate oxyanion and chloride anion being tethered via two and three hydrogen bonds, respectively, by analogy with the structure of **2**·2HCl (Fig. 4.6) [73].

Although it has not yet been definitively demonstrated that sapphyrin can act as a through-membrane carrier under physiological conditions (i.e. in cells), a rage of chemical evidence has accumulated that supports the notion that these or other oligopyrrolic systems could be useful in this regard. Much of this work has been reviewed [167].

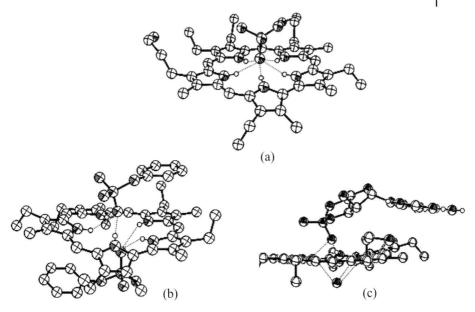

(a)

(b)   (c)

Figure 4.30 (a) Single-crystal X-ray structure of the 1:1 chelated complex formed between the diprotonated formed sapphyrin **36** and monobasic phosphoric acid. A second counteranion, present in the lattice, is not shown. (b) Single-crystal X-ray structure of the 1:2 chelated complex formed between sapphyrin **2** and bisphenylphosphate. (c) Single-crystal X-ray structure of the mixed salt formed between sapphyrin **36**, HCl, and the acid form of c-AMP. These structures were generated using information downloaded from the CCDC and correspond to structures originally reported in Refs. [73, 74].

The first test of sapphyrin as a potential anion carrier was made using a standard Pressman-type U-tube model membrane apparatus (Fig. 4.31). With this arrangement the anion transport ability of various putative sapphyrin-type anion receptors was measured by monitoring the movement of the targeted anions from a first aqueous phase (Aq I), through a layer of CH_2Cl_2 (Org), to a second receiving aqueous phase (Aq II). It was observed that the presence of sapphyrin **2** in the organic layer served to

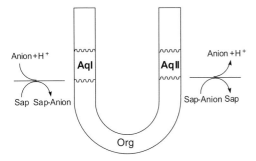

Figure 4.31 Schematic representation of the Pressman-type model membrane used to measure anion transport rates.

enhance the rate of fluoride anion transport from Aq I to Aq II over wide range of initial Aq I pH. Further, the rate of transport in the presence of monoprotonated sapphyrin **2** was found to be nearly 100 times greater than the background rate [68].

In similar model membrane experiments, the transport of adenosine 5'-monophosphate (5'-AMP) and guanosine 5'-monophosphate (5'-GMP) was achieved by using sapphyrin **2**. In this study it was found that reducing the pH of Aq I to below 4 (to provide a high concentration of the diprotonated sapphyrin) did not serve to enhance the rate of anion transport appreciably [67]. In subsequent work it was found that addition of triisopropylsilyl protected cytidine (C-Tips) enhanced the rate of sapphyrin-mediated 5'-GMP transport [168]. This latter finding led Sessler and coworkers to attach a nucleobase to the sapphyrin skeleton. The resulting combined systems, conjugates **37** and **38**, contain two putative recognition motifs – a sapphyrin-derived phosphate binding site and a Watson–Crick base-pairing element.

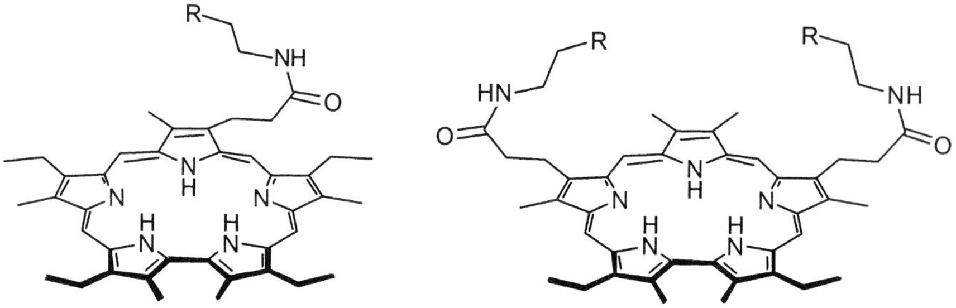

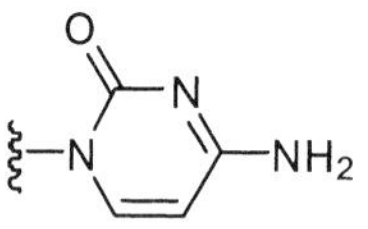

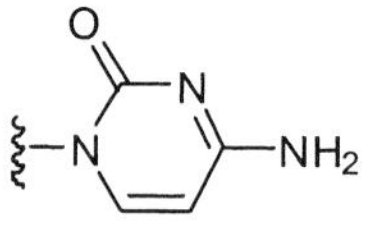

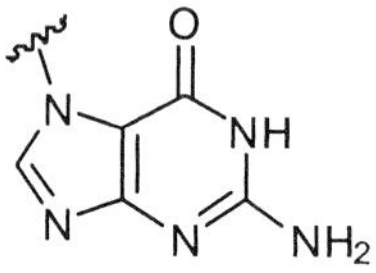

Tables 4.9–4.11 summarize the nucleotide phosphate transport rates observed for sapphyrins **37** and **38** under several different initial pH conditions (involving both Aq I and Aq II). The cytosine-bearing carriers **37a** and **38a** were found to enable selective transport of 5'-GMP over 5'-CMP (cytosine 5'-monophosphate) and 5'-AMP (by a factor of 8–100). These studies also revealed that better transport rates were always observed when the receiving phase (Aq II) is kept highly basic, as would be expected for a mechanism that involves combined anion + proton transport. Compound **37a**, which bears a single attached nucleobase only, was always found to have higher selectivity for 5'-GMP than **38a**, which contains two appended nucleobase sidechains [168]. Rationales proposed for these findings invoke a lower number of competing interactions for **37a**.

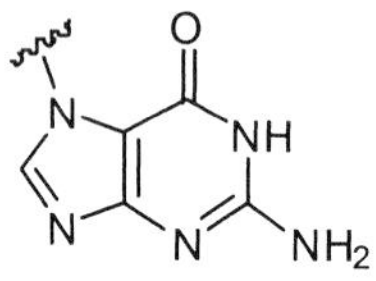

Table 4.9 Initial nucleotide-5'-monophosphate transport rates
(10^{-8} mol cm^{-2} h^{-1}) for carriers **37a** and **38a**. Data taken from Ref. [168].

Carrier	Aq I (pH)	Aq II	k_T 5'-GMP	$k_{5'\text{-GMP}}/k_{5'\text{-CMP}}$	$k_{5'\text{-GMP}}/k_{5'\text{-AMP}}$
37a	6.15	H_2O	1.201	101.7	7.66
37a	6.70	H_2O	0.287	42.9	8.87
37a	7.05	H_2O	0.001	20.1	9.49
37a	6.15	10 mM NaOH	1.423	26.3	2.73
37a	6.70	10 mM NaOH	1.228	40.8	4.36
37a	7.05	10 mM NaOH	0.708	43.3	9.60
38a	6.15	H_2O	0.101	6.2	1.38
38a	7.05	10 mM NaOH	0.115	23.7	3.18
None	7.00	H_2O	$<10^{-5}$		
2	7.00	H_2O	$<10^{-5}$		
2	7.00	10 mM NaOH	$<10^{-5}$		

Table 4.10 Initial rates of GMP isomer transport (10^{-8} mol cm^{-2} h^{-1})
for carriers **37a** and **38a**. Data taken from Ref. [169].

Carrier	Aq I (pH)	Aq II	k_T 2'-GMP	$k_{2'\text{-GMP}}/k_{5'\text{-GMP}}$	$k_{2'\text{-GMP}}/k_{3'\text{-GMP}}$
37a	6.70	H_2O	0.767	9.70	7.30
37a	6.70	1 mM NaOH	2.989	9.55	5.30
37a	7.00	H_2O	0.594	11.06	7.82
37a	7.20	H_2O	0.421	$>10^2$	$>10^2$
37a	7.35	H_2O	0.352	$>10^2$	$>10^2$
38a	6.70	1 mM NaOH	0.104	3.33	2.67
None	7.00	H_2O	$<10^{-5}$		
2	7.00	H_2O	2×10^{-5}		

Table 4.11 Initial nucleotide-5'-monophosphate transport rates
(10^{-8} mol cm^{-2} h^{-1}) for carriers **37b** and **38b**. Data taken from Ref. [169].

Carrier	Aq I (pH)	Aq II	k_T 5'-CMP	$k_{5'\text{-CMP}}/k_{5'\text{-GMP}}$	$k_{5'\text{-GMP}}/k_{5'\text{-AMP}}$
37b	6.70	H_2O	0.129	8.96	3.15
37b	6.70	1 mM NaOH	0.541	9.17	3.15
38b	6.70	1 mM NaOH	0.147	17.5	10.5

To study the regioselective preference of carrier **37a**, three different phosphate groups, 5'-GMP, 3'-GMP and 2'-GMP, were used. In most instances the 2'-GMP isomer was transported considerably faster than the 5' and 3' isomers (cf. Tab. 4.10). It was, on the other hand, observed that the association constants for formation of **37a**-5'-GMP and **37a**-2'-GMP were 8×10^3 and 2.2×10^4 M^{-1}, respectively, as inferred from UV–visible spectroscopic titrations conducted in methanol [169]. Taken together these findings support the conclusion that the rate of transport depends on (at least) two important processes, binding and release, and can thus be very sensitive to

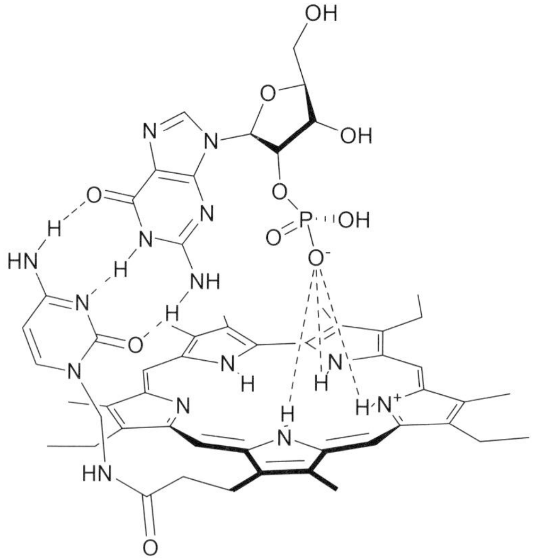

Figure 4.32 Structure of the proposed supramolecular complex formed between the monoprotonated form of sapphyrin **37a** and 2'-GMP.

small structural changes. Figure 4.32 illustrates the proposed complex formed between monoprotonated sapphyrin **37a** and 2'-GMP.

Transport selectivity for 5'-CMP was observed for the guanosine-bearing sapphyrins **37b** and **38b**. As emphasized by the data in Tab. 4.11, the relative rates of carrier induced, through-membrane transport were smaller for the monosubstituted system, **37b**, than for the bis-substituted system, **38b** [169].

The interactions between nucleotides, oligonucleotides, and other anions have also been probed by construction of a modified silica gel chromatography support. This system was produced by attaching a cytosine-substituted sapphyrin containing a carboxyl group to aminopropyl silica gel (Fig. 4.33). The resulting supports were found to enable HPLC-based separation of GMP, GDP, and GTP from a mixture of

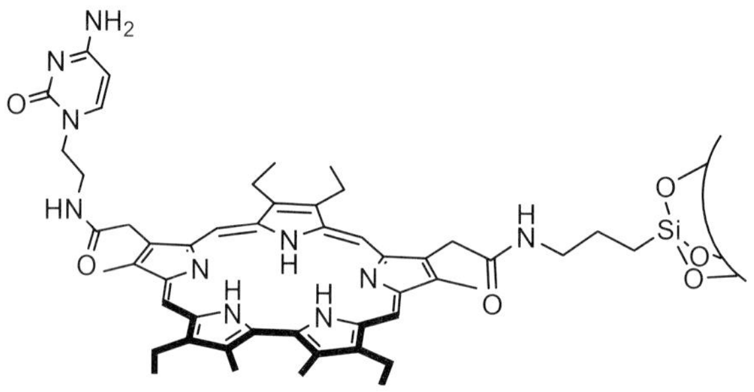

Figure 4.33 Schematic representation of the functionalized silica gel formed from a cytosine-bearing sapphyrin.

the mono-, di-, and triphosphates of cytidine, uridine, adenosine, and guanosine under isocratic conditions at pH 7 [166]. This selectivity was regarded as important, because it provided independent support for the proposed two-site recognition of anions invoked to rationalize the selective transport of nucleotides by sapphyrins. Although control experiments using a column packed with a sapphyrin-substituted silica gel [170] achieved separation of the mono-, di-, and triphosphates, no nucleoside-based selectivity was observed.

In an effort generate receptors for dianions, several sapphyrin dimers, trimer, and tetramers were prepared. In a first study, dimers **39** and **40** were found to be substantially more effective than monomers **2**, **37**, and **38** in enhancing the rate of transport of different dicarboxylate anions through a U-tube membrane model [171]. In addition, sapphyrin dimers linked by chiral linkers (**40**) were found to effect enantioselective recognition of N-protected amino acids [172]. The trimeric and tetrameric sapphyrins species **41–43** were prepared and studied as potential carriers for phosphorylated nucleotides. By using the standard U-tube arrangement it was shown that trimers **41** and **42** were indeed efficient carriers of nucleotide diphosphates, although they failed to effect the transport of nucleotide triphosphates. In contrast, tetramer **43** proved to be an effective carrier for mono-, di-, and trinucleotides [173].

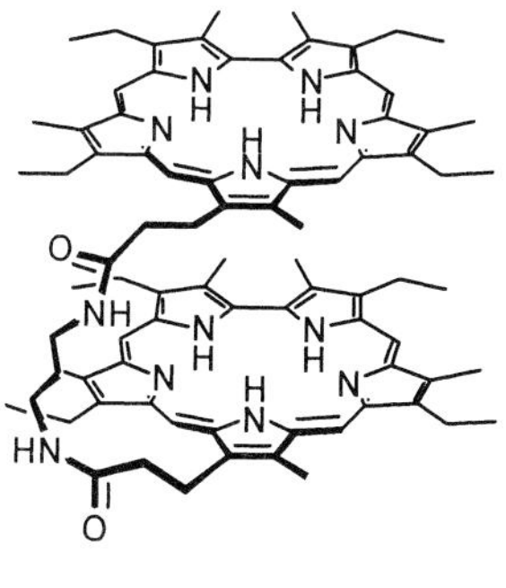

40a R =   **40b** R =

39

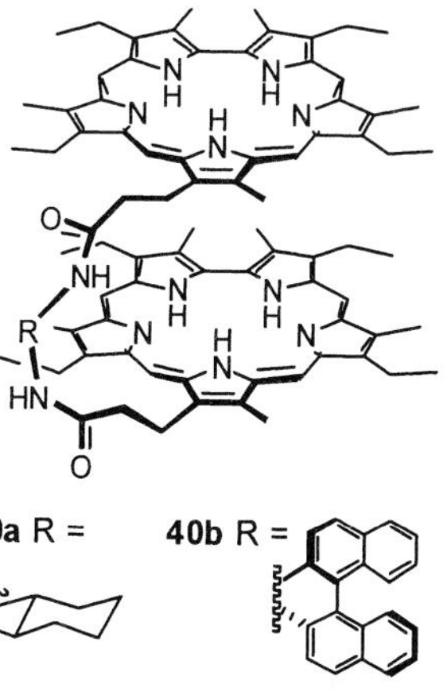

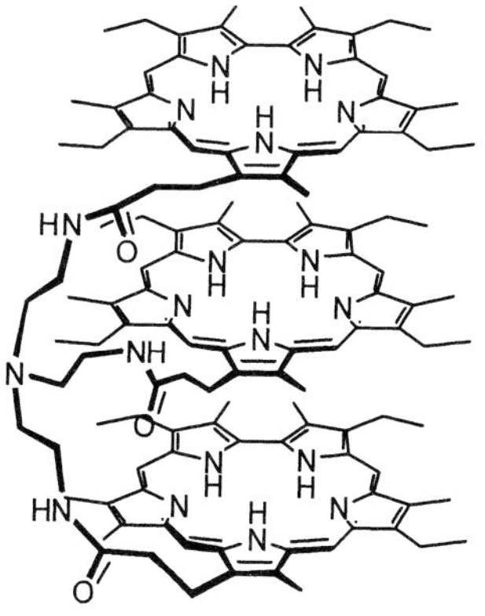

41

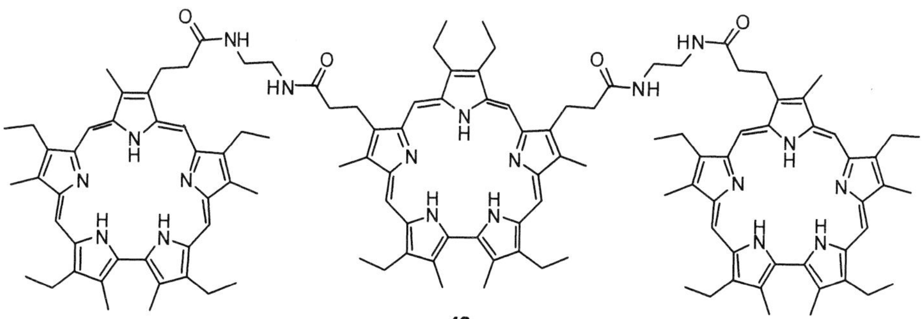

42

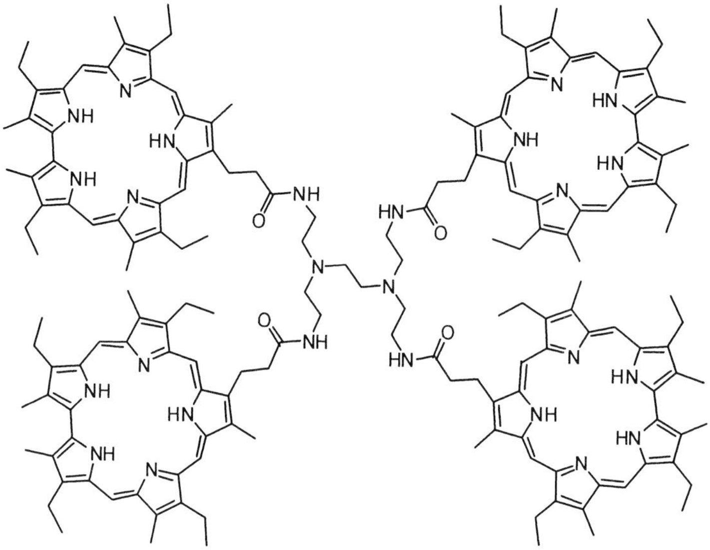

43

In related work the nucleoside-functionalized calix[4]pyrroles **44** and **45** were synthesized. As neutral receptors the calix[4]pyrroles have much lower phosphate anion-binding affinities than protonated sapphyrins. On the other hand, the sapphyrins are flat whereas the calixpyrroles adopt nonplanar conformations. What this means is that a nucleobase substituent, if appropriately connected to the calix[4]pyrrole framework, might be better able to produce a system capable of effecting the cooperative recognition of a complementary mononucleotide substrate. In an effort to test this hypothesis, calix[4]pyrroles **44** and **45** were studied as potential nucleotide carriers using a U-tube model membrane by analogy with experiments performed with the sapphyrin systems (*vide supra*). They were also studied as the key components in membrane-based ion-selective electrodes (ISE) as detailed in Sect. 4.6 [174].

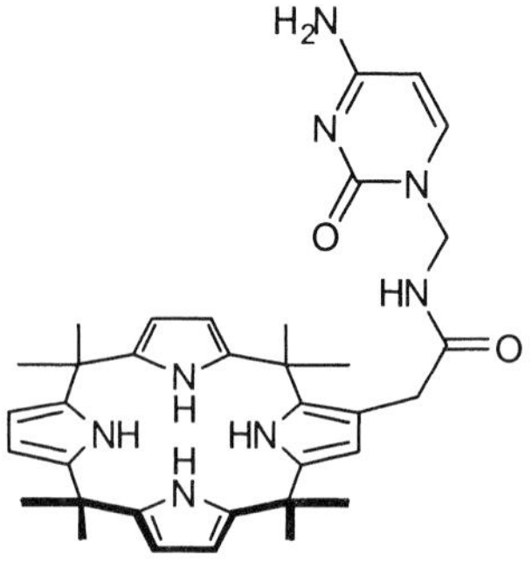

44

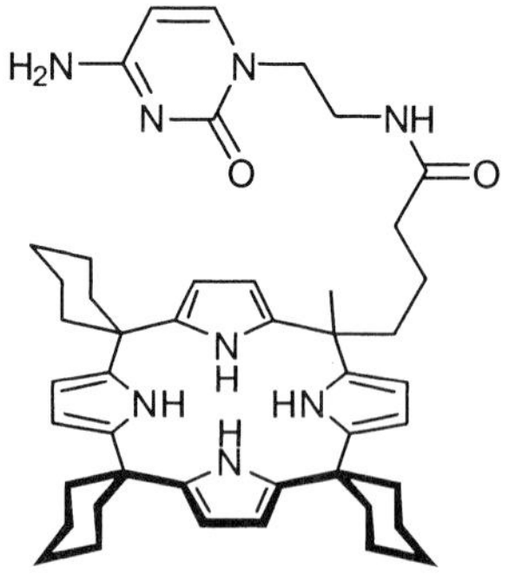

45

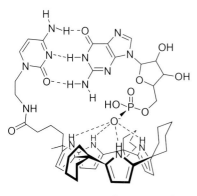

Figure 4.34 Structure of the proposed supramolecular complex formed between cytosine appended calix[4]pyrrole **45** and 5'-GMP.

When a U-tube model membrane containing tetrabutylammonium chloride in the organic phase (added to provide a hydrophobic, charge-neutralizing counter-cation) was used, good selectivity for 5'-GMP was observed for compound **45**, even though the absolute rate of transport (9.3×10^{-10} mol cm^{-2} h^{-1}) proved lower than that found for the sapphyrin-based carrier **37a** (1.42×10^{-8} mol cm^{-2} h^{-1}). The use of carrier **44** served to further reduce the rate of transport and led to selectivity for 5'-CMP rather than for 5'-GMP (Table 4.12) [174]. These results are consistent with carrier **45** acting as a *bona fide* ditopic receptor, capable of binding both the phosphate "head" and purine "tail" of 5'-GMP in a "two-point" fashion (Fig. 4.34). By contrast, it seems as if carrier **44** acts more as a nonselective "one-point" receptor, binding both the phosphate and nucleobase portions of nucleotide monophosphates at the calix[4]pyrrole NH hydrogen-bond donor site.

Table 4.12 Results of model through-membrane transport experiments conducted using receptors **44** and **45**. The concentration of the carriers in the organic phase was 0.1 mM, and the pH of the initial and receiving aqueous phases were 6.0 and 12.5, respectively. The organic phase also contained 0.1 M tetrabutylammonium perchlorate, a species added as a charge-neutralizing agent. The tabulated k_T values are in units of 10^{-11} mol cm^{-2} h^{-1}. Data taken from Ref. [174].

Carrier	Aq I (pH)	Aq II	k_T 5'-GMP	$k_{5'\text{-GMP}}/k_{5'\text{-CMP}}$	$k_{5'\text{-GMP}}/k_{5'\text{-AMP}}$
44	6.0	12.5	9.8	0.2	1.8
45	6.0	12.5	93	7.8	1.9

4.6
Anion Sensing

The synthesis of new receptor systems designed to sense and report the presence of an anion is a branch of supramolecular chemistry that is currently attracting much attention [2, 175–178]. It is likely that pyrrole-based systems will have a role to play in advancing work in this area. In one effort devoted to realizing this promise, metallocene units, electrochemically active "read out" species previously attached to calixarene moieties to generate electrochemical sensors for a variety of guest species [179–183], were attached to calixpyrroles by Gale and Sessler [184]. This gave the redox-active ferrocene-bearing systems 46 and 47.

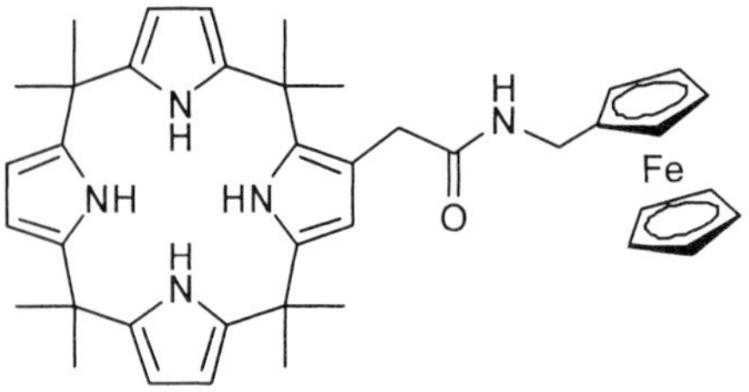

46

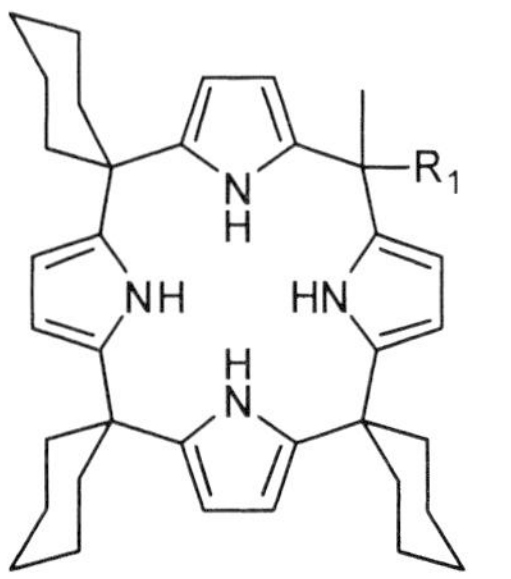

47a R =

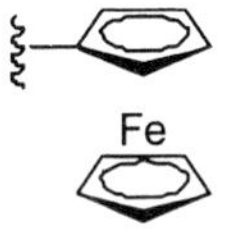

47b R =

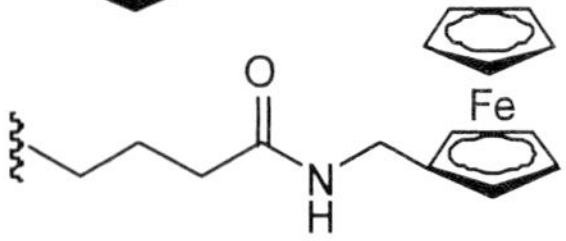

The solution-phase anion association constants of 46 and 47, studied in CD_2Cl_2 using standard ^1H NMR titration methods, are summarized in Table 4.13. Stability constants for receptor 46 were found to be of similar magnitude to those for other calix[4]pyrroles [102]. The association constants for receptor 47a also implied the same affinity trend $F^- > Cl^- > H_2PO_4^-$ usually observed in dichloromethane for other alkylated calix[4]pyrroles. The electrochemical properties of 46 and 47 were determined by use of cyclic voltammetry (CV). Receptors 46 and 47b were scanned between +600 and −100 mV and reversible ferrocene/ferrocenium waves at $E_{1/2}$ = 503 and 511 mV (relative to Ag/AgCl) were observed for 46 and 47a, respectively. When the CV of receptor 46 was scanned from +1.8 and −0.6 V, however, further oxidation waves were observed that were assigned to calix[4]pyrrole-centered processes. Although cathodic shifts of 31 and 63 mV were observed for the Fc/Fc⁺ couple of receptor 46 after addition of dihydrogen phosphate anion and fluoride anion, respectively, addition of chloride anion was found to engender an anodic shift of −22 mV. Further complicating a straightforward analysis, cathodic shifts of 14 and 207 mV, respectively, were observed after addition of fluoride and chloride anions to receptor **47b**, while addition of excess dihydrogen phosphate was found to lead to an anodic shift of −9 mV [184]. In contrast, the Fc/Fc$^+$ couple of receptor **47a** underwent an anodic shift after addition of fluoride, chloride, bromide, and dihydrogen phosphate anions [185].

In an effort to design other ferrocene–pyrrole conjugates, the ansa ferrocene **48a** containing two pyrrolic NH and two amidic NH groups was synthesized [186]. Quantitative ^{1}H NMR spectroscopic titration studies revealed that receptor **48a** had a high affinity for fluoride, chloride, and dihydrogen phosphate (K_a = >10^5, 9030, and 11,300 M^{-1} for F$^-$, Cl$^-$, and H$_2$PO$_4^-$, respectively, in CD$_3$CN). These results are consistent with those from electrochemistry studies and led to the suggestion that this receptor acts as an effective redox-based sensor for F$^-$ (80 mV) and H$_2$PO$_4^-$ (136 mV) [186]. To regulate the dihydrogen phosphate anion affinities and modulate the nature of the Fc/Fc$^+$-based electrochemical properties, three new ansa ferrocene systems containing bridging arms of different length and nature were synthesized. It was observed that the binding affinities for H$_2$PO$_4^-$ increased stepwise as the number of oxygen atoms in the linking bridge was increased from 0 to 2 (K_a = 4050, 13,200, 81,400 M^{-1}, respectively, for **48b**, **48c**, and **48d** in CD$_2$Cl$_2$ containing 2 % DMSO-d$_6$). This observation is consistent with the ether-type oxygen atoms being in-

Table 4.13 Affinity constants for interaction of **46** and **47** with anionic substrates and electrochemical data obtained from CV measurements. Data taken from Refs. [184, 185].

		No anion	F$^-$	Cl$^-$	Br$^-$	H$_2$PO$_4^-$	HSO$_4^-$
46[a]	K_a (M^{-1})	N/A	1496	444	–	40	–
	$E_{1/2}$ (mV)[c]	503	566	481	–	534	–
	ΔE (mV)	N/A	63	−22	–	31	–
47a[b]	K_a (M^{-1})	N/A	3375	3190	50	304	–[e]
	$E_{1/2}$ (mV)[d]	444	368	408	432	350	436
	ΔE (mV)	N/A	−76	−36	−12	−100	<10
47b[a]	K_a (M^{-1})	N/A	[f]	202	–	40[g]	–
	$E_{1/2}$ (mV)[c]	511	525	718	–	502	–
	ΔE (mV)	N/A	14	207	–	−9	–

[a] Measured in dichloromethane (deuterated solvent for K_a determinations)

[b] Measured in CH$_3$CN–DMSO, 9:1 (deuterated solvents for K_a determinations)

[c] Determined in dichloromethane containing 0.1 M n-Bu$_4$NPF$_6$ as the supporting electrolyte. The potentials were determined relative to Ag/AgCl

[d] Values obtained from square-wave voltammograms

[e] No shifts were observed

[f] The NMR signals became very broad under these conditions precluding accurate determination of this value

[g] The value was determined using the chemical shift of the pyrrole β-CH proton, because the pyrrole-NH signal became too broad to follow accurately over the course of the titration

volved directly in the anion binding process. On the other hand, the extent of the anion-induced cathodic shift in the Fc/Fc$^+$ potentials does not fully correlate with the number of oxygen atoms (ΔE = 128, 140, and 140 mV for **48b**, **48c**, and **48d**, respectively) [187].

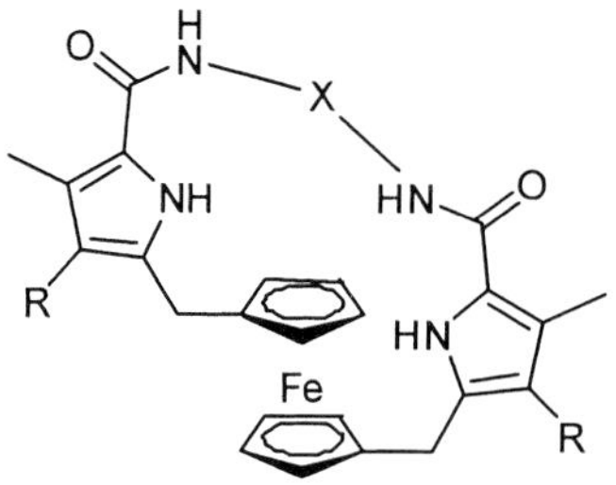

48a X = CH$_2$(CH$_2$OCH$_2$CH$_2$OCH$_2$)CH$_2$, R = C$_2$H$_5$
48b X = CH$_2$(CH$_2$)$_3$CH$_2$, R = CH$_3$
48c X = CH$_2$(CH$_2$OCH$_2$)CH$_2$, R = CH$_3$
48d X = CH$_2$(CH$_2$OCH$_2$)$_2$CH$_2$, R = CH$_3$

Quite recently, Gale and coworkers reported two new ferrocene-appended amidopyrroles with considerable promise as electrochemical anion sensors [188]. The electrochemical behavior of receptors **49** in the presence and absence of a variety of anions were determined in dichloromethane by CV methods using a platinum microdisk electrode. On the addition of fluoride anion to **49b**, a large anodic shift in Fc/Fc$^+$ was observed (ΔE = −130 mV). For **49a**, however, addition of fluoride anion produces the largest voltammetric shifts and yields two waves (ΔE = −125 and −255 mV). Interestingly, receptor **49a** was found to have strong affinity for benzoate anion (1820 M^{-1}) and addition of this anion produced a significant shift in the first voltammetric wave (ΔE = −120 mV). For **49a**, the most significant shifts in the voltammetric waves were observed with F$^-$ and C$_6$H$_5$CO$_2^-$. This is fully consistent with the proposal that the electrochemical oxidation of the receptors is affected by the bound anion.

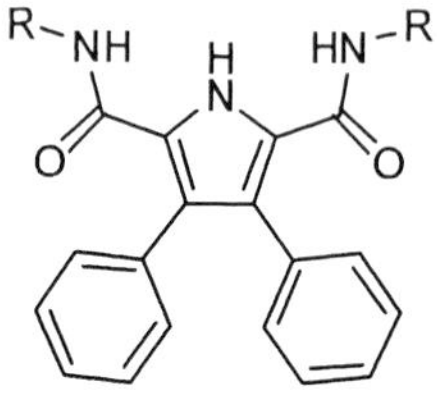

49a R =

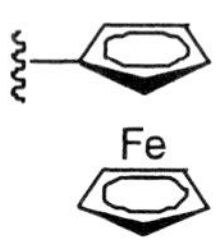

49b R =

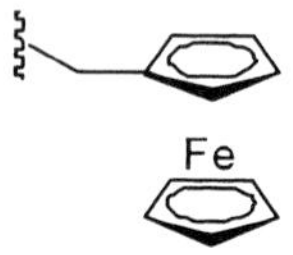

In 2003 Becher and coworkers reported the synthesis of a mono-tetrathiafulvalene (TTF) calix[4]pyrrole, **50**, obtained through the acid catalyzed condensation of pyrrole TTF and tripyrrane [189]. The affinity constant for bromide anion, measured by standard ^{1}H NMR titration methods, was found to be 7.6 × 10^3 M^{-1} in CD$_3$CN containing 0.5 % D$_2$O. For the chloride and fluoride anions the corresponding K_a values determined by competitive ^{1}H NMR spectroscopic titrations were calculated to be 1.2 × 10^5 and 2.1 × 10^6 M^{-1}, respectively. The electrochemical properties of receptor **50**, containing a TTF unit, were monitored by CV methods in acetonitrile solution. In the presence of anions anodic shifts are observed for this system; the values are −43 and −34 mV for chloride and bromide anions, respectively.

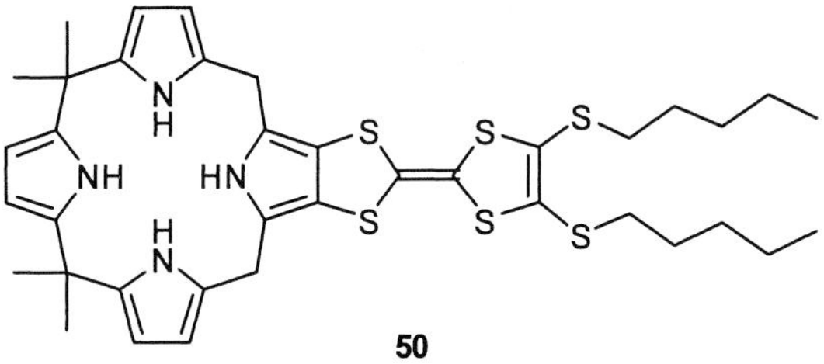

50

On consideration of the full collection of calixpyrrole-based electrochemical sensors prepared so far it becomes clear that significant electrochemical effects can be induced and that these are caused by addition of anions. The direction of the shifts varies from system to system, and within each system, however. Thus, in general, no useful or predictive rationalization of the anion-induced response can yet be proposed. This has made the electrochemical approach less effective than it perhaps could be.

In part because of limitations inherent in the direct electrochemical sensing approach, researchers have sought to prepare ion selective electrodes (ISE). The first ISE containing neutral pyrrole-based anion receptor elements were reported by Král, Gale, and Sessler in 1999 [190]. In this study, it was found that the potentiometric sensitivity toward anions of polyvinyl chloride membrane-type ISE (PVC membranes) based on calix[4]pyrrole **18a** and its pyridine analogs [191] (i.e. dichlorocalix[2]pyrrole[2]pyridine **51a** and tetrachlorocalix[4]pyridine **51b**) is pH-dependent. Whereas the ISE containing compound **18a** responded to the F$^-$ and HPO$_4^{2-}$ anions at pH 3.5, 5.5, and 9.0, the same system responded to the chloride and bromide cations at higher pH. Presumably, such findings reflect the high selectivity of **18a** for OH$^-$ and the rather low affinity for Cl$^-$ or Br$^-$ at higher pH. Consistent with this hypothesis a non-Hofmeister order of selectivity, namely Br$^-$ < Cl$^-$ < OH$^-$ ≈ F$^-$ < HPO$_4^{2-}$, was observed at pH 9.0. This result supports the notion that calix[4]pyrrole **18a** acts as a pure anion receptor at lower pH, whereas it has two different responses, reflecting both anion and counter cation binding, at higher pH.

At high pH the selectivity of ISEs derived from **51a** and **51b** is the complete opposite of that for an ISE based on **18a** (F$^-$ < OH$^-$ < Cl$^-$ < Br$^-$). At low pH, however, both systems (**51a** and **51b**) respond strongly when exposed to fluoride and phos-

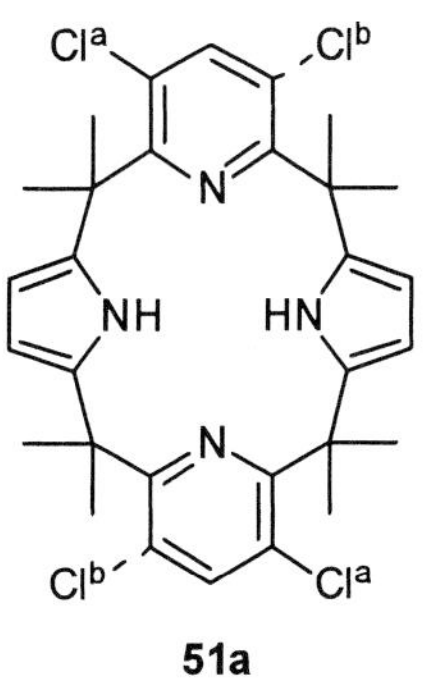

51a 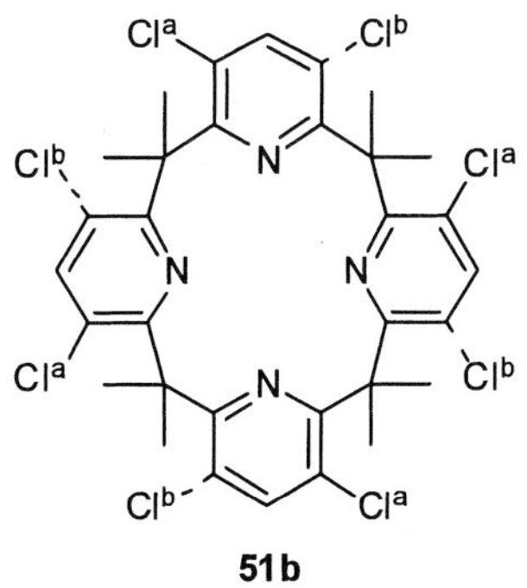   **51b**

phate anions. Such findings are consistent with the inherent anion-binding capacity of **51a** derived from neutral pyrrolic NH hydrogen-bonding interactions being augmented by electrostatic interactions involving the protonated pyridine units, at least at low pH.

In an effort to design nucleotide-specific ISEs, the nucleoside-functionalized cal-ix[4]pyrrole **44** and **45** were tested as membrane elements [174]. Before studying these systems, however, the potentiometric sensitivity and selectivity of ISEs containing the unfunctionalized analogous compounds (**18a** and **52**) without tridodecyl-methylammonium chloride (TDDMACl) were tested. At pH 6.6 (at which the phosphate moiety occurs mostly in its dianionic form) for both systems (i.e. those derived from both **18a** and **52**) the inherent selectivity, as judged from the observed anionic response, was 5'-AMP < 5'-GMP ≈ 5'-CMP < 5'-UMP ≈ 5'-TMP. On the other hand, it was found that system **18a** has a slightly different selectivity pattern (5'-AMP < 5'-UMP < 5'-CMP < 5'-GMP) when the electrically charged species TDDMACl is added to the PVC membranes (Table 4.14). That this additive has such an effect reinforces the conclusions drawn from the corresponding transport studies (discussed above), in which it proved necessary to add tetrabutylammonium perchlorate to obtain a decent transport rate. This, presumably, reflects the need to neutralize the negative charge present in the calix[4]pyrrole–nucleotide complex.

For the more hydrophobic cyclohexyl-substituted calix[4]pyrrole **52**, the selectivity pattern 5'-AMP (no carbonyl) < 5'-CMP (one carbonyl) < 5'-GMP < 5'-TMP (two carbonyl) ≈ 5'-UMP was observed when using a membrane prepared with TDDMACl. This result led to the suggestion that the ISE response selectivity is dom-

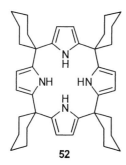

52

Table 4.14 Potentiometric selectivity of PVC membranes containing unfunctionalized cal-ix[4]pyrrole **18a** (log $K^{sel}_{5'\text{-AMP}/5'\text{-XMP}}$), the cytosine-functionalized calix[4]pyrroles **44** (log $K^{sel}_{5'\text{-UMP}/5'\text{-XMP}}$), and analog **45** (log $K^{sel}_{5'\text{-CMP}/5'\text{-XMP}}$). These membranes contained TDDMACl. Data taken from Refs. [174, 192].

	pH 6.6			pH 8.5		
	5'-AMP	5'-CMP	5'-GMP	5'-AMP	5'-CMP	5'-GMP
18a	0.00	+0.24	+0.76	–	–	–
44	−0.07	+0.97	+0.89	−0.20	+0.58	+0.10
45	+1.53	0.00	+2.23	−0.68	0.00	−1.23

inated by interactions between the nucleobase and the calix[4]pyrrole, even when the substrate is a nucleotide containing a negatively charged phosphate subunit that could be a source of an independent anionic response.

For ISEs prepared from the functionalized calix[4]pyrroles **44** and **45**, which have 5'-CMP and 5'-GMP selectivity in model membrane-transport experiments at pH 6.0 (initial Aq) and 12.5 (receiving Aq), as discussed above, it was considered important to measure the potentiometric response at both low and high pH. This is because the transport studies led to the inference that the pH of both aqueous phases could be a factor determining transport selectivity. Table 4.14 summarizes the potentiometric selectivity of PVC membranes derived from **44** and **45** and which also contain TDDMACl. For system **44** the order of selectivity was 5'-AMP < 5'-GMP < 5'-CMP at both pH 6.6 and pH 8.5. The selectivity sequence for system **45**, on the other hand, was found to be a more sensitive function of pH (viz. 5'-CMP < 5'-AMP < 5'-GMP at pH 6.6 and 5'-GMP < 5'-AMP < 5'-CMP at pH 8.5) [192]. The high selectivity for 5'-GMP at lower pH supports the notion that ditopic interactions play a key role in mediating receptor nucleotide binding events at or below pH 6.6. At high pH, on the other hand, the nucleobase is expected to exist as a dianion, making the selectivity advantage imparted by the Watson–Crick base-pairing interactions less significant. While not yet fully established, such rationales can help explain the low potentiometric selectivity for 5'-GMP that is, in fact, observed at pH 8.5. Less clear are the differences between the ISE and transport results, which are yet to be fully explained.

Another approach to pyrrole-based anion sensor development involves generation of species that enable direct optical detection of a bound anionic substrate, either via a visual (so-called "naked eye") read-out or via a spectroscopic change, such as a change in fluorescence intensity. Several groups have been working on this problem and within each group several alternative strategies have been pursued. The development of pyrrole-based optical anion sensors has therefore proceeded along several different paths.

One method used to prepare polypyrrole-based optical sensors has involved the covalent attachment of a chromophore or fluorophore to the polypyrrole skeleton. Perturbation of the electronic properties of these attached groups on anion complexation (either directly or indirectly) then produces a response detectable by the naked eye or through monitoring changes in the fluorescence spectrum, or both. Here, the first systems reported in the literature were covalently linked calixpyrroles bearing anthracene subunits (i.e. calix[4]pyrroles **53**). These systems were obtained by coupling of a calixpyrrole mono-acid with a variety of aminoanthracenes [193].

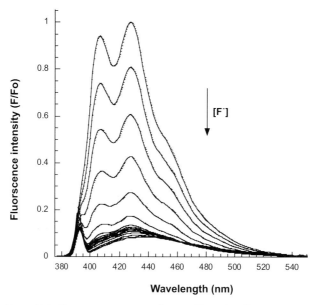

Figure 4.35 Fluorescence spectra of sensor **53a** in CH_2Cl_2 (0.05 mM) showing the changes induced on addition of increasing quantities of tetrabutylammonium fluoride. The excitation wavelength was 378 nm.

With systems **53** affinity constants for several representative anions were measured in organic solvents using ^1H NMR spectroscopic titrations and fluorescence-quenching methods. For the fluoride anion (studied, like the other anions, as its tetrabutylammonium salt) it proved impossible to determine a stability constant by ^1H NMR spectroscopic titration because of spectral broadening. The fluorescence of these sensors (i.e. **53a–53c**) was, on the other hand, shown to be quenched significantly in the presence of some anionic guests, with the extent of quenching depending on the specific choice of anion. The greatest quenching was seen for fluoride; the associated changes in the fluorescence spectrum for sensor **53a** are reproduced in Fig. 4.35. From these quenching experiments the affinity constants for fluoride, chloride, bromide, and dihydrogen phosphate anions were calculated and are listed in Table 4.15 for the full family of **53**.

In related work the second-generation calix[4]pyrrole anion sensors **54** were produced. In this case, dansyl, lissamine–rhodamine B, or fluorescein moieties were attached to the calix[4]pyrrole framework to serve as the fluorescent reporter groups, whereas an amide or thiourea residue was used as the covalent linker between these fluorophores and the calix[4]pyrrole backbone. This choice of linker serves to provide a second anion-binding site. The changes induced by anion binding, specifically the decrease in fluorescence emission intensity observed when sensor **54b** is titrated with an increasing concentration of tetrabutylammonium fluoride, is reproduced in Fig. 4.36 [194]. Quantitative assessment of the anion-induced spectral changes were

Table 4.15 Affinity constants for anion sensors **53** with different anions (studied as their n-Bu$_4$N$^+$ salts) in CH$_3$CN and CH$_2$Cl$_2$ as determined by fluorescence quenching analysis at 25 °C. Sensor **53a** was excited at 378 nm and emission was monitored at the λ_{max} of 446 nm; for sensors **53b** and **53c** the corresponding values were 393 nm and 429 nm, and 387 nm and 418 nm, respectively. Data taken from Ref. [193].

	log K_a in CH$_2$Cl$_2$			log K_a in CH$_3$CN		
	53a	**53b**	**53c**	**53a**	**53b**	**53c**
F$^-$	4.94	4.52	4.49	5.17	4.69	4.69
Cl$^-$	3.69	2.96	2.79	4.87	3.81	3.71
Br$^-$	3.01	–[a]	–[a]	3.98	2.86	–[a]
H$_2$PO$_4^-$	4.20	3.56	–[a]	4.96	3.90	–[a]

[a] Quenching insufficient to provide an accurate value

used to determine the affinity constants for this and other representative anions, and are summarized in Table 4.16. It was found that sensors **54** are very selective for dihydrogen phosphate and pyrophosphate anions relative to chloride anions. This selectivity was explained by the presence of the second anion binding group, i.e. the amide or thiourea link, and the favorable interactions it enables with nonspherical anions. These same ancillary interactions enable sensors **54** to operate successfully in the presence of a small amount of water at physiological pH.

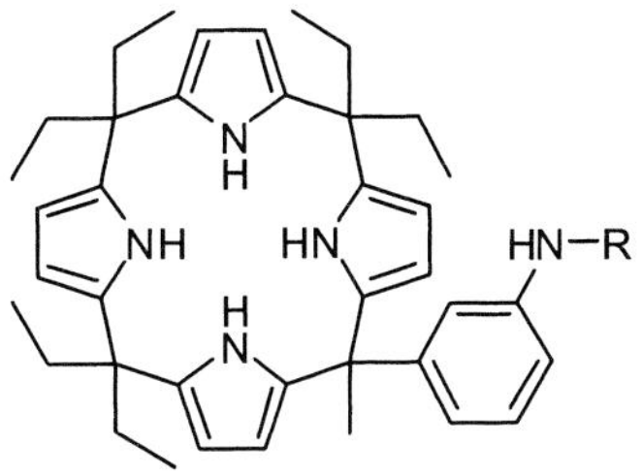

54a R =

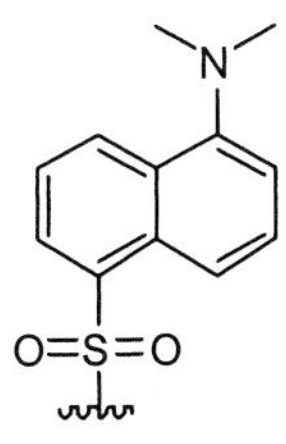

54b R =

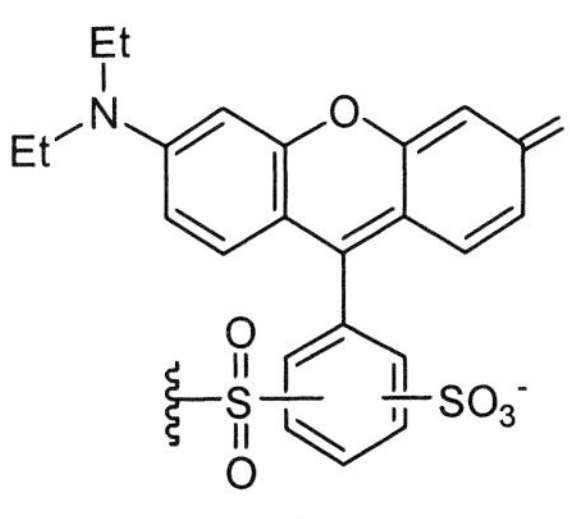

54c R =

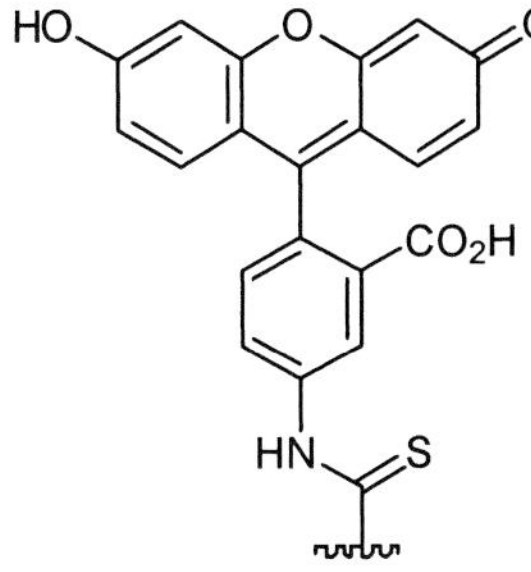

Table 4.16 Association constants (M^{-1}) for sensors **54** and representative anionic substrates (studied as their n-Bu_4N^+ salts) as determined by fluorescence emission quenching in acetonitrile (containing 0.01 % v/v water) for sensors **54a** and **54b** and in acetonitrile–water (94:4, pH 7.0 ± 0.1) for sensor **54c**. Data taken from Ref. [194].

	54a	54b	54c
F^-	222,500	>1,000,000	>2,000,000
Cl^-	10,500	18,200	<10,000
$H_2PO_4^-$	168,300	446,000	682,000
$HP_2O_7^{3-}$	131,000	170,000	>2,000,000

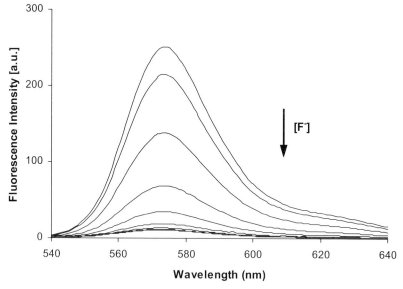

Figure 4.36 Decrease in fluorescence emission intensity observed when sensor **54b** (0.1 µM in acetonitrile containing 0.01 % v/v water) is titrated with increasing concentrations of tetrabutylammonium fluoride. From highest to lowest curve, $[F^-]$ = 0, 0.30, 0.76, 1.53, 2.30, 3.06, 3.83, and 4.60 µM.

In 2000, Sessler and coworkers reported the preparation of a set of powerful calix-pyrrole-based naked-eye sensors for selected anions such as F^-, Cl^-, and $H_2PO_4^-$ (cf. structures **55**). As shown in Fig. 4.37a, addition of tetrabutylammonium fluoride to dichloromethane solutions of receptor **55c** leads to a noticeable red-shift (λ_{max} from 441 to 498 nm) and broadening of the absorption maximum. Reflecting this anion-induced spectral difference, the color of the solution changes from yellow to red. Addition of either chloride and dihydrogen phosphate anions, on the other hand, caused the color of the solution to change from yellow to orange. For the mononitrobenzene-conjugated calixpyrrole **55b**, again in dichloromethane, the maximum absorption

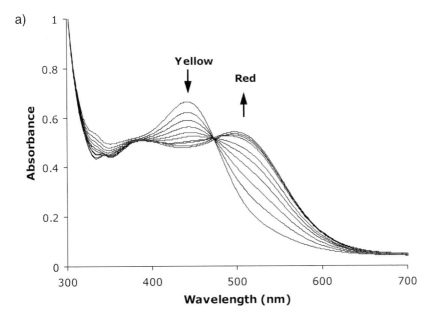

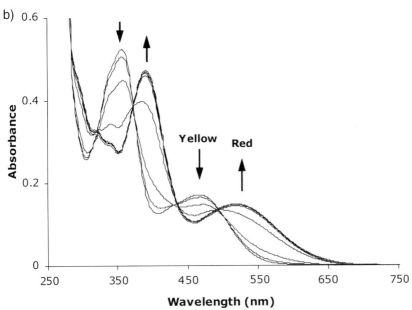

Figure 4.37 (a) Absorption spectra of sensor **55c** recorded in CH₂Cl₂ (0.05 mM) before and after addition of 0.2, 0.4, 0.6, 0.8, 1, 1.2, 1.4, 1.6, 1.8, and 2 equiv. tetrabutylammonium fluoride. (b) Absorption spectra of **55f** recorded in CH₂Cl₂ (0.05 mM) before and after addition of 0, 1, 2, 4, 6, 8, and 10 equiv. tetrabutylammonium fluoride.

peak shifted from 391 nm (pale yellow color) to 433 nm (intense yellow) on the addition of fluoride anion. Receptor **55a** is colorless and was used as a control. It was observed that the absorption maximum (λ_{max} = 308 nm) of this last species underwent a slight shift (from λ_{max} = 308 nm to λ_{max} max = 321 nm) on addition of 20 equiv. tetrabutylammonium fluoride, indicating that binding was occurring; no useful "naked eye" optical read out was observed, however [195]. Table 4.17 summarizes the affinity constants for interactions of receptors **55** with different anions, as determined from UV–visible titration studies.

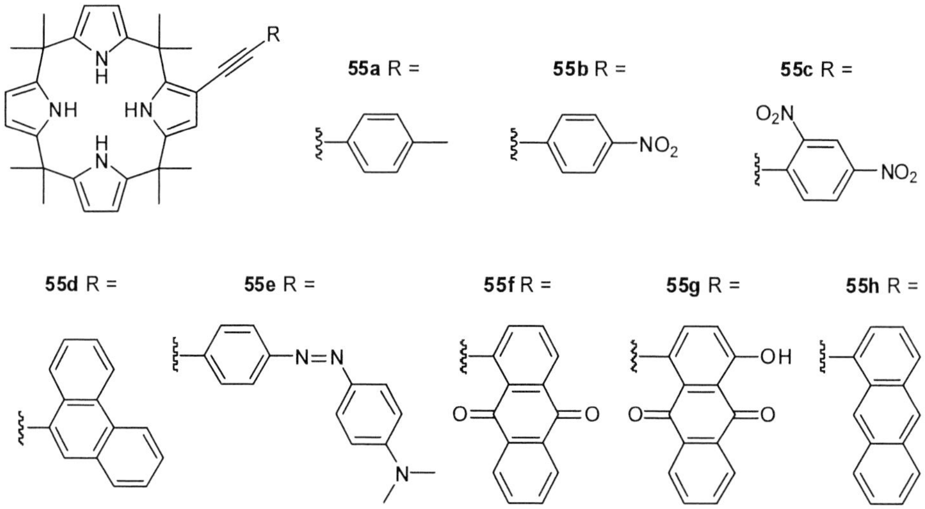

Particularly noteworthy color changes are observed with anthraquinone derivatives **55f** and **55g**. Figure 4.37b shows the changes in the absorption spectrum of sensor **55f** (in dichloromethane) observed on addition of 10 equiv. tetrabutylammonium fluoride. This addition caused the color of the solution to change from clear yellow to red. With this receptor, the addition of either chloride or phosphate anions induced a slightly less intense color change (to orange–red). The changes seen with sensor **55g** were even more dramatic – anion-induced color changes from red to blue, purple, or dark purple were observed on addition of fluoride, chloride, or dihydrogen phosphate anions. Such changes emphasize that sensors such as **55** not only enable facile colorimetric detec-

Table 4.17 Anion-binding affinity constants (log K_a) for sensors **55** and selected anionic substrates as deduced from UV–visible titration experiments in CH_2Cl_2.[a] Data taken from Refs. [195, 196].

	55a	55b	55c	55f	55g	55h
F⁻	4.04	4.23	4.51	3.71	3.43	3.66
Cl⁻	3.36	3.67	3.84	3.16	3.13	3.16
$H_2PO_4^-$	2.68	3.03	3.28	3.04	2.76	2.86

[a] All anions were studied as their tetrabutylammonium salts

tion of anions but also do it in a manner amenable to "color tuning". The one issue of note is that, as for other sensors, the effect is most pronounced for the fluoride anion.

In a new approach to anion sensor design Sessler and coworkers have explored dipyrrolylquinoxaline (DPQ) systems [197]. The first in a possibly generalized series of anion sensors in which two (or more) pyrrolic recognition elements are bridged by a rigid, aromatic chromophore, these acyclic systems were prepared in only two steps; specifically, they were made from simple compounds, namely pyrrole, oxalyl chloride, and a variety of phenylenediamines by using a modification of a literature procedure [198] first reported by Oddo [199]. In the absence of anions, the color of DPQ **56a**, **56b**, and **56c** in dichloromethane or DMSO solution were pale yellow and deep yellow. Addition of fluoride anion, as tetrabutylammonium fluoride, to these solutions resulted in immediate quenching of the fluorescence intensity and a colorimetric response that was reflected as a dramatic color change from yellow to orange–red for **56a** and **56c** and from yellow to dark purple for **56b** (Figs. 4.38a and b).

(a)

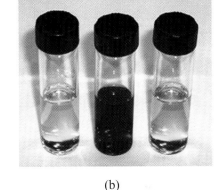

(b)

(c)

(d)

Figure 4.38 Color changes induced by addition of anions to a CH$_2$Cl$_2$ solution of receptors **56a**, **56b**, **65b**, and **66a**. (a) From left to right: **56a**; **56a** + F$^-$; **56a** + H$_2$PO$_4^-$. (b) From left to right **56b**; **56b** + F$^-$; **56b** + H$_2$PO$_4^-$. (c) From left to right **65b**; **65b** + malonate; **65b** + succinate. (d) From left to right **66a**; **66a** + F$^-$; **66a** + H$_2$PO$_4^-$.

In contrast, addition of chloride or phosphate anion did not induce appreciable fluorescence quenching or a noticeable change in color except for sensor **56c**; for this sensor phosphate anion induced a color change similar to that observed for fluoride, but only at substantially higher anion concentrations.

Anion-binding affinities were measured quantitatively by means of standard fluorescence quenching or UV–visible absorbance titration techniques. These analyses helped confirm that the observed colorimetric change did indeed reflect an anion-binding process. For example, as might be expected on the basis of its colorimetric behavior, receptor **56c** displayed a significantly enhanced affinity for the $H_2PO_4^-$ anion ($K_a = 1.73 \times 10^4$ M^{-1} for **56c** compared with 80 M^{-1} for **56b**; CH_2Cl_2, tetrabutylammonium salt) [128, 197]. Further, the more dramatic color changes seen for **56b** and **56c** in the presence of fluoride anion (tetrabutylammonium salt) are also reflected in higher K_a values (1.2×10^5 and 6.2×10^4 M^{-1} for **56b** and **56c**, respectively, in CH_2Cl_2, compared with 1.8×10^4 M^{-1} for **56a**). This is rationalized in terms of the increased NH hydrogen-bond-donor capacity of **56b** and **56c** which results from the electron-withdrawing groups present on the DPQ skeleton.

56a $R_1 = R_2 = H$
56b $R_1 = H, R_2 = NO_2$
56c $R_1 = F, R_2 = H$

The success of these initial systems led to the synthesis of the fused DPQ phenanthroline conjugates **57** and their metal complexes **58**. With the latter systems it was expected that appending cationic charges to a DPQ framework would withdraw electron density from NH bonds and, as is observed for **56b** and **56c**, increase the anion-binding affinity. Although the free phenanthroline DPQ system **57** had rather low affinity for the fluoride anion ($K_a = 440$ M^{-1} in DMSO for **57a** and 68,000 M^{-1} in CH_2Cl_2 containing 2 % CH_3CN for **57b**; tetrabutylammonium salts in both instances), presumably as a result of the additional electron density such appended heterocycles provide, the two Ru(II) complexes **58a** and **58c** showed a greater affinity for the fluoride anion ($K_a = 640,000$ M^{-1} in CH_2Cl_2 containing 2 % CH_3CN for **58a** and $K_a = 12,000$ M^{-1} in DMSO for **58c**) compared with DPQ **56a** itself ($K_a < 100$ M^{-1} in DMSO). With DPQ **58a** a red shift in the emission spectrum from 594 nm to 610 nm was also observed as a function of an increased CN^- concentration. The binding affinity for cyanide anion could thus be calculated and was found to be 428,000 M^{-1} in CH_2Cl_2 containing 2 % CH_3CN [200]. The fluoride anion affinity of the Co(III) complex **58b**, with its incrementally greater charge (+3 compared with +2 for **58c**), was one of the highest yet recorded in DMSO ($K_a = 54,000$ M^{-1}) [200, 201].

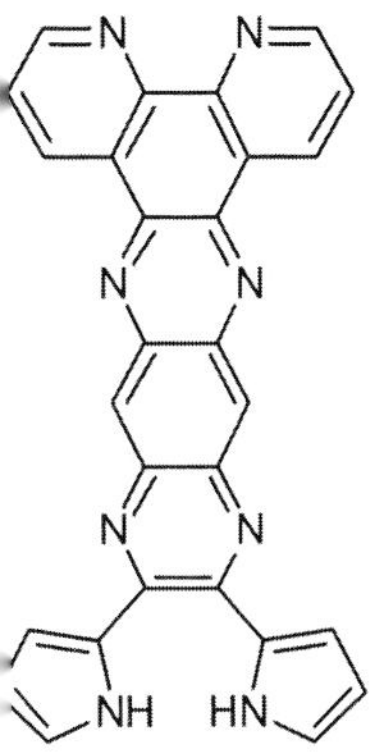

57a

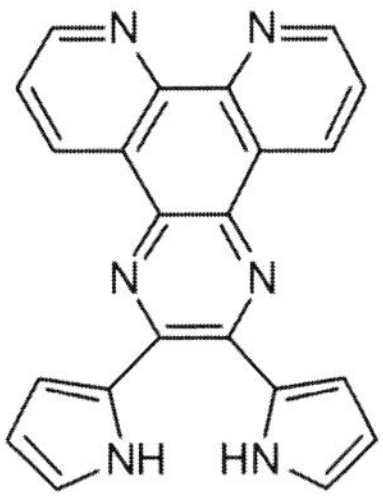

57b

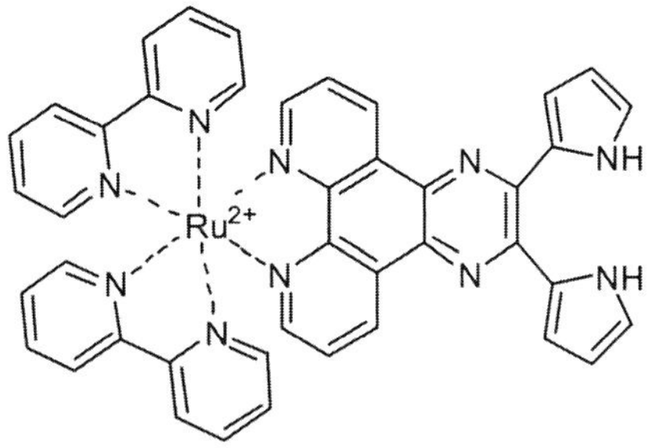

58a

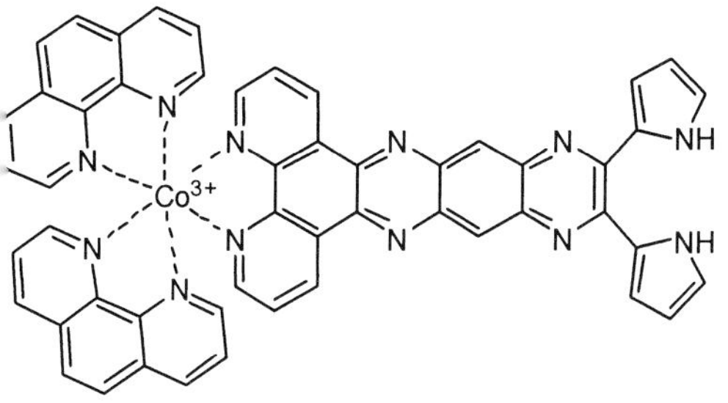

58b

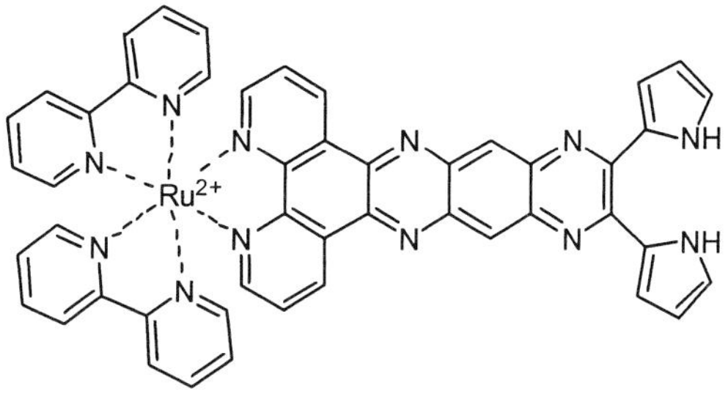

58c

In an elegant elaboration of this general approach Anzenbacher and coworkers reported the new DPQ systems **59** and **60**. These new systems contain aryl subunits and their fluorescent response is improved [202]. In the presence of fluoride and pyrophosphate anions (studied as their tetrabutylammonium salts in CH_2Cl_2) the intensity of the absorption bands in the 400–450 nm spectral region was reduced for both **59** and **60** and a strong new band appeared between 500 and 550 nm. Addition of these two anions also resulted in significant fluorescence quenching. Accompanying quantitative studies confirmed that these receptors had greater affinity than DPQ **56a** for fluoride and pyrophosphate anions in CH_2Cl_2.

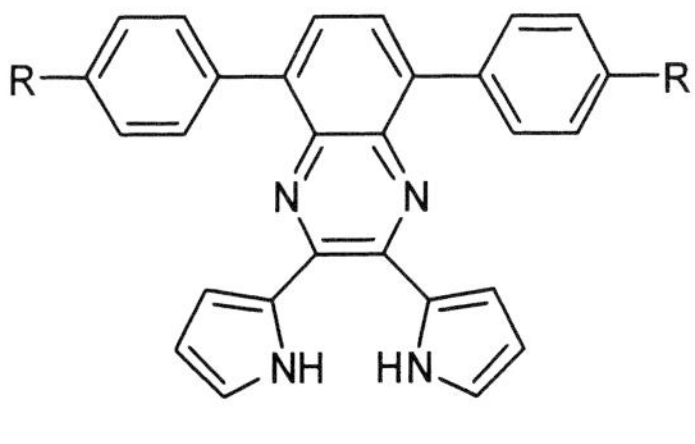

59a R = H
59b R = OCH$_3$
59c R = N(CH$_3$)$_2$

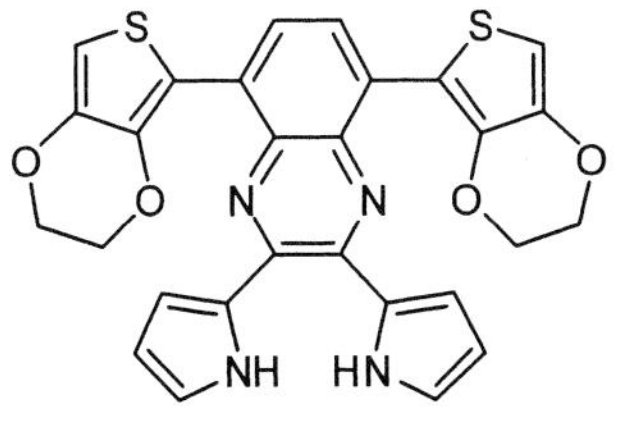

60

An alternative approach to the design of new pyrrole-based DPQ sensors involves modifying the basic chromophore-bridging element used to connect the two pyrrole recognition subunits. One recently, reported system that falls within this paradigm is 2,7-bis($1H$-pyrrol-2-yl)ethynyl-1,8-naphthyridine (BPN) **61** that contains two hydrogen-bonding donors (pyrrolic NH) linked by a dialkynyl naphthridine spacer [203]. In this system the specific structural details, involving an organized array of hydrogen bonding donor–acceptor–acceptor–donor subunits, produces a good saccharide receptor. The color of BPN in CH_2Cl_2 was found to change from cyanine to green on the addition of octyl β-D-glucopyranoside (OGU). The binding affinity was assessed quantitatively by use of three different methods – UV–visible absorption, fluorescence quenching, and ^1H NMR spectroscopic titrations. It was found that although results from absorption and fluorescence titrations measurements were almost coincident (K_a = 4.8 × 10^3 and 5.5 × 10^3 M^{-1}, respectively, in CH_2Cl_2), those from ^1H NMR titrations resulted in higher stability constants (K_a = 2.0 × 10^4 M^{-1} in $CDCl_3$). For octyl β-D-galactopyranoside (OGA) the corresponding K_a value is approximately one third that of OGU (K_a = 1600 M^{-1}, as inferred from fluorescence titration studies). These differences in K_a values were considered significant, because these two saccharides differ only in the orientation of the 4-hydroxyl group.

Another DPQ derivative containing an "alternative" bridging subunit is the pyrrole-linked system **62**. This system was originally obtained as the side-product during the iodination of pyrrole. It was then subsequently synthesized in excellent yield by use of a stepwise approach [204]. In the presence of fluoride (tetrabutylammonium salt in CH_2Cl_2), the color of solutions containing compound **62** changed from pale brown to yellow. Although the affinity of compound **62** for fluoride anion (K_a 1,8 × 10^5 M^{-1} in CH_2Cl_2) is approximately ten times higher than that of DPQ **56a** (K_a = 1.8 × 10^4 M^{-1} in CH_2Cl_2), its dihydrogen phosphate anion affinity is even more substantially enhanced (by factor of ca 220; K_a = 1.8 × 10^4 M^{-1} for **62** compared with 80 M^{-1} for **56a**, in CH_2Cl_2). Presumably, the presence of three electron-withdrawing ester substituents improves the anion affinity generally, whereas the geometry changes caused by replacing the quinoxaline bridge with a pyrrole serves to augment dihydrogen phosphate anion selectivity. Irrespective of such rationalizations, the differences seen leads to the suggestion that the inherent anion-binding affinity and selectivity of DPQ systems might be modulated by changing the nature of the bridging subunit.

61   **62**

Another strategy being developed in an effort to generalize the DPQ-based approach to anion sensing and, in particular, to produce systems with enhanced affinity, has involved appending additional pyrroles to the anion-binding portion of the DPQ backbone. So far, application of this appealing strategy has led to the synthesis of systems **63**, **64**, and **65**, all of which contain built-in pyrrole-derived NH "claws". Although the bowl-like conformations of DPQ **63** and **65** were expected to favor the binding of larger anions such as phosphate, acetate, oxalate, malonate, and succinate, affinity for fluoride anion was also expected to remain high. For **63a** and **65a** UV–visible spectroscopic titration studies performed in CH_2Cl_2 solution revealed that the absolute fluoride and phosphate anion affinity was increased relative to that of **56a** ($K_a(F^-)$ = 32,000 and >1,000,000 M^{-1} for **63a** and **65a**, respectively, and $K_a(H_2PO_4^-)$ = 4300 and 300,000 M^{-1} for **63a** and **65a**, respectively) [205]. More recently, color changes from dark yellow to purple or brown have been observed on addition of malonate or succinate anions to CH_2Cl_2 solutions of DPQ **65** (Fig. 4.38c).

UV–visible spectroscopic studies of mono- and dicarboxylate anion binding (tetrabutylammonium salts; CH_2Cl_2) confirmed that DPQ **65b** binds oxalate with good selectivity relative to malonate and succinate (the ratio of the respective K_a values is 450:40:1 in CH_2Cl_2). The same studies revealed that DPQ **65b** binds the monocarboxylate anion, acetate, even more effectively than these species, however. In contrast, the affinity of DPQ **65a** for the acetate anion is no greater than that for oxalate, malonate, or succinate [206].

63a R = -C₄H₄-
63b R = CN

64

65a R = -C₄H₄-
65b R = CN

To create potential "molecular cages" for anionic substrates, a first macrocyclic DPQ system **66** was prepared recently by Sessler and Furuta [207]. An example of what could emerge as a large class of quinoxaline-bridged porphyrinoids, this target was synthesized from formyl functionalized DPQ units. The absorption spectrum of **66a**, recorded in dichloromethane, is characterized by two bands at 367 and 427 nm. On addition of an increasing concentration of tetrabutylammonium fluoride, new peaks at 329 and 480 nm were seen to appear. It was also found that the color of CH_2Cl_2 solutions of **66a** changed from yellow to orange and to dark yellow in the presence of fluoride and dihydrogen phosphate anions, respectively (Fig. 4.38d). On the basis of quantitative analysis of these spectral changes it was deduced that receptor **66a** binds fluoride and phosphate anions in a cooperative 2:1 fashion and does so with rather high affinity ($K_a = 3 \times 10^5$ and 80 M^{-2} for fluoride and phosphate, respectively) [207].

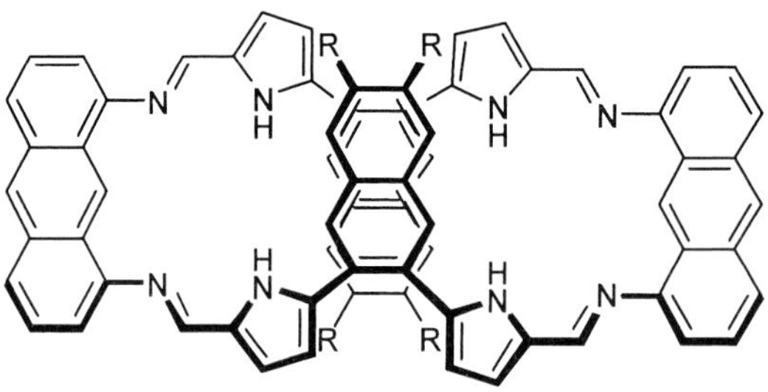

66a R = H
66b R = OCH_3
66c R = OCH_2CH_3

4.7
Guanidinium-based Anion Receptors

Although the focus of this chapter is on pyrrole-based anion receptors, it is nonetheless instructive to consider briefly other kinds of anion-binding system. Thus, for the sake of illustration, three classes of recognition motif will be presented – guanidiniums, amides, and ureas. The first of these, when incorporated into receptors, usually gives rise to charged anion-recognition systems, whereas the last two, as a rule, give rise to neutral receptors. The thought, therefore, is that these three motifs, when considered in concert, would provide readers with insights that would help them appreciate the diversity of approaches that can be used to effect anion recognition while enabling them to understand pyrrole-based anion receptors within the context of a more global perspective. By its nature, this overview is not designed to be comprehensive and readers are referred to several recent reviews for more complete treatment [1, 2, 6, 9, 10, 16, 208].

Since the early days of anion-recognition chemistry, guanidinium-based anion receptors have attracted attention because of their special combined ability to stabilize both hydrogen bonds and electrostatic interactions. They remain protonated over a wide pH range ($pK_a = 13.5$) [209, 210] and provide two protons that point in roughly the same direction and which can stabilize two parallel hydrogen bonds, as seen from the X-ray crystal structures of many guanidinium salts [211–223]. As a consequence, guanidinium-based receptors often display high affinities for oxoanion substrates, even in polar solvents and aqueous environments.

To quantify the binding of acetate anion, Tanford monitored the pK_a change of the acid on the formation of a complex with the anionic conjugate base of the acid and guanidinium. Under these conditions it was found that the association constant ($\log K_a$) was low, being less than approximately 0.5 [224]. This method was also used to measure the association constants for guanidinium complexes formed from several other carboxylates and from phosphates [225, 226].

In 1978, Lehn and coworkers reported the synthesis and anion-binding properties of the guanidium-based macrocycles **67** [227]. In the context of this work, three receptors were synthesized. They were prepared by conversion of the corresponding thiourea-containing systems into, first, their *S*-ethylthiouronium derivatives and then, after reaction with ammonia, into the desired target systems **67a–67c**.

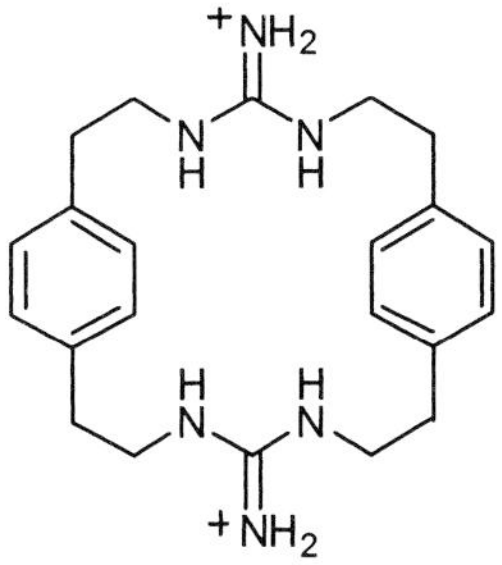

67a 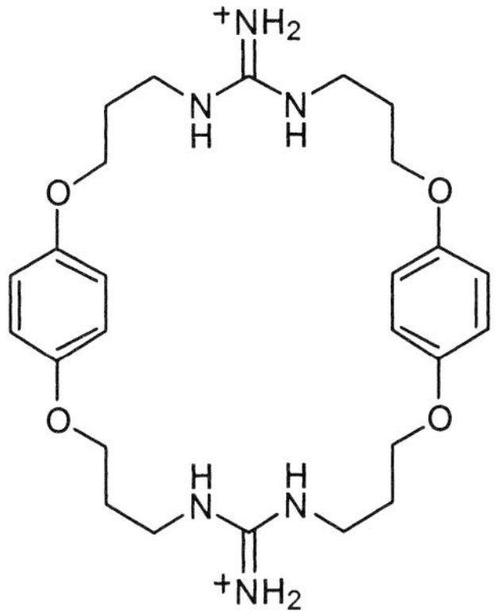   **67b** 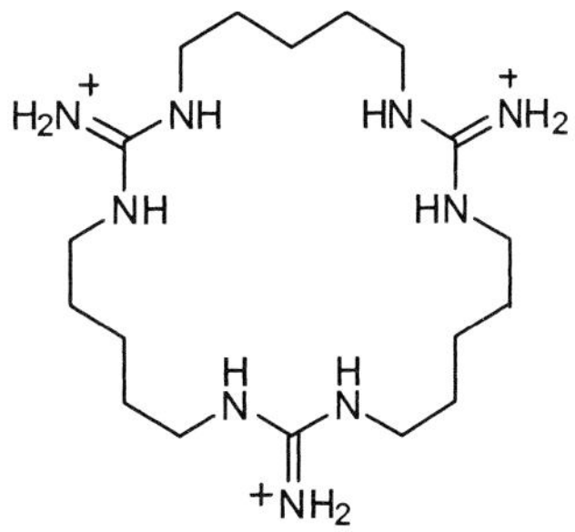   **67c**

The stability constants of macrocycles **67** were determined by analysis of the pH titration curves recorded in the presence and absence of a variety of putative anionic substrates. Taken in concert, these studies revealed that macrocycles **67a**, **67b**, and **67c** bind trianionic phosphate (PO_4^{3-}) in a 1:1 ratio and with affinity constants, log K_a = 3.1, 3.4, and 4.3, respectively, in methanol–water (9:1) solution at 20 °C [227]. That receptor **67c** had a higher stability constant than receptors, **67a** and **67b**, was considered consistent with the appealing notion that, at least within this series of ostensibly related macrocycles, PO_4^{3-} anion-binding affinity is related to the number of positively charged guanidium units available for interaction with the targeted anionic substrate.

Some years after Lehn's seminal studies, Hamilton and coworkers [228] prepared the bis-acylguanidium-based receptor **68**, an open chain bis-guanidinium system linked by an isophthalate spacer. This cleft-like receptor had the advantage that it could be prepared in a single step by reacting dimethyl isophthalate with guanidinium hydrochloride.

An intramolecular hydrogen-bond interaction between the NH and CO moieties, confirmed by proton NMR studies, is thought to endow receptor **68** with considerable rigidity and, as such, more preorganization than was present in the earlier systems of Lehn. The association constant for diphenylphosphate, studied in the form of its tetrabutylammonium salt, was measured by recording the change in the UV absorption intensity at 266 nm observed on diluting a 1:1 mixture of the host and guest in CH_3CN (K_a = 4,6 × 10^4 M^{-1}) [228]. The fact that strong binding behavior was observed led Hamilton and coworkers to apply receptor **68** to the problem of phosphodiester cleavage; they were able to show that it acted as an effective "artificial catalyst" of RNA hydrolysis [229, 230].

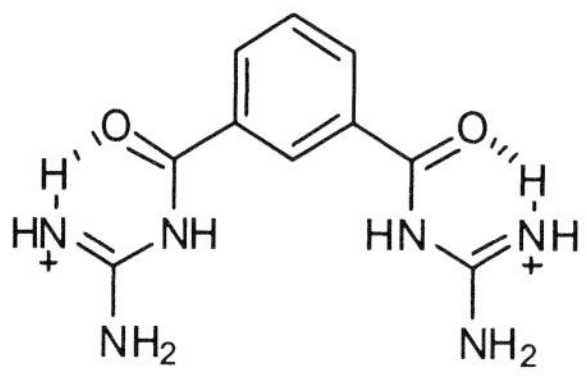

68

Another bisguanidium system, **69**, linked through a *trans*-decalin spacer, was introduced by Göbel in 1994. This receptor was found to bind the anionic form of a small cyclic phosphodiester, catechol cyclic phosphate, with a $K_a = 110$ M^{-1} in DMSO, as determined by ^{31}P NMR spectroscopy [231].

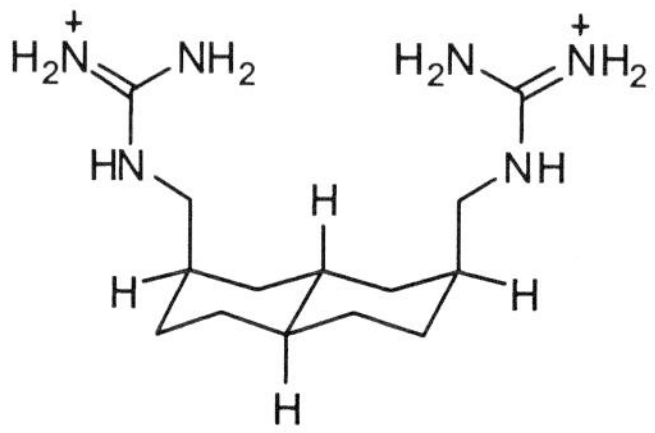

69

In an effort to improve the binding characteristics of this kind of anion receptor, the guanidinium group was incorporated into a bicyclic system. The geometrical advantage of such structures is that they provide two preorganized protons that point in the same direction. This enables receptors of this type to form complexes easily with a variety of anionic guests. The first bicyclic compounds containing a guanidinium group, systems with the generalized structure **70**, were actually reported by McKay and Kreling in the late 1950s [232]. It was not until Schmidtchen first improved the synthetic methodology leading to compounds such as **70** [233] and reported their anion binding properties in the 1980s, however, that bicyclic guanidium systems came to be appreciated as anion receptors [234].

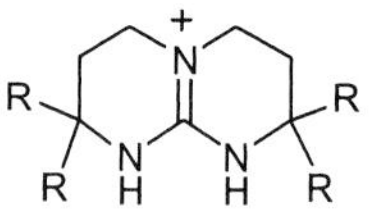

70a R = CH$_3$
70b R = CH$_2$CH=CH$_2$
70c R = CH$_2$CH$_2$CH$_2$OH

A single-crystal X-ray diffraction analysis of **70c** was conducted. It revealed a structure wherein two acetate anions are bound by the embedded guanidinium motifs of two separate molecules of **70c** by hydrogen-bond and electrostatic interactions in-

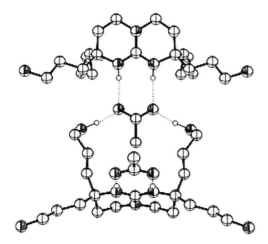

Figure 4.39 Single-crystal X-ray structure of the solid state 2:2 acetate anion complex formed from the bicyclic guanidinium receptor **70c**. This structure was generated using information downloaded from the CCDC and corresponds to a structure originally reported in Ref. [234].

volving the oxygen atoms of the acetate anions, the NH protons of the guanidinium groups, and additional hydrogen bonds between the OH moieties of a neighboring receptor and an acetate anion (Fig. 4.39). Inspired by the observation that acetate–guanidinium complexes could be stable in the solid state, the solution phase anion binding properties of **70b** were studied by using ^{1}H NMR spectroscopic titrations performed in two different solvents. The shapes of the resulting titration curve fits were consistent with the formation of a anion–receptor complex of 1:1 stoichiometry with *p*-nitrobenzoate (studied as the corresponding tetrabutylammonium salt) and from these curves affinity constants (K_a) of >1.0 × 10^4 and 1.4 × 10^5 M^{-1} were obtained from studies conducted in CDCN$_3$ and CDCl$_3$, respectively.

In work carried out in parallel with that of Schmidtchen, Lehn and de Mendoza prepared the first chiral bicyclic guanidinium receptor **71** [235]. It was found that this receptor is capable of extracting *p*-nitrobenzoate anion from aqueous media into organic solvents, lending credence to the notion it can bind oxyanions. Quantitative analysis of the binding constants by use of standard ^{1}H NMR titration methods gave a K_a of 1609 M^{-1} for the binding of *p*-nitrobenzoate (studied as the corresponding triethylammonium salt) in CDCl$_3$. Interestingly, the asymmetric nature of **71**-*SS* enabled enantioselective recognition of chiral carboxylate anions such as the sodium salts of (*S*)-mandelate and (*S*)-naproxenate, as judged from extraction experiments. Similarly, receptors **71**-*SS* and **71**-*RR* could be used to extract L- and D-*N*-acetyltryptophan from a racemic mixture of the L and D enantiomers. Supporting ^{1}H NMR spectroscopic titration studies, performed with the triethylammonium salt of *N*-acetyltryptophan in CDCl$_3$, revealed that receptor **71**-*SS* binds the L and D enantiomers with affinity constants of K_a = 1051 and 534 M^{-1}, respectively, values that match well with those from extraction experiments [235]. Thus, both methods support the conclusion that chiral guanidinium-based receptors such as **71** can be used to effect the enantioselective recognition of anionic substrates.

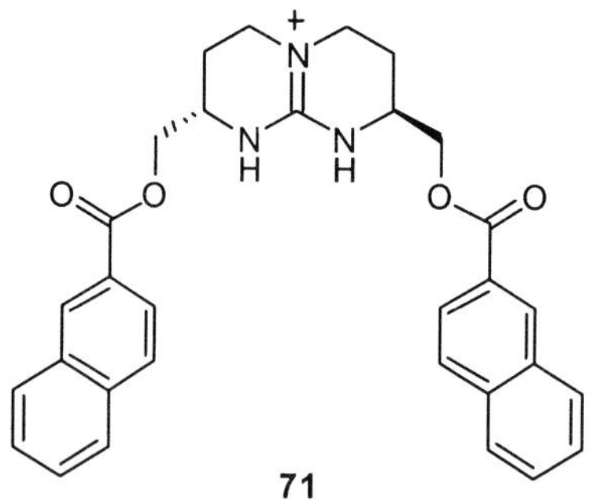

71

In 1993 Schmidtchen reported the bisguanidinium receptor **72**, in which bicyclic guanidiniums are linked via ether linkages, and showed that such systems could be used to effect dicarboxylate anion recognition [236]. Evidence for strong dicarboxylate anion binding was obtained from ^{1}H NMR spectroscopic titration analyses, from which K_a values of 2540, 16,500, 6060, and 14,500 M^{-1} (for 1:1 binding in CD_3OD) were obtained for oxylate, malonate, isophthalate, and *p*-nitroisophthalate anions. In contrast, no change in the ^{1}H NMR spectrum was observed when simpler monoanions, for example acetate and iodide, were added to CD_3OD solutions of **72**.

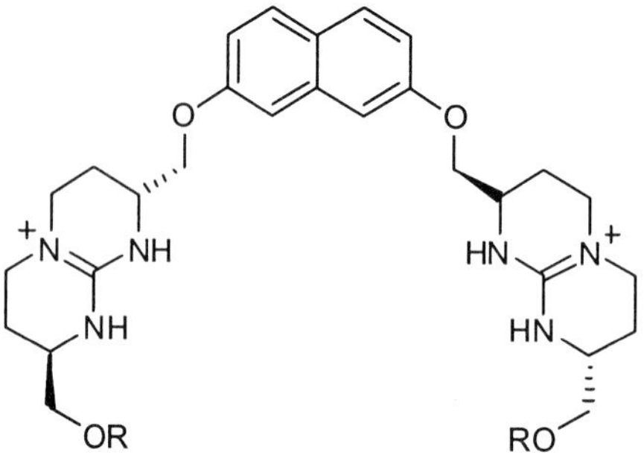

72 R = Si(C$_6$H$_5$)$_2$C(CH$_3$)$_3$

In an effort to design synthetic receptors with strong and selective anion binding in aqueous media, Anslyn and coworkers synthesized the polyguanidinium-based receptors **73**, **74**, and **75** [237–240]. The solid-state structure of the complex formed from receptor **73** and tricarballate was verified by an X-ray crystal diffraction analysis (Fig. 4.40). The selectivity of receptor **73** was greater for tricarboxylate anions (e.g. citrate and tricarballate) than for bis- and mono-carboxylates and phosphates in D_2O. With this receptor, association constants, corresponding to the formation of 1:1 receptor–carboxylate anion complexes were determined by use of standard ^{1}H NMR spectroscopic titration methods and yielded K_a values of 6.9×10^3, 7.3×10^3, 2.2×10^2, and <10 M^{-1} for citrate, tricarballate, glutarate, and acetate, respectively, in D_2O

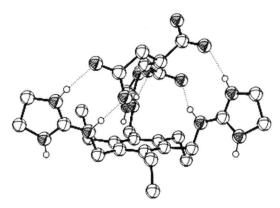

Figure 4.40 Single-crystal X-ray structure of the tricarballate anion complex of the tris-guanidinium receptor **73** as observed in the solid state. This structure was generated using information downloaded from the CCDC and corresponds to a structure originally reported in Ref. [237].

solutions adjusted to pD 7.4 [237]. A year later the citrate content of commercial drink was analyzed using receptor **73** and displacement of a fluorescence-based dye [241]. That this assay worked so well was considered further evidence that receptors such as **73** bind complementary polycarboxylate anions well, even in aqueous media.

73

74

75

New tri- and tetra-guanidinium-based receptors **74** and **75** containing a Cu²⁺ binding site were recently introduced by Anslyn [238–240]. Both receptors were found to occur in their preorganized forms after Cu²⁺ complexation and to have selective anion-binding behavior. It was observed that receptor **74** has high affinity and selectivity for phosphate-type anions in aqueous media at neutral pH. Measurement of specific anion-binding affinity, by monitoring the change in UV–visible absorbance intensity at 790 nm on addition of sodium-anion salts in aqueous solution (pH 7.4 TRIS buffer), gave K_a values of 1.5×10^4, 1.7×10^4, <100, and <100 M⁻¹ for HPO₄²⁻, HAsO₄²⁻, CH₃CO₂⁻, and Cl⁻, respectively [238]. The corresponding thermodynamic data were determined by use of ITC methods. For the HPO₄²⁻ anion, for example, calorimetry performed at pH 7.4 (HEPES buffer solution) yielded $\Delta H° = -3.8$ kcal mol⁻¹ and $\Delta G° = -5.3$ kcal mol⁻¹ [239].

Later, receptor **74** was used as a chemosensor for determining the concentration of inorganic phosphate in serum and saliva, again using the indicator–displacement assay [242].

In an effort to extend this work further, and to take advantage of the fact that Cu²⁺ centers can be used to generate highly preorganized binding motifs, receptor **75** was synthesized. This cyclophane-type system contains three bridging tetraguanidinium subunits and has the highest selectivity for 2,3-bisphosphoglycerate (2,3-BPG) among the various anions tested. The corresponding affinity constant, representing formation of a 1:1 receptor–anion complex, i.e. **75**-2,3-BPG, was determined by using a competitive spectrophotometric method ($K_a = 8 \times 10^8$ M⁻¹ in 1:1 water–methanol solution at pH 6.8) [240].

4.8
Amide-based Anion Receptors

It has long been known that proteins can bind anions not only by interactions involving positively charged residues but also via the neutral amide linkages that define their very essence. These latter subunits can serve as both hydrogen bond-donors (NH) and acceptors (oxygen lone pairs) [243]. For example, X-ray crystal structure analysis of the sulfate-binding protein of *Salmonella typhumurium* reveals that five of the seven identified hydrogen-bonding interactions involve amide NH groups of the polypeptide backbone (cf. Sect. 4.2) [27, 28]. Given the importance of amide-based binding motifs in Nature it is not surprising that much attention has been devoted to developing synthetic anion receptors that rely on similar kinds of interaction. Highlights of this chemistry are reviewed in this section; related approaches based on urea functionality are summarized in Sect. 4.9.

In 1981 Pascal and coworkers reported what can be regarded as the first amide-based synthetic anion receptor, the cage-like system **76** [244]. Single-crystal X-ray crystal structure analysis revealed that the cavity of **76** contains three amide motifs and is cylindrical in shape, approximately 4 Å in length and 3 Å in diameter (Fig. 4.41). As a result of these dimensions, it was considered to be of an appropriate size to serve as a receptor for fluoride and hydroxide anions. In fact, evidence of flu-

Figure 4.41 Single-crystal X-ray structure of receptor **76**. This structure was generated using information downloaded from the CCDC and corresponds to a structure originally reported in Ref. [244].

oride anion binding was obtained in DMSO-d$_6$ solution. Specifically, it was observed that peaks in the ^{1}H NMR and ^{19}F NMR spectra undergo shifts on addition of fluoride anion [244]. No results from a corresponding quantitative binding study were reported, however.

76

In 1997 the synthesis and anion-binding studies of the non-preorganized neutral acyclic receptors **77** were reported by Crabtree and coworkers [146, 147]. Important in its own right, this work is even more significant because it provided the inspiration for Gale's bis-amide pyrrole receptors described in Sect. 4.4.2 [148–153].

77a X = CH, Y = CO, R$_1$ = R$_2$ = H
77b X = CH, Y = CO, R$_1$ = *n*-Bu, R$_2$ = H
77c X = CH, Y = CO, R$_1$ = R$_2$ = CH$_3$
77d X = CH, Y = SO$_2$, R$_1$ = R$_2$ = H
77e X = N, Y = CO, R$_1$ = R$_2$ = H

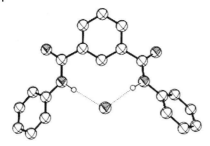

Figure 4.42 Single-crystal X-ray structure of the bromide anion complex of **77a**. This structure was generated using information downloaded from the CCDC and corresponds to a structure originally reported in Ref. [146].

The single-crystal X-ray structure of the bromide anion complex of the bis-amide system **77a** is shown in Fig. 4.42. It reveals the presence of two hydrogen-bond interactions between the amide NH groups and the bromide anion (N $\cdots$ Br = 2.39 and 2.68 Å) [146].

Because of the low solubility of the bis-amide **77a** in CD$_2$Cl$_2$, quantitative anion-binding studies were performed with analogs **77b–77e** [146, 147]. Table 4.18 shows the association constants (K_a) for **77b–77e**, as derived from ^{1}H NMR spectroscopic titration studies conducted in CD$_2$Cl$_2$. It was found that the anion-binding constants are related to the acidity of the amide NH proton. The sulfonamide **77d**, containing a more acidic NH unit (pK_a = 9–10), has a higher binding constant than either receptor **77b** or **77c** (pK_a = 14–15). The anion-binding affinity of **77b** is lower than that of **77c**, presumably because of a combination of lower acidity and higher steric encumbrance. Receptor **77d**, however, gives mixed 1:1 and 1:2 stoichiometry complexes. The highest selectivity for smaller halide anions was achieved by receptor **77e**, because of the rigidity of the hydrogen-bond interactions between the nitrogen of pyridine and the amide NH, and the steric effect of the lone pair electrons on nitrogen [147].

Table 4.18 Association constants K_a (M^{-1} or M^{-2}) for receptors **77b–77e** and selected anionic substrates as deduced from ^{1}H NMR titration experiments in CD$_2$Cl$_2$ at 19.2 °C, using the *n*-Bu$_4$N$^+$ salts of the indicated anions.[a] Data taken from Refs. [146, 147].

Anion	77b	77c	77d	77e
F$^-$	30,000	7500	55,000, 1000[b]	24,000
Cl$^-$	61,000	5300	1000	1500
Br$^-$	7100	1400	4600	57
I$^-$	460	220	1200	<20
AcO$^-$	19,800	2800	21,000, 300[b]	525

[a] The association constants are calculated according to a 1:1 binding profile

[b] The association constants are calculated assuming a 1:2 (host:guest) complex stoichiometry

Hamilton recently reported the synthesis of a new class of amide-containing receptor, **78**, designed to bind tetrahedral anions strongly [245–247]. These rigid systems have three hydrogen bond-donating amide NH groups constrained within the interior of a macrocyclic core. MM2 force-field calculations provided support for the assumption that the cavity of receptors **78** would be well matched in terms of size and shape to several targeted tetrahedral oxyanions. More direct experimental evidence came from quantitative anion-binding studies conducted using standard ^{1}H NMR and fluorescence spectral titration methods.

For *p*-tosylate, studied as the corresponding triethylammonium salt in 2 % DMSO-d_6–CDCl$_3$, the formation of a 1:1 complex with both receptors **78a** and **78b** was inferred from ^{1}H NMR spectroscopic titration studies. From these, association constants (K_a) of 2.6×10^5 and 2.1×10^5 M^{-1} were calculated for **78a** and **78b**, respectively [245]. The other halides, nitrate, hydrogen sulfate and dihydrogen phosphate anions, however, were found to bind to receptors **78a** and **78b** in a 2:1 fashion (host:guest ratio) at low anion concentrations and in a 1:1 ratio after ≥0.5 equiv. of the anion in question had been added. On the basis of this stoichiometry and other considerations it was proposed that the complex formed at low anion concentrations (relative to the receptor) exists in the form of a sandwich-like structure wherein the anion sits between two host molecules [245, 247].

In pure DMSO-d_6 only 1:1 binding behavior was observed for receptor **78b** and the association constants for halide, nitrate, *p*-tosylate anions were found to be dramatically decreased. On the other hand, a high binding affinity for hydrogen sulfate and dihydrogen phosphate anions was observed and the formation of 1:1 complexes was inferred ($K_a = 1.7 \times 10^3$ and 1.5×10^5 M^{-1} for HSO$_4^-$ and H$_2$PO$_4^-$, respectively) [245].

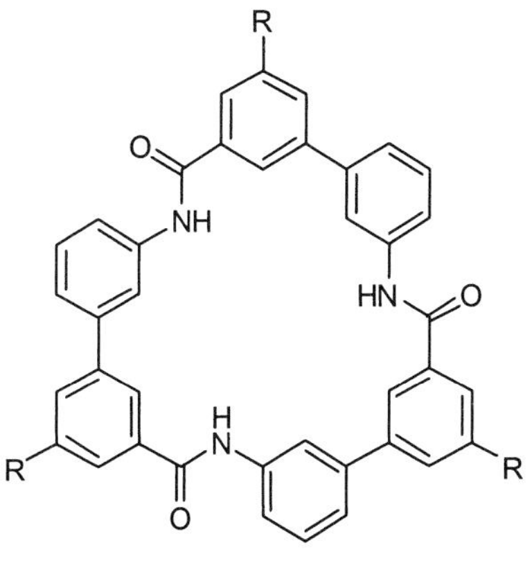

78a R = CO$_2$C$_2$H$_5$
78b R = NHBoc

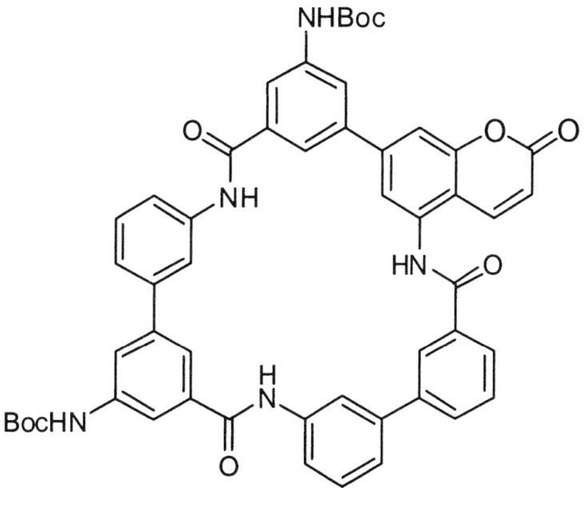

78c

System **78c**, designed to act as a fluorescence chemosensor, had selectivity similar to that observed for receptors **78a** and **78b**; again affinity was high for tetrahedral oxyanions. The association constants were measured by monitoring changes in the maximum emission intensity observed after addition of tetrabutylammonium-anion salts in 1:1 DMSO–1,4-dioxane. The results of these fluorescence titration studies yielded affinity constants of 2.0×10^6, 6.3×10^5, and 3.8×10^3 M^{-1} for dihydrogen phosphate, phenyl phosphate, and *p*-tosylate, respectively [246]. For **78c**, as for **78a** and **78b**, the selectivity for dihydrogen phosphate was considered to reflect the presence of three convergent amide NH hydrogen-bond-donor subunits.

In 2001 Jurczak reported the anion-binding properties of two neutral macrocyclic polylactam receptors **79a** and **79b**, synthesized from a 2,6-pyridinedicarboxylate precursor and 1,2-diaminoethane [248, 249]. The X-ray crystal structure of the fluoride complex formed from tetralactam **79a** is shown in Fig. 4.43; it supports the notion that the size of the cavity is too small to accommodate chloride anion easily but provides a good fit for the smaller fluoride anion. Specifically, the bound fluoride anion is found to be held in place by four inward-pointing hydrogen bonds.

79a n = 1
79b n = 3

To verify the anion-binding properties of these receptors, ^{1}H NMR spectroscopic titrations were performed with a variety of tetrabutylammonium-anion salts in DMSO-d_6. The results of these studies confirmed that the greatest anion-binding affinity and highest selectivity were for the acetate and dihydrogen phosphate anions (K_a = 2640 and 1680 M^{-1}, for $CH_3CO_2^-$ and $H_2PO_4^-$, respectively) and also served to establish that the K_a for fluoride is higher than that for chloride (830 and 65 M^{-1} for F^- and Cl^-, respectively) [248].

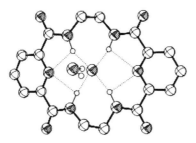

Figure 4.43 Single-crystal X-ray structure of the fluoride anion complex of tetralactam **79a**, showing one bound water molecule. This structure was generated using information downloaded from the CCDC and corresponds to a structure originally reported in Ref. [248].

In independent work Bowman-James synthesized another ostensibly similar tetraamide macrocycle, receptor **80**, specifically designed to bind tetrahedral anions such as sulfate and phosphate with high specificity [250–252]. In contrast with **79a**, receptor **80** not only has four amide subunits to provide neutral NH hydrogen-bond-donor functionality, but also two tertiary amine groups that, when protonated, were expected to provide additional anion-binding sites, along with two positive charges within the binding core. An X-ray crystal structure of the sulfate complex formed from **80** revealed a sandwich complex in which a single sulfate anion is surrounded by two macrocyclic receptors, at least in the solid state (Fig. 4.44).

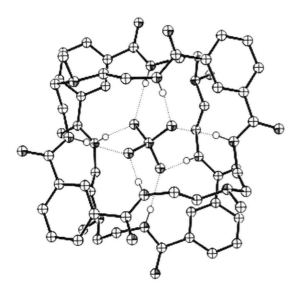

80a X = O, Y = CH
80b X = O, Y = N
80c X = S, Y = CH
80d X = S, Y = N

Figure 4.44 Single-crystal X-ray structure of the 1:2 (anion to receptor) solid-state sulfate anion complex formed from receptor **80a**. This figure was generated using information downloaded from the CCDC and corresponds to a structure originally reported in Ref. [250].

Quantitative solution-phase anion-binding studies, performed by monitoring changes in the ^{1}H NMR spectrum on addition of different tetrabutylammonium salts to the neutral form of **80a** in CDCl$_3$ solution, revealed selectivity for hydrogen sulfate and dihydrogen phosphate anions, species containing at least one acidic proton (Table 4.19) [250]. Presumably, these results reflect both an increase in the effective negative charge of these two anions as the result of proton transfer to the two amine units and the enhanced electrostatic interactions (between the bound anions and the protonated amines) that would result from this proton transfer [250].

Table 4.19 Association constants log K_a (M^{-1}) for receptors **80a–80d** and selected anionic substrates, as deduced from ^{1}H NMR titration experiments using the *n*-Bu$_4$N$^+$ salts of the anions.^{a} Data taken from Refs. [250–252].

	80a		**80b**	**80c**	**80d**
Solvent	CDCl$_3$	DMSO-d$_6$	DMSO-d$_6$	DMSO-d$_6$	DMSO-d$_6$
H$_2$PO$_4^-$	4.66	2.92	4.05	4.97^b	4.63^c
HSO$_4^-$	4.50	2.89	2.03	3.15	4.99^c
F$^-$	–	2.63	2.61	2.85	4.11^c
Cl$^-$	1.32	1.39	2.69	2.02	2.60
Br$^-$	–	1.30	2.71	2.00	1.40
I$^-$	2.13	<1.0	<1.0	1.44	<1.0

[a] Estimated deviations less than 5 % unless otherwise noted
[b] Estimated deviation 10 %
[c] Slow exchange; estimated error less than 15 %

Bowman-James also helped pioneer the use of thioamide motifs to effect anion recognition. In the context of this work, thioamide and pyridine subunits were introduced into macrocyclic frameworks in an effort to enhance the affinity of receptors such as **80b**, **80c**, and **80d** [251, 252]. In general, it was observed that anion-binding constants in DMSO-d$_6$ were higher for the pyridine analogs (receptors **80b** and **80d**) than for the corresponding phenyl analogs (receptors **80a** and **80c**). Presumably, this result reflects the pyridine-assisted pre-organization of two of the four amide NH donor groups [251, 252]. Another inference that can be drawn from Tab. 4.19 is that in DMSO-d$_6$ the bis-thioamide receptors **80c** and **80d** have higher binding affinity for dihydrogen phosphate, hydrogen sulfate, and fluoride anions than the bis-amide receptors **80a** and **80b**. The increased acidities of the thioamide NH is thought to enhance the relative anion-binding affinity of receptors **80c** and **80d** [252].

It has long been a goal in the anion-recognition community to develop neutral synthetic anion receptors effective in polar solvents such as water. In recent years Kubik and coworkers have succeeded in preparing a set of cyclic hexapeptides, **81**, that contain L-proline and 6-aminopicolinoic acid subunits. They also determined their anion-binding affinities in aqueous media [253, 254]. In the context of this work these workers determined the solid-state structure of the 2:1 (host:guest) iodide complex (Fig. 4.45). They found a sandwich-type structure in which one iodide anion was bound between two cyclic peptide receptors **81a**, being stabilized there by six hydrogen bonds involving the NH functionality of the macrocycle-derived amides.

The solution phase anion binding properties of **81a** and **81b** were determined in aqueous media by monitoring shifts in the NH peaks seen in the ^{1}H NMR spectrum upon the addition of different anion-containing salts; the result are summarized in Table 4.20. For receptor **81a**, the 1:2 (anion:host) binding stoichiometry inferred from the X-ray structure of the iodide complex was confirmed by use of Job plots. On the other hand, only a 1:1 anion:host stoichiometry was observed for receptor **81b**, which differs from **81a** only in that it contains additional hydroxyl groups on the pyrrolidine rings. Although these hydroxyl groups are not close to the binding site, they presumably serve to prevent aggregation between the macrocycles. Such considerations not withstanding, it is important to note that the K_{a1} values of receptor **81a** corresponding to the first binding event match well the K_a values of receptor **81b**. Furthermore, the product $K_{a1} \times K_{a2}$, which can be considered to reflect the absolute anion-binding ability, reveals that receptor **81a** is a very effective anion receptor for large anions such as sulfate and iodide in aqueous media [254].

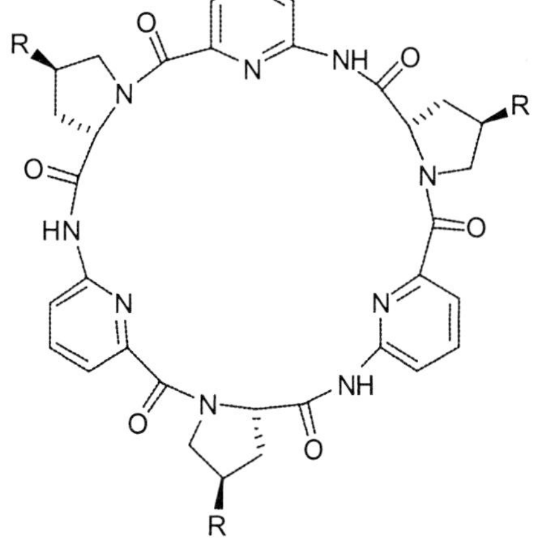

81a R = H
81b R = OH

Table 4.20 Anion-binding affinities (M^{-1} or M^{-2}) for receptors **81a** and **81b** and selected anionic substrates, as deduced from ^{1}H NMR titration experiments in 80 % D_2O–CD_3OD. Data taken from Ref. [254].

		NaCl	NaBr	NaI	KI	N(CH$_3$)$_4$I	Na$_2$SO$_4$
81a	K_{a1}	5	16	22	24	24	96
	K_{a2}	6770	6820	7380	6660	7470	1270
	$K_{a1} \times K_{a2}$	3.4×10^4	1.1×10^5	1.6×10^5	1.6×10^5	1.8×10^5	1.2×10^5
81b	K_a	8	13	19	–	–	95

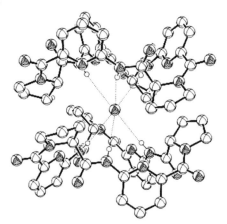

Figure 4.45 Single-crystal X-ray structure of the 1:2 (anion to receptor) iodide anion complex stabilized by receptor **81a** in the solid state. This structure was generated using in- formation downloaded from the CCDC and corresponds to a structure originally reported in Ref. [253].

Inspired by the results obtained with receptor **81a**, Kubik developed a set of second-generation dimeric receptors, **82**, linked by spacers of different lengths. These systems were found to have substantial selectivity for sulfate anion in aqueous solution [255, 256]. The stoichiometry of the different anion complexes formed with receptors **82**, in-ferred from electrospray ionization mass spectrometry in the gas phase and from Job plots in 50 % D_2O–CD_3OD solution, provided strong support for the notion that this re-ceptor forms 1:1 complexes with a range of different anions. This finding led the au-thors to suggest that the two cyclopeptide units present in **82a**–**82c** snap down and bind the anionic guest like a "molecular oyster" pinching a chemical "pearl".

82a
X =

82b
X =

82c
X =

The first quantitative binding studies involving receptor **82a** were conducted in 50 % D_2O–CD_3OD using standard ^1H NMR titration methods. It was observed that receptor **82a** has high selectivity for sulfate anion (K_a = 3.5 × 10^5, 8.9 × 10^3, and 5.3 × 10^3 M^{-1} for Na_2SO_4, NaI, and NaBr, respectively, in this solvent mixture) [255]. One year after this initial report appeared a second quantitative anion-binding study was performed using ITC in 2:1 (*v*/*v*) acetonitrile–water solution at 298 K. It was observed that receptors **82b** and **82c** bind sulfate and iodide anions one order of magnitude more strongly than receptor **82a**. For sulfate, the association constants (K_a) of receptors **82a**, **82b**, and **82c** are 2.0 × 10^5, 5.4 × 10^6, and 6.7 × 10^6 M^{-1}, respectively. For iodide anion, the corresponding values were found to be 3.3 × 10^3, 2.9 × 10^4, and 5.6 × 10^4 M^{-1} for receptors **82a**, **82b**, and **82c**, respectively [256].

In 1997 Anslyn and coworkers reported results of studies involving an anion-binding polyamide cage-type receptor, **83** [257]. This receptor, the first to be investigated since Pascal's initial synthesis of his trisamide cyclophane **76**, contains a rather rigid framework that provides six inward-pointing amide NH hydrogen bond-donor subunits. X-ray crystal structure analysis of the acetate complex formed from receptor **83** revealed 1:1 binding stoichiometry in the solid state. It also shows that the acetate anion is encapsulated within the cavity by four hydrogen bonds (Fig. 4.46). On the basis of standard ^1H NMR titrations, performed in CD_2Cl_2–CD_3CN (1:3, *v*/*v*), affinity constants (K_a) were determined for $CH_3CO_2^-$, NO_3^-, Cl^-, and $H_2PO_4^-$ (studied in the form of their tetrabutylammonium salts) and were found to be 770, 300, 40, and 25 M^{-1}, respectively, for these four anionic substrates [257]. Taken together, these results were believed to support the proposition that the amide functional group, when properly oriented, can be used to bind Y-shaped anions, for example acetate, effectively. Although, the highest binding constant was observed for the acetate anion, a relatively high affinity for nitrate was observed, reflecting once again what was thought to be a good geometric match. Two years later, anion binding constants of receptor **83** were re-investigated by monitoring the UV–visible spectral changes induced by addition of substrate under conditions of competitive indicator-displacement. Association constants for NO_3^-, Br^-, and ClO_4^- (studied in the form of their sodium salts), calculated in this way, were found to be 380, 220, and 130 M^{-1} in CH_3OH–CH_2Cl_2 (1:1) and 500, 190, and 70 M^{-1} in CH_3CN–CH_2Cl_2 (3:1) [258].

83

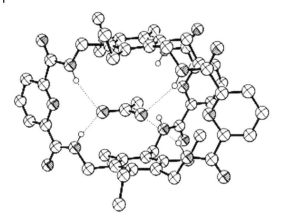

Figure 4.46 Single-crystal X-ray structure of the acetate anion complex of receptor **83**. This structure was generated using information downloaded from the CCDC and corresponds to a structure originally reported in Ref. [257].

Quite recently a new type of polyamide cryptand-type receptor, represented by systems **84a** and **84b**, were synthesized by Bowman-James and coworkers [252, 259]. Receptor **84** has six amide NH donor subunits and two tertiary amine groups. X-ray diffraction analysis revealed that the fluoride anion complex of receptor **84a** is stabilized by six hydrogen bonds (N $\cdots$ F = 2.84 and 2.89 Å) as shown in Fig. 4.47. The anion-binding properties of receptors **84a** and **84b** in solution were measured by ^{1}H NMR spectral titration experiments using the corresponding tetrabutylammonium salts. The association constants (log K_a) calculated in this way were F$^-$ 5.90, Cl$^-$ 3.48, H$_2$PO$_4^-$ 3.31, HSO$_4^-$ 1.83, Br$^-$ 1.60, and I$^-$ 1.30 for amide cryptand **84a** and F$^-$ 4.50, Cl$^-$ 1.54, H$_2$PO$_4^-$ 3.40, HSO$_4^-$ 1.69, Br$^-$ <1.0, and I$^-$ <1.0 for thioamide cryptand **84b**. Both cryptand-type receptors **84a** and **84b** have higher affinity for fluoride anion and lower affinity for chloride, acetate, and dihydrogen phosphate anions than the two-dimensional receptors **80a–80d**.

84a X = O
84b X = S

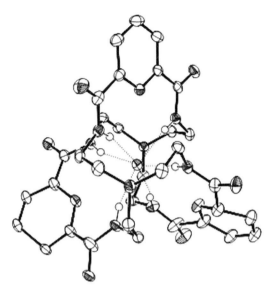

Figure 4.47 Single-crystal X-ray structure of the fluoride anion complex of receptor **84a**. This structure was generated using informa- tion downloaded from the CCDC and corre- sponds to a structure originally reported in Ref. [259].

4.9
Urea-based Anion Receptors

The urea and thiourea functional groups are among the most popular binding motifs used to prepare neutral anion-binding receptors. They provide two hydrogen bond-donor groups that point in the same direction and which are spaced appropriately to interact with a range of anionic substrates, including Y-shaped anions such as carboxylates.

The first use of urea and thiourea functional groups for anion recognition was reported by Wilcox in 1992 [260]. He and his coworkers reported the synthesis and anion-recognition properties of the acyclic systems **85**. Here, an initial indication that receptor **85a** was capable of binding anions came from the large downfield shift seen for the NH signals in the ^{1}H NMR spectrum of **85a** in CDCl$_3$ when tetrabutylammonium benzoate and tosylate were added. Accompanying UV–visible titrations, performed in chloroform, enabled the association constants (K_a) for benzoate and p-tosylate (studied as their tetrabutylammonium salts) to be calculated for this system; they were found to be 2.7×10^4 and 6.1×10^3 M^{-1}, respectively [260].

In 1994, Kelly designed the bisurea-based anion receptor **86** for dicarboxylate-type anions [261]. In this work a predicative study of space-filling models was performed and provided support for the proposal that receptor **86** was of an appropriate size and shape to bind m-dinitro-substituted benzene substrates, for example TNT. Quantitative anion-binding studies, performed by means of ^{1}H NMR spectral titrations conducted in DMSO-d$_6$ with a variety of putative ditopic substrates, were found to be consistent with this supposition. In particular, it was found that the association constants (K_a) for isophthalate, terephthalate, and *m*-dinitrobenzene (studied as the corresponding triethylammonium salts) were 63,000, 745, and 86 M^{-1}, respectively [261]. Receptor **86** had higher affinity for isophthalate than terephthalate; this was taken as a further confirmation that geometry optimization plays a key role in enhancing host–guest binding interactions.

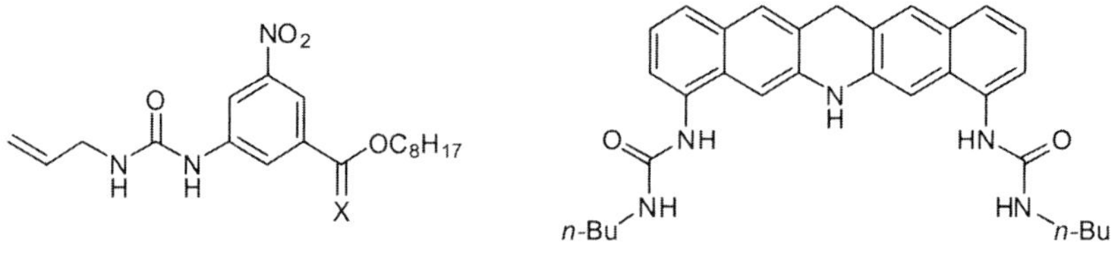

85a X = O
85b X = S

86

In an effort to increase the anion-binding ability of urea-based neutral anion receptors, two open-chain systems, **87a** and **87b**, with four thiourea units linked via two different spacers, *p*- and *m*-xylylene, were synthesized by Benito et al. [262]. With these systems, addition of tetrabutylammonium glutarate caused the NH signal in the ^{1}H NMR spectrum to shift. Such shifts were ascribed to the presence of hydrogen-bonding interactions involving the receptor and the presumably bound glutarate anion. For receptor **87a** the binding stoichiometry and association constants were determined from Job plots and from ^{1}H NMR spectroscopic titrations, respectively. In the context of these studies it was found that glutarate was bound too well to enable an accurate affinity constant to be determined in either CDCl$_3$ or DMSO-d$_6$, because of the detection limits of ^{1}H NMR spectroscopy. Analogous titrations were therefore conducted in aqueous solution (10 % D$_2$O–DMSO-d$_6$). From these an association constant of 10^3 M^{-1} was obtained for binding of glutarate by **87a** under these conditions.

In contrast with the behavior of **87a**, the stoichiometry for receptor **87b**–glutarate was observed as 1:2 (host:guest), as inferred from a Job plot carried out in DMSO-d$_6$.

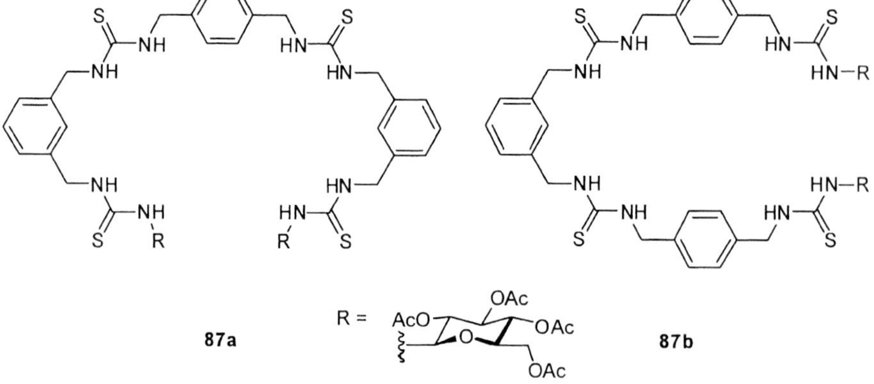

87a   R =   **87b**

Also, results from a ^{1}H NMR spectroscopic titration revealed $K_{a1} = 10^5$ M^{-1} and $K_{a2} = 10^2$ M^{-1} [262]. Nonetheless, the fact that the anion-binding patterns are so different for two ostensibly similar receptors **87a** and **87b**, with different spacers, provides support for the notion that geometry optimization plays a critical role in anion recognition.

In 2000, two C_3-symmetric metacyclophane-type anion receptors, **86a** and **88b**, containing three thiourea subunits were reported by Hong [263]. The conformationally more rigid receptor **88b**, made by "attaching" three ethyl groups to the receptor scaffold, was found to have higher binding affinity than **88a**, as judged from quantitative NMR spectroscopic studies. These studies also revealed that **88b** is selective for acetate anion over dihydrogen phosphate anion and chloride anion in DMSO-d_6 (K_a = 5300, 1600, and 95 M^{-1} for $CH_3CO_2^-$, $H_2PO_4^-$, and Cl^-, respectively; tetrabutylammonium salts for $CH_3CO_2^-$ and $H_2PO_4^-$, tetraethylammonium salt for Cl^-). The order of selectivity is different for the more flexible system **88a** and its overall affinity for anions is lower than that of **88b** (K_a = 800 M^{-1}, 320 M^{-1} and 40 M^{-1} for $H_2PO_4^-$ $CH_3CO_2^-$, and Cl^-, respectively, in DMSO-d_6) [263].

Essentially contemporaneous with the above report, but actually slightly before its publication, the urea-linked macrocycles **89a** and **89b** were described by Reinhoudt and coworkers [264]. The anion-binding affinity of both receptors **89a** and **89b** were investigated using standard ^{1}H NMR spectral titrations, performed in DMSO-d_6. Tetrabutylammonium-anion salts were used as the source of the anions. The more flexible receptor **89a** was found to bind dihydrogen phosphate and chloride anions with K_a values of 4000 and <50 M^{-1}, respectively, and with 1:1 stoichiometry. This system thus has relatively high $H_2PO_4^-/Cl^-$ selectivity ($K_{rel} > 100$) [264]. This contrasts with the behavior of the more rigid receptor **89b**, for which the corresponding K_a values are 2500 and 500 M^{-1} for dihydrogen phosphate and chloride anion, respectively. These findings presumably reflect the fact that **89b** is not structurally optimized for dihydrogen phosphate binding and lacks the flexibility needed to accommodate this relatively large anion.

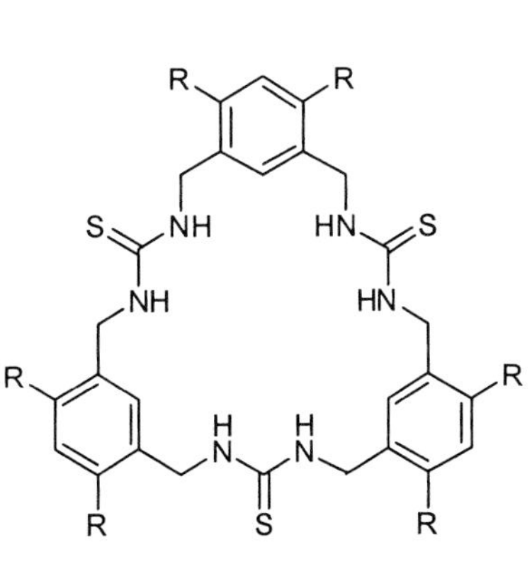

88a R = H
88b R = C_2H_5

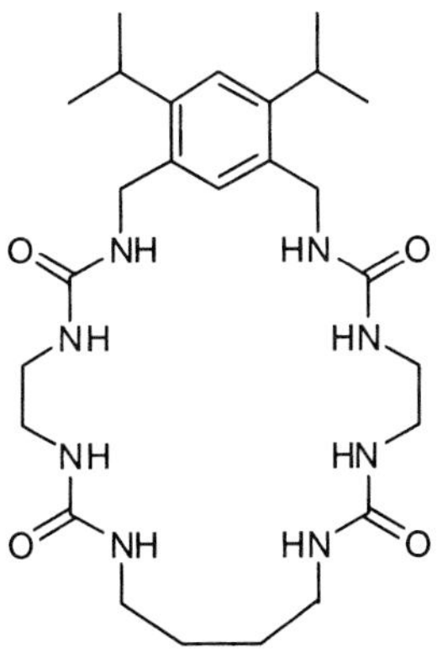

89a

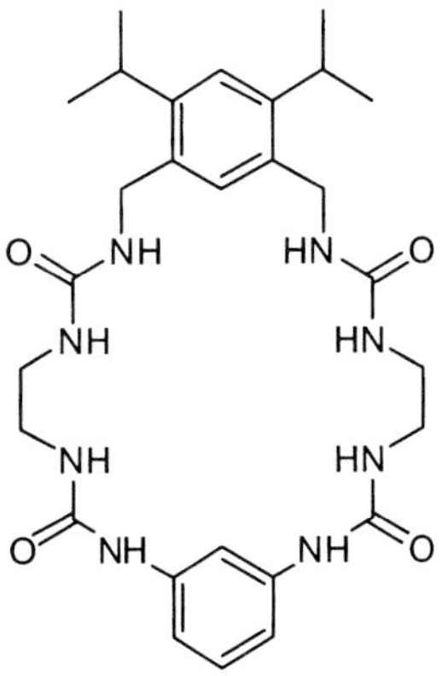

89b

Before introducing systems **89a** and **89b**, Reinhoudt demonstrated that thiourea-containing calix[6]arenes, of general structure **90**, could serve as useful anion and cation receptors [265]. Receptors **90a** and **90b** were both found to bind halide and carboxylate anions with 1:1 stoichiometry and have high selectivity for tricarboxylate anions. Quantitative analyses of the anion-binding behavior of **90a** and **90b** were performed by recording changes in the NH chemical shifts seen upon the addition of tetrabutylammonium-anion salts in $CDCl_3$.

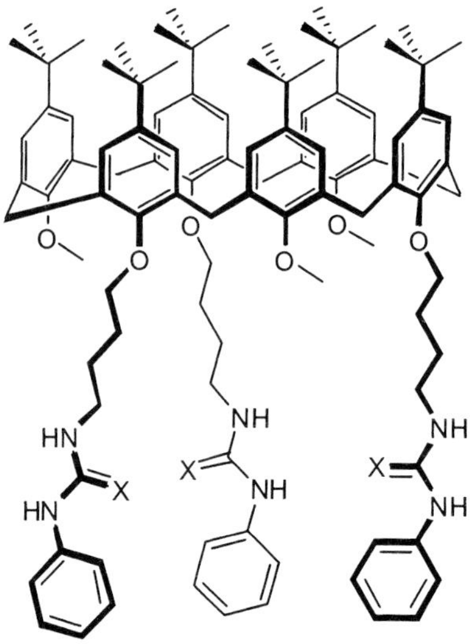

90a X = O
90b X = S

As a result of these latter studies it was concluded that receptor **90a**, containing urea-based binding units, is more selective for *cis*-1,3,5-cyclohexanetricarboxylate ($K_a = 1.0 \times 10^5$ M^{-1}) than for 1,3,5-benzenetricarboxylate ($K_a = 8.7 \times 10^4$ M^{-1}), isophthalate ($K_a = 6.9 \times 10^4$ M^{-1}), and benzoate ($K_a = 1.6 \times 10^4$ M^{-1}). In contrast, for receptor **90b**, with three thiourea units in place of the urea linkages, the selectivity order is slightly different – 1,3,5-benzenetricarboxylate ($K_a = 2.9 \times 10^5$ M^{-1}) > *cis*-1,3,5-cyclohexanetricarboxylate ($K_a = 2.9 \times 10^4$ M^{-1}) > isophthalate ($K_a = 6.4 \times 10^3$ M^{-1}) > benzoate ($K_a = 1.4 \times 10^3$ M^{-1}) [265]. For both receptors, however, the threefold symmetry of these functionalized calix[6]arene receptors leads to a preference for tricarboxylate anions.

Quite recently, Kilburn and coworkers reported the synthesis and anion-binding properties of a calix[4]arene-based anion receptor **91**, characterized by the presence of a thiourea-based strap [266]. The association constants (M^{-1}) for acetate, phenylphosphinate, and diphenylphosphate (studied as their respective tetrabutylammonium salts) were found, from quantitative ^1H NMR titrations, to be 11,000, 24,000, and 1800 M^{-1}, respectively, in CD_3CN [266]. When considered together, these results provide support for the conclusion that receptor **91** binds tetrahedral phenylphosphinate anions more strongly than the Y-shaped acetate anion.

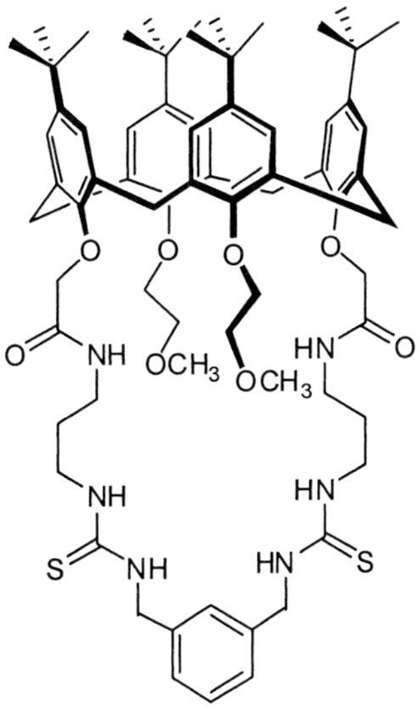

91

4.10
Conclusions

The results presented in this chapter, focused on pyrrole-based receptors but including discussion of alternative systems, help make clear the central role currently played by pyrrole-based recognition strategies in the design of anion receptors. The diversity of structures constructed to date, and the range of applications these structures have enabled researchers to target, reflect the many advantages of pyrrole as an anion-binding motif. Existing in its neutral form over a wide range of pH (in contrast with the readily protonated polyamines or the inherently charged guanidiniums), the unoxidized form of pyrrole acts as a strong NH-derived hydrogen bond-donor unencumbered by concerns associated with potential nearby hydrogen bond-acceptor functionality (as are, for instance, receptors based on, e.g., amides and ureas). Pyrrole as a receptor-generating "building block" has the further advantage that it can be functionalized readily and thus incorporated into an almost infinite number of very elaborate receptor systems. When such systems enable oxidation of pyrrole, imine-like sp^2-hybridized nitrogen centers are produced that, when protonated, provide a source for both NH-based hydrogen-bonding recognition and more generalized cation–anion electrostatic interactions. Thus, depending on how it is used, pyrrole can provide the basis for either neutral or cationic anion receptors. This duality provides a richness of function that is essentially without parallel within the world of anion recognition. It is thus safe to predict that pyrrole-based systems will continue to play a prominent part in the evolving story of anion recognition.

Acknowledgment

This work was supported by the NIH (GM 58907)

References

1 A. Bianchi, K. Bowman-James, E. García-España, *Supramolecular Chemistry of Anions*, Wiley–VCH, New York, **1997**.

2 P. D. Beer, P. A. Gale, *Angew. Chem., Int. Ed.* **2001**, *40*, 486.

3 P. D. Beer, J. Cadman, *Coord. Chem. Rev.* **2000**, *205*, 131.

4 C. H. Park, H. E. Simms, *J. Am. Chem. Soc.* **1968**, *90*, 2431.

5 P. A. Gale, *Coord. Chem. Rev.* **2003**, *240*, 1.

6 M. D. Best, S. L. Tobey, E. V. Anslyn, *Coord. Chem. Rev.* **2003**, *240*, 3.

7 J. L. Sessler, S. Camiolo, P. A. Gale, *Coord. Chem. Rev.* **2003**, *240*, 17.

8 J. M. Llinares, D. Powell, K. Bowman-James, *Coord. Chem. Rev.* **2003**, *240*, 57.

9 C. R. Bondy, S. J. Leob, *Coord. Chem. Rev.* **2003**, *240*, 77.

10 K. Choi, A. D. Hamilton, *Coord. Chem. Rev.* **2003**, *240*, 101.

11 T. J. Wedge, F. M. Hawthorne, *Coord. Chem. Rev.* **2003**, *240*, 111.

12 T. N. Lambert, B. D. Smith, *Coord. Chem. Rev.* **2003**, *240*, 129.

13 A. P. Davis, J.-B. Joos, *Coord. Chem. Rev.* **2003**, *240*, 143.

14 M. W. Hosseini, *Coord. Chem. Rev.* **2003**, *240*, 157.

15 P. D. Beer, E. J. Hayes, *Coord. Chem. Rev.* **2003**, *240*, 167.

16 P. A. Gale, *Coord. Chem. Rev.* **2003**, *240*, 191.

17 F. P. Schmidtchen, M. Berger, *Chem. Rev.* **1997**, *97*, 1609.

18 B. Alberts, D. Bray, J. Lewis, M. Raff, K. Roberts, J. D. Watson, *Molecular Biology of the Cell. 3rd Ed*, **1995**.

19 M. P. Anderson, R. J. Gregory, S. Thompson, D. W. Souza, S. Paul, R. C. Mulligan, A. E. Smith, M. J. Welsh, *Science* **1991**, *253*, 202.

20 D. B. Simon, R. S. Bindra, T. A. Mansfield, C. Nelson-Williams, E. Mendonca, R. Stone, S. Schurman, A. Nayir, H. Alpay, A. Bakkaloglu, J. Rodriguez-Soriano, J. M. Morales, S. A. Sanjad, C. M. Taylor, D. Pilz, A. Brem, H. Trachtman, W. Griswold, G. A. Richard, E. John, R. P. Lifton, *Nature Genetics* **1997**, *17*, 171.

21 O. Devuyst, P. T. Christie, P. J. Courtoy, R. Beauwens, R. V. Thakker, *Human Molec. Genetics* **1999**, *8*, 247.

22 D. A. Scott, R. Wang, T. M. Kreman, V. C. Sheffield, L. P. Karniski, *Nature Genetics* **1999**, *21*, 440.

23 A. Yoshida, S. Taniguchi, I. Hisatome, I. E. Royaux, E. D. Green, L. D. Kohn, K. Suzuki, *J. Clin. Endocrinol. Metab.* **2002**, *87*, 3356.

24 U. Kornak, D. Kasper, M. R. Bosl, E. Kaiser, M. Schweizer, A. Schulz, W. Friedrich, G. Delling, T. J. Jentsch, *Cell* **2001**, *104*, 205.

25 R. S. Kaplan, *J. Membr. Biol.* **2001**, *179*, 165.

26 H. Xu, N. Sträter, W. Schröder, C. Böttcher, K. Ludwing, W. Saenger, *Acta Crystallogr.* **2003**, *D59*, 815.

27 J. W. Pflugrath, F. A. Quiocho, *Nature* **1985**, *314*, 257.

28 J. W. Pflugrath, F. A. Quiocho, *J. Mol. Biol.* **1988**, *200*, 163.

29 L. Chantalat, J. M. Nicholson, S. J. Lambert, A. J. Reid, M. J. Donovan, C. D. Reynolds, C. M. Wood, J. P. Baldwin, *Acta Crystallogr.* **2003**, *D59*, 1395.

30 R. Elizabeth, R. J. P. Williams, *Biochem. J.* **1987**, *245*, 641.

31 H. Alexandre, M. Geuskens, *Arch. Biol.* **1984**, *95*, 55.

32 G. J. Quigley, M. M. Teeter, A. Rich, *Proc. Natl. Acad. Sci. USA* **1978**, *75*, 64.

33 I. Labadi, E. Jenei, R. Lahti, H. Lonnberg, *Acta Chem. Scand.* **1991**, *45*, 1055.

34 E. Lahti, H. Lonnberg, *Biochem. J.* **1989**, *259*, 55.

35 W. H. Voige, R. I. Elliott, *J. Chem. Ed.* **1982**, *59*, 257.

36 H. Ohishi, I. Nakanishi, K. Inubushi, G. v. d. Marel, J. H. v. Boom, A. Rich, A. H. J. Wang, T. Hakoshima, K.-I. Tomita, *FEBS Lett.* **1996**, *391*, 153.

37 H. Ohishi, N. Terasoma, I. Nakanishi, G. v. d. Marel, J. H. v. Boom, A. Rich, A. H. J. Wang, T. Hakoshima, K.-I. Tomita, *FEBS Lett.* **1996**, *398*, 291.

38 B. P. Orner, A. D. Hamilton, *J. Incl. Phenom.* **2001**, *41*, 141.

39 J. C. Manimala, E. V. Anslyn, *Eur. J. Org. Chem.* **2002**, 3909.

40 R. P. Dioxn, S. J. Geib, A. D. Hamilton, *J. Am. Chem. Soc.* **1992**, *114*, 365.

41 V. Jubian, R. P. Dixon, A. D. Hamilton, *J. Am. Chem. Soc.* **1992**, *114*, 1120.

42 P. Schiebl, F. P. Schmidtchen, *J. Org. Chem.* **1994**, *59*, 509.

43 G. Muller, J. Riede, F. P. Schmidtchen, *Angew. Chem., Int. Ed.* **1988**, *27*, 1516.

44 A. Gleich, F. P. Schmidtchen, P. Mikulcik, G. Muller, *J. Chem. Soc., Chem. Commun.* **1990**, 55.

45 R. Dutzler, E. B. Campbell, M. Cadene, B. T. Chait, R. MacKinnon, *Nature* **2002**, *415*, 287.

46 P. H. Schlesinger, R. Ferdani, R. Pajewski, J. Pajewska, G. W. Gokel, *Chem. Commun.* **2002**, 840.

47 P. H. Schlesinger, R. Ferdani, J. Liu, J. Pajewska, R. Pajewski, M. Saito, H. Shabany, G. W. Gokel, *J. Am. Chem. Soc.* **2002**, *124*, 1848.

48 G. Srinivasan, C. M. James, J. A. Krzycki, *Science* **2002**, *296*, 1459.

49 B. Hao, W. Gong, T. K. Ferguson, C. M. James, J. A. Krzycki, M. K. Chan, *Science* **2002**, *296*, 1462.

50 F. Wrede, *Z. Physiol. Chem.* **1932**, *210*, 125.

51 V. F. Wrede, A. Rothhaas, *Z. Physiol. Chem.* **1933**, *222*, 203.

52 F. Wrede, A. Rothhaas, *Z. Physiol. Chem.* **1933**, *219*, 267.

53 F. Wrede, A. Rothhaas, *Z. Physiol. Chem.* **1933**, *215*, 67.

54 F. Wrede, A. Rothhaas, *Z. Physiol. Chem.* **1934**, *226*, 95.

55 A. Fürstner, *Angew. Chem., Int. Ed.* **2003**, *42*, 3582.

56 K. Harashima, T. Tanaka, J. Nagatsu, *Agric. Biol. Chem.* **1967**, *31*, 481.

57 H. H. Wasserman, G. C. Rodgers, D. D. Keith, *Tetrahedron* **1976**, *32*, 1851.

58 S. B. Han, H. M. Kim, Y. H. Kim, C. W. Lee, E.-S. Jang, K. H. Son, S. U. Kim, Y. K. Kim, *Int. J. Immunopharmacol.* **1998**, *20*, 1.

59 B. Montaner, R. Perez-Tomas, *Life Sciences* **2001**, *68*, 2025.

60 C. Yamamoto, H. Takemoto, K. Kuno, D. Yamamoto, A. Tsubura, K. Kamata, H. Hirata, A. Yamamoto, H. Kano, T. Seki, K. Inoue, *Hepatology* **1999**, *30*, 894.

61 D. Yamamoto, Y. Uemura, K. Tanaka, K. Nakai, C. Yamamoto, H. Takemoto, K. Kamata, H. Hirata, K. Hioki, *Int. J. Cancer* **2000**, *88*, 121.

62 C. Diaz-Ruiz, B. Montaner, R. Perez-Tomas, *Histoll. Histopathol.* **2001**, *16*, 415.

63 B. Montaner, S. Navarro, M. Pique, M. Vilaseca, M. Martinell, E. Giralt, J. Gil, R. Perez-Tomas, *Brit. J. Pharmacol.* **2000**, *131*, 585.

64 J. L. Sessler, D. Seidel, *Angew. Chem., Int. Ed.* **2003**, *42*, 5134.

65 J. L. Sessler, M. J. Cyr, V. Lynch, E. McGhee, J. A. Ibers, *J. Am. Chem. Soc.* **1990**, *112*, 2810.

66 M. Shionoya, H. Furuta, V. Lynch, A. Harriman, J. L. Sessler, *J. Am. Chem. Soc.* **1992**, *114*, 5714.

67 H. Furuta, M. J. Cyr, J. L. Sessler, *J. Am. Chem. Soc.* **1991**, *113*, 6677.

68 J. L. Sessler, D. A. Ford, M. J. Cyr, H. Furuta, *J. Chem. Soc., Chem. Commun.* **1991**, 1733.

69 J. L. Sessler, M. J. Cyr, A. K. Burrell, *Synlett* **1991**, 127.

70 J. L. Sessler, T. D. Mody, D. A. Ford, V. M. Lynch, *Angew. Chem., Int. Ed.* **1991**, *31*, 452.

71 J. L. Sessler, T. Morishima, V. M. Lynch, *Angew. Chem., Int. Ed.* **1991**, *30*, 977.

72 V. J. Bauer, D. L. J. Clive, D. Dolphin, J. B. Paine, III, F. L. Harris, M. M. King, J. Loder, S. W. C. Wang, R. B. Woodward, *J. Am. Chem. Soc.* **1983**, *105*, 6429.

73 V. Král, H. Furuta, K. Shreder, V. Lynch, J. L. Sessler, *J. Am. Chem. Soc.* **1996**, *118*, 1595.

74 B. L. Iverson, K. Shreder, V. Král, P. Sansom, V. Lynch, J. L. Sessler, *J. Am. Chem. Soc.* **1996**, *118*, 1608.

75 J. L. Sessler, M. Cyr, H. Furuta, V. Král, T. Mody, T. Morishima, M. Shionoya, S. Weghorn, *Pure Appl. Chem.* **1993**, *65*, 393.

76 J. L. Sessler, J. M. Davis, V. Lynch, *J. Org. Chem.* **1998**, *63*, 7062.

77 S. J. Weghorn, V. Lynch, J. L. Sessler, *Tetrahedron Lett.* **1995**, *36*, 4713.

78 H. Furuta, T. Morishima, V. Král, J. L. Sessler, *Supramol. Chem.* **1993**, *3*, 5.

79 K. Umezawa, K. Tohda, X. M. Lin, J. L. Sessler, Y. Umezawa, *Anal. Chim. Acta* **2001**, *426*, 19.

80 J. L. Sessler, S. J. Weghorn, Y. Hiseada, V. Lynch, *Chem. Eur. J.* **1995**, *1*, 56.

81 T. Köhler, D. Seidel, V. Lynch, F. O. Arp, Z. Ou, K. M. Kadish, J. L. Sessler, *J. Am. Chem. Soc.* **2003**, *125*, 6872.

82 T. Köhler, D. Seidel, V. Lynch, F. O. Arp, Z. Ou, K. M. Kadish, J. L. Sessler, *Abstracts of Papers, 226th ACS National Meeting*, New York, NY, US, September 7–11, 2003, ORGN.

83 T. Köhler, Z. Ou, J. T. Lee, D. Seidel, V. Lynch, K. M. Kadish, J. L. Sessler, *Angew. Chem., Int. Ed.* **2004**, in press.

84 J. L. Sessler, S. J. Weghorn, V. M. Lynch, M. R. Johnson, *Angew. Chem., Int. Ed.* **1994**, *33*, 1509.

85 E. Vogel, M. Bröring, J. Fink, D. Rosen, H. Schmickler, J. Lex, K. W. K. Chan, Y.-D. Wu, D. A. Plattner, M. Nendel, K. N. Houk, *Angew. Chem., Int. Ed.* **1995**, *34*, 2511.

86 J. L. Sessler, D. Seidel, V. M. Lynch, *J. Am. Chem. Soc.* **1999**, *121*, 11257.

87 D. Seidel, V. Lynch, M., J. L. Sessler, *Angew. Chem., Int. Ed.* **2002**, *41*, 1422.

88 J. L. Sessler, W.-S. Cho, unpublished results.

89 J. L. Sessler, T. D. Mody, D. A. Ford, V. Lynch, *Angew. Chem., Int. Ed.* **1992**, *31*, 452.

90 J. L. Sessler, W.-S. Cho, S. P. Dudek, L. Hicks, V. M. Lynch, M. T. Huggins, *J. Porphyrins Phthalocyanines* **2003**, *7*, 97.

91 V. Král, J. L. Sessler, R. S. Zimmerman, D. Seidel, V. Lynch, B. Andrioletti, *Angew. Chem., Int. Ed.* **2000**, *39*, 1055.

92 C. Bucher, D. Seidel, V. Lynch, V. Král, J. L. Sessler, *Org. Lett.* **2000**, *2*, 3103.

93 C. Bucher, R. S. Zimmerman, V. Lynch, V. Král, J. L. Sessler, *J. Am. Chem. Soc.* **2001**, *123*, 2099.

94 M. S. Melvin, J. T. Tomlinson, G. Park, C. S. Day, G. R. Saluta, G. L. Kucera, R. A. Manderville, *Chem. Res. Toxicol.* **2002**, *15*, 734.

95 A. Fürstner, E. J. Grabowski, *Chem-BioChem* **2001**, *2*, 706.

96 R. D'Alessio, A. Bargiotti, O. Carlini, F. Colotta, M. Ferrari, P. Gnocchi, A. Isetta, N. Mongelli, P. Motta, A. Rossi, M. Rossi, M. Tibolla, E. Vanotti, *J. Med. Chem.* **2000**, *43*, 2557.

97 M. S. Melvin, M. W. Calcutt, R. E. Noftle, R. A. Manderville, *Chem. Res. Toxicol.* **2002**, *15*, 742.

98 A. Fürstner, J. Grabowski, C. W. Lehmann, T. Kataoka, K. Nagai, *Chem-BioChem* **2001**, *2*, 60.

99 W.-S. Cho, J. L. Sessler, L. R. Eller, G. D. Pantos, B. R. Peterson, *Abstracts of Papers, 225th ACS National Meeting*, New Orleans, LA, US, March 23–27, 2003, ORGN.

100 J. L. Sessler, S. J. Weghorn, V. Lynch, K. Fransson, *J. Chem. Soc., Chem. Commun.* **1994**, 1289.

101 A. Baeyer, *Dtsch. Chem. Ges.* **1886**, *19*, 2184.

102 P. A. Gale, J. L. Sessler, V. Král, V. Lynch, *J. Am. Chem. Soc.* **1996**, *118*, 5140.

103 S. Camiolo, S. J. Coles, P. A. Gale, M. B. Hursthouse, J. L. Sessler, *Acta Crystallogr.* **2001**, *E57*, o816.

104 P. A. Gale, J. L. Sessler, W. E. Allen, N. A. Tvermoes, V. Lynch, *Chem. Commun.* **1997**, 665.

105 F. P. Schmidtchen, *Org. Lett.* **2002**, *4*, 431.

106 J. L. Sessler, W.-S. Cho, J. A. Shriver, D. E. Gross, unpublished results.

107 J. L. Sessler, D. An, W.-S. Cho, V. M. Lynch, *J. Am. Chem. Soc.* **2003**, *125*, 13646.

108 J. L. Sessler, D. An, W.-S. Cho, V. Lynch, *Angew. Chem., Int. Ed.* **2003**, *42*, 2278.

109 C.-H. Lee, H.-K. Na, D.-W. Yoon, D.-H. Won, W.-S. Cho, V. M. Lynch, S. V. Shevchuk, J. L. Sessler, *J. Am. Chem. Soc.* **2003**, *125*, 7301.

110 S. Camiolo, P. A. Gale, *Chem. Commun.* **2000**, 1129.

111 L. Bonomo, E. Solari, G. Toraman, R. Scopelliti, C. Floriani, M. Latronico, *Chem. Commun.* **1999**, 2413.

112 C. J. Woods, S. Camiolo, M. E. Light, S. J. Coles, M. B. Hursthouse, M. A.

King, P. A. Gale, J. W. Essex, *J. Am. Chem. Soc.* **2002**, *124*, 8644.

113 P. Anzenbacher, Jr., K. Jursíková, V. M. Lynch, P. A. Gale, J. L. Sessler, *J. Am. Chem. Soc.* **1999**, *121*, 11020.

114 M. Dukh, P. Dra?ar, I. Cern?, V. Pouzar, J. A. Shriver, V. Král, J. L. Sessler, *Supramol. Chem.* **2002**, *14*, 237.

115 P. A. Gale, J. W. Genge, V. Král, M. A. McKervey, J. L. Sessler, A. Walker, *Tetrahedron Lett.* **1997**, *38*, 8443.

116 B. Turner, M. Botoshansky, Y. Eichen, *Angew. Chem., Int. Ed.* **1998**, *37*, 2475.

117 B. Turner, A. Shterenberg, M. Kapon, K. Suwinska, Y. Eichen, *Chem Commun* **2001**, 12.

118 J. L. Atwood, S. G. Bott, S. Harvey, P. C. Junk, *Organometallics* **1994**, *13*, 4151.

119 A. Alvanipour, J. L. Atwood, S. G. Bott, P. C. Junk, U. H. Kynast, H. Prinz, *J. Chem. Soc., Dalton Trans.* **1998**, 1223.

120 B. Turner, A. Shterenberg, M. Kapon, Y. Eichen, K. Suwinska, *Chem. Commun.* **2001**, 13.

121 G. Cafeo, F. H. Kohnke, G. L. La Torre, A. J. P. White, D. J. Williams, *Angew. Chem., Int. Ed.* **2000**, *39*, 1496.

122 G. Cafeo, F. H. Kohnke, M. F. Parisi, R. P. Nascone, G. L. La Torre, D. J. Williams, *Org. Lett.* **2002**, *4*, 2695.

123 G. Cafeo, F. H. Kohnke, G. L. La Torre, A. J. P. White, D. J. Williams, *Chem. Commun.* **2000**, 1207.

124 G. Cafeo, F. H. Kohnke, G. L. La Torre, M. F. Parisi, R. P. Nascone, A. J. P. White, D. J. Williams, *Chem. Eur. J.* **2002**, *8*, 3148.

125 P. A. Gale, J. L. Sessler, V. Král, *Chem. Commun.* **1998**, 1.

126 J. L. Sessler, P. Anzenbacher, Jr., J. A. Shriver, K. Jursíková, V. M. Lynch, M. Marquez, *J. Am. Chem. Soc.* **2000**, *122*, 12061.

127 J. A. Shriver, Ph.D. Dissertation, The University of Texas at Austin, **2002**.

128 P. Anzenbacher, Jr., A. C. Try, H. Miyaji, K. Jursíková, V. M. Lynch, M. Marquez, J. L. Sessler, *J. Am. Chem. Soc.* **2000**, *122*, 10268.

129 T. G. Levitskaia, M. Marquez, J. L. Sessler, J. A. Shriver, T. Vercouter, B. A. Moyer, *Chem. Commun.* **2003**, 2248.

130 D.-W. Yoon, H. Hwang, C.-H. Lee, *Angew. Chem., Int. Ed.* **2002**, *41*, 1757.

131 C. H. Lee, J.-S. Lee, D.-W. Yoon, S.-K. Park, W.-S. Cho, B. Rubin, J. L. Sessler, unpublished results.

132 J. L. Sessler, M. C. Hoehner, D. W. Johnson, A. Gebauer, V. M. Lynch, *Chem. Commun.* **1996**, 2311.

133 O. D. Fox, T. D. Rolls, P. D. Beer, M. G. B. Drew, *Chem. Commun.* **2001**, 1632.

134 C. Bucher, R. S. Zimmerman, V. Lynch, J. L. Sessler, *J. Am. Chem. Soc.* **2001**, *123*, 9716.

135 C. Bucher, R. S. Zimmerman, V. Lynch, J. L. Sessler, *Chem. Commun.* **2003**, 1646.

136 C. Schmuck, L. Geiger, *Current Organic Chemistry* **2003**, *7*, 1485.

137 C. Schmuck, M. Heil, *Org. Lett.* **2001**, *3*, 1253.

138 C. Schmuck, *Chem. Eur. J.* **2000**, *6*, 1279.

139 C. Schmuck, *Tetrahedron* **2001**, *57*, 3063.

140 C. Schmuck, J. Lex, *Eur. J. Org. Chem.* **2001**, 1519.

141 C. Schmuck, J. Lex, *Org. Lett.* **1999**, *1*, 1779.

142 C. Schmuck, *Eur. J. Org. Chem.* **1999**, 2397.

143 C. Schmuck, *Chem. Commun.* **1999**, 843.

144 C. Schmuck, W. Wienand, *J. Am. Chem. Soc.* **2003**, *125*, 452.

145 C. Schmuck, V. Bickert, *Org. Lett.* **2003**, *5*, 4579.

146 K. Kavallieratos, S. R. de Gala, D. J. Austin, R. H. Crabtree, *J. Am. Chem. Soc.* **1997**, *119*, 2325.

147 K. Kavallieratos, C. M. Bertao, R. H. Crabtree, *J. Org. Chem.* **1999**, *64*, 1675.

148 S. Camiolo, P. A. Gale, M. B. Hursthouse, M. E. Light, *Tetrahedron Lett.* **2002**, *43*, 6995.

149 S. Camiolo, P. A. Gale, M. B. Hursthouse, M. E. Light, *Org. Biomol. Chem.* **2003**, *1*, 741.

150 P. A. Gale, S. Camiolo, C. P. Chapman, M. E. Light, M. B. Hursthouse, *Tetrahedron Lett.* **2001**, *42*, 5095.

151 I. E. D. Vega, S. Camiolo, P. A. Gale, M. B. Hursthouse, M. E. Light, *Chem. Commun.* **2003**, 1686.

152 P. A. Gale, S. Camiolo, G. J. Tizzard, C. P. Chapman, M. E. Light, S. J. Coles, M. B. Hursthouse, *J. Org. Chem.* **2001**, *66*, 7849.

153 K. Navakhun, P. A. Gale, S. Camiolo, M. E. Light, M. B. Hursthouse, *Chem. Commun.* **2002**, 2084.

154 S. Ohkuma, T. Sato, M. Okamoto, H. Matsuya, K. Arai, T. Kataoka, K. Nagai, H. H. Wasserman, *Biochem. J.* **1998**, *334*, 731.

155 T. Sato, H. Konno, Y. Tanaka, T. Kataoka, K. Nagai, H. H. Wasserman, S. Ohkuma, *J. Biol. Chem.* **1998**, *273*, 21455.

156 T. Kataoka, M. Muroi, S. Ohkuma, T. Waritani, J. Magae, A. Takatsuki, S. Kondo, M. Yamasaki, K. Nagai, *FEBS Lett.* **1995**, *359*, 53.

157 M. S. Melvin, K. E. Wooton, C. C. Rich, G. R. Saluta, G. L. Kucera, N. Lindquist, R. A. Manderville, *J. Inorg. Biochem.* **2001**, *87*, 129.

158 M. S. Melvin, J. T. Tomlinson, G. R. Saluta, G. L. Kucera, N. Lindquist, R. A. Manderville, *J. Am. Chem. Soc.* **2000**, *122*, 6333.

159 M. S. Melvin, D. C. Ferguson, N. Lindquist, R. A. Manderville, *J. Org. Chem.* **1999**, *64*, 6861.

160 R. A. Manderville, *Curr. Med. Chem. – Anti-Cancer Agents* **2001**, *1*, 195.

161 G. Park, J. T. Tomlinson, M. S. Melvin, M. W. Wright, C. S. Day, R. A. Manderville, *Org. Lett.* **2003**, *5*, 113.

162 A. Levin, L. Stevens, P. A. McCullough, *Postgrad. Med.* **2002**, *111*, 53.

163 B. F. Culleton, M. G. Larson, P. W. Wilson, J. C. Evans, P. S. Parfrey, D. Levy, *Kidney Int.* **1999**, *56*, 2214.

164 K. Záruba, Z. Tománková, D. Sýkora, J. Charvátová, I. Kavenová, P. Bour, P. Matejka, J. Fähnrich, K. Volka, V. Král, *Anal. Chim. Acta* **2001**, *437*, 39.

165 K. Tohda, R. Naganawa, M. L. Xiao, M. Tange, K. Umezawa, K. Odashima, Y. Umezawa, H. Furuta, J. L. Sessler, *Sens. Actuators, B* **1993**, *14*, 669.

166 J. L. Sessler, J. W. Genge, V. Král, B. L. Iverson, *Supramol. Chem.* **1996**, *8*, 45.

167 J. L. Sessler, W. E. Allen, *Chemtech* **1999**, *29*, 16.

168 V. Král, J. L. Sessler, H. Furuta, *J. Am. Chem. Soc.* **1992**, *114*, 8704.

169 V. Král, J. L. Sessler, *Tetrahedron* **1995**, *51*, 539.

170 B. L. Iverson, R. E. Thomas, V. Král, J. L. Sessler, *J. Am. Chem. Soc.* **1994**, *116*, 2663.

171 V. Král, A. Andrievsky, J. L. Sessler, *J. Am. Chem. Soc.* **1995**, *117*, 2953.

172 J. L. Sessler, A. Andrievsky, V. Král, V. Lynch, *J. Am. Chem. Soc.* **1997**, *119*, 9385.

173 V. Král, A. Andrievsky, J. L. Sessler, *J. Chem. Soc., Chem. Commun.* **1995**, 2349.

174 J. L. Sessler, V. Král, T. V. Shishkanova, P. A. Gale, *Proc. Natl. Acad. Sci. USA* **2002**, *99*, 4848.

175 P. A. Gale, *Coord. Chem. Rev.* **2000**, *199*, 181.

176 P. A. Gale, *Coord. Chem. Rev.* **2001**, *213*, 79.

177 C. Suksai, T. Tuntulani, *Chem. Soc. Rev.* **2003**, *32*, 192.

178 R. Martínez-Máñez, F. Sancenón, *Chem. Rev.* **2003**, *103*, 4419.

179 P. D. Beer, A. D. Keefe, M. G. B. Drew, *J. Organomet. Chem.* **1988**, *353*, C10.

180 P. D. Beer, A. D. Keefe, M. G. B. Drew, *J. Organomet. Chem.* **1989**, *378*, 437.

181 P. D. Beer, A. D. Keefe, A. M. Z. Slawin, D. J. Williams, *J. Chem. Soc., Dalton Trans.* **1990**, 3675.

182 P. D. Beer, Z. Chen, M. G. B. Drew, P. A. Gale, *J. Chem. Soc., Chem. Commun.* **1995**, 1851.

183 P. D. Beer, C. G. Crane, J. P. Danks, P. A. Gale, J. F. McAleer, *J. Organomet. Chem.* **1995**, *490*, 143.

184 J. L. Sessler, A. Gebauer, A. Gale Philip, *Gazz. Chim. Ital.* **1997**, *127*, 723.

185 P. A. Gale, M. B. Hursthouse, M. E. Light, J. L. Sessler, C. N. Warriner, R. S. Zimmerman, *Tetrahedron Lett.* **2001**, *42*, 6759.

186 M. Scherer, J. L. Sessler, A. Gebauer, V. Lynch, *Chem. Commun.* **1998**, 85.

187 J. L. Sessler, R. S. Zimmerman, G. J. Kirkovits, A. Gebauer, M. Scherer, *J. Organomet. Chem.* **2001**, *637–639*, 343.

188 G. Denuault, P. A. Gale, M. B. Hursthouse, M. E. Light, C. N. Warriner, *New J. Chem.* **2002**, *26*, 811.

189 K. A. Nielsen, J. O. Jeppesen, E. Levillain, J. Becher, *Angew. Chem., Int. Ed.* **2003**, *42*, 187.

190 V. Král, J. L. Sessler, T. V. Shishkanova, P. A. Gale, R. Volf, *J. Am. Chem. Soc.* **1999**, *121*, 8771.

191 V. Král, P. A. Gale, P. Anzenbacher, Jr., K. Jursíková, V. Lynch, J. L. Sessler, *Chem. Commun.* **1998**, 9.

192 V. Král, M. Valík, V. Shishkanova Tatiana, J. L. Sessler, *Dekker Encyclopedia of Nanoscience and Nanotechnology* **2004**, in press.

193 H. Miyaji, P. Anzenbacher, Jr., J. L. Sessler, E. R. Bleasdale, P. A. Gale, *Chem. Commun.* **1999**, 1723.

194 P. Anzenbacher, Jr., K. Jursíková, J. L. Sessler, *J. Am. Chem. Soc.* **2000**, *122*, 9350.

195 H. Miyaji, W. Sato, J. L. Sessler, V. M. Lynch, *Tetrahedron Lett.* **2000**, *41*, 1369.

196 H. Miyaji, W. Sato, J. L. Sessler, *Angew. Chem., Int. Ed.* **2000**, *39*, 1777.

197 C. B. Black, B. Andrioletti, A. C. Try, C. Ruiperez, J. L. Sessler, *J. Am. Chem. Soc.* **1999**, *121*, 10438.

198 B. Oddo, *Gazz. Chim. Ital.* **1911**, *41*, 248.

199 B. Lindstom, *Acta Chem. Scand.* **1973**, *27*, 2411.

200 P. Anzenbacher, Jr., D. S. Tyson, K. Jursíková, F. N. Castellano, *J. Am. Chem. Soc.* **2002**, *124*, 6232.

201 T. Mizuno, W.-H. Wei, L. R. Eller, J. L. Sessler, *J. Am. Chem. Soc.* **2002**, *124*, 1134.

202 P. Anzenbacher, Jr., D. Aldakov, *Chem. Commun.* **2003**, 1394.

203 J.-H. Liao, C.-T. Chen, H.-T. Chou, C.-C. Cheng, P.-T. Chou, J.-M. Fang, Z. Slanina, T. J. Chow, *Org. Lett.* **2002**, *4*, 3107.

204 S. Shevchuk, V., V. M. Lynch, J. L. Sessler, *Tetrahedron* **2004**, *60*, 11283.

205 J. L. Sessler, H. Maeda, T. Mizuno, V. M. Lynch, H. Furuta, *Chem. Commun.* **2002**, 862.

206 J. L. Sessler, G. D. Pantos, E. Katayev, V. M. Lynch, *Org. Lett.* **2003**, *5*, 4141.

207 J. L. Sessler, H. Maeda, T. Mizuno, V. M. Lynch, H. Furuta, *J. Am. Chem. Soc.* **2002**, *124*, 13474.

208 R. J. Fitzmaurice, G. M. Kyne, D. Douheret, J. D. Kilburn, *J. Chem. Soc., Perkin Trans. 1*, **2002**, 841.

209 T. H. Wirth, N. Davidson, *J. Am. Chem. Soc.* **1964**, *86*, 4325.

210 R. Schwesinger, *Chimia* **1985**, *39*, 269.

211 R. M. Curtis, R. A. Pasternak, *Acta Crystallogr.* **1955**, *8*, 1955.

212 F. A. Cotton, V. W. Day, E. E. J. Hazen, S. Larsen, *J. Am. Chem. Soc.* **1973**, *95*, 4834.

213 J. M. Adams, R. W. H. Small, *Acta Crystallogr.* **1974**, *B30*, 2191.

214 J. M. Adams, R. W. H. Small, *Acta Crystallogr.* **1976**, *B32*, 832.

215 J. M. Adams, R. G. Pritchard, *Acta Crystallogr.* **1976**, *B32*, 2438.

216 M. Cygler, M. J. Grabowski, A. Stepień, E. Wajsman, *Acta Crystallogr.* **1976**, *B32*, 2391.

217 E. Wajsman, M. Cygler, M. J. Grabowski, A. Stepień, *Roczniki Chemii* **1976**, *50*, 1587.

218 A. Stepień, M. J. Grabowski, *Acta Crystallogr.* **1977**, *B33*, 2924.

219 J. M. Adams, V. Ramdas, *Acta Crystallogr.* **1978**, *B34*, 2150.

220 J. M. Adams, *Acta Crystallogr.* **1978**, *B34*, 1218.

221 A. Katrusiak, *Acta Crystallogr.* **1994**, *C50*, 1161.

222 A. Waśkowska, *Acta Crystallogr.* **1997**, *C53*, 128.

223 T. Kolev, H. Preut, P. Bleckmann, V. Radomirska, *Acta Crystallogr.* **1997**, *C53*, 805.

224 C. Tanford, *J. Am. Chem. Soc.* **1954**, *76*, 945.

225 J. I. Watters, S. Matsumoto, *J. Am. Chem. Soc.* **1964**, *86*, 3961.

226 B. Springs, P. Haake, *Bioorg. Chem.* **1977**, *6*, 181.

227 B. Dietrich, T. M. Fyles, J.-M. Lehn, L. G. Pease, D. Fyles, *J. Chem. Soc., Chem. Commun.* **1978**, 934.

228 R. P. Dixon, S. J. Geib, A. D. Hamilton, *J. Am. Chem. Soc.* **1992**, *114*, 365.

229 V. Jubian, R. P. Dixon, A. D. Hamilton, *J. Am. Chem. Soc.* **1992**, *114*, 1120.

230 V. Jubian, A. Veronese, R. P. Dixon, A. D. Hamilton, *Angew. Chem., Int. Ed.* **1995**, *34*, 1237.

231 R. Gross, J. W. Bats, M. W. Göbel, *Liebigs Ann. Chem.* **1994**, 205.

232 A. F. McKay, M. E. Kreling, *Can. J. Chem.* **1957**, *35*, 1438.

233 F. P. Schmidtchen, *Chem. Ber.* **1980**, *113*, 2175.

234 G. Müller, J. Riede, F. P. Schmidtchen, *Angew. Chem., Int. Ed.* **1988**, *27*, 1516.

235 A. Echavarren, A. Galán, J.-M. Lehn, J. de Mendoza, *J. Am. Chem. Soc.* **1989**, *111*, 4994.

236 P. Schiebl, F. P. Schmidtchen, *Tetrahedron Lett.* **1993**, *34*, 2449.

237 A. Metzger, V. M. Lynch, E. V. Anslyn, *Angew. Chem., Int. Ed.* **1997**, *36*, 862.

238 S. L. Tobey, B. D. Jones, E. V. Anslyn, *J. Am. Chem. Soc.* **2003**, *125*, 4026.

239 S. L. Tobey, E. V. Anslyn, *J. Am. Chem. Soc.* **2003**, *125*, 14807.

240 Z. Zhong, E. V. Anslyn, *Angew. Chem., Int. Ed.* **2003**, *42*, 3005.

241 A. Metzger, E. V. Anslyn, *Angew. Chem., Int. Ed.* **1998**, *37*, 649.

242 S. L. Tobey, E. V. Anslyn, *Org. Lett.* **2003**, *5*, 2029.

243 D. Voet, J. G. Voet, C. W. Pratt, *Fundamentals of Biochemistry*, Wiley, New York, **1999**.

244 R. A. J. Pascal, J. Sepergel, D. V. Engen, *Tetrahedron Lett.* **1986**, *27*, 4099.

245 K. Choi, A. D. Hamilton, *J. Am. Chem. Soc.* **2001**, *123*, 2456.

246 K. Choi, A. D. Hamilton, *Angew. Chem., Int. Ed.* **2001**, *40*, 3912.

247 K. Choi, A. D. Hamilton, *J. Am. Chem. Soc.* **2003**, *125*, 10241.

248 A. Szumma, J. Jurczak, *Eur. J. Org. Chem.* **2001**, 4031.

249 A. Szumma, J. Jurczak, *Helv. Chim. Acta* **2001**, *84*, 3760.

250 M. A. Hossain, J. M. Llinares, D. Powell, K. Bowman-James, *Inorg. Chem.* **2001**, *40*, 2936.

251 M. A. Hossain, S. O. Kang, D. Powell, K. Bowman-James, *Inorg. Chem.* **2003**, *42*, 1397.

252 M. A. Hossain, S. O. Kang, J. M. Llinares, D. Powell, K. Bowman-James, *Inorg. Chem.* **2003**, *42*, 5043.

253 S. Kubik, R. Goddard, R. Kirchner, D. Nolting, J. Seidel, *Angew. Chem., Int. Ed.* **2001**, *40*, 2648.

254 S. Kubik, R. Goddard, *Proc. Natl. Acad. Sci. USA* **2002**, *99*, 5127.

255 S. Kubik, R. Kirchner, D. Nolting, J. Seidel, *J. Am. Chem. Soc.* **2002**, *124*, 12752.

256 S. Otto, S. Kubik, *J. Am. Chem. Soc.* **2003**, *125*, 7804.

257 A. P. Bisson, V. M. Lynch, M.-K. C. Monahan, E. V. Anslyn, *Angew. Chem., Int. Ed.* **1997**, *36*, 2340.

258 K. Niikura, A. P. Bisson, E. V. Anslyn, *J. Chem. Soc., Perkin Trans. 2,* **1999**, 1111.

259 S. O. Kang, J. M. Llinares, D. Powell, D. VanderVelde, K. Bowman-James, *J. Am. Chem. Soc.* **2003**, *125*, 10152.

260 P. J. Smith, M. V. Reddington, C. S. Wilcox, *Tetrahedron Lett.* **1992**, *33*, 6085.

261 T. Ross Kelly, M. H. Kim, *J. Am. Chem. Soc.* **1994**, *116*.

262 J. M. Benito, M. Gómez-García, J. L. Jiménez Blanco, C. Ortiz Mellet, J. M. García Fernández, *J. Org. Chem.* **2001**, *66*, 1366.

263 K. H. Lee, J.-I. Hong, *Tetrahedron Lett.* **2000**, *41*, 6083.

264 B. H. M. Snellink-Ruël, M. M. G. Antonisse, J. F. J. Engberson, P. Timmerman, D. N. Reinhoudt, *Eur. J. Org. Chem* **2000**, 165.

265 J. Scheerder, J. F. J. Engberson, A. Casnati, R. Ungaro, D. N. Reinhoudt, *J. Org. Chem.* **1995**, *60*, 6448.

266 G. Tumcharern, T. Tuntulani, S. J. Coles, M. B. Hursthouse, J. D. Kilburn, *Org. Lett.* **2003**, *5*, 4971.

5
Molecular Containers in Action

Dmitry M. Rudkevich

5.1
Introduction

Binding sites of enzymes have convergent and concave surfaces; synthetic cavities – so-called *molecular containers* – mimic these features [1]. They also act as "nanometer-size" cups, flasks, boxes, and bottles, which are so important in our daily life. The interest of organic chemists in three-dimensional, cavity-containing hosts dates back to the work of Cramer on the inclusion complexes of cyclodextrins in the 1950s [2]. Since then, especially over the last decade, the desirability of controlling the size and shape of the cavity has led to successful synthesis of a variety of molecular containers. Among the major types are *cavitands*, *carcerands*, and *self-assembling capsules*. We will start this chapter with a brief introduction to their structural features and the dynamics of their "molecule-within-molecule", or *encapsulation complexes*. Despite obvious differences among their design, size, and molecular architecture, they all have cavities and entrap guests. Many excellent reviews on this subject have been written [1, 3–12], but we will introduce here a different approach – applications of molecule-within-molecule complexes will be discussed. Although the field is relatively young, these applications are rather broad and include such areas as chemical reactivity and catalysis, sensors, separations, new materials, and biologically relevant studies. In contrast with alternative cavity-forming structures employed in chemistry and chemical biology, for example imprinted polymers, dendrimers, liposomes, zeolites, and catalytic antibodies [13], the properties and behavior of molecular containers and their encapsulation complexes can be tuned and monitored on the level of individual, relatively small, molecules by using the conventional approaches and techniques of organic chemistry, although the potential of applied encapsulation reaches far beyond this.

Functional Synthetic Receptors. Edited by T. Schrader, A. D. Hamilton
Copyright © 2005 WILEY-VCH Verlag GmbH & Co. KGaA, Weinheim
ISBN: 3-527-30655-2

5.2
Variety of Molecular Containers

Typically, molecular containers are built from calixarenes, resorcinarenes, and their relatives. These compounds have had a large impact on the history of supramolecular chemistry [12]. Their rigid, concave surface, commercial availability, and rich synthetic chemistry have made them extremely popular platforms for elaboration.

Cavitands are synthetic host-molecules with open-ended enforced cavities large enough to bind complementary organic molecules and ions [6, 10, 14]. Concave inner surfaces of cavitands of a variety of shapes have been designed with either integrated or appended binding sites. Complexes of cavitands are called *caviplexes*. The most popular strategy used to build cavitands involves extension of the aromatic walls of the resorcinarene. For example, condensation of resorcinarene with 2,3-dichloroquinoxaline results in cavitand **1** with extended walls and a shielded inner cavity (Fig. 5.1) [15]. According to X-ray analysis, molecule **1** has a lipophilic interior 7.2 Å

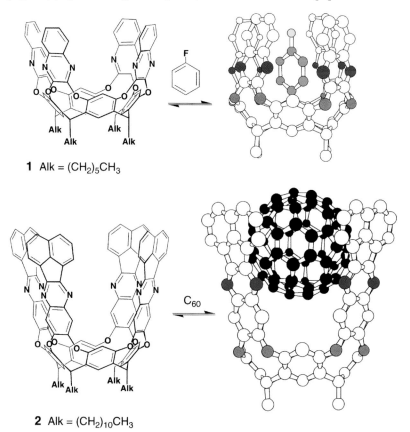

1 Alk = $(CH_2)_5CH_3$

2 Alk = $(CH_2)_{10}CH_3$

Figure 5.1 Top: cavitand **1** and X-ray structure of complex **1**•fluorobenzene [15]. Bottom: deep cavitand **2** and its 1:1 complex with fullerene C_{60} (energy-minimized representation) [16].

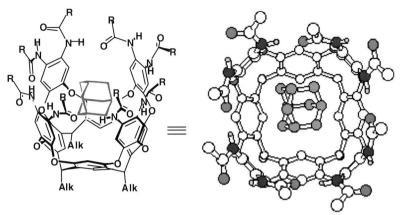

3 • Adamantane

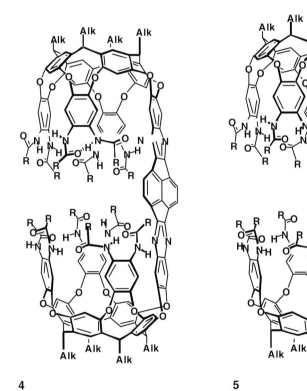

4   5

Figure 5.2 Top: self-folding caviplex **3**•adamantane and its energy-minimized representation [17]. Here and subsequently in the legends CH hydrogen atoms and long alkyl chains are omitted for viewing clarity. A $\Delta G^{\#}$ barrier of ~17 kcal mol^{-1} was measured for exchange of adamantane (EXSY; p-xylene-d$_{10}$) and a $\Delta G^{\#}$ barrier of ~17.5 kcal mol^{-1} was measured for collapse of the hydrogen bonding seam in **3** (VT ^{1}H NMR; toluene, p-xylene). Bottom: nanoscale self-folding containers **4** and **5** [18, 19].

wide and 8.3 Å deep which is large enough to accommodate aromatic molecules. Indeed, fluorobenzene has been seen inside cavitand **1** in the solid state. The complexation–decomplexation process in solution was rapid on the NMR time-scale, and only time-averaged chemical shifts were seen when binding occurred. For aromatic guests binding energies of $-\Delta G^{298}$ = 2.0 to 3.1 kcal mol^{-1} were measured for a 1:1 complex with **1** in acetone-d$_6$.

Similarly constructed cavitand **2** is much deeper and has a nanoscale cavity 14 Å deep and 12 Å wide (Fig. 5.1) [16]. According to molecular modeling, the cavity can accommodate up to three benzene- or chloroform-sized molecules. More interestingly, the π-electron rich interior of **2** resulted in considerable attraction to electron-deficient fullerene C$_{60}$ in toluene solution; the binding energy $-\Delta G^{293}$ = 4.0 ± 0.2 kcal mol^{-1} was determined for a 1:1 complex.

The open-ended nature of cavitands is clearly a problem – it results in modest affinity toward guests. Constant flow of guest/solvent molecules in and out of the cavity obviously occurs – not only through the upper neck but also through the wide holes between the aromatic walls. Hydrogen bonding has been successfully used to narrow the gaps and stabilize the vase-conformation in extended cavitands. In molecule **3** the vicinal secondary amides on the upper rim form a seam of head-to-tail intramolecular hydrogen bonds which results in a self-folding deepened vase with internal dimensions of 8×10 Å and *higher* kinetic stability of the caviplexes (Fig. 5.2) [17]. Exchange between bound and free guest species becomes slow on the NMR time-scale. Apparently, the circle of hydrogen bonds resists opening of the wall, thus creating a kinetic barrier to escape of the guest. At the same time, the thermodynamic stability of complexes with **3** remains rather low: $-\Delta G^{298} \leq 2$ kcal mol^{-1}. The guest exchange rate in self-folding caviplexes **3** is $k = 2 \pm 1$ s^{-1} at 295 K. These unique kinetic features were subsequently used in the design of larger molecular containers.

For example, cylindrical structure **4** combines two cavities preorganized for cooperative binding (Fig. 5.2) [18]. Long (~18 Å) and rigid guests such as diaryl-substituted adamantanes readily form kinetically stable 1:1 inclusion complexes. Association constants K_{ass} up to 500 M^{-1} ($-\Delta G^{295}$ = 3.6 kcal mol^{-1}) were obtained (toluene-d$_8$, ^{1}H NMR experiments). These are an order of magnitude higher than the values observed for complexes of smaller cavitands **3** with related adamantanes. A guest exchange rate constant of $k \approx 0.5 \pm 0.3$ s^{-1} at 295 K was obtained from ROESY experiments. More recently, the synthesis and characterization of self-folding cavitand-porphyrin **5** were completed [19]. This features a huge unimolecular cavity containing two cavitands attached to the Zn-porphyrin wall (Fig. 5.2). Its dimensions, ~10 × 25 Å place it among the largest synthetic hosts yet prepared [10]. The metalloporphyrin fragment introduces additional binding functionality for several heterocyclic compounds. As a result, not only were kinetically stable caviplexes obtained but unprecedentedly high thermodynamic stability of the complexes was observed for guests able to interact with both sites of **5** – the self-folding cavities and the metalloporphyrin. As with containers **4**, intramolecular hydrogen bonds at the upper rims of the cavitands resist the unfolding of the inner cavities and thereby increase the energy barrier to guest exchange. The exchange is slow on the NMR time scale (at ≤300 K). When the cavities and metalloporphyrins participate simultaneously in the binding event, very high affinities for guests are observed ($-\Delta G^{295}$ up to 10 kcal mol^{-1} in toluene), to which the porphyrin fragments contribute significantly ($-\Delta G^{295}$ up to 6 kcal mol^{-1}). The quantitative pairwise selection of two different guests by molecular container **5** has been reported [19]. The porphyrin functions also raise the possibility of metal-catalyzed reactions – alkane hydroxylations or alkene epoxidations – in these nanoscale containers.

Carcerands are close-surface host-molecules with enforced inner cavities which are large enough to incarcerate smaller organic molecules, but with portals too narrow to allow the guest to escape without breaking covalent bonds [7, 20]. Complexes of carcerands – *carceplexes* – are an extreme form of host–guest complex stability. Carcerands hold guests permanently. In fact, most carceplexes include guests (or solvent molecules) during their synthesis – on covalent shell-closure. Guests can often template the synthetic process. More interesting for applications are *hemicarceplexes* – carceplexes with larger portals which allow the entrapped guest to escape at high temperature; the guest remains incarcerated under the conditions of isolation and characterization at ordinary temperatures [7, 21]. Guest-free hemicarceplexes are called *hemicarcerands*.

Hemicarcerands are usually divided into two structural types. In molecules **6** one of the bridges, connecting two hemispheres, is omitted, thus opening a wider window for a guest to enter (Fig. 5.3). The other structures, **7**, have all four bridges but these are long and conformationally flexible; this generates temporary portals large enough to accept the guest. Hemicarceplexes are usually stable at ambient temperatures and release the entrapped guest at higher temperatures. The shielding effects of the cavity's aromatic walls cause a large upfield shift ($\Delta\delta \approx$ 3–5) of the guest's ^{1}H NMR signals on encapsulation. The unique host–guest chemistry of hemicarcerands has been thoroughly discussed in numerous reviews [3, 7]. Here, we mention some examples.

Hemicarcerand **8** has four (CH$_2$)$_4$ bridges and crystallizes with six H$_2$O molecules entrapped within its interior (Fig. 5.3) [23]. The H$_2$O molecules are hydrogen bonded to one another and also to the walls of the host. This arrangement is approximately octahedral and complementary to the inner shape of **8**. The O•••O distances and the O•••O•••O bond angles of the encapsulated H$_2$O are comparable with the crystallographic values of ordinary ice. We will see more of the unique features of container **8** later.

First-generation methylene-bridged carcerands (for example, **7**, X = X' = OCH$_2$O) typically entrap only one DMF-sized molecule. Recently prepared molecular container **9** encapsulates three DMF molecules (Fig. 5.3) [24]. The complex is effectively sealed – no DMF was lost even on heating in nitrobenzene at 160 °C for 6 h.

Enlarging the portals of the cavity often leads towards more flexible hemicarcerands, which in turn reduces the kinetic stability of the hemicarceplexes [3]. To avoid this, rigid portals were designed. Thus, hemicarcerand **10** was constructed from two cavitand hemispheres and four rigid OCH$_2$C≡C–C≡CCH$_2$O linkers [25]. It has huge, thirty-membered portals (Fig. 5.3). 1,3,5-Triethylbenzene, 1,3,5-tri-*i*-propylbenzene, and 1,3-dimethyladamantane produced 1:1 hemicarceplexes with **10** upon heating. Very bulky 1,3,5-tri-*t*-butylbenzene could not enter the portals. It was used

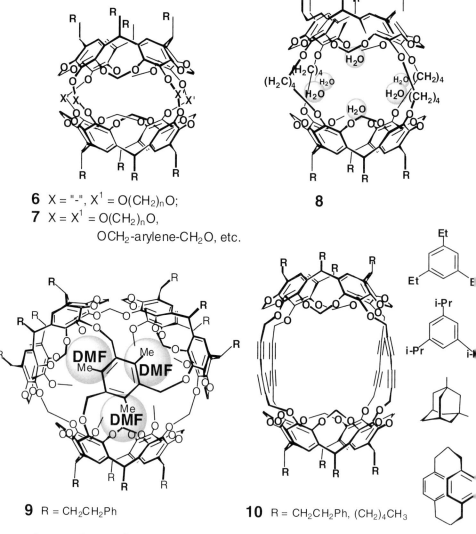

6 X = "-", X^1 = O(CH$_2$)$_n$O;
7 X = X^1 = O(CH$_2$)$_n$O,
  OCH$_2$-arylene-CH$_2$O, etc.

8

9 R = CH$_2$CH$_2$Ph

10 R = CH$_2$CH$_2$Ph, (CH$_2$)$_4$CH$_3$

Figure 5.3 Top: typical structures **6** and **7** of Cram's hemicarcerands. Encapsulation of multiple H$_2$O molecules by hemicarcerands 8 [23]. Bottom: larger carceplex **9**·3DMF [24] and large hemicarcerand **10** [25] and its guests.

as a solvent in which 1:1 hemicarceplexes of **10** were prepared with nine different [*m.n*]paracyclophanes. The largest encapsulated guest was [3.3]paracyclophane. The half-lives for decomplexation with these large guests were extremely high – 1608 h for **10**·1,3,5-tri-*i*-propylbenzene and 960 h for **10**·1,3,5-triethylbenzene (CDCl$_3$, 25°C). The portals could not, however, retain smaller molecules.

Both the cavity and the portal's dimensions and their conformational rigidity have a very significant effect on the association and dissociation activation energies of the complex. As already mentioned, enlarging the cavity easily leads to greater flexibility of the host; the two hemispheres often undergo twists and turns relative to each other. This was seen in the X-ray structures of Cram's larger hemicarcerands [26, 27]. The kinetic stability of the complexes decreases when the portals are widely open, enabling the guest to escape easily. At the same time, twists and turns of the linkers can significantly reduce the cavity size and close the portals after the complex is formed. Very rigid and preorganized portals prevent the guest from entering. Competition with the multiple solvent molecules situated inside the cavity can also be a problem.

Both the size and shape of the guest are crucial. Occupancy factors, or packing coefficients, of molecule-within-molecule complexes in solution are usually 55 ± 9 % [28]. In the absence of additional intermolecular forces, recognition-through-encapsulation is largely determined by the host and guest volumes. This is also believed to be a feature of natural biological cavities, for example those of enzymes.

The differences between cavitands, carcerands, and hemicarcerands are their ease of guest uptake and release – the exchange rates. More recently, *self-assembling capsules* have emerged. Self-assembly is the spontaneous noncovalent association of two or more molecules under equilibrium conditions into stable, well-defined aggregates [29]. Intermolecular forces are capable of effectively organizing multicomponent supramolecular assemblies reversibly and accurately, using error-correcting mechanisms. In Nature, self-assembly is a ubiquitous strategy involved in the formation of cell membranes, double-stranded nucleic acids, and viruses. In chemistry, self-assembly provides a novel, rapid means of constructing complex nanostructures, receptor systems, catalysts, and materials. Self-assembling capsules have enclosed cavities formed as a result of reversible noncovalent interactions between two or more subunits [9]. In the same way as carcerands and hemicarcerands, capsules completely surround their guests but can release them without breaking covalent bonds. Indeed, the encapsulation occurs under subtle control, because formation–dissipation of the capsule occurs under equilibrium conditions. Capsules form reversibly on time-scales of 10^{-3} to 10^3 s, between those of diffusion complexes $\sim 10^{-10}$ s) and carceplexes ($\sim 10^{10}$ s) [9].

Early generations of self-assembling capsules include Rebek's sport balls (Fig. 5.4). The "tennis ball" **11** assembles as a result of formation of eight C=O•••H–N hydrogen bonds between the self-complementary glycoluril NH and C=O groups [30]. The inner volume of **11** is ~ 50 Å^3, which enables encapsulation of small molecules only, for example CH_4, C_2H_6, and Noble gases. Larger bisglycoluril-containing balls contain extended spacers and are capable of encapsulating two benzene-sized molecules. For example, the enclosed cavity of the "soft ball" **12** is 300 Å^3 [31].

Dimeric calix[4]arene tetraurea capsules **13**, which were introduced shortly after the balls, are probably the most studied (Fig. 5.5) [32, 33]. In dimers **13** a seam of sixteen intermolecular C=O•••H–N hydrogen bonds is formed at the upper rim. Eight urea moieties, four from each hemisphere, assemble head-to-tail. This results in a rigid cavity of ~ 200 Å^3 which can accommodate a benzene-sized guest molecule. Ex-

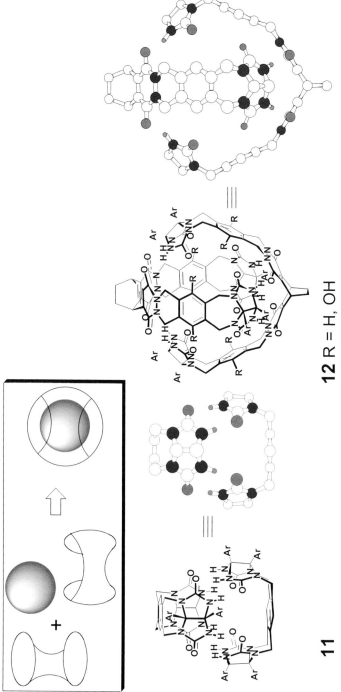

12 R = H, OH

11

Figure 5.4 Rebek's tennis ball **11** [30] and soft ball **12** [31] and their molecular models.

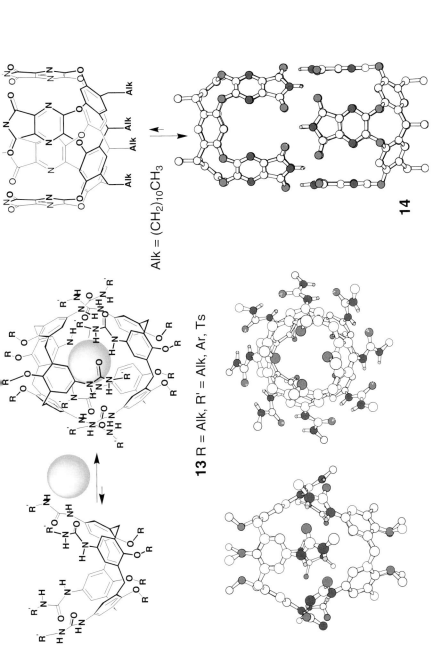

Alk = (CH₂)₁₀CH₃

13 R = Alk, R' = Alk, Ar, Ts

14

Figure 5.5 Self-assembling calixarene capsules **13** [32] and **14** [37]. Bottom: X-ray structure of **13** (left) [32c] and MacroModel representation of **14**.

change of the encapsulated guest is slow on the NMR time-scale. Capsules **13** include benzene, toluene, halobenzenes, and polycyclic aliphatic compounds such as camphor derivatives and 3-methylcyclopentanone. Because of their high kinetic stability, the encapsulation complexes can be conveniently studied by ^{1}H NMR spectroscopy. The encapsulated benzene molecule, for example, is observed at ~4 ppm in deuterated *p*-xylene. The CH protons of encapsulated fluorobenzene are observed between 5.5 and 3 ppm and the para and ortho hydrogen atoms are separated by more than $\Delta\delta$ = 2 ppm. Encapsulated aliphatic guests are seen upfield of 0 ppm. From NOESY experiments with the lower symmetry capsule **13** a rate constant of 0.47 ± 0.1 s^{-1} was obtained for benzene exchange [34]. The guest exchange time can, however, be significantly prolonged by introducing sterically bulky substituents (e.g. 2,6-diisopropylphenyl, 3,5-di-*t*-butylphenyl, trityl, *p*-tritylphenyl) to the urea moieties. Some of these capsules are stable even in polar solvents [35].

When two different calixarene tetraureas are mixed, heterodimers form with the corresponding homodimers. From a mixture of tetraurea and tetrasulfonyl urea, heterodimers form exclusively [36]. The increased acidity of the –SO$_2$NH urea proton might be the reason for such selection.

Electrospray ionization mass spectrometry (ESI MS) was successfully used to characterize capsules **11–13** in solution and in the gas phase. In this work ion labeling was achieved by encapsulation of quaternary ammonium ions in CHCl$_3$.

Larger capsules entrap more than one guest. Self-complementary cyclic tetraimides built up on the resorcinarene platform dimerize by hydrogen bonding into cylindrical capsule **14** (Fig. 5.5) [37]. Its estimated internal volume is 460 Å^3 and the internal dimensions are 6×15 Å. In deuterated mesitylene, which is too bulky to fit inside, capsule **14** dissociates, but addition of benzene or toluene immediately results in complex **14**•2×guest – two benzene or toluene molecules were quantitatively encapsulated inside **14** (^{1}H NMR). The complexes are kinetically stable at ambient temperatures and the encapsulation processes are slow on the NMR time-scale.

Metal-induced capsular self-assembly **15** (Fig. 5.6) of two deep, resorcinarene-based cavitands was described by Dalcanale and coworkers [38]. Two tetracyanocavitands were connected through four square-planar Pd(II) or Pt(II) entities in CH$_2$Cl$_2$, CHCl$_3$, and acetone. Evidence of encapsulation of one triflate anion on dimerization was obtained by ^{19}F NMR and X-ray analysis. For Pt(II) the assembly process was shown to be reversible – Et$_3$N dissociated the capsule **15** whereas addition of trifluoroacetic acid restored it.

Examples of Co(II) and Fe(II) resorcinarene-based cages **16** came from Harrison's group (Fig. 5.6) [39]. These are water-soluble and formed by combining CoCl$_2$ or FeCl$_2$ in aqueous solution at pH > 5 with the cavitand, functionalized with four iminodiacetic acid moieties. On synthesis, six water molecules initially occupy the inner cavity. When the appropriate guest molecule is present in the reaction mixture it usually goes inside, replacing the water. Benzene, halobenzenes, chlorinated hydrocarbons, (cyclo)alkanes, and alcohols were thus trapped in aqueous solution. At lower pH the ligand becomes protonated. The metal ion can no longer be coordinated to the cavitand and the assembly falls apart.

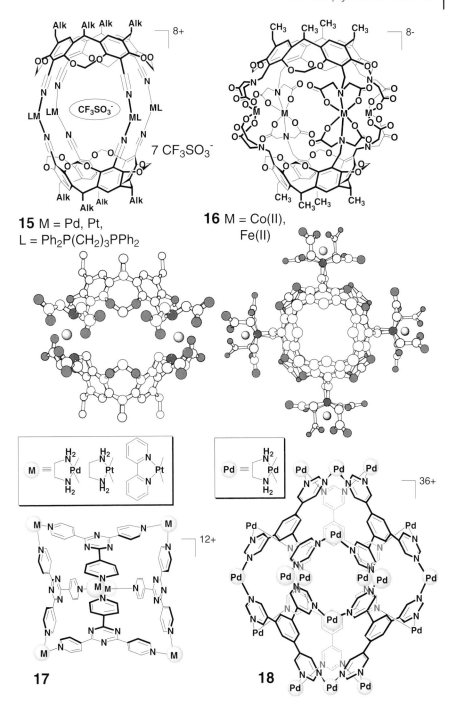

15 M = Pd, Pt,
L = Ph₂P(CH₂)₃PPh₂

16 M = Co(II),
Fe(II)

17

18

Figure 5.6 Metal-containing self-assembling capsules
15–18 [38–40]. Middle row: the X-ray structure of **16** [M = Co(II)].

Fujita and coworkers elegantly combined triangular aza-heterocyclic ligands with cis-protected square-planar Pd and Pt entities into highly symmetric capsules **17** and **18** (Fig. 5.6) [40]. The water solubility of these capsules was ensured by the ionic metal centers, and their hydrophobic interiors were could accommodate a wide variety of organic guests. In capsule **17**, Pd or Pt atoms occupy each corner of an octahedron formed by four tris(4-pyridyl)triazine molecules; the longest metal–metal separation is 19 Å and the internal volume is ~500 Å^3 (Fig. 5.6). The Pt-based assembly **17** is remarkably stable even in HNO$_3$ and aqueous K$_2$CO$_3$. Cavities **17** are large enough to encapsulate *four* guests the size of adamantane, two *cis*-azobenzenes, or two *cis*-stilbenes. Mixing six tris(pyrimidyl)benzenes with eighteen Pd complexes resulted in the spectacular hexahedral capsule **18**. Unlike **17**, each face of the hexahedral capsule **18** is completely enclosed by the planar threefold-symmetric ligands (Fig. 5.6). The volume of the inner cavity is considerably larger (~900 Å). This means that guests such as C$_{60}$ fit easily inside. Only water or oxygen molecules can pass through the narrow pores (2 × 2 Å) in the structure, however.

By similar design, homooxacalix[3]arenes bearing 4-pyridyl groups at the upper rim form capsules **19** with Pd(II)(Ph$_2$P(CH$_2$)$_3$PPh$_2$)(OTf)$_2$ in a 2:3 ratio in CH$_2$Cl$_2$ (Fig. 5.7) [41]. A kinetically stable complex between capsule **19** and fullerene C$_{60}$ has been detected (^{1}H and ^{13}C NMR) in Cl$_2$CDCDCl$_2$. Complexation of the Li$^+$ cation at the lower rims of the calixarene induced a more flattened conformation, and this seemed to be even more suitable for C$_{60}$ inclusion. A K_{ass} value as high as 2100 M^{-1} (30 °C) was obtained.

Spectacular trimeric assemblies **20** of cavitands functionalized with four dithiocarbamate units were obtained on mixing with Zn(II) and Cd(II) salts in EtOH–H$_2$O (Fig. 5.7) [42]. In apolar solvents the ^{1}H NMR spectra were broad, but the complexes crystallized readily from pyridine–H$_2$O. They were characterized by X-ray crystallography. Trimers **20** form very stable 1:1 complexes with fullerene C$_{60}$ in benzene and toluene solutions; log K_{ass} values are within the range 3.5–6 (UV–visible spectroscopy).

Six molecules of resorcinarene (R = Me or *n*-undecyl) and eight H$_2$O molecules self-assemble into unique spherical nanocavity **21** [43]. Its diameter is 17.7 Å, and the internal volume reaches ~1375 Å^3 (Fig. 5.7). *Sixty* hydrogen bonds hold the nanocavity together. In wet CDCl$_3$ hexamer **21** entraps bulky tetraalkylammonium and tetraalkylphosphonium cations [44]. Molecular modeling suggests **21** is large enough to accommodate even fullerenes or porphyrins.

Some Comments on Binding Within Molecular Containers: Conventional molecular recognition complexes depend on stereoelectronic complementarity between a receptor and a substrate and are held together by hydrogen bonding, metal–ligand attractions, ion–ion, ion–dipole, CH–π, and π–π interactions, and van der Waals and solvophobic forces [45]. In contrast, carceplexes, hemicarceplexes, self-assembling capsules, and, to a lesser extent, caviplexes rely primarily on so-called *constrictive binding*. The stability of these complexes is mostly provided *not* through the intrinsic host–guest attraction but through mechanical inhibition of the decomplexation process. According to Cram [22], constrictive binding is:

$$\Delta G^{\#}_{constrictive} = \Delta G^{\#}_{assoc} = \Delta G^{\#}_{dissoc} - (-\Delta G^{\circ}),$$

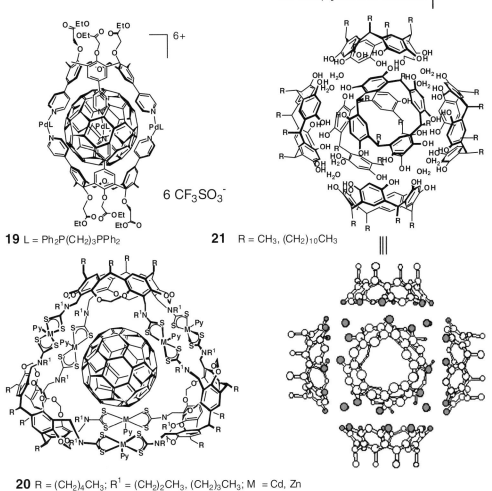

19 L = Ph$_2$P(CH$_2$)$_3$PPh$_2$

21 R = CH$_3$, (CH$_2$)$_{10}$CH$_3$

20 R = (CH$_2$)$_4$CH$_3$; R^1 = (CH$_2$)$_2$CH$_3$, (CH$_2$)$_3$CH$_3$; M = Cd, Zn

Figure 5.7 Capsules **19** and **20** entrap fullerene C$_{60}$ [41, 42].
Right: Atwood's nanocavity assembles from six molecules of
resorcinarene and eight H$_2$O molecules [43].

where $\Delta G^{\#}_{assoc}$ and $\Delta G^{\#}_{dissoc}$ are the free energies of activation for complex formation and dissociation, respectively, and $-\Delta G°$ is the intrinsic host–guest binding energy. For a medium-sized hemicarceplex it was demonstrated experimentally that most of the constrictive binding comes from structural reorganization of the host [22]. Both the inner cavity dimensions and the size and rigidity of the opening/portal have a significant effect on activation energies $\Delta G^{\#}_{assoc}$ and $\Delta G^{\#}_{dissoc}$. Enlarging the cavity often leads to higher flexibility of the host. The kinetic stability of the complexes decreases when the portals are wide open, allowing the guest to escape easily. At the same time very rigid and preorganized portals can prevent the guest entering.

The size and shape of the guest both reflect the binding ability of the hosts. In the absence of additional intermolecular forces, recognition-through-encapsulation is largely determined by the host and guest volumes. Competition with solvent molecules inside the cavity might be one important reason why entry of the guest is blocked. Accordingly, the $\Delta G^{\#}_{assoc}$ barriers can be reduced by using a solvent with minimum solvating ability. In addition, "very large" solvent molecules, which cannot fit inside the cavity or are simply too bulky to pass through the portals, can be used. And, finally, introduction of additional binding sites within the interior increases the $-\Delta G°$ intrinsic binding free energy. So far, however, this remains a challenging task in the synthetic chemistry of molecular containers.

Another type of multiple assembly is the so-called *double-rosette*. These are composed of complementary melamines and barbiturates/(iso)cyanurates, which form hydrogen bonds. Thus, calix[4]arenes diametrically substituted at their upper rim with two melamine moieties assemble into well-defined, box-like, double-rosette structures **22** on addition of two equivalents of barbiturates/isocyanurates (Fig. 5.8) [46]. These assemblies consist of nine components and are held together by 36 hydrogen bonds. They are stable in apolar solvents at ≥0.1 mM concentrations. In addition to high-resolution ^{1}H NMR spectroscopy and, in some spectacular instances, X-ray crystallography, matrix-assisted laser-desorption ionization time-of-flight (MALDI-TOF) mass spectroscopy has been used extensively to characterize the double-rosettes.

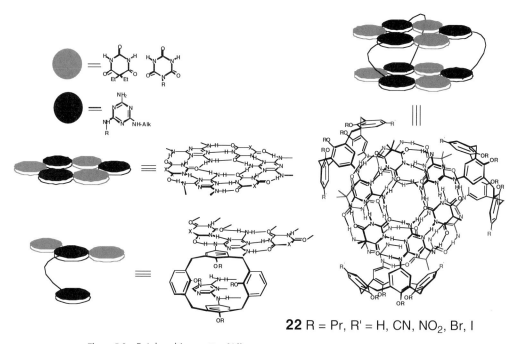

22 R = Pr, R' = H, CN, NO$_2$, Br, I

Figure 5.8 Reinhoudt's rosettes [46].

When 5,5-disubstituted barbituric acid derivatives (e.g. 5,5-diethylbarbiturate, DEB) were employed, only the staggered D_3 symmetrical assemblies form, with antiparallel orientation of the two rosette motifs (Fig. 5.8). These are chiral and occur as a racemate in the absence of other chiral sources. When chiral melamines were attached to the calixarene upper rim, complete induction of supramolecular chirality was achieved [47].

By mixing different calixarene bismelamines with DEB, libraries of self-assembled double-rosettes were generated. These are *dynamic* libraries, formed under thermodynamic control [48]. When the dynamic combinatorial libraries were formed their covalent capture was effectively achieved by a ring-closing metathesis reaction [49]. More complex assemblies (for example, **23**), consisting of 15 components held together by 72 hydrogen bonds, were constructed from calixarene-supported tetramelamines and four equivalents of DEB (Fig. 5.9) [50, 51].

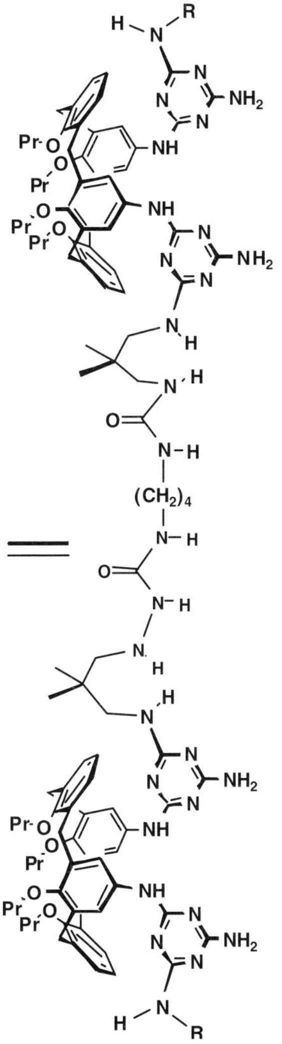

23

Figure 5.9 Rosette assemblies **23** combine 15 components held together by 72 hydrogen bonds [50].

5.3
Chemistry Inside Capsules

Molecular containers have been effectively used to stabilize reactive and otherwise unfavorable species within the interior, to control and change the reaction rates and regiochemistry, and to amplify and catalyze reactions.

5.3.1
Observing Unusual Species Through Encapsulation

Cram and coworkers impressively pioneered this field, showing as early as 1991 that cyclobutadiene can be synthesized and stored inside hemicarcerand **6** (X' = OCH$_2$O) [52]. Since then, in less than a decade, results from a wide range of experiments have been published describing how capsules can stabilize unusual molecular entities [11]. We discuss here some key results.

o-Benzyne (see **24**, Fig. 5.10) was effectively stabilized by encapsulation in a hemicarcerand [53]. *o*-Benzyne is very important in physical organic chemistry; it is, how-

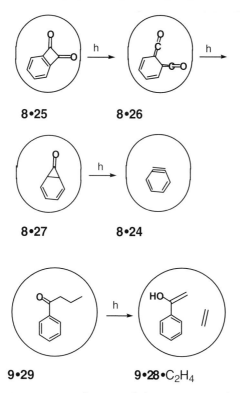

8•25   **8•26**

8•27   **8•24**

9•29   **9•28•**C$_2$H$_4$

Figure 5.10 Top: formation of *o*-benzyne **24** inside hemicarcerand **8** [53]. Bottom: kinetic stabilization of enol **24** in Sherman's large carcerand **9** [54]. Here and subsequently in the legends molecular containers and capsules are depicted as circles or ellipses for simplicity and viewing clarity.

ever, highly elusive and could only be characterized at extremely low temperatures (~20 K). Inside hemicarcerand **8**, *o*-benzyne seemed to be rather stable and was studied by ^{1}H and ^{13}C NMR spectroscopy at much higher temperatures (175–198 K). First, the hemicarceplex with benzocyclobutenedione **8•25** was prepared. Its photolysis at λ > 400 nm produced hemicarceplex **8•26** containing encapsulated bisketene **26**, and then the hemicarceplex with benzocyclopropenone **8•27**. Further photolysis at $\lambda \approx 280$ nm resulted in extrusion of CO and generated *o*-benzyne **24** inside the capsule. The half-life of *o*-benzene inside **8** is 205 s at –75 °C and ~7 ms at room temperature. Under these conditions it readily reacts with the interior wall in an "innermolecular" Diels–Alder reaction.

Kinetic stabilization of an enol has also been achieved [54]. Typically, enols are unstable and tend to ketonize. Acids, bases, protic solvents, and water are known to accelerate this process. Sherman and coworkers obtained stable acetophenone enol **28** inside carcerand **9**. First, carceplex **9•29** with butyrophenone was prepared (Fig. 5.10). Its photolysis (λ = 300 nm, 5–6 h) in benzene resulted in ~85 % yield of **9•28•ethylene**. Enol **28** was studied by conventional NMR spectroscopy and its stability toward water- and TFA-catalyzed ketonization was estimated. TFA cannot pass through the portals and no isomerization was seen. In contrast, water can enter the cavity. As a result, ketonization of **28** occurred ($t_{1/2}$ = 78 min at 100 °C; $t_{1/2}$ = 3 years at 25 °C, in H_2O-saturated nitrobenzene). At the same time, no ketonization was observed in the absence of water even after 26 days at 100 °C (nitrobenzene, crushed 4-Å molecular sieves).

Discovery of self-assembling capsules, which allow easier exchange of guests in and out broadened the scope of this chemistry. For example, Fujita used self-assembling containers **17** for stabilization of unfavorable geometrical isomers. Treatment of an aqueous solution of capsule **17** with a solution of 4,4-dimethylazobenzene (cis/trans, 1:6) in hexane resulted in the formation of an unusual complex within the capsule walls; the capsule selectively entrapped two equivalents of the cis isomer (2D NMR) [55]. Accordingly, the *cis*-azobenzene molecules were stabilized within this encapsulation complex (Fig. 5.11). Exposing the solution to visible light for several weeks did not result in the production of any of the thermodynamically favored *trans*-azobenzene. Interestingly, inside **17** the guest exists as its dimeric hydrophobic complex. It is too large to penetrate and the compound enters as a single species.

Another example of stabilization by encapsulation was demonstrated with capsule **14**. On encapsulation within **14** the tertiary anilide *p*-[*N*-methyl-*N*-(*p*-tolyl)]toluamide, **30**, is fixed in its *Z* configuration (Fig. 5.11) [56]. In bulk solution such anilides typically exist as mixtures of *E* and *Z* isomers, with the *E* configuration favored. The *E* isomer, however, cannot fit inside capsule **14** and its inherent stability is overcome by the CH–π, van der Waals, and dipolar interactions offered by the interior surface of the capsule.

Enclosed cavities in self-assembling capsules can stabilize otherwise labile molecules formed *in situ* by reaction of smaller molecular components. Thus, the condensation reaction of trimethoxysilanes in the cavity of **17** led exclusively to cyclic, trimeric silanol **31**, which had never before been isolated [57]. In a typical reaction, phenyltrimethoxysilane was suspended in a D_2O solution of **17** at 100 °C (Fig. 5.11).

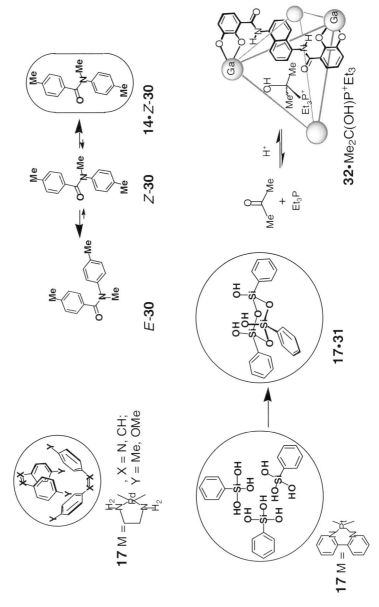

Figure 5.11 Top: unusual isomers can be observed inside self-assembling capsules **17** and **14** [55, 56]. Bottom: unfavored molecular species are stabilized in capsules **17** and **32** [57, 59].

After 1 h the ^{1}H NMR spectrum showed the exclusive presence of only one complex **17**·**31**. The formation of **17**·**31** was also apparent from ESI-MS and single-crystal X-ray crystallography. Not only were cyclic trimers **31** formed in a "ship-in-a-bottle" fashion, they were protected by the cavity. They were very stable even in acidic aqueous solutions and isolable as pure clathrate compounds.

Capsule **14** was used to entrap such useful chemical reagents as dicyclohexylcarbodiimide (DCC) and dibenzoyl peroxide (DBPO) [58]. The chemical stability of DCC and DBPO was greatly improved by complexation inside capsule **14**. For example, inside the capsule DBPO is stable for at least 3 days at 70 °C in mesitylene-d_{12} solution whereas in the absence of **14** it decomposes within 3 h under those conditions. Because of the reversible nature of the hydrogen bonds stitching **14**, competitive guest molecules or polar solvents can release the encapsulated DCC and DBPO.

The tetrahedral gallium catecholate $Na_{12}[Ga_4L_6]$ cluster **32** encapsulated and stabilized highly reactive $Me_2C(OH)P^+Et_3$ species in water and methanol (Fig. 5.11) [59]. In the absence of **32**, molecules $Me_2C(OH)P^+Et_3$ form from acetone and PEt_3 in the presence of a proton source and quickly decompose in aqueous solution. Inside **32**, however, these species are stable for several hours in D_2O and for a day in methanol-d_4; ^{1}H and ^{31}P NMR spectroscopy and ESI-MS evidence was provided.

5.3.2
Changing Reaction Rates by Encapsulation

For self-assembling capsules, acceleration of reaction rates and even catalysis by encapsulation have been established. Initially, pronounced (~200-fold) rate acceleration was observed in the Diels–Alder reaction between benzoquinone and cyclohexadiene in the presence of self-assembling capsule **12** (Fig. 5.12) [60]. Two molecules of benzoquinone were first encapsulated and clearly seen by ^{1}H NMR spectroscopy, but the rate acceleration was a consequence of a "mixed encapsulation" complex **33**. This is the Michaelis complex. No rate acceleration was observed for electronically similar reactants which could not fit inside **12**. In this reaction, however, the Diels–Alder product **34** seemed to be a good guest for **12**, and strong product inhibition prevented turnover. True catalysis was observed for the reaction between 2,5-dimethylthiophene dioxide and benzoquinone in capsule **12**. It was shown that the reaction product **35** could be easily ejected from **12** by benzoquinone, which resulted in catalytic turnover [61]. Tenfold rate enhancement compared with the control was observed for the reaction inside **12**.

The Michaelis complex **36** in the Diels–Alder reaction between 1,4-naphthoquinone and 1,3-cyclohexadiene was finally seen by Fujita and coworkers (Fig. 5.13) [62]. Initially, capsule **17** was left to extract two equivalents of naphthoquinone into aqueous solution, after which cyclohexadiene was added. The Michaelis complex **36** formed (^{1}H NMR, D_2O) and, as a consequence, the Diels–Alder reaction was accelerated 21-fold compared with the reaction without capsule **17** added. Furthermore, when isoprene (2-methyl-1,3-butadiene) was employed instead of cyclohexadiene, a 113-fold acceleration was observed. The corresponding Michaelis complex was seen in this reaction also.

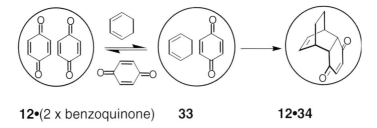

12•(2 x benzoquinone)   **33**   **12•34**

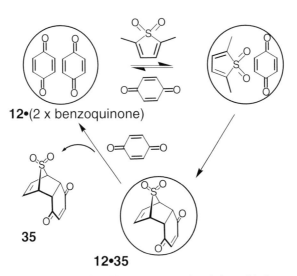

12•(2 x benzoquinone)

35

12•35

Figure 5.12 The Diels–Alder reactions inside Rebek's soft ball **12**
(R = OH) [60, 61].

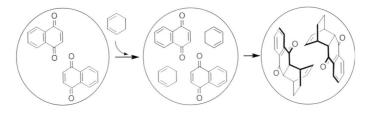

17• 2 x naphthoquinone   **36**

M =

Figure 5.13 The Michaelis complex **36** in the Diels–Alder reaction
between 1,4-naphthoquinone and 1,3-cyclohexadiene was seen by
Fujita [62].

The origin of these rate accelerations and catalysis still needs more investigation. Whether they are because of the increased apparent concentrations of reactants in the inner phase or because of stabilization of the transition states is difficult to evaluate experimentally [63].

5.3.3
Encapsulated Reagents

Encapsulation of reactants/reagents directly affects their physical and chemical properties. Among the challenges of modern synthetic chemistry are to control kinetics and selectivity of reactions and to create reagents which modify a portion of a molecule without the need to protect other reactive sites. Molecule-within-molecule complexes have unique features in this area also.

Through-shell borane reduction and methylation with incarcerated benzocyclobutenedione have been studied [64]. Heating hemicarceplex **8•26** under reflux with BH$_3$–THF in THF solution resulted in 85–90 % incarcerated monoreduction product **37**, but none of the corresponding bis alcohol was detected (Fig. 5.14). Addition of excess MeLi to a solution of benzocyclobutenedione hemicarceplex **8•26** in THF–Et$_2$O at –78 °C, then quenching with water, resulted in 65 % formation of complex **8•38** with 8-hydroxy-8-methylbenzocyclobuten-7-one **38**. This is the product of addition of MeLi to one carbonyl group only. Despite the excess of MeLi used, the corresponding bis adduct has not been detected. These results are clearly the consequence of the inaccessibility of one of the carbonyl groups to the bulk-phase reagents. Functional groups of the encapsulated species, which are located deep inside, are less reactive.

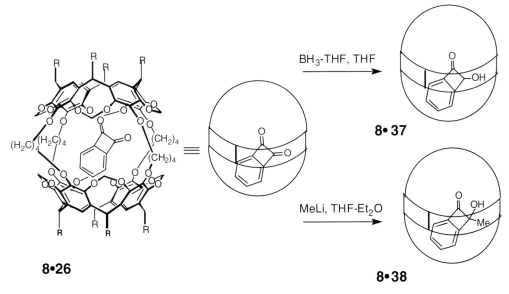

Figure 5.14 Through-shell borane reduction and methylation with incarcerated benzocyclobutenedione **29** [64].

Molecular containers can freely entrap and release highly reactive guest species, under subtle chemical or physical control. *Encapsulated reagents* can thus be created [65]. These are highly reactive species, reversibly entrapped within the cavity of the host, which can be released to the reaction mixture under subtle control. The cavity gives protection from the bulk environment and thus controls the rate of reaction. Chemical transformations with encapsulated reagents can occur not only within the cavity interior [64], but also outside, on release. As far as delicate, noncovalent forces holding the molecule-within-molecule complex together are concerned, temperature, solvent polarity, and substrate-cavity size–shape complementarity are the critical factors responsible for reagent release so reaction can occur.

Terakis-*O*-alkylated calix[4]arenes **39** and **40** have been shown to interact reversibly with NO_2/N_2O_4 gases and entrap highly reactive NO^+ cation within their cavities (Fig. 5.15). NO^+ is generated from N_2O_4, which is known to disproportionate to $NO^+NO_3^-$ on exposure to aromatic compounds. Stable nitrosonium complexes **39**•NO^+ and **40**•NO^+ were quantitatively isolated upon addition of a stabilizer – Lewis acidic $SnCl_4$ [66].

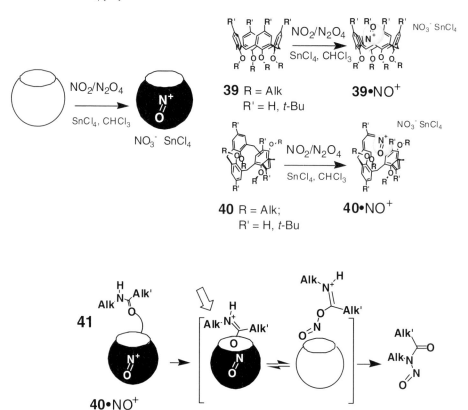

Figure 5.15 Top: calixarenes **39**, **40** convert NO_2/N_2O_4 gases into stable calix–nitrosonium complexes [66]. Bottom: reactions of amides **41** with encapsulated nitrosating reagent **40**•NO^+ [65].

Addition of H_2O or alcohols resulted in the dissociation of **39**•NO^+ and **40**•NO^+ and recovery of **39** and **40**. Complexes of calixarenes with R' = H decompose instantly, but it takes minutes to release NO^+ when R' = *t*-Bu. Apparently, bulky *t*-Bu groups at the upper rim of **39** and **40** protect the encapsulated NO^+. Such stability of the arene–NO^+ complex is without precedent.

Complexes **40**•NO^+ act as encapsulated nitrosating species (Fig. 5.15) [65]. Thus, secondary amides **41** reacted with **40**•NO^+ (R' = *t*-Bu), with remarkable selectivity. Amides with N-CH_3 substituents were smoothly nitrosated (298 K, $CHCl_3$; 50–95 %) whereas bulkier amides did not react. Stereochemistry was responsible – N-alkyl groups larger than CH_3 do not enable the substrate to enter the cavity. In the proposed mechanism the dimensions and shapes of the N-alkyl substituent are crucial. The amide substrate approaches the cavity facing it with the carbonyl oxygen; this places the N-alkyl group in a close proximity to the calixarene bulky rims. For sizeable N-alkyl groups this situation is sterically unfavorable, so that the substrate C=O and the encapsulated NO^+ cannot reach each other. Such stability and selectivity is not easily achieved for other nitrosating agents – NO^+ salts, $NaNO_2/H_2SO_4$, NO/O_2, N_2O_3, and NO_2/N_2O_4.

A recent paper from the Rebek's laboratory introduces another novel concept – autocatalysis with encapsulated reagents [67]. In detail, when DCC was trapped by capsule **14** (Fig. 5.16), the kinetic behavior of the reaction between *p*-toluic acid and *p*-eth-

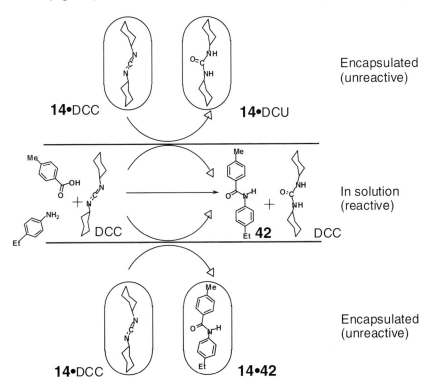

Figure 5.16 Autocatalysis with encapsulated DCC [67].

ylaniline was very interesting. Trace amounts of DCC free in solution promoted the initial formation of an amide bond between the acid and the amine. The products of this reaction were anilide **42** and dicyclohexylurea (DCU), both of which are better guests for capsule **14** than DCC. Accordingly, increasing amounts of DCC are displaced from the capsule by emerging **42** and DCU, so the rate of the reaction increases as the reaction proceeds. The kinetics are sigmoidal in character. Such nonlinear kinetics can be viewed as an emergent property of the system as a whole, with the partnership of compartmentalization and molecular recognition giving rise to chemical amplification.

5.4
Storage of Information Inside Capsules

The use of molecules and molecular assemblies for information processing, storage, and transmission is one of the most appealing perspectives in nanotechnology and molecular electronics [68]. Molecular and supramolecular species capable of existing in different states which can be interconverted by external factors is a topic of great interest for molecular computing. Molecular switching and logic gate devices based on polymeric molecular wires, layered materials, and interlocked molecules, for example rotaxanes and catenanes, have been successfully introduced. These devices can be scaled down to real molecular dimensions and might be very thermodynamically efficient. Molecule-within-molecule complexes are alternative, still unexplored but very promising, information/memory storage sources. The information content accumulated within the capsule's interior is unique. In some carceplexes a tight fit between the host and its single guest mechanically restricts molecular motion to the extent that different isomers can be distinguished on the basis of different orientations of guest molecules inside a host. Reinhoudt's supramolecular switch in diastereomeric carceplexes **43**•DMF is early example of this science, called supramolecular isomerism (Fig. 5.17) [69]. Self-assembling monolayers of caviplexes **43**•DMF on gold have subsequently been produced.

When two guests are encapsulated, the volume of structural information inside the cavity increases significantly.

An early example of pairwise guest selection by encapsulation was described for cylindrical capsule **14**. When both benzene and *p*-xylene were added in 1:1 ratio to a mesitylene-d$_{12}$ solution of capsule **14**, an unsymmetrically filled complex formed (Fig. 5.17) [37, 56, 58]. Such pairwise selection of the guests was also observed for benzene with *p*-trifluoromethyltoluene, *p*-chlorotoluene, 2,5-lutidine, and *p*-methylbenzyl alcohol, giving new species with one of each guest inside. Complexation of smaller hydrogen-bonded pairs, for example 2-pyridone–2-hydroxypyridine dimer, benzamide and benzoic acid dimers, *trans*-1,2-cyclohexanediol dimers, and some others is, moreover, observed with capsule **14**.

When equal amounts of α-, β-, and γ-picolines were added to a mesitylene-d$_{12}$ solution of **14**, in addition to the corresponding homocapsules, nonsymmetrically filled heterocapsules were also formed [70]. The chemical shift differences between guests

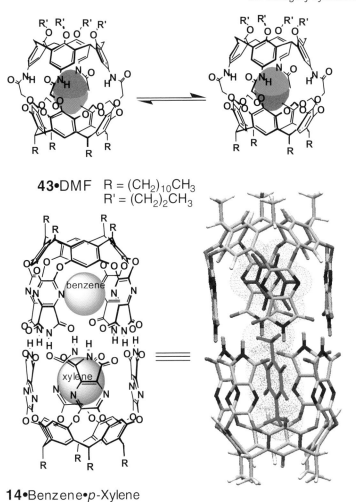

43•DMF R = (CH$_2$)$_{10}$CH$_3$
             R' = (CH$_2$)$_2$CH$_3$

14•Benzene•*p*-Xylene

Figure 5.17 Top: Reinhoudt's supramolecular isomerism involving carceplex **43•**DMF [69]. Bottom: pairwise selection of benzene and *p*-xylene by capsule **14** [37].

in the bulk solution and those in capsule **14** are related to their positions in the dimeric assembly. In mesitylene-*d*$_{12}$ solution, α-, β- and γ-picolines gave complexes with two identical guests per capsule – *homo*-capsules. When γ-picoline is encapsulated, the chemical shift of the methyl groups in the NMR spectrum places them near the ends of the cavity – at δ = –2.79. The dipole involving the nitrogen, therefore, prefers the middle of the cavity, and at the methyl ends the change in shift is maximum –Δδ = 4.55. For the encapsulated β-picoline the methyl chemical shift is δ = –1.65 (Δδ = 3.43) whereas with α-picoline the methyl group is seen at δ = 0.58 (Δδ = 1.33). These guests spin along the long axis of the capsule but apparently do not tumble about other axes. When equal amounts of two different picolines were added to a mesitylene-d$_{12}$ solution of **14**, in addition to the corresponding homo-capsules, non-symmetric *hetero*-capsules were also formed, which were filled with two different guests (Fig. 5.18). Again, these guests are too large to move past each other within the capsule, and they can exchange their positions only by leaving the capsule and reentering it – a process that is slow on the NMR timescale. Accordingly, the capsule gives two sets of signals (two different ends) when two different picolines are inside. Moreover, when all three picolines were added to **14**, two homo-capsules (γ-picoline and β-picoline filled) and three hetero-capsules were clearly detected (Fig. 5.18).

More of these peculiar stereochemical relationships are likely to emerge as capsules become larger and able to accommodate more guest molecules. The control of guest orientation should lead to enhanced interactions, and even reactions inside these cylindrical capsules.

When racemic guests are added to capsule **14**, two diastereomeric host–guest complexes result. For example, when *trans*-1,2-cyclohexanediol was employed ¹H NMR

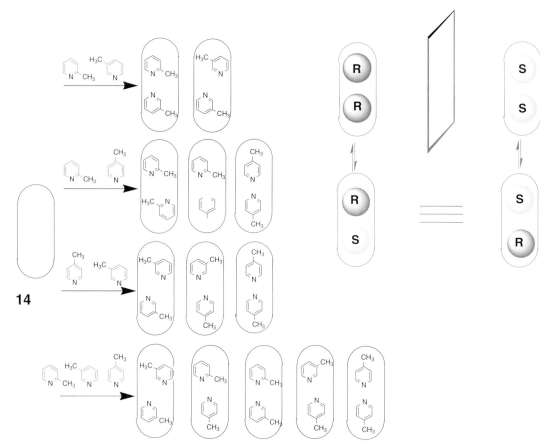

Figure 5.18 Information storage inside capsules. Left: variety of supramolecular isomers formed on encapsulation of picolines by cylindrical capsule **14** [70]. Right: chirality expression on encapsulation of racemic guests by **14** [56].

revealed the presence of two capsules; these were filled with enantiopure and racemic pairs of guest molecules (Fig. 5.18) [56, 71].

Nanoscale bis-cavitands **5** were also shown to complex two different guests, one per cavity, simultaneously and highly selectively [19].

These effects can be readily achieved by using a limited number of molecules only. The results might be very useful in nanotechnology – for design of switchable surfaces [69] and information storage and processing by molecular encapsulation. As more guests are encapsulated, the volume of information increases further and leads to many more exciting possibilities.

5.5
Materials and Sensors by Encapsulation

5.5.1
Molecular Containers as Sensors and Sensing Materials

Encapsulation phenomena can be applied to the sensing of small molecules. Because molecule-within-molecule complexes often give rise to rather delicate, chemically unique and specific interactions between a host and a guest, selectivity and specificity might well be expected. This field is still not well explored, but some spectacular results have already been achieved. For example, self-assembling capsules **44** were prepared with fluorophore donor and acceptor groups for fluorescence resonance energy transfer (FRET) [72]. On dimerization the donor and acceptor are situated within 20 Å, which is suitable for FRET (Fig. 5.19). As already mentioned, capsules **44** can form only in the presence of a suitable guest molecule. When the solvent is not a good fit the FRET system can be employed for detection of small molecules. For example, although FRET is not observed for **44** in large *p*-xylene, addition of 3-methylcyclopentanone triggers energy transfer. Quantitatively, at a capsule concentration of 150 nM the threshold concentration for energy transfer is ~0.1 mM 3-methylcyclopentanone. The assembly is 50% complete when the guest concentration is ~4 mM.

Water-soluble Co(II) and Fe(II) resorcinarene-based cages **16** trap benzene, halobenzenes, chlorinated hydrocarbons, (cyclo)alkanes, and alcohols in aqueous solution at pH > 5. These guests can thus be detected and monitored by conventional spectroscopic techniques [39]. Because of their paramagnetic nature, complexes of **16** act as NMR shift reagents, because they cause very substantial upfield isotropic hydrogen shifts ($\Delta\delta \approx 25-40$) in the guest molecules with a signal separation of $\delta \approx 12$.

As already mentioned, on exposure to NO_2/N_2O_4 calix[4]arenes **39** and **40** reversibly entrap NO^+ cation within their cavities (Fig. 5.15). Stable charge-transfer complexes ($K_{ass} > 10^6$ M^{-1} in CDCl$_3$) are deeply colored [66]. The UV–visible spectra contained broad charge-transfer bands with λ_{max} ranging between 520 to 600 nm. This can be used for NO_x sensing – calixarenes conveniently transmit information about NO^+ binding via visible light signals. Importantly, NO^+ can be generated not only from NO_2/N_2O_4 but also from other environmentally toxic NO_x, especially $NO-O_2$ and NO–air mixtures, and N_2O_3. The charge-transfer interactions are chemically unique

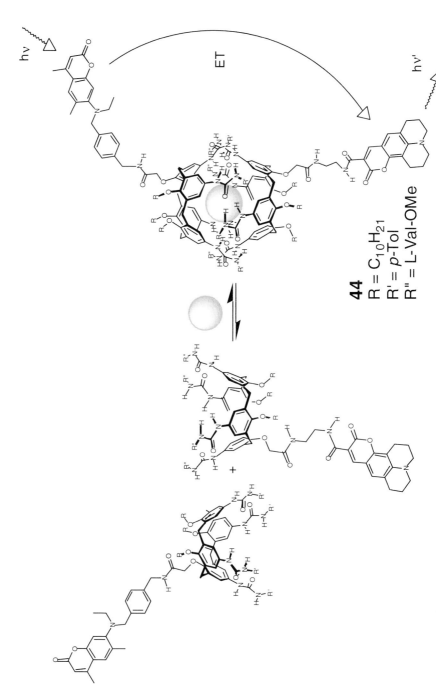

Figure 5.19 Self-assembling capsules **44** have fluorophore donor and acceptor groups and can be used for sensing small molecules by means of FRET [72].

44
R = C₁₀H₂₁
R' = *p*-Tol
R'' = L-Val-OMe

and would guarantee detection of NO$_\chi$ in the presence of such gases as H$_2$O, O$_2$, HCl, SO$_x$, NH$_3$, and even neat NO. On the other hand, Kochi and coworkers showed that calix[4]arenes, first oxidized to their cation-radicals, act as colorimetric sensors for NO, both in solution and in the solid state [73]. Complexes similar to **39** • NO$^+$ and **40** • NO$^+$ are formed.

NO$_2$-sensitive calix[4]arenes **39** have been modified and immobilized on a solid support. The silica gel-based material **45** produced can be used for sensing and entrapment of NO$_2$/N$_2$O$_4$ and other NO$_x$ gases (Fig. 5.20) [66]. When a stream of NO$_2$/N$_2$O$_4$ was passed through small columns loaded with silica gel **45**, either dry or

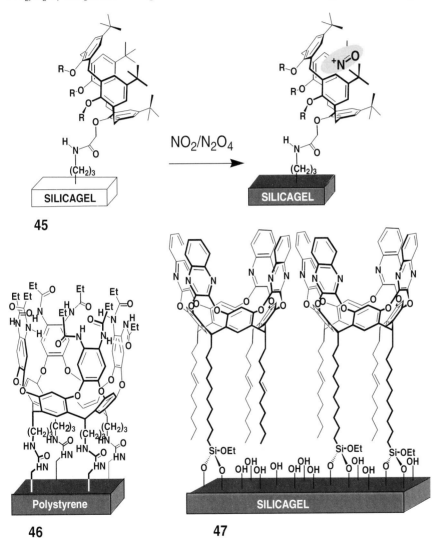

Figure 5.20 Cavity-containing materials for sensing and separation. Calixarene-silica gel **45** [66] and solid-supported cavitands **46** and **47** [74, 75].

wetted with $CHCl_3$, the material instantly turned dark purple, indicating NO^+ complexation. The color of the wetted material **45** $\cdot (NO^+)_n$ seemed to be deeper and persisted for hours. Although dry **45** $\cdot (NO^+)_n$ bleached within minutes, this result clearly demonstrated that the gas sensing is possible, even without solvent, on the gas–solid interphase. Silica gel **45** may be used as "supramolecular filters" for purification of, for instance, biomedical grade NO.

Phosphonate-bridged resorcinarene-based cavitands have excellent selectivity for and sensitivity to ($\sim 10^2$ ppm) short-chain alcohols (e.g., MeOH, EtOH, etc.), because of hydrogen bonding between the analyte and P=O groups and also CH–π interactions within the π-electron-rich cavity. Immobilized on a quartz surface these hosts act as mass sensors for alcohols [74].

Polymer-bound cavitands have also been prepared (Fig. 5.20). Thus, size–shape selective guest uptake and release was observed for self-folding cavitands **46** immobilized on polystyrene and poly(ethylene glycol) (PEGA) [75]. Adamantane-based guests were employed, and the binding events were followed by UV–visible spectroscopy and, in the spectacular case of soluble polymers, by ^{1}H NMR spectroscopy. The entrapped guests were clearly seen in the shielded, upfield region of the spectra. Quinoxaline-based cavitands **47** were grafted on to silica gel and had impressive sorption properties for airborne aromatic hydrocarbons (BTX) at ppb levels [76].

These findings clearly show that practical application of molecular containers is within sight. Besides sensing, their immobilization promises success in selective chromatographic separations and eventually in preparative organic synthesis and catalysis.

5.5.2
Supramolecular Polymers

Another emerging area of research involving capsules is the science of reversibly formed polymers. When two calix[4]arene tetraureas are covalently linked at their lower rims (**48**, Fig. 5.21), hydrogen bonding results in a polymer chains of capsules, or *polycaps* [77]. Polycaps are a novel class of polymer, recently emerged, in which monomeric units are held together by reversible forces. These self-assembling, supramolecular polymers combine features of conventional polymers with properties resulting from bonding reversibility. Structural and dynamic properties of supramolecular polymeric materials, for example their 2D and 3D architectures, degree of polymerization, strength, and liquid crystallinity, can be adjusted or switched "on–off" by assembly–dissociation processes.

The polycaps form only when a guest of proper size and shape is present. In competitive solvents, for example DMSO or MeOH, the polycaps dissociate. By direct analogy with parent capsules **13**, the formation of heterodimeric systems was also explored. Specifically, the polycap rapidly broke down to a dumbbell-shaped assembly when treated with an excess of the simple dimeric capsule (e.g. **13**). Combination of aryl- and sulfonylurea biscalixarenes effectively yielded heteromeric polycaps [36].

To fill the space between polymer chains, the monomers were functionalized with long alkyl groups capable of forming a liquid-like sheath about them. At high con-

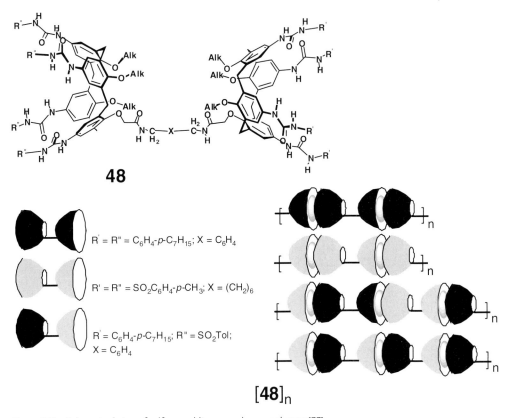

48

R' = R" = C$_6$H$_4$-*p*-C$_7$H$_{15}$; X = C$_6$H$_4$

R' = R" = SO$_2$C$_6$H$_4$-*p*-CH$_3$; X = (CH$_2$)$_6$

R' = C$_6$H$_4$-*p*-C$_7$H$_{15}$; R" = SO$_2$Tol; X = C$_6$H$_4$

[48]$_n$

Figure 5.21 Polymeric chains of self-assembling capsules, or polycaps [77].

centrations the resulting polycaps self-organize into polymeric liquid crystals [78]. Lyotropic, nematic liquid-crystalline phases were usually observed. Molecules like difluorobenzene and nopinone were readily encapsulated in these liquid-crystalline phases. Further characterization by X-ray diffraction showed peaks at 2.4 and 1.6 nm that match well the repeat distances and the dimensions of the polycaps.

Switchable polycaps were recently prepared, which change their aggregation properties without dissociation. For example, pH-switchable supramolecular polymer **49** takes advantage of the side-chain acid-sensitive functionalities (Fig. 5.22) [79]. Addition of small quantities (5–7 equiv.) of trifluoroacetic acid (TFA) to a benzene solution of **49** results in instant precipitation of **49** • *n*TFA. Obviously, the α-NH$_2$ groups in **49** become protonated, and the insoluble TFA salt is formed. It was determined that at ≤20 equiv. TFA per capsule, no significant dissociation occurs (≤30 %). Accordingly, a pH-switch can be provided at low concentrations of acids without breaking the self-assembling polymeric chain. Capsules **49** • *n*TFA, however, dissociate to give the corresponding monomers (as TFA salt) in DMSO-d$_6$. On the other hand, addition of Et$_3$N to a suspension of **49** • *n*TFA in benzene quickly regenerates **49**, which dissolves without dissociation of the capsules.

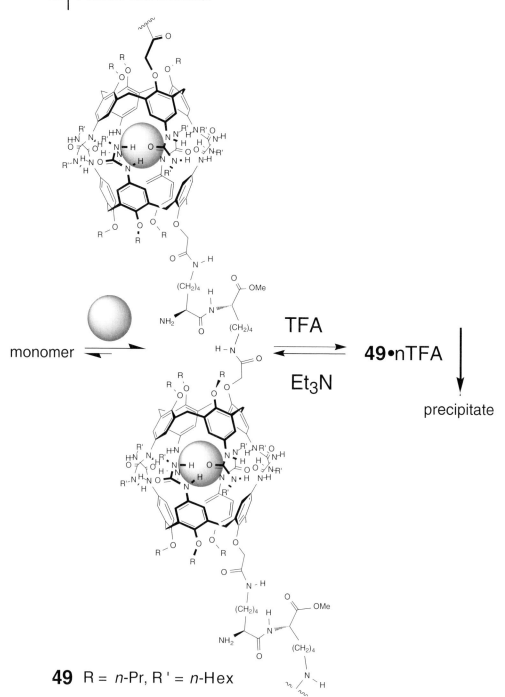

49 R = *n*-Pr, R ' = *n*-Hex

Figure 5.22 pH-Switchable supramolecular polymers **49** [79].

The most interesting feature of material **49** is, of course, its capsules. These are already preformed in apolar solution, but precipitate as salt **49** • nTFA only on protonation. While precipitating they captivate and store guest/solvent molecules in the solid state. Thus on formation of the capsules polymer **49** entraps benzene and toluene already in solution and then precipitates with it as **49** • nTFA. The guests are protected inside the capsules, but escape can be arranged even without dissociation. Indeed, when redissolved in $CDCl_3$ or $(CDCl_2)_2$, polymer **49** • nTFA releases benzene simply because the solvent now competes for the cavities.

These findings thus provide an opportunity to construct materials which reversibly trap, store, and then deliver chemicals to reaction mixtures.

Similarly, CO_2 reacts with amino-functionalized calix[4]arene capsules **50•50** in aprotic solvent with the formation of supramolecular polymeric chains **51** (Fig. 5.23) [80]. These are reversible and can dissipate either by hydrogen bond dissociation or a thermal release of CO_2. These can also entrap and store guests within the polymeric self-assembling capsules.

5.6
Biologically Relevant Encapsulation

In this section we will discuss biologically related molecule-within-molecule complexes. After all, the inspiration for molecular cavities came from Nature, specifically from mimicking enzyme–substrate complexes.

5.6.1
Entrapment of Biologically Active Guests

Self-folding cavitands **3-5** complex various adamantine derivatives (in chlorinated and aromatic solvents), which have a wide spectrum of biological activity [17–19]. More interesting, however, are water-soluble versions of **3**, which form thermodynamically and kinetically stable complexes in water [81]. Polymer-immobilized self-folding cavitands **46** have also been prepared [75].

Triethylene glycol footed deep cavitands **52** (Fig. 5.24) have four m-amidinium groups at the upper rim and effectively bind nucleotides and dinucleotides in buffered aqueous and methanol solutions [82]. For adenine nucleotides association constant (K_{ass}) values in the range 1.4×10^3 to 6.6×10^5 M^{-1} were determined by ^{1}H NMR spectroscopy (D$_2$O, pH 8.3, 300 K). In addition to charged interactions and hydrogen bonding the cavity also contributes to the overall binding process. Analysis of NMR chemical shifts and molecular modeling strongly suggest that the adenine fragment is situated deep inside host **52**.

Acetyl choline and choline were recently entrapped within the hydrophobic cavity of host **53** in water (Fig. 5.24) [83]. The complexes seemed to be thermodynamically stable; association constant, K_{ass}, values of ~2×10^4 M^{-1} were determined (^{1}H NMR, isothermal titration calorimetry, ITC). The Me$_3$N$^+$ groups of acetyl choline and choline were found deeply trapped inside the hydrophobic pocket of the host, which was evident from a large $\Delta\delta \approx 4.5$ ppm upfield shift of the methyl protons of the guest. Free and encapsulated guests were seen simultaneously by NMR, from which a rather high kinetic barrier for guest exchange of 16 kcal mol^{-1} was estimated (release rate ~8 s^{-1}).

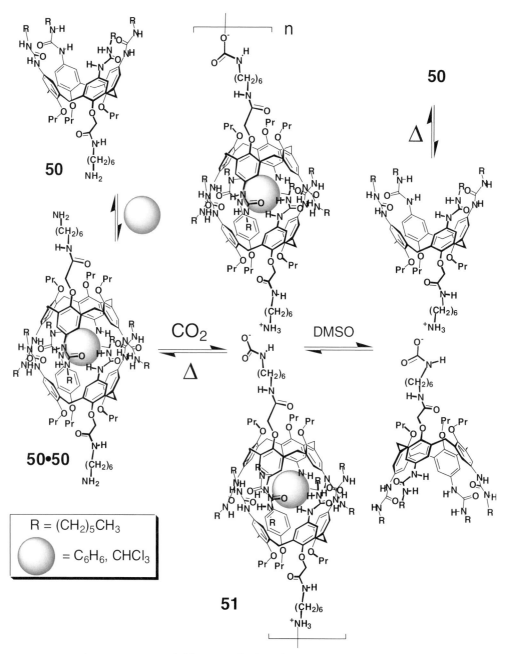

Figure 5.23 CO$_2$-Switchable supramolecular polymers **51** [80].

Self-assembling capsules have also been tested for biological recognition. Thus, cylindrical capsule **14** traps N-α-protected amino acid esters such as BOC-L-alanine alkyl esters and BOC-L-alanine alkyl esters in mesitylene [84]. In this complex the size and shape of the guest determined the binding affinities.

Self-assembled tetrarosettes **23** have been shown to bind glucopyranosides. Interestingly, one of the two enantiomers of a racemic mixture of **23** stereoselectively recognizes and binds the chiral carbohydrate guests, with the formation of one diastereomeric complex (^{1}H NMR, CD-spectroscopy). Assemblies **23** have six urea functionalities positioned in the central cavity for saccharide recognition [85].

In general, understanding of self-assembling cavities and cages might facilitate the design of more complex, reversibly formed nanostructures, binding features of which mimic antibodies [86]. This would require, first of all, full control over the structure and stereochemistry of the receptor sites, capable of organizing themselves spontaneously into well-defined nanostructures. It should then be possible to prepare dynamic, noncovalent libraries of such structures. On screening of these libraries, template effects of the guest can select the strongest binding receptor under thermodynamic control. After that, covalent post-modification can be performed. Fig. 5.25 sketches this idea for rosettes, but other structural motifs could work as well. This research is at the very early stage, however.

5.6.2
Encapsulation of Gases

Encapsulation of biologically and environmentally relevant gases within molecular containers might lead to important medical applications. We have already discussed application of calix[4]arenes (**39**, **40**, and **45**, Figs. 5.15 and 5.20) for sensing and separation of NO_x–NO, NO_2, and their relatives [65, 66]. Reinhoudt and coworkers targeted O_2. They prepared water-soluble models of heme–protein active sites [87]. These are self-assembling, heterodimeric capsules of cationic Co(II) porphyrin **54** and tetrasulfonato calix[4]arene **55**. For assembly of **54**·**55**, log $K = 3.8$, $\Delta H° = -42$ kJ mol^{-1}, and $\Delta S° \approx -71$ J mol^{-1} K were determined at pH 10. A Millipore VMWP membrane loaded with an aqueous solution of **54** (2 mM), **55** (6 mM), and 1-methylimidazole (20 mM) facilitated transport of O_2 with a facilitation factor of 1.15 and the O_2/N_2 selectivity of 2.3–2.4. For a membrane loaded with H_2O only the facilitation factor was 1.0 and O_2/N_2 selectivity was 2.0.

Studies with cyclodextrins and later with Colett's cryptophanes showed that it is possible to use ^{129}Xe NMR spectroscopy to obtain quantitative information about complexation processes [88]. NMR data are very sensitive to the environment in which Xe is located, and this can be used to probe structural and dynamic properties of host–guest complexes both in the solid state and in solution. Specifically, in NMR studies of proteins weak Xe–protein interactions influence the ^{129}Xe chemical shift, depending on the accessible protein surface, and enable monitoring of protein void space and conformations. Pines and coworkers recently attached cryptophane-A to biotin and used ^{129}Xe NMR spectroscopy to detect biotin–avidin binding (Fig. 5.26) [89].

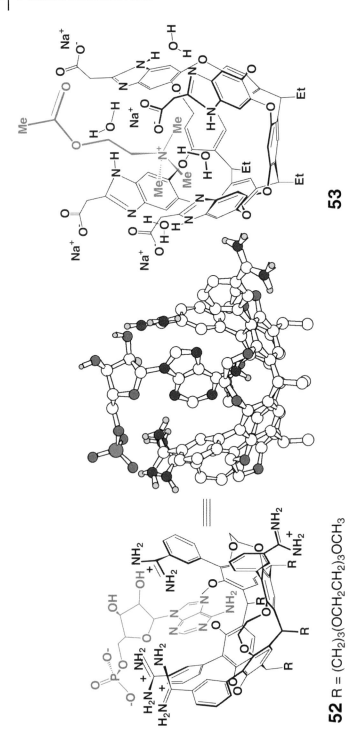

Figure 5.24 Left: Diederich's complex of cavitand **52** with AMP [82]. Right: energy-minimized structure of **52** · AMP in H$_2$O. Right: caviplex **53** · acetyl choline [83].

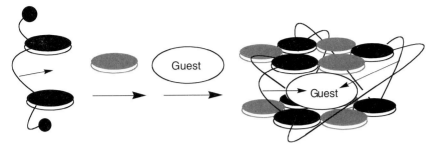

Figure 5.25 Mimicking antibodies with self-assembling cavities and cages [86].

Free Xe in water generates a peak at δ = 193 ppm whereas the **56**·Xe peak is located at δ = 70 ppm. On addition of avidin, a third peak appears 2.3 ppm downfield of the **56** • Xe peak; this was assigned to the protein-bound **56** • Xe complex. When the concentration of avidin was increased the intensity of this peak increased. The proposed method suggests the possibility of attaching different ligands to different cages, thus forming Xe sensors associated with distinct resolved chemical shifts.

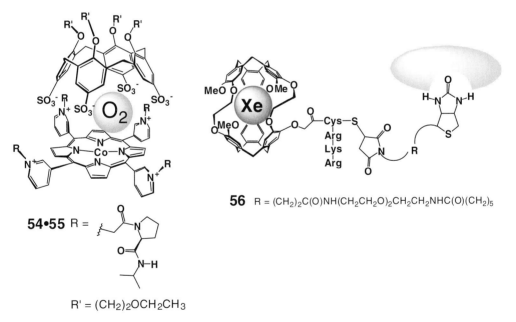

56 R = (CH$_2$)$_2$C(O)NH(CH$_2$CH$_2$O)$_2$CH$_2$CH$_2$NHC(O)(CH$_2$)$_5$

54•55 R =

R' = (CH$_2$)$_2$OCH$_2$CH$_3$

Figure 5.26 Encapsulation of gases. Self-assembling carrier **54•55** for O$_2$ [87]. Cryptophane-A–biotin conjugate **56** and its Xe complex enable monitoring of biotin–avidin binding [89].

5.7
Concluding Remarks

The notion of molecular containers emerged several decades ago. At that time, naturally occurring cyclodextrins were the only cavity-containing hosts known to be able to accommodate smaller guests. The ambitious desire of chemists for totally synthetic molecules enabling great structural variation and control of cavity size and shape led to the discovery of first man-made hollow containers in the early 1980s. Even then, inclusion complexes with small molecules were observed experimentally [90–92]. One of the earliest stable cryptophane complexes, prepared by Collet, had a maximum internal dimension of ~8 Å and a cavity volume of 95 Å^3, and entrapped just one CH_2Cl_2 molecule [90]. Today, a wide variety of nanoscale molecular containers is available. Their internal volumes are as large as 1500 Å^3 and they can entrap guest molecules and ions up to 20 Å long and more than 700 Da in mass. They can also hold several, up to ten, guest species, which can even be different from each other [10]. Both covalent bonds and noncovalent interactions have been used effectively to control size and shape of the interior nanocavities. The kinetic stability of the inclusion complexes can vary significantly but can be tuned by synthesis. Guest exchange can occur within milliseconds or can take hours or days. The uptake and release of guest species sometimes involves breaking the covalent bonds and sometimes just folding and unfolding of the host, which can be achieved by varying either solvent polarity or temperature, or both. It is now also possible to incorporate additional binding and catalytically active sites within the interior. The thermodynamics of the complexes can also be improved.

The impact of molecular containers in chemistry and related fields has been impressive and is expected to grow. Chemists have already made an enormous intellectual contribution to the field of molecular recognition, aggressively attacking problems of intermolecular interactions and generating new supramolecular functions. The functions came with cavities – recognition and binding, then sensing, separation, stabilization, and controlled release of active reagents. Separation and sensing are the first visible applications. We have seen that capsules and cages have been successfully used to detect and monitor industrially important substances such as gases, organic vapor, fullerenes, and other substances. Biologically significant molecules and even some drugs have been complexed; examples with nucleotides, carbohydrates, adamantane derivatives have been mentioned.

For preparative organic chemistry, active reagents can safely be stored inside the capsule for a prolonged time and released easily. Various problems of physical organic chemistry are also being approached. Use of capsules as NMR shift reagents is promising. Molecule-within-molecule complexes can be used for structural information/memory storage. Stabilization of reactive intermediates is another important area under investigation and when two or more guests are encapsulated, reaction/catalytic chambers can be created. In biochemistry and molecular biology, molecular containers are being developed for drug encapsulation, active transport of drugs through cell membranes, and drug delivery. Approaches toward polymeric cavities and capsules and even novel, cavity-based materials have been made. Further modification and application of these materials offer even wider perspectives.

Acknowledgment

I am grateful to the Alfred P. Sloan Foundation for financial support. This work is dedicated to C. David Gutsche.

References

1 (a) D. J. Cram, J. M. Cram, Science 1974, 183, 803–809; (b) D. J. Cram, Science 1983, 219, 1177–1183; (c) C. D. Gutsche, Acc. Chem. Res. 1983, 16, 161–170; (d) D. J. Cram, Angew. Chem., Int. Ed. Engl. 1988, 27, 1009–1020; (e) D. J. Cram, Nature 1992, 356, 29–36.

2 F. D. Cramer, *Rev. Pure Appl. Chem.* **1955**, 5, 143–164.

3 D. J. Cram, J. M. Cram, Container Molecules and their Guests; Royal Society of Chemistry: Cambridge, 1994.

4 Review collection on cyclodextrins: Chem. Rev. 1998, 98, pp. 1741–2076.

5 F. Diederich, *Cyclophanes*; Royal Society of Chemistry: Cambridge, **1991**.

6 Cavitands: D. M. Rudkevich, J. Rebek, Jr., *Eur. J. Org. Chem.* **1999**, 1991–2005.

7 Carcerands and hemicarcerands: (a) A. Jasat, J. C. Sherman, *Chem. Rev.* **1999**, 99, 931–967; (b) R. Warmuth, J. Yoon, *Acc. Chem. Res.* **2001**, 34, 95–105. Structural classification: (c) L. R. MacGillivray, J. L. Atwood, *Angew. Chem. Int. Ed.* **1999**, 38, 1019–1033.

8 Cryptophanes: A. Collet, J.-P. Dutasta, B. Lozach, J. Canceill, *Top. Curr. Chem.* **1993**, 165, 103–129.

9 Self-assembling capsules: (a) F. Hof, S. L. Craig, C. Nuckolls, J. Rebek, Jr., *Angew. Chem. Int. Ed.* **2002**, 41, 1488–1508; (b) M. M. Conn, J. Rebek, Jr., *Chem. Rev.* **1997**, 97, 1647–1668.

10 Nanoscale containers: D. M. Rudkevich, *Bull. Chem. Soc. Jpn.* **2002**, 75, 393–413.

11 Reactions inside cavities: R. Warmuth, *Eur. J. Org. Chem.* **2001**, 423–437.

12 (a) C. D. Gutsche, *Calixarenes Revisited*; Royal Society of Chemistry: Cambridge, **1998**; (b) V. Böhmer, *Angew. Chem., Int. Ed. Engl.* **1995**, 34, 713–745; (c) P. Timmerman, W. Verboom, D. N. Reinhoudt, *Tetrahedron* **1996**, 52, 2663–2704; (d) S. Ibach,

V. Prautzsch, F. Vögtle, C. Chartroux, K. Gloe, *Acc. Chem. Res.* **1999**, 32, 729–740.

13 (a) Imprinted polymers: B. Sellergren, *Angew. Chem. Int. Ed.* **2000**, 39, 1031–1037; (b) Dendrophanes: D. K. Smith, F. Diederich, *Top. Curr. Chem.* **2000**, 210, 183–227; (c) Liposomes: D. D. Lasic, D. Papahadjopoulos, *Science* **1995**, 267, 1275–1276 and literature therein; (d) A. Corma, H. Garcia, *J. Chem. Soc., Dalton Trans.* **2000**, 1381–1394; (e) Antibodies: P. G. Schultz, *Proc. Natl. Acad. Sci. USA* **1998**, 95, 14590–14591 and references therein.

14 (a) J. R. Moran, S. Karbach, D. J. Cram, *J. Am. Chem. Soc.* **1982**, 104, 5826–5828; (b) R. C. Helgeson, M. Lauer, D. J. Cram, *J. Chem. Soc., Chem. Commun.* **1983**, 101–103.

15 (a) P. Soncini, S. Bonsignore, E. Dalcanale, F. Ugozzoli, *J. Org. Chem.* **1992**, 57, 4608–4612; (b) M. Vincenti, C. Minero, E. Pelizzetti, A. Secchi, E. Dalcanale, *Pure Appl. Chem.* **1995**, 67, 1075–1084.

16 F. C. Tucci, D. M. Rudkevich, J. Rebek, Jr., *J. Org. Chem.* **1999**, 64, 4555–4559.

17 (a) D. M. Rudkevich, G. Hilmersson, J. Rebek, Jr. *J. Am. Chem. Soc.* **1998**, 120, 12216–12225; (b) A. Shivanyuk, K. Rissanen, S. K. Körner, D. M. Rudkevich, J. Rebek, Jr., *Helv. Chim. Acta* **2000**, 83, 1778–1790.

18 U. Lücking, F. C. Tucci, D. M. Rudkevich, J. Rebek, Jr., *J. Am. Chem. Soc.* **2000**, 122, 8880–8889.

19 S. D. Starnes, D. M. Rudkevich, J. Rebek, Jr., *J. Am. Chem. Soc.* **2001**, 123, 4659–4669.

20 D. J. Cram, S. Karbach, Y. H. Kim, L. Baczynskyj, G. W. Kalleymeyn, *J. Am. Chem. Soc.* **1985**, 107, 2575–2576.

21 (a) M. E. Tanner, C. B. Knobler, D. J. Cram, *J. Am. Chem. Soc.* **1990**, 112, 1659–1660;

(b) D. J. Cram, M. E. Tanner, C. B. Knobler, *J. Am. Chem. Soc.* **1991**, *113*, 7717–7727.

22 D. J. Cram, M. T. Blanda, K. Paek, C. B. Knobler, *J. Am. Chem. Soc.* **1992**, *114*, 7765–7773.

23 T. A. Robbins, C. B. Knobler, D. R. Bellew, D. J. Cram, *J. Am. Chem. Soc.* **1994**, *116*, 111–122.

24 N. Chopra, J. C. Sherman, *Angew. Chem. Int. Ed.* **1999**, *38*, 1955–1957.

25 D. J. Cram, R. Jaeger, K. Deshayes, *J. Am. Chem. Soc.* **1993**, *115*, 10111–10116.

26 H.-J. Choi, D. Bühring, M. L. C. Quan, C. B. Knobler, D. J. Cram, *J. Chem. Soc., Chem. Commun.* **1992**, 1733–1735.

27 C. von dem Bussche-Hünnefeld, D. Bühring, C. B. Knobler, D. J. Cram, *J. Chem. Soc., Chem. Commun.* **1995**, 1085–1087.

28 S. Mecozzi, J. Rebek, Jr., *Chem. Eur. J.* **1998**, *4*, 1016–1022.

29 (a) D. Philp, J. F. Stoddart, *Angew. Chem., Int. Ed. Engl.* **1996**, *35*, 1154–1196; (b) G. M. Whitesides, B. Grzybowski, *Science* **2002**, *295*, 2418–2421; (c) L. J. Prins, D. N. Reinhoudt, P. Timmerman, *Angew. Chem. Int. Ed.* **2001**, *40*, 2382–2426.

30 N. Branda, R. Wyler, J. Rebek, Jr., *Science* **1994**, *263*, 1267–1268.

31 (a) R. S. Meissner, J. Rebek, Jr., J. de Mendoza, *Science* **1995**, *270*, 1485–1488; (b) R. Meissner, X. Garcias, S. Mecozzi, J. Rebek, Jr. *J. Am. Chem. Soc.* **1997**, *119*, 77–85.

32 (a) K. D. Shimizu, J. Rebek, Jr., *Proc. Natl. Acad. Sci. USA* **1995**, *92*, 12403–12407; (b) O. Mogck, V. Böhmer, W. Vogt, *Tetrahedron* **1996**, *52*, 8489–8496. Review: J. Rebek, Jr., *Chem. Commun.* **2000**, 637–643. X-ray: (c) O. Mogck, E. F. Paulus, V. Böhmer, I. Thondorf, W. Vogt, *Chem. Commun.* **1996**, *52*, 2533–2534.

33 D. M. Rudkevich, In: *Calixarenes 2001*; Z. Asfari, V. Böhmer, J. M. Harrowfield, J. Vicens, Eds., Kluwer Academic Publishers, Dordrecht, **2001**, p.155–180.

34 O. Mogck, M. Pons, V. Böhmer, W. Vogt, *J. Am. Chem. Soc.* **1997**, *119*, 5706–5712.

35 M. O. Vysotsky, I. Thondorf, V. Böhmer, *Angew. Chem. Int. Ed.* **2000**, *39*, 1264–1267.

36 R. K. Castellano, J. Rebek, Jr., *J. Am. Chem. Soc.* **1998**, *120*, 3657–3663.

37 T. Heinz, D. M. Rudkevich, J. Rebek, Jr., *Nature* **1998**, *394*, 764–766.

38 F. Fochi, P. Jacopozzi, E. Wegelius, K. Rissanen, P. Cozzini, E. Marastoni, E. Fisicaro, P. Manini, R. Fokkens, E. Dalcanale, *J. Am. Chem. Soc.* **2001**, *123*, 7539–7552.

39 O. D. Fox, J. F.-Y. Leung, J. M. Hunter, N. K. Dalley, R. G. Harrison, *Inorg. Chem.* **2000**, *39*, 783–790.

40 M. Fujita, K. Umemoto, M. Yoshizawa, N. Fujita, T. Kusukawa, K. Biradha, *Chem. Commun.* **2001**, 509–518.

41 A. Ikeda, M. Yoshimura, H. Udzu, C. Fukuhara, S. Shinkai, *J. Am. Chem. Soc.* **1999**, *121*, 4296–4297.

42 O. D. Fox, E. J. S. Wilkinson, P. D. Beer, M. G. B. Drew, *Chem. Commun.* **2000**, 391–392.

43 L. R. MacGillivray, J. L. Atwood, *Nature* **1997**, *389*, 469–472.

44 A. Shivanyuk, J. Rebek, Jr., *Proc. Natl. Acad. Sci. U. S. A.* **2001**, *98*, 7662–7665.

45 (a) J.-M. Lehn, *Supramolecular Chemistry: Concepts and Perspectives*; VCH: Weinheim, **1995**; (b) J. W. Steed, J. L. Atwood, *Supramolecular Chemistry*, John Wiley and Sons, Chichester, **2000**.

46 P. Timmerman, R. H. Vreekamp, R. Hulst, W. Verboom, D. N. Reinhoudt, K. Rissanen, K. A. Udachin, J. Ripmeester, *Chem. Eur. J.* **1997**, *3*, 1823–1832.

47 (a) L. J. Prins, J. Huskens, F. de Jong, P. Timmerman, D. N. Reinhoudt, D. N. *Nature* **1999**, *398*, 498–502; (b) L. J. Prins, K. A. Jolliffe, R. Hulst, P. Timmerman, D. N. Reinhoudt, *J. Am. Chem. Soc.* **2000**, *122*, 3617–3627.

48 M. Crego Calama, R. Hulst, R. Fokkens, N. M. M. Nibbering, P. Timmerman, D. N. Reinhoudt, *Chem. Commun.* **1998**, 1021–1022.

49 F. Cardullo, M. Crego Calama, B. H. M. Snellink-Ruël, J.-L. Weidmann, A. Bielejewska, R. Fokkens, N. M. M. Nibbering, P. Timmerman, D. N. Reinhoudt, *Chem. Commun.* **2000**, 367–368.

50 K. A. Jolliffe, P. Timmerman, D. N. Reinhoudt, *Angew. Chem., Int. Ed. Engl.* **1999**, *38*, 933–937.

51 Polymeric rod-like nanostructures were obtained from calix4arene bismelamines and biscyanurates, see: H.-A. Klok, K. A. Jolliffe, C. L. Schauer, L. J. Prins, J. P. Spatz, M. Möller, P. Timmerman, D. N. Reinhoudt, *J. Am. Chem. Soc.* **1999**, *121*, 7154–7155.

52 D. J. Cram, M. E. Tanner, R. Thomas, *Angew. Chem., Int. Ed. Engl.* **1991**, *30*, 1024–1027.

53 (a) R. Warmuth, *Angew. Chem., Int. Ed. Engl.* **1997**, *36*, 1347–1350; (b) R. Warmuth, *Chem. Commun.* **1998**, 59–60.

54 D. A. Makeiff, K. Vishnumurthy, J. C. Sherman, *J. Am. Chem. Soc.* **2003**, *125*, 9558–9559.

55 T. Kusukawa, M. Fujita, *J. Am. Chem. Soc.* **1999**, *121*, 1397–1398.

56 T. Heinz, D. M. Rudkevich, J. Rebek, Jr., *Angew. Chem. Int. Ed.* **1999**, *38*, 1136–1139.

57 M. Yoshizawa, T. Kusukawa, M. Fujita, K. Yamaguchi, *J. Am. Chem. Soc.* **2000**, *122*, 6311–6312.

58 S. K. Körner, F. C. Tucci, D. M. Rudkevich, T. Heinz, J. Rebek, Jr., *Chem. Eur. J.* **2000**, *6*, 187–195.

59 M. Ziegler, J. L. Brumaghim, K. N. Raymond, *Angew. Chem. Int. Ed.* **2000**, *39*, 4119–4121.

60 J. Kang, G. Hilmersson, J. Santamaria, J. Rebek, Jr., *J. Am. Chem. Soc.* **1998**, *120*, 3650–3656.

61 J. Kang, J. Santamaria, G. Hilmersson, J. Rebek, Jr., *J. Am. Chem. Soc.* **1998**, *120*, 7389–7390.

62 T. Kusukawa, T. Nakai, T. Okano, M. Fujita, *Chem. Lett.* **2003**, *32*, 284–285.

63 S. P. Kim, A. G. Leach, K. N. Houk, *J. Org. Chem.* **2002**, *67*, 4250–4260.

64 R. Warmuth, E. F. Maverick, C. B. Knobler, D. J. Cram, *J. Org. Chem.* **2003**, *68*, 2077–2088.

65 G. V. Zyryanov, D. M. Rudkevich, *Org. Lett.* **2003**, *5*, 1253–1256.

66 G. V. Zyryanov, Y. Kang, D. M. Rudkevich, *J. Am. Chem. Soc.* **2003**, *125*, 2997–3007.

67 J. Chen, S. Körner, S. L. Craig, D. M. Rudkevich, J. Rebek, Jr., *Nature* **2002**, *415*, 385–386.

68 A. R. Pease, J. F. Stoddart, *Struct. Bonding* **2001**, *99*, 189–236.

69 (a) P. Timmerman, W. Verboom, F. C. J. M. van Veggel, J. P. M. van Duynhoven, D. N. Reinhoudt, *Angew. Chem. Int. Ed. Engl.* **1994**, *33*, 2345–2348. Self-assembling monolayers of carceplexes on gold: B.-H. Huisman, D. M. Rudkevich, A. FarrÁn, W. Verboom, F. C. J. M. van Veggel, D. N. Reinhoudt, *Eur. J. Org. Chem.* **2000**, 269–274.

70 F. C. Tucci, D. M. Rudkevich, J. Rebek, Jr., *J. Am. Chem. Soc.* **1999**, *121*, 4928–4929.

71 A. Scarso, A. Shivanyuk, O. Hayashida, J. Rebek, Jr., *J. Am. Chem. Soc.* **2003**, *125*, 6239–6243.

72 R. K. Castellano, S. L. Craig, C. Nuckolls, J. Rebek, Jr., *J. Am. Chem. Soc.* **2000**, *122*, 7876–7882.

73 (a) R. Rathore, S. V. Lindeman, K. S. S. Rao, D. Sun, J. K. Kochi, *Angew. Chem. Int. Ed.* **2000**, *39*, 2123–2127; (b) S. V. Rosokha, J. K. Kochi, *J. Am. Chem. Soc.* **2002**, *124*, 5620–5621; (c) S. V. Rosokha, S. V. Lindeman, R. Rathore, J. K. Kochi, *J. Org. Chem.* **2003**, *68*, 3947–3957.

74 (a) M. Suman, M. Freddi, C. Massera, F. Ugozzoli, E. Dalcanale, *J. Am. Chem. Soc.* **2003**, *125*, 12068–12069; (b) R. Paolesse, C. Di Natale, S. Nardis, A. Macagnano, A. D'Amico, R. Pinalli, E. Dalcanale, *Chem. Eur. J.* **2003**, *9*, 5388–5395.

75 A. Rafai Far, A.; Y. L. Cho, A. Rang, D. M. Rudkevich, J. Rebek, Jr., *Tetrahedron* **2002**, *58*, 741–755.

76 F. Bianchi, R. Pinalli, F. Ugozzoli, S. Spera, M. Careri, E. Dalcanale, *New J. Chem.* **2003**, *27*, 502–509.

77 R. K. Castellano, D. M. Rudkevich, J. Rebek, Jr., *Proc. Natl. Acad. Sci. USA* **1997**, *94*, 7132–7137. Review on supramolecular polymers: A. T. ten Cate, R. P. Sijbesma, *Macromol. Rapid Commun.* **2002**, *23*, 1094–1112.

78 (a) R. K. Castellano, C. Nuckolls, S. H. Eichhorn, M. R. Wood, A. J. Lovinger, J. Rebek, Jr., *Angew. Chem., Int. Ed. Engl.* **1999**, *38*, 2603–2606; (b) R. K. Castellano, R. Clark, S. L. Craig, C. Nuckolls, J. Rebek, Jr., *Proc. Natl. Acad. Sci. USA* **2000**, *97*, 12418–12421.

79 H. Xu, S. P. Stampp, D. M. Rudkevich, *Org. Lett.* **2003**, *5*, 4583–4586.

80 H. Xu, E. M. Hampe, D. M. Rudkevich, *Chem. Commun.* **2003**, 2828–2829.

81 T. Haino, D. M. Rudkevich, A. Shivanyuk, K. Rissanen, J. Rebek, Jr., *Chem. Eur. J.* **2000**, *6*, 3797–3805.

82 L. Sebo, F. Diederich, V. Gramlich, *Helv. Chim. Acta* **2000**, *83*, 93–113.

83 F. Hof, L. Trembleau, E. C. Ullrich, J. Rebek, Jr., *Angew. Chem. Int. Ed.* **2003**, *42*, 3150–3153.

84 O. Hayashida, L. Sebo, J. Rebek, Jr., *J. Org. Chem.* **2002**, *67*, 8291–8298.

85 T. Ishi-i, M. A. Mateos-Timoneda, P. Tim-
    merman, M. Crego-Calama, D. N. Rein-
    houdt, S. Shinkai, *Angew. Chem. Int. Ed.*
    2003, *42*, 2300–2305.

86 P. Timmerman, D. N. Reinhoudt, *Adv.
    Mater.* **1999**, *11*, 71–74.

87 R. Fiammengo, K. Wojciechowski,
    M. Crego-Calama, P. Timmerman,
    A. Figoli, M. Wessling, D. N. Reinhoudt,
    Org. Lett. **2003**, *5*, 3367–3370.

88 (a) K. Bartik, M. Luhmer, J.-P. Dutasta,
    A. Collet, Reisse, *J. Am. Chem. Soc.* **1998**,
    120, 784–791; (b) T. Brotin, A. Lesage,
    L. Emsley, A. Collet, *ibid.* **2000**, *122*,
    1171–1174; (c) T. Brotin, J.-P. Dutasta, *Eur.
    J. Org. Chem.* **2003**, 973–984.

89 M. M. Spence, S. M. Rubin, I. E. Dimitrov,
    E. J. Ruiz, D. E. Wemmer, A. Pines, S. Q.
    Yao, F. Tian, P. G. Schultz, *Proc. Natl.
    Acad. Sci. USA* **2001**, *98*, 10654–10657.

90 First cryptophanes: (a) J. Gabard, A. Collet,
    J. Chem. Soc., Chem. Commun. **1981**,
    1137–1139; (b) J. Canceill, M. Cesario,
    A. Collet, J. Guilhem, C. Pascard, *J. Chem.
    Soc., Chem. Commun.* **1985**, 361–363 (X-ray
    studies); (c) J. Canceill, L. Lacombe,
    A. Collet, *J. Am. Chem. Soc.* **1985**, *107*,
    6993–6996 (solution studies).

91 First speleands: J. Canceill, A. Collet,
    J. Gabard, F. Kotzyba-Hibert, J.-M. Lehn,
    Helv. Chim. Acta **1982**, *65*, 1894–1897.

92 For an early study of calixarene complexes
    in solution, see: L. J. Bauer, C. D. Gutsche,
    J. Am. Chem. Soc. **1985**, *107*, 6063–6069.

6
Formation and Recognition Properties of Dynamic Combinatorial Libraries

Andrew D. Hamilton, Debarati M. Tagore, and K. Ingrid Sprinz

6.1
Introduction

Combinatorial chemistry was first introduced as a way of synthesizing a large number of structurally-related compounds and then quickly screening them against a target of interest. The technology has now evolved to be a critical component of drug-discovery processes. A combinatorial library by definition consists of building blocks linked together to generate diverse combinations of library members. The libraries are generally produced by two methods – split-mix and parallel synthesis (Fig. 6.1A and B). In the split-mix technique, diversity is generated by sequential steps of separation, reaction, and recombination to obtain a large library whereas in parallel syntheses the library members are made in separate containers and can be screened individually or pooled and subjected to a target of interest [1].

To circumvent the time and extensive synthetic chemistry involved in the development of these libraries, dynamic combinatorial chemistry (DCC) was developed; in this method relatively few building block components can form reversible linkages and thus interconvert to form multiple distinct species in a library (Fig. 6.1C+D).

DCC can be described as combinatorial chemistry in which target recognition can be used to achieve selectivity and amplification of selected species [2].

Dynamic combinatorial libraries (DCL) have been developed for many applications including the screening and isolation of active peptide inhibitors of antibodies [3]. The library designs have further evolved to enable considerable success in the discovery of drug molecules. DCL are based on the concept that the library members can participate in reversible interactions and therefore many combinations can be made from the initial building blocks. Three important aspects of a DCL are: first, it uses organic fragments that can form reversible linkages with the other components; second, the reaction conditions must be conducive for the fragments to form linkages to each other; and, third, the library must be subjected to selection by a target of interest. Reversibility in a dynamic library can be introduced either by covalent or nonco-

Functional Synthetic Receptors. Edited by T. Schrader, A. D. Hamilton
Copyright © 2005 WILEY-VCH Verlag GmbH & Co. KGaA, Weinheim
ISBN: 3-527-30655-2

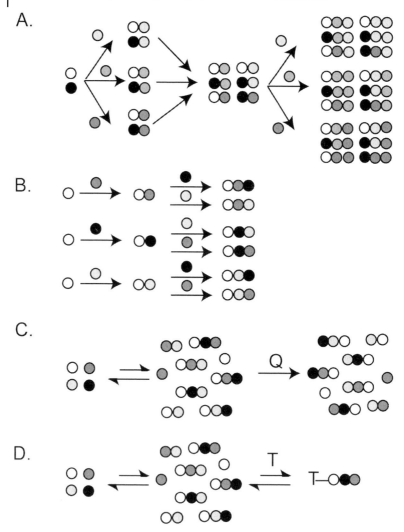

Figure 6.1 Differences between a combinatorial library (A, B) and a dynamic combinatorial library (C, D). A. Split-mix technique. B. Parallel synthesis. C. Dynamic combinatorial chemistry where Q is a quenching agent to transform the dynamic library to a static library. D. Template-assisted synthesis amplifies the best binding species from the library because the species is removed from the pool, thus causing the pool to re-equilibrate, making more of the binding species.

valent interactions. Covalent interactions suitable for DCLs include: disulfide bond formation, transimination, transesterification, oxime and hydrazone exchange, and olefin metathesis. Appropriate noncovalent forces include: metal coordination and hydrogen-bonding interactions. Another aspect that is critical to a dynamic library is the ability not only to control reversible exchange between the library members but also to "quench" and produce a "static" library by altering reaction conditions

(Fig. 6.1C). Compatibility with biological targets often requires aqueous conditions and the need for linkages that are stable in the presence of water and salts. DCL formed by transimination and acyl hydrazone exchange, disulfide interchange, olefin cross metathesis, and, to a lesser extent, boronate transesterification are all compatible with biological systems. Rearrangement of library members can be triggered by changing the pH, oxidation/reduction conditions, or by use of a catalyst. Many types of macromolecules have been used as targets, including enzymes, nucleic acids, and carbohydrates.

The term "virtual combinatorial library" (VCL) is frequently used to describe the theoretical number of different combinations that can potentially be obtained from a library of building blocks in dynamic equilibrium [4]. When, however, a "target" molecule is added (for example, a biopolymer or a small molecule), the products that bind strongly with the target will be removed from the equilibrium, thus driving further formation of the strong binders [5–7]. Unlike conventional combinatorial libraries in which one preformed molecule binds to its target, the dynamic and adaptive nature of VCL results in an increase in the molecular diversity. The library of compounds generated can be utilized in three different ways.

One possible use of a library involves selection from an interconverting mixture followed by further rearrangement and reselection of the unselected members, thus enabling evolution and amplification. This approach is seen in configurational/ conformational isomerization systems which are discussed later in this chapter (Sect. 6.4). Another strategy involves the recognition of small molecule targets through a process called "molding" in which the target acts as a scaffold around which suitable library components will assemble to form a receptor for the target (Fig. 6.2). Examples of this method are discussed later in this chapter (Sect. 6.5). In a

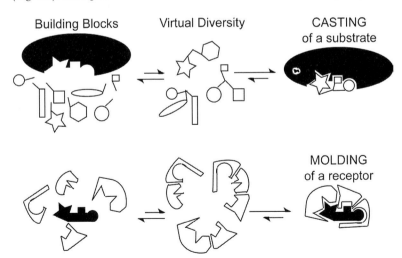

Figure 6.2 Schematic representation of the casting and molding processes. In the casting process (top), the receptor selects the best binding species from a library of diverse com- binations. In the molding process (bottom), a templating substrate induces the formation of specific complementary receptors.

process called "casting," the building blocks are assembled trough dynamic exchange to form a good fit for the target. For example, substrates for biopolymer active sites are designed where the product that will be amplified is the one that fits into the binding site of the biopolymer.

6.2
Covalent Interactions Used in DCC Design

6.2.1
Acyl Hydrazone and Imine Exchange

6.2.1.1 Biological Applications
An example of "casting" was reported by Huc and Lehn who created a dynamic library of imines, from a series of amines and aldehydes, as inhibitors of the Zn^{2+} metalloenzyme carbonic anhydrase II (CAII) [5]. The amines were used in excess compared with the aldehydes to compensate for side reactions with lysine residues on the enzyme. CAII contains a hydrophobic pocket above the zinc-binding site and previous studies have shown that aryl derivatives containing a sulfonamide group at the para-position are good inhibitors of CAII because the sulfonamide coordinates the Zn^{2+}. To facilitate the characterization of the best binders from the library, the imines were reduced with $NaBH_3CN$ to prevent any hydrolysis, thus the dynamic interconversion was "quenched". The components of the library were designed to have comparable polarities and reactivity. The final products, after reduction of the imine with $NaBH_3CN$, absorbed at 220–240 nm and hence detection was straightforward. All the library components were incubated in the absence and presence of the enzyme and the two spectral traces were compared. The library's highest affinity molecule for CAII fits the binding pocket based on earlier structural details of CAII inhibitors.

Because potent sulfonamide inhibitors of CAII had previously been identified, the protein was investigated for its potential to template the synthesis of its own high affinity inhibitors [8]. The library was generated by reacting thiol-containing sulfonamides at the para-position with different alkyl halides. The inhibition constants were determined individually by use of a colorimetric assay. Template-directed synthesis was performed with the library in the absence and presence of the enzyme where reversed-phase high performance liquid chromatographic (RP-HPLC) analysis showed that with the enzyme an increase in the strongest inhibitor concentration occurred. The thiol presumably binds to the active site of the enzyme and recruits the binding partner that best fits the hydrophobic binding area above the active site. The irreversible thiol alkylation reaction led to enhancement of certain library species and clearly selected for the different affinities of the library components.

Hochgurtel et al. explored the idea of making a virtual dynamic library in which hits are generated only in the presence of a target [9]. This was tested using influenza A virus neuraminidase as the target of an imine library. This library comprised an amine scaffold and many alkyl- and aryl-substituted aldehydes, based on the structure of a commercially available drug for influenza (Fig. 6.3).

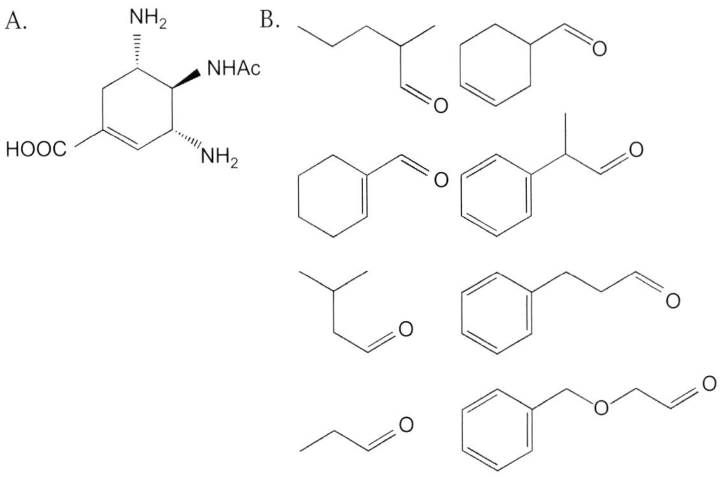

Figure 6.3 A. Structure of the amine scaffold used in the generation of the library. B. Aldehyde building blocks used in the library.

A transient library was formed in the presence of the enzyme, and consisted of rapidly equilibrating imine and hemiaminal species. Reduction of the imines with tetrabutylammonium cyanoborohydride (TBC) resulted in the irreversible formation of stable amines. Because equilibration occurs in the presence of the target, reduction gives rise to a biased library enriched in the binding species. Detection of components is often difficult when the library is large and diverse; hence conditions were explored to reduce the library of imines slowly so that significant amounts of any one combination were not reduced at any time. Because not all the components of the library could be detected by HPLC–MS, this represented a virtual combinatorial library. Enrichment of the binding species was calculated to be 120-fold in the presence of the protein as compared to the absence of the protein. The amplified products had structural properties similar to those of a commercially available inhibitor of neuraminidase (Fig. 6.4).

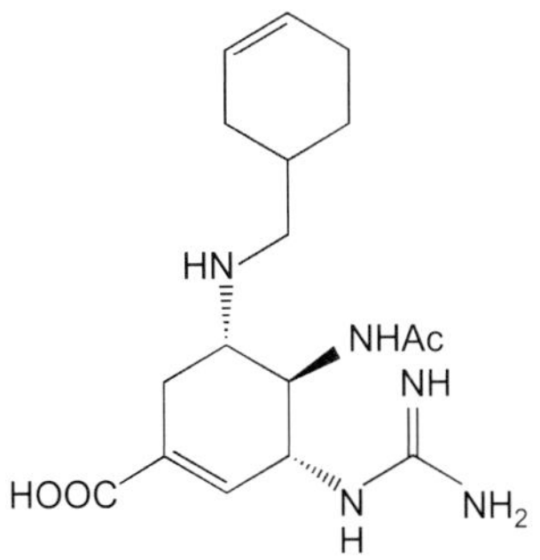

Figure 6.4 The product obtained after selection by neuraminidase.

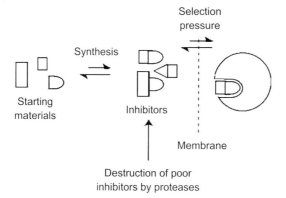

Figure 6.5 Schematic representation of the amplification of compounds binding with high affinity to the target by proteolytically removing the weak binders from solution.

In the generation and screening of DCL members, the detection process would be facilitated if there were a way of eliminating the weak binders from the library. Cheeseman et al. have tested such an approach by incubating their library of dipeptides with CAII and then proteolytically destroying compounds with little or no affinity (Fig. 6.5) [10].

Five dipeptides were synthesized so that each dipeptide contains a 4'-sulfonamidophenylalanine at the N-terminus. These were tested individually to determine the affinity of each compound. To demonstrate that the enzyme could recruit the best binders, an experiment was devised in which a membrane separated two chambers, one containing CAII and the other containing a buffer solution. The membrane allowed transport of small molecules but prevented the enzyme from crossing the barrier. Both chambers contained equal concentrations of good and poor binding molecules. With time, the concentration of the strong binding molecules increased in the chamber containing CAII whereas the concentration of the poor binding molecules did not vary appreciably. This was further continued by screening all five dipeptides. The poor binders were destroyed by use of a protease (pronase) that cleaved all five dipeptides at a reasonable rate. The same experiment using a membrane between two compartments was repeated but using pronase in the chamber without CAII. If the enzyme could selectively bind the good inhibitors there would be an increased concentration of these dipeptides in the chamber containing CAII and hence a decrease in their cleavage by pronase. A decrease in the proteolysis rate of binding molecules that bound strongly to CAII was clearly demonstrated.

To extend this approach, Corbett et al. used compartmentalization to synthesize, select and amplify dipeptides that bind to CAII [11]. A vessel, divided into three chambers by semipermeable membranes, was used in the process of generating the dipeptides, screening for binding to CAII, and then destruction of non-binders. The dipeptides were synthesized in the first chamber and then allowed to diffuse into the second chamber containing CAII. The nonbinding dipeptides passed through this second chamber and into the third chamber which contained pronase. The cleaved dipeptides diffused back into the synthesis chamber and the process was repeated resulting in amplification of CAII binding dipeptides.

Bunyapaiboonsri et al. created a system to generate inhibitors of acetylcholinesterase (AChE), to efficiently screen the VCL, and to identify lead compounds [12]. AChE hydrolyzes acetylcholine and hence terminates the transmission of nerve impulses to the brain. The active site contains a deep hydrophobic gorge lined by aromatic residues with another binding site in proximity. Reversible acyl hydrazone connections between a series of aldehydes and hydrazide components were used to generate the DCL. The building blocks had quaternary ammonium groups reflective of key features in previous inhibitors. The functionalities used for protein recognition were varied along with their separation and relative orientation by implementing linkers with different flexibilities. After DCL formation in the presence of the enzyme, a method of dynamic deconvolution was used to identify the building blocks from the protein-binding species. The components contained both ammonium and pyridinium groups which were reacted in the presence of dialdehyde linkers. At acidic pH, rapid interchange occurred compared with neutral and basic conditions. To efficiently evaluate and deconvolute the binding components from the pool, one building block was removed at a time causing the library to re-equilibrate. A decrease in inhibition was detected if a component of the inhibiting species was removed. Using this strategy, a bis-pyridinium inhibitor (1) with a low nM affinity for AChE was identified (Fig. 6.6).

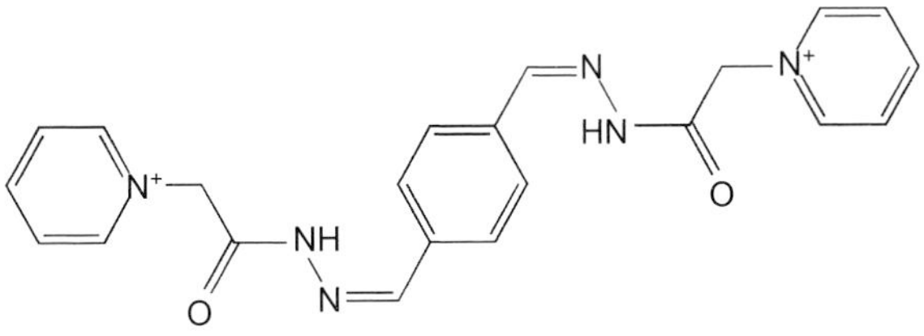

Figure 6.6 High-affinity compound **1** obtained from a DCL of acyl hydrazones that targets AChE with low nM affinity.

The main problem with the design of imine combinatorial libraries is that they are susceptible to hydrolysis. The presence of an enzyme with imine reductase activity can, however, convert the imines to hydrolytically-stable amines, which shifts the equilibrium toward enrichment of the particular substrate that gave the reduced amine. Enzyme-catalyzed reduction of a library of imines to secondary amines has been studied by Li et al. in work in which an anaerobic bacterium, *A. woodii*, was screened for "imine reductase" activity because of its ability to reduce carbon–carbon double bonds [13]. Incubation of the imine library with *A. woodii* gave two reduced amines in small quantities when the bacterium had been induced by caffeate (Fig. 6.7). One of the two amines was also obtained in the absence of caffeate, possibly indicating the occurrence of another imine reductase activity. Even though the amount of product formed was very low and side-products were also seen, this system has the potential to generate novel enzyme activity.

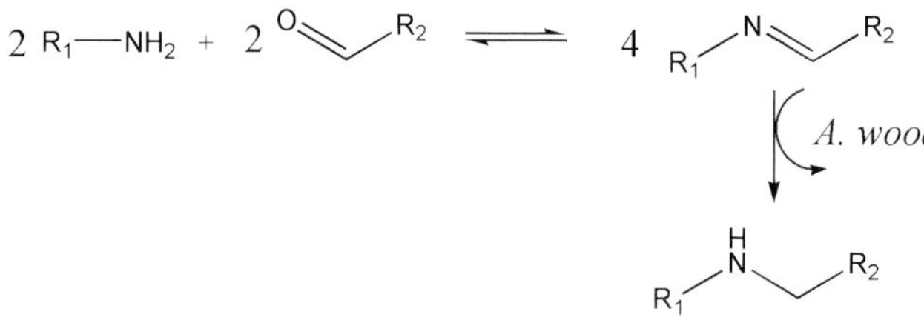

Figure 6.7 Reaction catalyzed by an enzyme with "imine reductase" activity.

The chemistry of DNA replication is of vital interest to biologists and if a reaction resembles the template-directed synthesis of nucleic acids, insight into the biological process can be obtained [14]. Goodwin et al. synthesized a piece of DNA, three nucleotides long with a 3'-allyl-3'-deoxythymidine molecule tethered at the 3' terminus. A separate DNA trimer was modified with 5'-amino-5'deoxythymidine at the 5'- end. Reaction of the allyl-modified oligonucleotide with OsO$_4$ and NaIO$_4$ generates an aldehyde that can then react with the amine-modified trimer to give an imine hexamer product. Imine formation was performed with and without the complementary hexamer. Imine instability was circumvented by reduction with NaBH$_3$CN to give an amine-tethered hexamer which was detected by high-performance liquid chromatography (HPLC). At different temperatures and in the presence of complementary hexamer, the product was amplified. At 0 °C, the product increased by 65% which correlates with DNA duplex stability at lower temperatures. The results were also compared with a parallel reaction that contained a non-complementary amine-modified tetramer competing with the 5'-amine trimer by reacting with the 3'-aldehyde trimer. Hexamer formation as a result of base complementarity was favored over non-complementary heptamer formation.

Generation of a dynamic library of carbohydrates using enzyme catalysis was shown by Lens et al., who used N-acetylneuraminic acid (NANA) aldolase to form different carbohydrate species that were screened against a plant lectin, wheat germ aglutinin (WGA) [15]. The reaction was catalyzed by NANA aldolase which converts N-acetylneuraminic acid to N-acetylmannosamine and pyruvate. WGA binds N-acetylneuraminic acid with millimolar affinity and hence it was envisaged that the enzyme would "amplify" N-acetylmuraminic acid over other library members. This was tested using N-acetylmannosamine and D-mannose as substrates. An enzyme-catalyzed aldol reaction in the presence of sodium pyruvate resulted in equal formation of the aldol products. When the products were incubated with WGA and analyzed by HPLC it was observed that, with time, WGA suppressed the generation of the non-binding component in favor of N-acetylmuraminic acid, which binds to WGA. Two more analogs of N-acetylmuraminic acid, D-lyxose, and D-galactose, were tested against WGA. The aldol products generated from all four compounds were incubated with WGA and the enzyme selectivity amplified N-acetylmuraminic acid from the equilibrium mixture.

6.2.1.2 **Chemical Applications**

Hydrazone reactions can also be templated by metal cations or crown ethers. In one study, Furlan et al. found that when compound **2** was mixed with trifluoroacetic acid (TFA) as catalyst, multiple oligomers were detected by ESI-MS and HPLC but when the crown ether [18]crown-6 was added to the mixture, a formerly undetectable compound **3** became the major product formed (67%) (Fig. 6.8) [16].

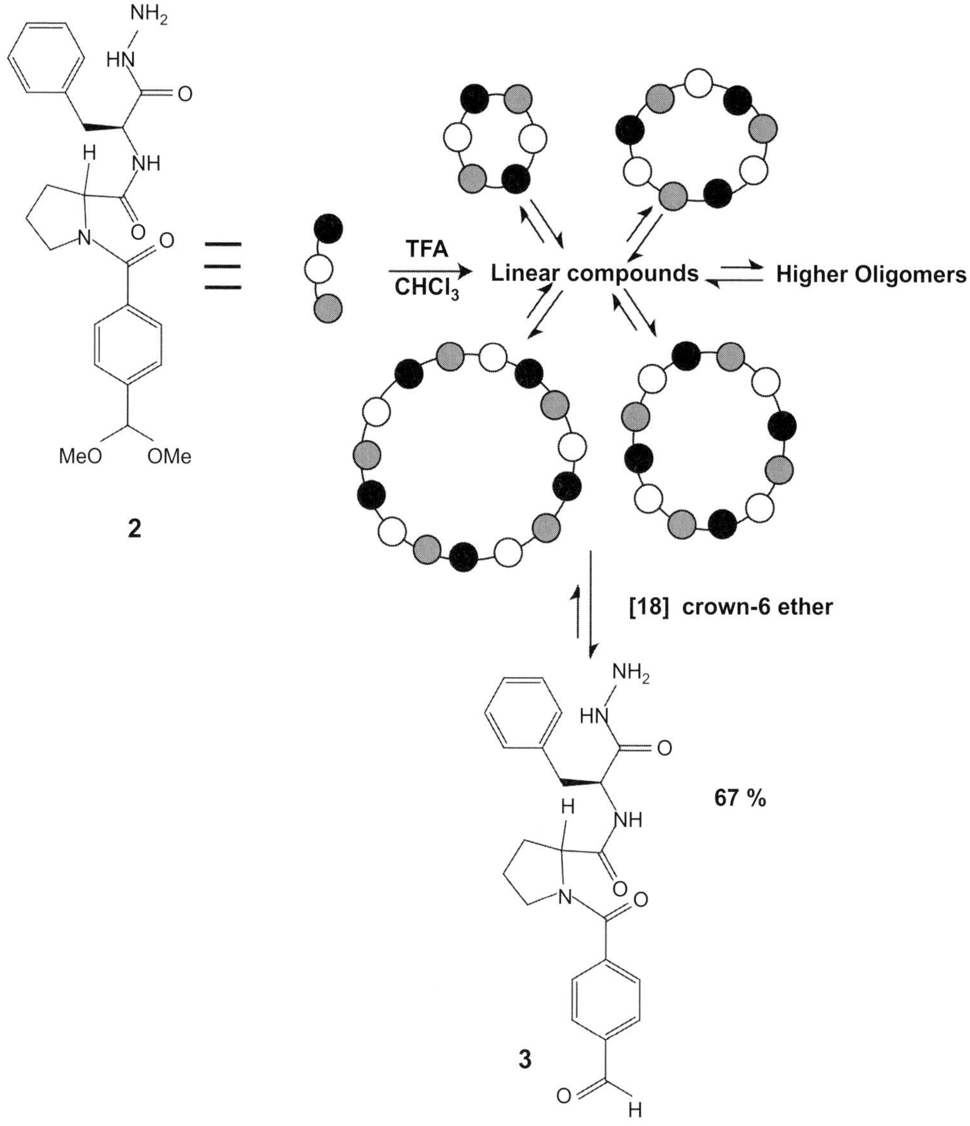

Figure 6.8 Generation of higher-order macrocycles of **2** in presence of TFA. Product **3**, is detected and amplified only in the presence of [18] crown-6-ether template.

The Sanders group then took a peptide-hydrazone DCL of interchanging macrocycles and demonstrated the selective amplification of certain members in the presence of Li$^+$ and Na$^+$ ions [17–19]. In the absence of metal ions, formation of dimers, trimers, tetramers, and higher-order macrocycles was observed for some monomers but addition of LiI or NaI led to the predominant formation of cyclic trimers in two cases and of cyclic dimers in another. Complex formation was analyzed by nuclear magnetic resonance (NMR), electrospray ionization mass spectrometry (ESI-MS), isothermal calorimetry (ITC), HPLC and, infrared spectroscopy (IR) and amplification was shown to occur when the macrocycle binds the metal ion, thus causing the geometry of the complex to change. Formation and amplification of either dimer or trimer depends on the coordination of the metal ion to the carbonyl groups in the macrocycles. In another study, the group found that addition of two different monomers in TFA produced 75 different macrocycles [20, 21]. Reversibility of the hydrazone exchange reaction and the dynamic nature of the macrocyclic library were shown when two different solutions of monomers were mixed to produce oligomers which lacked free monomer. Combination of the two oligomeric solutions resulted in the formation of mixed macrocycles which occur if the preexisting oligomers undergo fragmentation and rearrangement. The library was converted from "dynamic" to "static" by adding base (Et$_3$N) to quench the TFA in the solution and hence stop the hydrazone exchange. Characterization of

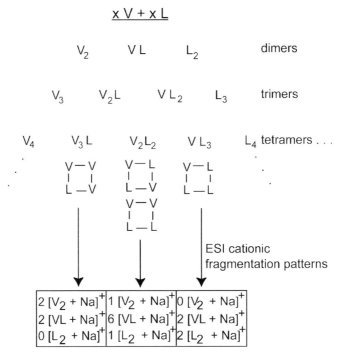

Figure 6.9 All possible combinations from multiple ligands and all possible cationic fragments generated by ESI. Other than the parent ion, the fragment ions are generated by dimers. V$_2$L$_2$ can exist as two different sequence isomers.

the macrocyclic library was found to be more accurate using electrospray ionization Fourier-transform ion cyclotron resonance tandem mass spectrometry (ESI-FTICR-MS). A library of macrocycles generated from monomers V and L should have a composition $V_x L_y$, V_x, L_y (x, y are the number of monomer units in the macrocycle). From tetramers formed in the library, those with the composition $V_2 L_2$ can exist as sequence isomers with the fragmentation pattern shown in Fig. 6.9.

The presence of sequence isomers in the macrocyclic library was demonstrated using the sensitive FTICR technique. Sanders and colleagues have also shown the amplification of a single compound from a DCL of different equilibrating components, in response to N-methyl alkylammonium salts [16, 22]. A pseudo-peptide cyclic hydrazone library was generated by adding TFA to a solution of 4 in CH_2Cl_2 (Fig. 6.10).

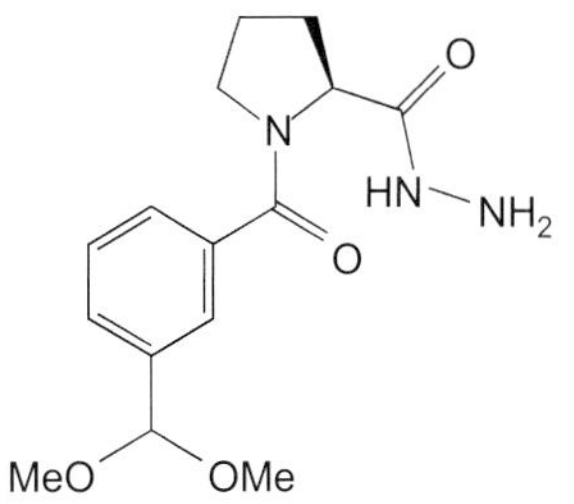

Figure 6.10   Monomer 4 used for the generation of the macrocycles.

Hydrazone exchange gave many different kinetic products, but over time the reaction came to thermodynamic equilibrium and only the cyclic dimer was produced.

Cyclization of 4 in TFA and in the presence of N-methyl quinuclidinium (NMQ) iodide, however, led to the formation of the cyclic trimers in 56% yield. When acetylcholine was used as the source of cations, the cyclic trimer was 86% of the total species formed. The template effect was verified by mass spectrometry (MS) and NMR studies and it is believed that the C=O bond of the pseudo-peptide and the methylene group of the ammonium salts associate by hydrogen bonding thus driving product formation and molecular amplification. To further ensure that final product distribution was a result of thermodynamic equilibrium, a crown ether was used to complex the hydrazinium terminus of the linear macrocyclic hydrazone molecules when they are interconverting between the linear and cyclic forms during the equilibrium process. The complexation was detected by HPLC and ESI-MS and it was seen that the library was dynamic and interconverting.

Two important criteria of DCL are that the different members of the library should be present in comparable amounts and there should be random distribution of the different components. Nazarpack-Kandlousy et al. have generated a library of oxime ethers that have the potential to be a DCL (Fig. 6.11) [23].

Figure 6.11   Oxime exchange reaction.

They also studied reversible imine exchange of O-aryl and O-alkyl oximes and investigated the thermodynamics and kinetics of the reaction. Reversible imine exchange reactions are ideal for a DCL because the rate at which equilibration is reached in these compounds is low, hence they can be isolated easily. The rate of hydrolysis of oximes is negligible under the conditions used for generating the equilibrating mixture.

6.2.2
Transesterification

Sanders created a "living" macrolactonization library of macrocycles under equilibrium conditions using reversible transesterification [24]. The library was "living", because on addition of new building blocks or templates components exchanged to form new template-directed ring sizes. By making a covalently linked supramolecular macrocyclic library and letting the library interact with a predetermined template, the size of the ring system could be changed. As a result of increased stabilization the equilibrium will be pushed towards the template-directed products over the assembly of other products, thus amplifying the desired macrocycle. Macrocycles are interesting compounds to perform binding experiments, because of the geometry of functional groups and the increased rigidity of the structure. The yield is usually fairly low for this reaction, so templated steps can help to increase the synthetic efficiency. Adding a template can bring the components of the macrocycle close enough, thus enabling the ends to react. Brady and Sanders made a series of macrocycles composed of cholates linked by ester groups to form different oligomers [25]. If the monomeric cholic acid derivatives are subjected to transesterification conditions under thermodynamic control, a variety of linear and cyclic esters would be obtained. To circumvent this, the design was based on ester-exchange reactions described in the equations:

$$RCO_2Me + R'OH \rightarrow RCO_2R' + MeOH$$
$$RCO_2R' + R''OH \rightleftharpoons RCO_2R'' + R'OH$$

Removal of methanol drives the formation of oligomers in the presence of sodium methoxide as the transesterification catalyst. The precise ratio of different oligomers depends on the nature of substituents at the C7 and C12 positions of the cholate. Furthermore, a templating effect has been demonstrated using alkali metal cations. The templating approach allows a change in the thermodynamic equilibrium because only the components that bind the template best are favored.

The problem with large receptor libraries is that only a few rigid components can be used which leads to self-assembly and a decrease in the level of diversity. The Sanders group extended their library by using building blocks with unique detection features and concave shapes [26]. These building blocks can react through base-catalyzed transesterification to give diverse library members. They had previously found that individual solutions of quinine- and cinchondine-derived components underwent transesterification and each formed macrocyclic trimers exclusively. In con-

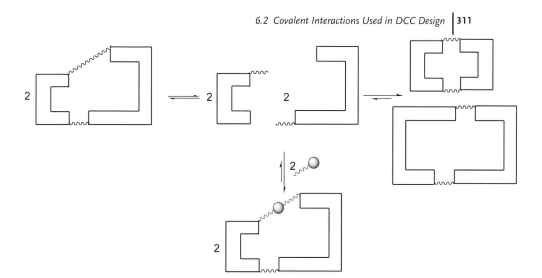

Figure 6.12 Amplification of the heterodimer in the presence of linker. In the absence of linker only homodimer formation occurs, because of incompatible bite size.

trast, the quinidine-derived monomers only formed dimers (Fig. 6.12). Systematic mixing of two different components gave different results ranging from self-assembly to diverse combinations within the macrocycles. This effect of self-sorting shows that even though the two building blocks are the same shape, the "bite size" is sufficiently different such that linking them causes too much strain whereas the two building blocks with the same bite size match, allowing macrocyclization to occur. To avoid self-association and therefore to increase diversity, small, flexible monomers were added to the mixture of larger monomers, with the expectation that they would ease the strained connection by bridging the different sized building blocks. As a result, a range of oligomeric species was detected.

6.2.3
Disulfides

Disulfide exchange reactions are well suited for the generation of dynamic libraries of covalently linked compounds, because of their compatibility with aqueous solutions, the mild reaction conditions, and the ability to control the extent of scrambling of library members. Disulfide exchange can be initiated at slightly basic pH and can be quenched at an acidic pH. Hioki and Still used a library of immobilized tripeptides and screened them for binding to a disulfide-linked small molecule receptor A-SS-A to find the receptor that can selectively bind peptides [27]. To test the evolution of disulfides in the presence of a target peptide, a mixed disulfide was prepared by reacting A-SH and thiophenol. (Fig. 6.13, left, and Fig. 6.14).

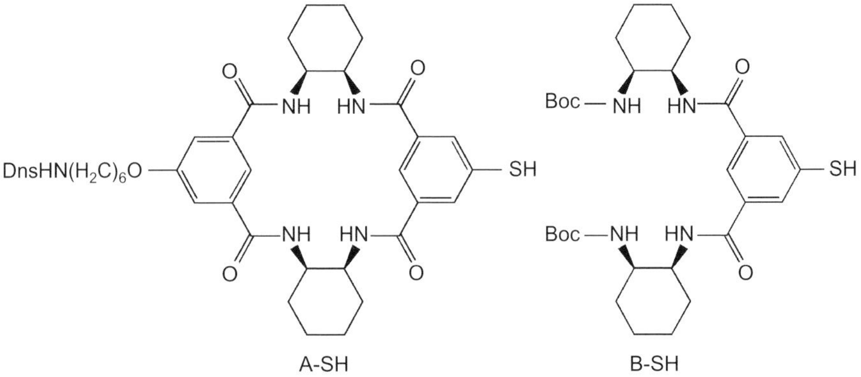

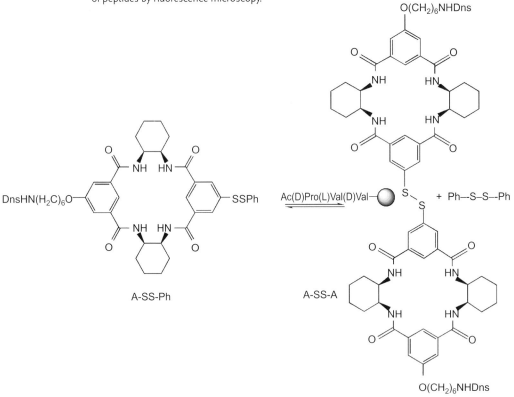

Figure 6.13 The building block thiols used for generation of disulfides. The dansyl group (Dns) was attached to detect binding of peptides by fluorescence microscopy.

Figure 6.14 Amplification of symmetric disulfides by a tripeptide on a solid support.

The equilibrium concentrations of A-SS-A, A-SS-Ph, and Ph-SS-Ph were obtained in the absence and presence of the immobilized tripeptide, when it was seen that the tripeptide amplifies the symmetric A-SS-A and Ph-SS-Ph disulfides.

$$2 \text{ (A-SS-Ph)} \rightleftharpoons \text{A-SS-A} + \text{Ph-SS-Ph}$$

Similarly, treating a mixed disulfide A-SS-B with excess B-SH resulted in amplification of the symmetrical disulfides in the presence of the tripeptide ligand (Fig. 6.13, right).

DCL is limited by the availability of methods to generate libraries from the building blocks, because reversible exchange has to occur under conditions that are not too harsh and where the building block identities are conducive to easy deconvolution. Otto et al. generated a disulfide exchange library in which the products are characterized by FTICR ESI-MS [28]. The thiols were chosen in such a way that each combination generates a disulfide that has distinct mass and hence is easy to identify.

A thiol–disulfide exchange reaction has been used to screen a library of disulfide-linked carbohydrates against the plant lectin concanavalin A (ConA) [29]. ConA binds to a trimannoside unit such that most of the interaction is centered on two peripheral mannoside groups. A library of 21 disulfide combinations was generated from six different thiol-derivatized carbohydrates of different chain lengths to help span the binding sites of ConA. Library member selection was performed in the presence of immobilized ConA and the binding interaction of bis-mannoside disulfide led to its subsequent amplification.

6.2.4
Olefin Metathesis

Olefin metathesis [30] is an important carbon–carbon bond forming reaction that can be used for the generation of DCLs. Brandli and Ward have studied the metathesis of linear internal olefins using Grubbs' catalyst [31]. The self-metathesis of each individual olefin gave six compounds which were characterized by gas chromatography–mass spectrometry (GC–MS) analysis after ozonolysis. Cross-metathesis of the olefins proceeded smoothly to give the expected number of products in quantitative yield. The dynamic nature of the metathesis library was probed by adding an external olefin to an equilibrium mixture of two olefins after metathesis using Grubbs' catalyst. The product distribution reached thermodynamic equilibrium after incorporating the external olefin, proving that it played a role in generating a DCL.

The problem with the generation of DCLs by olefin metathesis is the need for a catalyst tolerant to a wide range of functionality [32]. Catalysts stable to a variety of functional groups and which catalyze the metathesis of sterically-hindered olefins have recently been developed [33–35]. Hamilton et al. have used reversible olefin metathesis coupled with the donor–acceptor interaction of π-electron-rich aromatic diethers with π-electron poor aromatic diimides to generate [2]catenanes under thermodynamic control. A [2]catenane consists of two macrocycles that are interlocked. Ring-closing metathesis (RCM) of diimide derivatives with 1,5-dinaphthol[38]crown-10 and Grubbs' catalyst led to the formation of different isomers of [2]catenane which upon hydrogenation yielded a single product. X-ray studies showed that electron-rich aromatic diethers and electron-poor aromatic diimides were stacked alternately, with each individual [2]catenane stacked one upon another. While performing RCM in the absence of the crown ether did not lead to formation of any product, addition of the crown ether to the mixture and subsequent hydrogenation resulted in the formation of [2]catenane in comparable yield when all three components were together.

Olefin metathesis has been used by Nicolaou et al. to generate a library of van-comycin dimers [36]. Vancomycin is an antibiotic used against gram-positive bacteria resistant to other antibiotics. The emergence of vancomycin-resistant *enterococci* (VRE) and vancomycin intermediate susceptible *Staphylococcus aureus* (VISA) strains has led to the search for vancomycin analogs that are not affected by VRE or VISA. Vancomycin binds to the terminal Lys-D-Ala-D-Ala on a growing peptidoglycan chain during bacterial cell wall biosynthesis. Back-to-back dimerization of vancomycin leads to binding to its target with greater affinity. Niclolaou et al. generated van-comycin derivatives containing either terminal olefins or thiols attached to the sugar moiety by different length tethers (Fig. 6.15). The vancomycin derivatives were sub-jected to olefin metathesis in the absence and presence of the target which was ex-pected to facilitate self-association with either disulfide bond formation or olefin metathesis to lead to the covalent dimerization of the most stable product. A combi-natorial library of vancomycin derivatives with different tether lengths formed dimers in the presence of the target, with shorter tether lengths being more active as dimers against VRE. The first amino acid residue (NMeLys) in the target tripeptide was also found to play an important role in binding to vancomycin. From the combi-natorial library, three dimers were amplified that have strong activity against VRE and VISA.

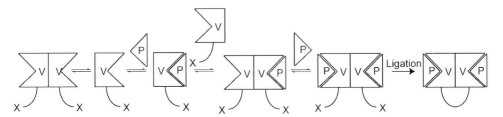

Figure 6.15 Schematic representation of selection of ligands for vancomycin dimers. Vancomycin was modified with a linker and dimerization of vancomycin monomers was enhanced in the presence of the ligand. The modified linkers are then coupled together. V = vancomycin, P = peptide L-Lys-L-Ala-D-Ala. The monomers are coupled either by disulfide bond formation or by olefin metathesis.

Giger et al. used a technique called receptor-assisted combinatorial synthesis (RACS) as a novel way to expand the size of a library [37]. If the components are al-lowed to form reversible bonds in the presence of a protein, to make a protein-spe-cific inhibitor, the equilibrium of the components will shift so that more of the bind-ing species is made (Fig. 6.16). A 1,2-diol group was used to link components through

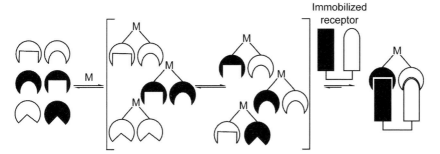

Figure 6.16 Schematic diagram of the how the best boronate ester is amplified in the presence of a receptor.

its borate esters. The borate–diol complexes do not cross-react with the protein and are stable under physiological conditions. The borate esters that best bind the protein are then converted into spiroketals. Olefin metathesis was used to generate alkenes which are dihydroxylated to give the library of 1,2-diols.

6.3
Noncovalent Interactions Used in DCC Design

6.3.1
Metal Ligand Coordination

Enrichment of DNA binding ligands from an equilibrating library of coordination complexes has been successfully demonstrated by Klekota et al., who started with six salicylaldimines and added Zn^{2+} ions to generate a library of 36 different zinc–bis(salicylaldimine) complexes (Fig. 6.17) [38].

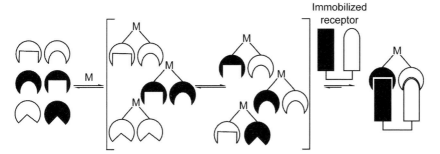

Figure 6.17 Schematic diagram of the generation of a library of interconverting metal-coordinated salicylaldimines, and amplification of the metal-ligand DCC in the presence of a poly d(A•T) DNA affinity column.

The library members were screened against an affinity column of d(A•T) DNA immobilized on cellulose resin, in the absence and presence of zinc ions. From this it was established that the homodimer of compound **5** with Zn^{2+} had the highest affinity for d(A•T) of the members of the library (Fig. 6.18).

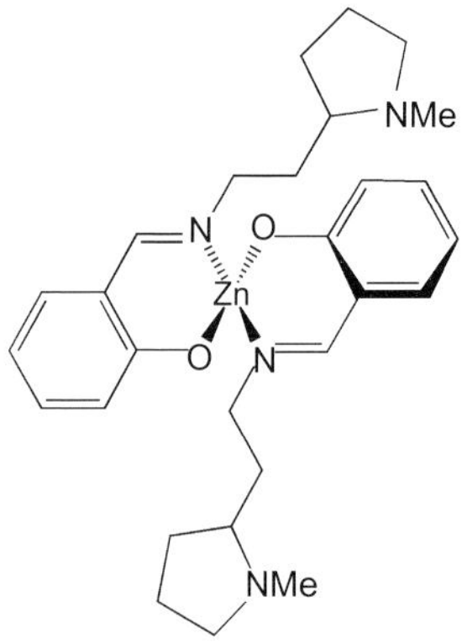

Figure 6.18 The bis(salicylaldimino)–Zn complex (**5**-Zn-**5**) was shown to bind to an oligo d(A•T) affinity column.

Self-assembly can also occur in the presence of metal ions which dictate distinct distance and orientation patterns of appropriately designed ligands [39]. Two ways to control the product distribution of self-assembly is in the design of the metal-binding ligand and variation of the metal ion. The reversibility of a DCL allows various inorganic supramolecular combinations to form a single complex of interest by changing the reaction conditions. When Baxter et al. added nitromethane to ligand **6** in the presence of copper at room temperature, ESI-MS identified three oligomeric species: $[Cu_n 6_n]^{n+}$ (where $n = 2, 3, 4$) in the ratio 6:4:3, respectively (Fig. 6.19).

When the same nitromethane solution was layered with benzene, crystallization occurred leaving only the $[Cu_2 6_2]^{2+}$ complex in the solid state. By crystallizing the mixture, a single complex was isolated. They found that formation was dictated by the preferred coordination geometry of the metal and the number of ligand metal binding sites.

Albrecht et al. carried this idea further by looking at "hetero"-recognition [40] in which the complex is made up of different ligands [41]. In particular, the ability of alkyl-bridged bis-(catechols) **7** and **8** to form dinuclear titanium(IV) architectures was investigated. The distribution of tetra-anionic complexes was affected by selective countercation binding. Addition of two equivalents of $Ti(OMe)_4$ to a 1:1 mixture of two alkyl-bridged bis-(catechol) ligands with alkali metal carbonate as base can result in the formation of the two homoleptic dinuclear complexes A1 and A2, the heteroleptic dinuclear complexes B1 and B2, the homoleptic oligomeric complexes C1 and C2, and the heteroleptic oligomeric complexes D (Fig. 6.20).

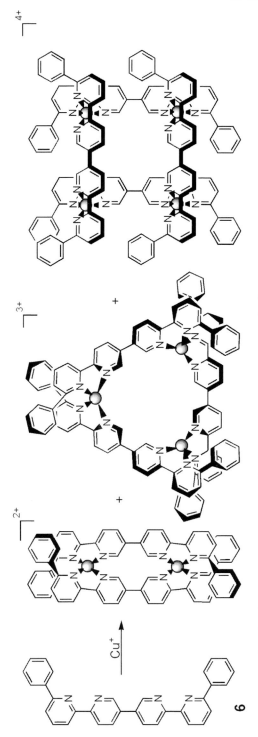

Figure 6.19 Product distribution obtained by addition of nitromethane to **6** in the presence of Cu²⁺.

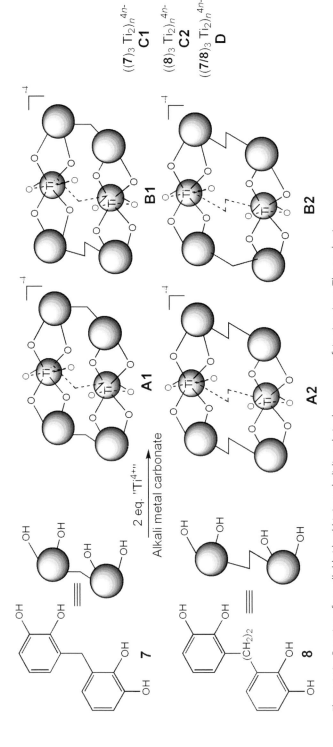

Figure 6.20 Reaction of two alkyl-bridged bis(catechol) ligands in the presence of titanium ions. The production of homoleptic and heteroleptic species varies depending on the presence of different salts.

Under different conditions, the species produced varied dramatically, depending on the countercation present. In the presence of sodium carbonate or a mixture of alkali metal cations, homoleptic complexes A1 and A2 were the only species observed. In the presence of lithium carbonate the heteroleptic complex B1 was observed with A1 and A2. Although formation of A and B complexes would be entropically favored compared with oligomeric complexes, the oligomeric homoleptic complex C1 was observed in addition to A1 when potassium carbonate was added as the base. This evidence suggested that the ligands are sorted and self-assembled, because A2 and C1 are both homoleptic. Albrecht et al. exploited a clever self-assembling interaction between linear ligands and metal ions to form a series of spontaneously generated metalla-cryptates [42]. On addition of a guest molecule, however, only one metalla-cryptate species was produced that was specific to the templating molecule. Earlier work from the same group had looked at helicate-type metalla-cryptates formed by self-association of three ligands in the presence of two gallium(III) ions and potassium cations. Without alkali metal counterions an insoluble product was generated that forms the correct self-assembled structure when KCl was added. Two neutral 8-hydroxyquinoline derivatives were investigated to show that in the presence of gallium(III) ions, supramolecular libraries were formed. In the presence of cationic guest molecules, however, amplification of the most stable metalla-cryptate was achieved. The cryptand is a flexible architecture that can adjust to a variety of cation sizes.

DCL are built on the concept of reversibility that can cause scrambling of the library components. Goral et al. have extended this concept to the generation of a dynamic library by two different reversible reactions that are independent of each other [43]. First, coordination of two terpyridine ligands to Co^{2+} was used to form an octahedral complex. Exchange of ligands can readily occur if there is an excess of free ligand present in the solution and can be controlled by oxidizing the metal to Co^{3+}, because this undergoes ligand exchange at a much slower rate (Fig. 6.21).

The terpyridine ligands have aldehyde substituents which can be reacted with hydroxylamines or hydrazines to give imines that, in turn, can be exchanged by use of an excess of free amines at acidic pH. Imine interconversion generates a second level of diversity, because ligand exchange and imine exchange are independent of each other. Pairwise combination of two terpyridine ligands generated six complexes and scrambling of imines led to a double-level library of 21 compounds, in which 15 members had unique masses distinguishable by ESI-MS. In an alternative system, Stultz et al. also generated and amplified a library of mixed metal porphyrin cages consisting of zinc(II) porphyrin donors and rhodium(II) or ruthenium(II) porphyrin acceptors by using the templating molecules 4,4'-bpy, 3,3'-dimethyl-4,4'-bipyridine, and benzo[*lmn*]-3,8-phenanthroline [44].

When screening a library against a target the optimum compound should not only have the correct recognition groups but should also be in the proper orientation and conformation relative to the target [45]. *Vicia villosa* B_4 lectin binds multiple 2-acetamido-2-deoxy-α-D-galactopyranose (GalNAc) residues. To recreate the multivalent nature of this binding, Sakai et al. generated a synthetic bipyridine linked to GalNAc. Reaction of the functionalized bipyridine with Fe(II) generated four diastereomeric octahedral complexes at equilibrium, each with three bipyridine-modified GalNAc

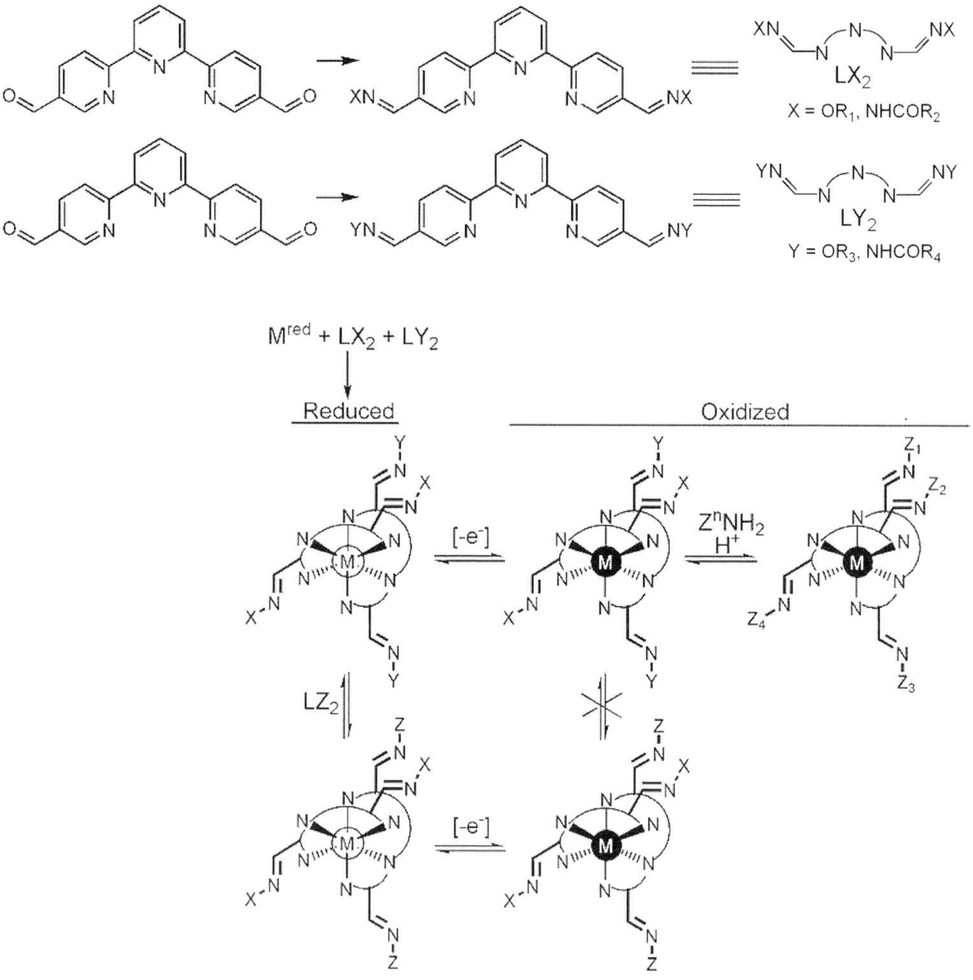

Figure 6.21 Coordination of the tridentate ligands (LX_2, LY_2) with the metal ion results in an octahedral complex that can undergo rapid ligand exchange in the presence of free ligand, LZ_2. Oxidation of the metal ion makes the library "static" because ligand exchange is very slow. Imine exchange can be initiated at an acidic pH by using free amines (Z^nNH_2).

groups. In the presence of the B_4 lectin, the Λ-mer isomer was amplified indicating a shift in the equilibrium because of the selective binding of Λ-mer Fe^{II}(bipy-GalNAc)$_3$.

Karan et al. developed coordination complexes that recognize RNA by reacting Cu^{2+} with a variety of salicylamide–amino acid building blocks [46]. The copper-coordinated amino acids were dialyzed in the presence and then absence of RNA or DNA hairpin templates. The molecules selected by the nucleic acids were collected and analyzed to show that the salicylamide–histidine derivatives had greater affinity for RNA than for the DNA hairpin.

6.3.2
Hydrogen Bonding

A DCL comes from reversible connection between building blocks under thermodynamic equilibrium to make all potential combinations. The inherent selectivity of one binding member of the library for the target will increase its concentration because of target-directed enrichment. Huc et al. made a selective library templated by chloride ions [47]. The building blocks self-assemble into a pentameric circular helicate. On addition of chloride ions, a certain shape and size will distinguish the binding cavity. The building blocks in this study contain hydrogen-bonding groups and can also coordinate metal ions. Those building blocks that coordinate and best complement a Janus molecule (a template molecule with two hydrogen bond recognition faces) will be preferentially selected from those that are less complementary (Fig. 6.22).

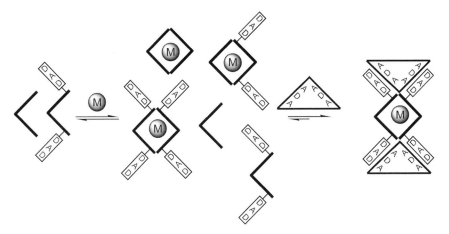

Figure 6.22 Dimer formation is initiated in the presence of the metal. The dimers formed are treated with a complementary Janus molecule to amplify the molecule that forms the maximum number of H-bonds.

The library created contained a mixture of molecules, some with two donor–acceptor–donor (DAD) components, some with one DAD motif and some without any DAD motifs. The metal ion acted to recruit two ligands and present, if possible, two hydrogen-bonding groups for recognition. Those molecules presenting two DAD components were complementary to the Janus molecule which contains two (ADA) faces that enable a maximum of 12 hydrogen bonds. The complexes with maximum hydrogen bonding were preferred over the less complementary combinations. Palladium and copper ions were investigated and it was observed that with different Janus molecules selectivity changed between two functionalized ligands.

The linear oligo-isophthalamide **9** is a sequence of four 2,6-diaminopyridines which are joined by three isophthalate groups and form a sequence of four donor–acceptor–donor (DAD) hydrogen-bonding subunits [48] (Fig. 6.23).

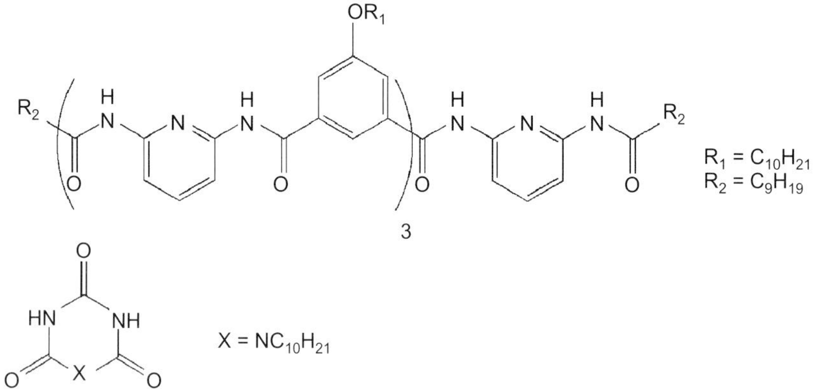

Figure 6.23 The structure of the linear oligo-isophthalamide **9** and the cyanurate **10**.

Compound **9** can exist in 36 possible rotameric forms as a result of rotation around the aryl–CO bonds. Addition of a double-faced, ADA/ADA cyanurate template **10** induces curvature in **9**, leading to the formation of a helical conformer in which two molecules of the cyanurate bind to one molecule of **9**. The binding stoichiometry was obtained from a Job plot and the helical nature of the complex was verified by ^{1}H NOE experiments on a model system. It was further shown that the **9:10$_2$** complexes can self-assemble by helical stacking to form fibers. Fiber formation depended on solvent composition, concentration, and temperature and can therefore potentially be fine-tuned for applications in material science.

Most DCL are based on reversible covalent bonds between a guest templating molecule and different building blocks to amplify the best binding combinations for a target of interest. Cardullo et al. have looked at weak, reversible noncovalent interactions involving multiple hydrogen bonds [49]. Hydrogen bonds have not been extensively explored for this purpose, because detection is hard when using large libraries. A four-membered library was produced in which the dynamic combinations were a result of hydrogen-bond exchange but the assemblies were then reacted to form covalently-linked analogs that could be characterized by HPLC and MALDI-TOF mass spectrometry. Three hydrogen-bonded calix[4]arene members were turned into covalently-linked macrocycles via RCM.

The use of a template to assist in the assembly of library components resembles biological strategies in which molecular recognition is often essential to produce a specific product. Because reversible covalent interactions such as disulfide bonds, cis–trans isomerization, and transesterification are widely used to form thermodynamically and kinetically controlled libraries, there has been interest in dynamic noncovalent interactions as alternatives. Multiple hydrogen-bonding complexes have been used to create a library of 5,5-diethylbarituric acid (DEB) assemblies [50]. Three dimelamine molecules are used to sandwich six barbiturate DEB molecules and give an aggregate composed of nine components with 36 hydrogen bonds (Fig. 6.24). By

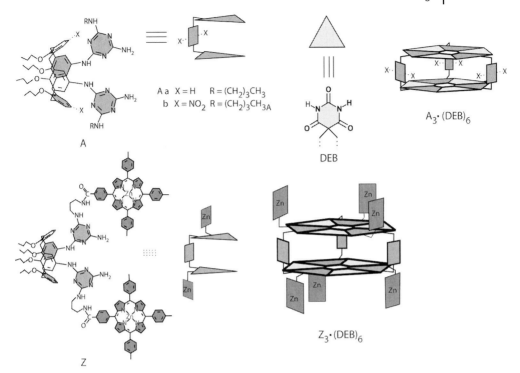

Figure 6.24 Diagrammatic representation showing how 5,5-diethylbarituric acid (DEB) molecules form alternating (DEB)$_3$A$_3$ or (DEB)$_3$Z$_3$ in the presence of A (dimelamine) or Z (zinc porphyrin dimelamine), respectively (from M.C. Calama, P. Timmerman, D.N. Reinhoudt, *Angew. Chem. Int. Ed. Engl.* **2000**, *39*, 755–758).

producing ten analogs of the dimelamine component, a diverse library of 220 unique noncovalent species was formed in statistical amounts under thermodynamic conditions. A collection of DEB components with dimelamine molecules that project two zinc porphyrin groups was also prepared. Because these library members were formed in statistical proportions without a template present, the authors tested for the selection of a receptor from the noncovalent dynamic library with dimelamine and zinc porphyrin-linked dimelamine. The literature suggested that zinc porphyrins have an affinity for pyrimidine derivatives; affinity is, however, even greater for cyclic multiple porphyrins. Mixing three components: DEB, dimelamine (A), and zinc porphyrin dimelamine (Z) led to formation of four equilibrating assemblies: (DEB)$_6$A$_3$, (DEB)$_6$A$_2$Z, (DEB)$_6$AZ$_2$, and (DEB)$_6$Z$_3$ (Fig. 6.25). Addition of tripyridine as a template led to a library that evolved to contain only one species, (DEB)$_6$Z$_3$, which formed the most stable interactions with the tripyridine. The ratio of (DEB)$_6$Z$_3$ to tripyridine was shown to be 1:2, because (DEB)$_6$Z$_3$ has two binding faces of the three zinc porphyrins where each face was able to bind three pyridine groups.

Hof et al. have used thermodynamically-controlled conditions to noncovalently link self-assembling components together into multi-component capsule-like species [51].

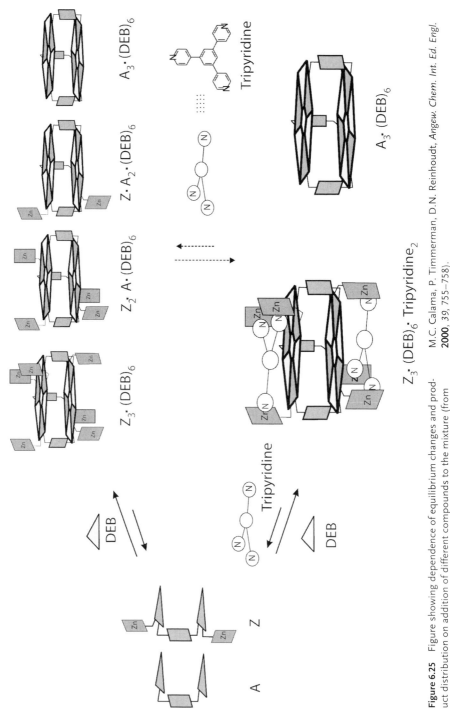

Figure 6.25 Figure showing dependence of equilibrium changes and product distribution on addition of different compounds to the mixture (from M.C. Calama, P. Timmerman, D.N. Reinhoudt, *Angew. Chem. Int. Ed. Engl.* **2000**, *39*, 755–758).

These reversibly-associated capsules act as receptors when a target molecule is added to the pool of assembled subunits. The component design included several criteria: the ability to reversibly associate, the presence of curvature in the molecule, and the potential for hydrogen bonding complementarity. Empty capsules did not form, thus ensuring that only one size and shape was produced when a target molecule was added. In particular, a tetrameric capsule was formed by hydrogen bonding between a cyclic sulfamide and a glycoluril group. By incorporating different substituents on the spacer between the two hydrogen-bonding groups, the size, shape and surface functionality of the capsule changed. The capsule components often had different affinities for a range of target molecules. ESI-MS was used to characterize the equilibrating pool of capsule products as the self-associating components were varied. Addition of distinctive alkyl groups to each component enabled detection of the resulting tetrameric capsules by mass spectrometry. The potential library of tetrameric capsules formed from two components gives six capsule combinations. Having equimolar concentrations of the two components in the presence of an ethyltrimethylammonium cation gave different amounts of each capsule type. On addition of a methylquinuclidinium cation to the same equimolar mixture, a different ratio of capsule products was formed. Therefore, the receptor capsule that best fits the target molecule formed spontaneously in solution, because of the increased stability, and was the principal assembled species detected. A larger library of 70 capsule varieties was prepared and mass spectrometry was used to establish that the most complementary receptor was amplified within the equilibrium mixture.

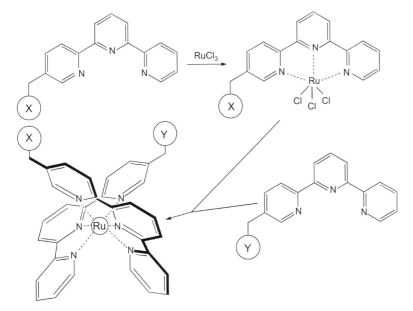

Figure 6.26 Formation of metal-templated libraries projecting a variable substrate-binding region (X and Y) on a constant metal binding template.

In the past, libraries of small nucleic acids, peptides, and synthetic agents, have been screened to search for active site-binding molecules. Antibodies are examples in which Nature uses combinatorial sorting of light and heavy chains to give large numbers of selective antigen-binding molecules. The light and heavy chains both contain a constant and variable region in which the constant domains enable the variable regions to position the hypervariable loops to produce an antigen binding site. Goodman et al. replicated the design of an antibody by constructing a synthetic library in which the constant region was a metal-binding ligand attached to the variable region which contained the substrate binding site [52]. The metal binding domain was based on a terpyridine group which was reacted with $RuCl_3$ to give an octahedral Ru with two different binding domains (Fig. 6.26).

This coordination of two different terpyridine groups leaves the two substrate binding groups in fixed positions. By using terpyridine analogs linked to different functional domains, a library of 15 combinations was developed. Microcalorimetry was used to estimate the enthalpic contribution to binding when the library was screened against bis(tetrabutylammonium) pimelate. The best binding molecule was titrated against the polar substrate using H^1 NMR to obtain a $K_a > 10^4$ M^{-1}. The metal has two effects – it recruits two thiourea hydrogen bonding sites and stabilizes electrostatic interactions by means of its positive charge.

Oligosaccharide clusters have been found in many biological processes as mediators of molecular recognition, as seen in lipid rafts [53]. Oligodeoxynucleotides (ODN) have been investigated for application in DNA-computing and molecular machines. Matsuuri et al. have exploited the self-assembly characteristics of DNA to form glycoclusters. Galactosylated-ODN were synthesized and designed to be "half sliding" complementary ODN (hsc-ODN) with each Gal-ODN design regulating the carbohydrate distance and position. Clusters with 18-, 20-, 22-mers Gal-ODN hybridized to hsc-ODN were produced in this way. In particular, the different spatial relationships of the galactose groups were assessed for their importance in controlling the binding affinity to lectin.

6.4
Conformational/Configurational Isomerization

Molecular evolution has three components: first, there must be a pool of components with different characteristics; second, there must be a method of selection to differentiate, select, and amplify certain species in the pool; and third, the structure of the compounds must be able to change without bias. Elizeev and Nelen used an approach that enables amplification of the highest affinity binding receptors to a target [54]. They developed a simple chemical system based on Darwinian mutagenesis and selection. Their design used an anionic receptor for arginine selected from a dynamic equilibrating receptor mixture. A dicarboxylate compound was used in the experiments because it can form three isomers under ultra violet (UV) light irradiation: trans–trans, trans–cis, and cis–cis (Fig. 6.27).

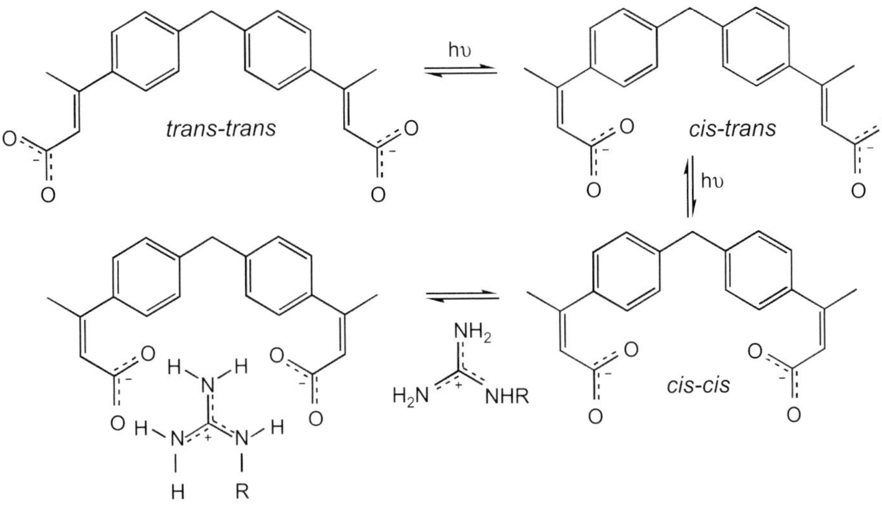

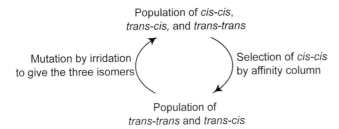

Figure 6.27 The photochemical equilibration of trans–trans, trans–cis, cis–cis dicarboxylate compounds, followed by cycles of guanidinium selection.

The cis–cis isomer was predicted to have the optimum geometry for binding the quanidinium moiety of arginine with two salt bridges. In the other two isomers the carboxylates are too far apart for efficient interaction. An ethanol mixture of the isomer salts was put into an experimental setup that enables recycling of mixture components (Fig. 6.28).

Population of *cis-cis*,
trans-cis, and *trans-trans*

Mutation by irridation
to give the three isomers

Selection of *cis-cis*
by affinity column

Population of
trans-trans and *trans-cis*

Figure 6.28 Diagram depicting the molecular evolution process. Irradiation of the dialkene species generates different photoisomers that are passed through an affinity column which binds one isomer preferentially. The recycling of the other two isomers by irradiation makes them available for another round of affinity chromatography.

The mixture was subjected to UV light irradiation thus mutating the solution to contain a range of the isomers in the ratio cis–cis:cis–trans:trans–trans, 3:28:69. The isomer solution was pumped on to an arginine-immobilized affinity column for selection. Most of the cis–cis isomer became attached to the arginine affinity column. Further irradiation of the remaining solution led to increased amounts of the cis–cis conformer that was added to the column, further amplifying the cis–cis concentration. The solution was photochemically isomerized 30 times to give the isomer ratio cis–cis:cis–trans:trans–trans, 48:29:23. Elution of the attached species from the solid support and analysis gave an isomer ratio of cis–cis:cis–trans:trans–trans, 85:13:2 by HPLC. Another test was performed in which a sample was irradiated continuously for eight hours and the flow through was collected to give a mixture with an isomer ratio cis–cis:cis–trans:trans–trans, 52:31:17 but elution from the column gave the cis–cis isomer only. These experiments showed the ability of a designed system to amplify, select, and isolate true binding receptors.

Whether a potential DCL is composed of reversible covalent or noncovalent linked components, is real or virtual, or consists of molecular or supramolecular components, the requirement of synthesizing all possible combinations is removed, because the target is left to select the best binders from the mixture [55]. Berl et al. investigated creation of a library that, on binding, can change conformation by isomerization, whether by site inversion or ring inversion. By allowing the molecules to change conformation on binding, an induced fit was ensured. Specifically, an equilibrium mixture of 5,5-dimethyl-1,3-cyclohexanedione and 2-hydrazinopyridine condensation products was generated. In addition to the three dihydrazone isomer products (trans–trans, cis–cis, and cis–trans), cis and trans isomers were obtained from the monohydrazone monoketone, and its hydrazine–enone tautomer when the reaction was performed in a 1:2 ratio in chloroform. To reduce the diversity of products produced during equilibration, dibutylbarbiturate was added until only one species was apparent in the NMR spectrum. This single species was found to be the cis–cis hydrazone isomer, which strongly hydrogen bonds to a dibutylbarbiturate molecule. The barbiturate molecule induced the formation of the cis–cis isomer from the equilibrium mixture, because of the stable binding interaction. This work is an example of the molding process, because it shows that a receptor can be chosen from an equilibrating mixture. Adaptive chemistry can be explored by using DCL systems able to change conformation or configuration depending on the target molecule.

6.5
Receptor-based Screening, Selection, and Amplification

The next step in combinatorial chemistry is to increase the efficiency of creation and optimization of lead compounds and to reduce the dependence on synthesis and screening. If the target of interest is left to direct the screening, selection and evolution of true binding molecules from a library of building blocks, this would be a significant advance. This can be achieved by using a library of building blocks with complementary reactive functionalities. Simultaneous binding of the complementary ligands in the active site of the enzyme will enable them to react, thus linking them. Lewis et al. have exploited this approach by using Huisgen's 1,3-dipolar cycloaddition of azides and acetylenes to generate 1,2,3-triazoles (Fig. 6.29) [56].

Dipolar cycloaddition is compatible with a variety of building blocks, tolerates aqueous conditions, results in minimum formation of undesired products, and depends on the imposed proximity and orientation of the reactants. The Sharpless lab-

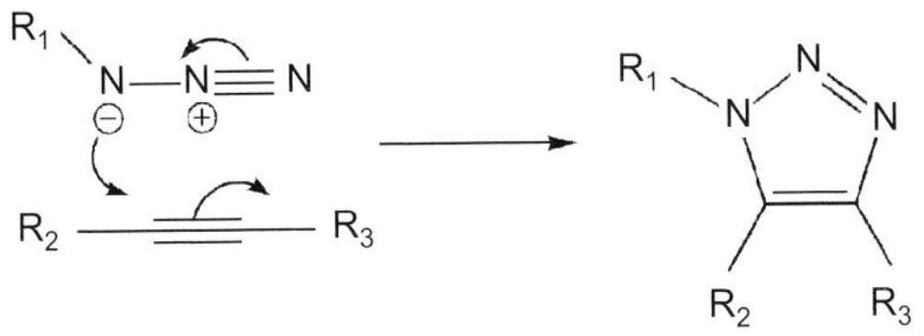

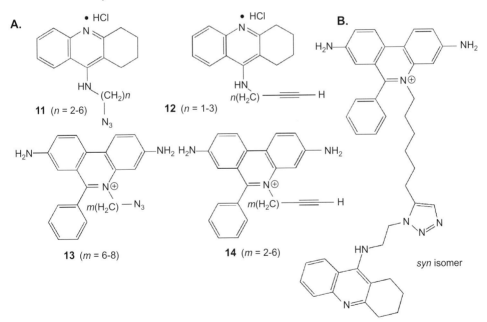

Figure 6.29 1,3-dipolar cycloaddition.

oratory used AChE to direct the inhibitor screening and selection. Two inhibitors known to bind the active site and two inhibitors known to bind the peripheral site were chosen as members of the library. The features of AChE binding and active sites have been discussed previously (Sect. 6.2.1.1). The respective AChE inhibitors were derivatized to contain either azide or acetylene functional groups of different chain length. On incubation with the enzyme, the component inhibitors react by dipolar cycloaddition to form a tightly binding divalent inhibitor. DIOS mass spectrometry was used to examine 49 reactions between **11**, **12**, **13**, and **14** and only one combination (**11** + **14**) produced a hit (Fig. 6.30).

The **11** + **14** syn isomer was found to be the predominant product and was 100-fold ($K_d = 77$ fM) more potent than the anti isomer ($K_d = 720$ fM) when tested against *T. californica*. Each individual component of the syn product bound to the enzyme with much lower affinity.

Figure 6.30 A. Azide and acetylene components used to make bidentate inhibitors of AChE. B. The *syn*-triazole inhibitor of AChE obtained from the library of azides and acetylene building blocks.

In the past, natural products have given rise to compounds that medicinal chemists have used as leads [57]. Recently, biopolymer libraries have been used in combinatorial approaches, whether for phage display or peptide epitope libraries. The peptides used in these studies have been made by solid-state methods and have then been evaluated against a protein target while the peptide is still attached to the solid support. Immobilized peptides have several disadvantages: the terminal amino or carboxyl group is sequestered into a covalent bond; there are potential steric issues; and peptide structure is hard to identify. Swann et al. used proteases to increase peptide diversity by shifting the enzyme activity from hydrolysis and toward synthesis of longer peptides. The direction of the reaction can be shifted by using a less polar organic solvent, increasing the concentration of one reactant, producing an insoluble product, or trapping the product. Swann et al. used nonspecific proteases to create diversity among a mixture of peptides. As proof of principle, two peptides YGG and FL were incubated with the protease thermolysin. HPLC and capillary zone electrophoresis (CZE) detection of the library of short peptides showed the formation of the desired YGGFL peptide in 0.1% yield. The ability of a receptor to sequester the peptide products was demonstrated by using a monoclonal antibody (3E7) specific to the amino terminus of β-endorphin with a 7.1 nM affinity for YGGFL. The antibody was separated from the thermolysin peptide synthesis vessel by means of a semipermeable membrane through which only the peptide substrates could diffuse; this prevented proteolysis of the antibody (Fig. 6.31). The peptides were synthesized by the protease and picked up by the antibody without being subjected further to protease activity. The equilibrium was pushed toward increased peptide synthesis to replace those removed by the antibody. Indeed, binding and amplification of YGGFL peptides was detected in the presence of 3E7 antibody.

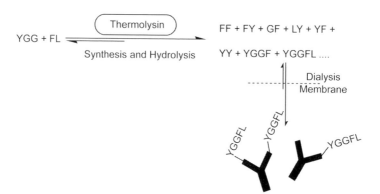

Figure 6.31 A peptide library was generated in the presence of thermolysin and the peptide (YGGFL) with affinity for the antibody target was subsequently amplified.

6.6
Conclusions

We have described several examples of the formation and application dynamic combinatorial libraries. These libraries are based on rapidly interconverting and exchanging components whose equilibrium composition can be influenced by addition of a potential target. Specific binding to the target will lead to removal or enrichment of one or more members of the library. Critical to the success of the dynamic approach is the choice of the connecting motif that links the different components. The connection must undergo rapid exchange, must be compatible with the target and any necessary medium restrictions (water, pH, salts), and ideally should be convertible into a locked and stable form for ready structural characterization. Particular success has been achieved by use of disulfide bond formation, transimination, transesterification, oxime, and hydrazone exchange, olefin metathesis, metal coordination, and hydrogen-bonding interactions.

Acknowledgment

We thank the NIH for financial support of our work described in the review.

References

1 J.-M. Lehn, *Chem. Eur. J.* **1999**, *5*, 2455–2463.

2 J.-M. Lehn, A. V. Eliseev, *Science* **2001**, *291*, 2331–2332.

3 O. Ramstrom, T. Bunyapaiboonsri, S. Lohmann, J. M. Lehn, *Biochim. Biophys. Acta.* **2002**, *1572*, 178–186.

4 B. Hasenknopf, J.-M. Lehn, B. O. Kneisel, G. Baum, D. Fenske, *Angew. Chem. Int. Ed. Engl.* **1996**, *35*, 1838–1840.

5 I. Huc, J. M. Lehn, *Proc. Natl. Acad. Sci. U S A* **1997**, *94*, 2106–2110.

6 G. R. Cousins, S. A. Poulsen, J. K. M. Sanders, *Curr. Opin. Chem. Biol.* **2000**, *4*, 270–279.

7 S. Otto, R. L. Furlan, J. K. M. Sanders, *Curr. Opin. Chem. Biol.* **2002**, *6*, 321.

8 R. Nguyen, I. Huc, *Angew. Chem. Int. Ed. Engl.* **2001**, *40*, 1774–1776.

9 M. Hochgurtel, H. Kroth, D. Piecha, M. W. Hofmann, C. Nicolau, S. Krause, O. Schaaf, G. Sonnenmoser, A. V. Eliseev, *Proc. Natl. Acad. Sci. U S A* **2002**, *99*, 3382–3387.

10 J. D. Cheeseman, A. D. Corbett, R. Shu, J. Croteau, J. L. Gleason, R. J. Kazlauskas, *J. Am. Chem. Soc.* **2002**, *124*, 5692–5701.

11 A. D. Corbett, J. D. Cheeseman, R. J. Kazlauskas, J. L. Gleason, *Angew. Chem. Int. Ed. Engl.* **2004**, *43*, 2432–2436.

12 T. Bunyapaiboonsri, O. Ramstrom, S. Lohmann, J. M. Lehn, L. Peng, M. Goeldner, *ChemBioChem* **2001**, *2*, 438–444.

13 H. Li, P. Williams, J. Micklefield, J. M. Gardiner, G. Stephens, *Tetrahedron* **2004**, *60*, 753–758.

14 J. T. Goodwin, D. G. Lynn, *J. Am. Chem. Soc.* **1992**, *114*, 9197–9198.

15 R. J. Lins, S. L. Flitsch, N. J. Turner, E. Irving, S. A. Brown, *Tetrahedron* **2004**, *60*, 771–780.

16 R. L. E. Furlan, Y. F. Ng, G. R. L. Cousins, J. E. Redman, J. K. M. Sanders, *Tetrahedron* **2002**, *58*, 771–778.

17 R. L. E. Furlan, G. R. L. Cousins, J. K. M. Sanders, *Chem. Commun.* **2000**, 1761–1762.

18 R. L. E. Furlan, Y. F. Ng, S. Otto, J. K. M. Sanders, *J. Am. Chem. Soc.* **2001**, *123*, 8876–8877.

19 S. L. Roberts, R. L. E. Furlan, S. Otto, J. K. M. Sanders, *Org. Biomol. Chem.* **2003**, *1*, 1625–1633.

20 G. R. L. Cousins, S. Poulsen, J. K. M. Sanders, *Chem. Commun.* **1999**, 1575–1576.

21 S. A. Poulsen, P. J. Gates, G. R. Cousins, J. K. M. Sanders, *Rapid Commun. Mass Spectrom.* **2000**, *14*, 44–48.

22 G. R. Cousins, R. L. Furlan, Y. F. Ng, J. E. Redman, J. K. M. Sanders, *Angew. Chem. Int. Ed. Engl.* **2001**, *40*, 423–428.

23 N. Nazarpack-Kandlousy, J. Zweigenbaum, J. Henion, A. V. Eliseev, J. Comb. Chem. 1999, 1, 199–206.

24 P. A. Brady, R. P. Bonar-Law, S. J. Rowan, C. J. Suckling, J. K. M. Sanders, *Chem. Commun.* **1996**, 319–320.

25 P. A. Brady, J. K. M. Sanders, *J. Chem. Soc. Perkin Trans. 1*, **1997**, 3237–3254.

26 S. J. Rowan, P. S. Lukeman, D. J. Reynolds, J. K. M. Sanders, *New J. Chem.* **1998**, *10*, 1015–1018.

27 H. Hioki, W. C. Still, *J. Org. Chem.* **1998**, *63*, 904–905.

28 S. Otto, R. L. E. Furlan, J. K. M. Sanders, *J. Am. Chem. Soc.* **2000**, *122*, 12063–12064.

29 O. Ramstrom, J. M. Lehn, *ChemBioChem* **2000**, *1*, 41–48.

30 R. H. Grubbs, S. Chang, *Tetrahedron* **1998**, *54*, 4413–4450.

31 C. Brandli, T. R. Ward, *Helv. Chim. Acta.* **1998**, *81*, 1616–1621.

32 D. G. Hamilton, N. Feeder, S. J. Teat, J. K. M. Sanders, *New J. Chem.* **1998**, *22*, 1019–1021.

33 M. Scholl, T. M. Trnka, J. P. Morgan, R. H. Grubbs, *Tetrahedron Lett.* **1999**, *40*, 2247–2250.

34 J. Huang, E. D. Stevens, S. P. Nolan, J. L. Petersen, *J. Am. Chem. Soc.* **1999**, *121*, 2674–2678.

35 T. Weskamp, F. J. Kohl, W. Hieringer, D. Gleich, W. A. Herrmann, *Angew. Chem. Int. Ed. Engl.* **1999**, *38*, 2416–2419.

36 K. C. Nicolaou, R. Hughes, S. Y. Cho, N. Winssinger, C. Smethurst, H. Labischinski, R. Endermann, *Angew. Chem. Int. Ed. Engl.* **2000**, *39*, 3823–3828.

37 T. Giger, M. Wigger, S. Audétat, S. A. Benner, *Syn. Lett.* **1998**, *6*, 688–691.

38 B. Klekota, M. H. Hammond, B. L. Miller, *Tetrahedron Lett.* **1997**, *38*, 8639–8642.

39 P. N. W. Baxter, J.-M. Lehn, K. Rissanen, *Chem. Commun.* **1997**, 1323–1324.

40 B. Hasenknopf, J. M. Lehn, G. Baum, D. Fenske, *Proc. Natl. Acad. Sci. U S A* **1996**, *93*, 1397–1400.

41 M. Albrecht, M. Schneider, H. Röttele, *Angew. Chem. Int. Ed. Engl.* **1999**, *38*, 557–559.

42 M. Albrecht, O. Blau, R. Frohlich, *Chem. Eur. J.* **1999**, *5*, 48–56.

43 V. Goral, M. I. Nelen, A. V. Eliseev, J. M. Lehn, *Proc. Natl. Acad. Sci. U S A* **2001**, *98*, 1347–1352.

44 E. Stulz, S. M. Scott, A. D. Bond, S. J. Teat, J. K. M. Sanders, *Chem. Eur. J.* **2003**, *9*, 6039–6048.

45 S. Sakai, Y. Shigemasa, T. Sasaki, *Tetrahedron Lett.* **1997**, *38*, 8145–8148.

46 C. Karan, B. L. Miller, *J. Am. Chem. Soc.*, **2001**, *123*, 7455–7456.

47 I. Huc, M. J. Krische, D. P. Funeriu, J.-M. Lehn, *Eur. J. Inorg. Chem.* **1999**, 1415–1420.

48 V. V. Berl, M. J. Krische, I. I. Huc, J. M. Lehn, M. Schmutz, *Chem. Eur. J.* **2000**, *6*, 1938–1946.

49 F. Cardullo, M. Crego Calama, B. H. M. Snellink-Ruël, J.-L. Weidmann, A. Bielejewska, R. Fokkens, N. M. M. Nibbering, P. Timmerman, D. N. Reinhoudt, *Chem. Commun.* **2000**, 367–368.

50 M. C. Calama, P. Timmerman, D. N. Reinhoudt, *Angew. Chem. Int. Ed. Engl.* **2000**, *39*, 755–758.

51 F. Hof, C. Nuckolls, J. Rebek, Jr., *J. Am. Chem. Soc.* **2000**, *122*, 4251–4252.

52 M. S. Goodman, V. Jubian, B. Linton, A. D. Hamilton, *J. Am. Chem. Soc.* **1995**, *117*, 11610–11611.

53 K. Matsuura, M. Hibino, T. Ikeda, Y. Yamada, K. Kobayashi, *Chem. Eur. J.* **2004**, *10*, 352–359.

54 A. V. Eliseev, M. I. Nelen, *J. Am. Chem. Soc.* **1997**, *119*, 1147–1148.

55 V. Berl, I. Huc, J.-M. Lehn, A. DeCian, J. Fischer, *Eur. J. Org. Chem.* **1999**, 3089–3094.

56 W. G. Lewis, L. G. Green, F. Grynszpan, Z. Radic, P. R. Carlier, P. Taylor, M. G. Finn, K. B. Sharpless, *Angew. Chem. Int. Ed. Engl.* **2002**, *41*, 1053–1057.

57 P. G. Swann, R. A. Casanova, A. Desai, M. M. Frauenhoff, M. Urbancic, U. Slomczynska, A. J. Hopfinger, G. C. Le Breton, D. L. Venton, *Peptide Sci.* **1996**, *40*, 617–625.

7
Synthetic Molecular Machines

Euan R. Kay and David A. Leigh

7.1
Introduction

The widespread use of molecular-level motion in key natural processes suggests that great rewards could come from bridging the gap between the present generation of synthetic molecular systems – which usually rely on electronic and chemical effects to perform their functions – and the machines of the macroscopic world, which utilize the synchronized movements of smaller parts to perform particular tasks. In recent years it has proved relatively straightforward to design synthetic molecular systems in which positional changes of submolecular components occur by moving energetically downhill, but what are the structural features necessary for molecules to use directional displacements to do work? How can we make a synthetic molecular machine that pumps ions against a gradient, for example, or moves itself energetically uphill along a track? Artificial compounds that can do such things have yet to be realized; the field of synthetic molecular machines is still in its infancy and only the most basic systems – mechanical switches and slightly more sophisticated, but still rudimentary, molecular rotors – have been made thus far. In this chapter we outline the early successes in taming molecular-level motion and the progress made toward utilizing synthetic molecular structures to perform mechanical tasks. We also highlight some of the challenges and problems that must still be overcome.

The path toward synthetic molecular machines starts nearly two centuries ago, with the discovery of the random nature of molecular-level motion. In 1827, the Scottish botanist Robert Brown noted through his microscope the incessant, haphazard motion of tiny particles within translucent pollen grains suspended in water. Subsequent investigations of both biological and inorganic materials led to the realization that all objects are subject to constant buffeting by their surroundings, as a result of the motions which intrinsically occur at the molecular level. A complete explanation of the phenomenon – now known as Brownian motion or movement – had to wait for Einstein nearly 80 years later, but ever since scientists have been fascinated by the stochastic nature of molecular-level motion and its implications. In exciting develop-

Functional Synthetic Receptors. Edited by T. Schrader, A. D. Hamilton

ISBN: 3-527-30655-2

ments over the past decade, theoretical physics has explained how random, directionless fluctuations can cause directed motion of particles [1, 2] which successfully accounts for the general principles behind biological motors [3]. The chemist's interest in the creation, behavior, and control of molecular structures means that an understanding of the physics involved can now help him/her to design artificial structures which perform mechanical operations at the molecular-level and transpose those effects to the macroscopic world. Nature provides proof that such an approach is possible in the form of ion pumps, motor proteins, photoactive proteins, retinal, and many other natural products [4]. Indeed, biological structures have already been incorporated into semi-synthetic biomaterials which can perform "unnatural" mechanical tasks [5]; now the synthetic chemist is trying to create functional molecular machines from scratch. Initially, such systems will probably be simpler and less effective than their biological counterparts, yet there is no reason to suggest that one day they might not be just as powerful or ubiquitous.

7.1.1
Molecular-level Machines and the Language Used to Describe Them

It has been suggested that problems arise as soon as a scientific definition is established. In fact, of course, language scholars tell us that precisely the opposite is true. Language – especially scientific language – must be suitably defined and correctly used to accurately convey concepts in a field. Nowhere is the need for accurate scientific language more apparent than in the discussion of the ideas and mechanisms by which nanoscale machines could – and do – operate. Much of the terminology used comes from phenomena observed by physicists and biologists, but unfortunately their findings and descriptions have sometimes been misunderstood and misapplied by chemists. Perhaps inevitably in a newly emerging field, there is not even clear agreement in the literature about what constitutes molecular machines and what differentiates them from other molecular devices [6]. Initially, categorization of molecules as machines was purely iconic – the structures "looked" like pieces of machinery – or they were so-called because they carried out a function that in the macroscopic world would require a machine to perform it. Many of the chemical systems first likened to pistons and other machines were simply host–guest complexes in which the binding could be switched "on" or "off" by external stimuli such as light. Although these early studies were unquestionably the key to popularizing the field, with hindsight consideration of the effects of scale tells us that supramolecular decomplexation events have little in common with the motion of a piston (the analogy is better within a rotaxane architecture, because the components are still kinetically associated after decomplexation) and that a photosensitizer is not phenomenologically related to a "light-fuelled motor". In fact, it is probably most useful differentiate "device" and "machine" on the basis that the etymology and meaning of "machine" implies mechanical movement – i.e. at the molecular level a net nuclear displacement – which causes something useful to happen. This leads to the definition that "molecular machines" are a subset of molecular devices (functional molecular systems) in which a stimulus triggers the controlled, large-amplitude mechanical motion of one component relative to another (or of a substrate relative to the machine) which results in a net task being performed. In this chapter we shall not discuss the larger field of supramolecular devices but will limit our discussion to approaches to machines, systems that actually feature some control over molecular-level motion. The examples given illustrate this point and help demonstrate the requirements for mechanical task performance at the molecular-level.

7.1.2
Principles of Motion at the Molecular Level – the Effects of Scale

The random thermal fluctuations experienced by molecules dominate mechanical behavior in the molecular world. Even the most efficient nanoscale machines – the motor proteins found in Nature – are swamped by its effect. A typical motor protein consumes ATP fuel at a rate of 100–1000 molecules every second, corresponding to a maximum possible power output in the region 10^{-16} to 10^{-17} W per molecule. When compared with the random environmental buffeting of ~10^{-8} W experienced by molecules in solution at room temperature, it seems remarkable that *any* form of controlled motion is possible [1].

The constant presence of Brownian motion is not the only distinction between motion at the molecular-level and in the macroscopic world. Because the physics which govern mechanical dynamic processes in the two regimes are completely different, vastly different mechanisms are required for controlled transport or propulsion. In the macroscopic world the equations of motion are governed by inertial terms (dependent on mass). Viscous forces (dependent on surface areas) dampen motion by converting kinetic energy into heat and objects do not move until provided with specific energy to do so. In a macroscopic machine this is often provided through a directional force when work is done to move mechanical components in a particular way. As objects become less massive and smaller in dimension, inertial terms decrease in importance and viscous terms begin to dominate. A parameter which quantifies this effect is Reynolds number – essentially the ratio of inertial to viscous forces – given by Eq. (1) for a particle of length dimension a, moving at velocity v, in a medium with viscosity η and density ρ [7]:

$$R = \frac{a v \rho}{\eta} \tag{1}$$

Size affects modes of motion long before we reach the nanoscale. Even at the mesoscopic level of bacteria (length dimensions ~10^{-5} m), viscous forces dominate. At the molecular level, Reynolds number is extremely low (except at low pressures in the gas phase) and the result is that molecules, or their components, cannot be given a one-off "push" in the macroscopic sense – momentum has become irrelevant. The motion of a molecular-level object is determined entirely by the forces acting on it at that particular instant – whether they be externally applied forces, viscosity or random thermal perturbations and Brownian motion. In more general terms, this analysis points to a central tenet – although the macroscopic machines we encounter in everyday life may provide the inspiration for what we might like molecular machines to achieve, drawing too close an analogy for their modes of operation is a poor design strategy. The "rules of the game" at different length scales are simply too different. Two basic principles *must* be followed for any molecular device to be able to carry out a mechanical function:

• First, the movement of the kinetically-associated molecules or their components must be controlled by employing interactions which restrict the natural tendency for three-dimensional random motion and, somehow, bias, rectify, or direct the motion along the required vectors.

• Second, the Second Law of Thermodynamics tells us that no machine can continually operate solely using energy drawn from the thermal bath, so an external input of energy is required to perform a mechanical task with any synthetic molecular machine.

Learning how to make successful designs based on these two principles is the subject of the rest of the chapter.

7.2
Controlling Conformational Changes

Consideration of the restriction of thermal motion in chemistry first arose in regard to a fundamental question of molecular stereochemistry [8]. The rotational triple energy minima about C–C single bonds follows directly from the tetrahedral geometry of saturated carbon centers, yet it was not proven experimentally until 1936 [9]. In the 1960s and 70s sterically crowded systems were synthesized which included examples of different atropisomers of constitutionally identical chemical structures and characterization of rotamerization about sp²–sp³ and sp³–sp³ linkages, elegantly illustrated by the triptycenes investigated by Ōki and coworkers (for example **1**, Fig. 7.1) [10].

These studies inspired the first molecular analogs of macroscopic machine parts – propellers and gears [11]. Linking 9-triptycyl units through an sp³ linker, Iwamura and Mislow independently created "molecular gears" such as **2** in the early 1980s. In **2**, the blades of each triptycyl group are tightly intermeshed so that rotation of one unit is inextricably linked to the other – if **A** rotates clockwise, **B** must rotate counter-

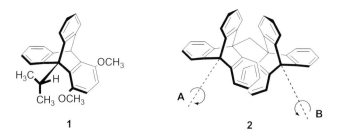

Figure 7.1 Triptycene derivatives **1** and **2** with hindered and correlated rotation around C–C bonds.

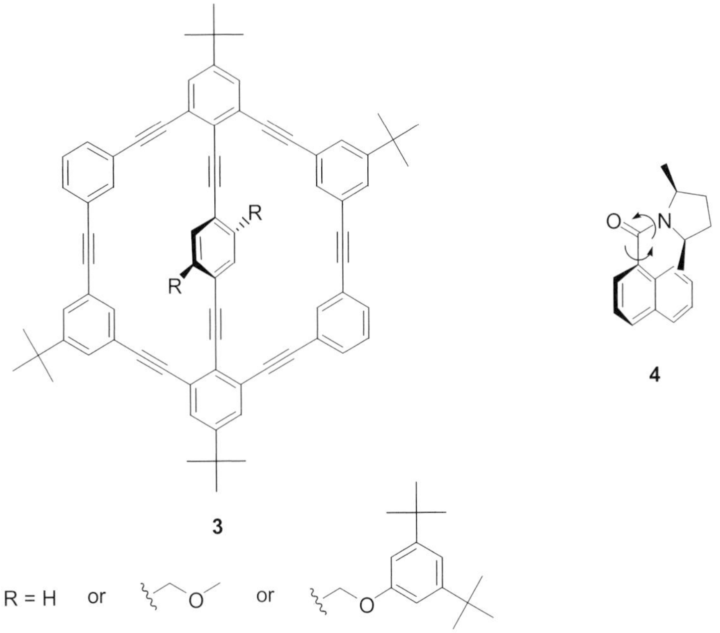

3

R = H   or   or

4

Figure 7.2 Molecular structures of **3** and **4** in which restricted rotation around single bonds has been observed.

clockwise and vice versa. The disrotatory correlated motion mirrors the operation of a macroscopic bevel gear. Experimental and theoretical studies confirm that the barrier to the conrotatory process – analogous to gear slippage – is large. Significantly, barriers to net disrotatory motion are negligible (0.2 kcal mol^{-1} for **2**), indicating that such motion encounters almost no friction – an early illustration of the differences between motion at the molecular and macroscopic levels.

Later, a "molecular turnstile", **3**, (which subsequently inspired "molecular gyroscopes") was created in which the rate of rotation of the central phenyl ring could be tuned by varying substitutions [12]; correlated rotational motion has also been observed around hindered amides such as **4** (Fig. 7.2) [13].

Although these studies clearly demonstrate the role steric interactions can play, the submolecular motions are, of course, non-directional even within a partial rotational event. Simply restricting the thermal rotary motion of one unit by a larger blocking group or by the similarly random motion of another unit cannot, in itself, lead to directionality. A molecular machine requires some form of external modulation over the dynamic processes.

As a first step toward achieving controlled and externally initiated rotation around C–C single bonds, Kelly combined triptycene structures (Fig. 7.1) with a molecular recognition event [14]. In the resulting "molecular brake", **5** (Fig. 7.3) [15], free rotation of a triptycyl group could be halted on application of Hg^{2+} ions which caused a conformational change in the appended bipyridyl unit – effectively putting a "stick"

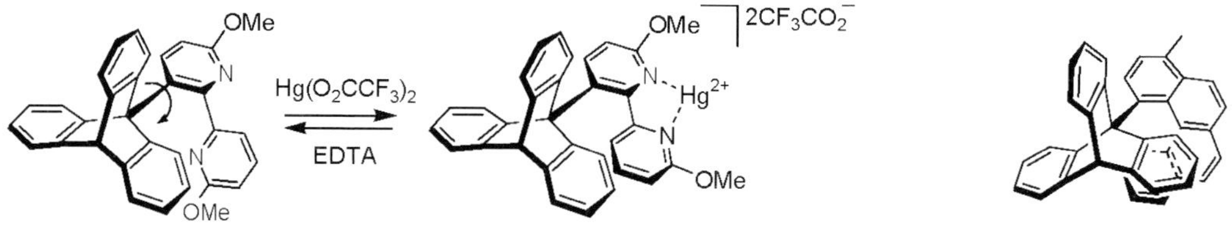

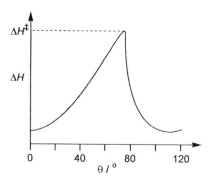

Figure 7.3 A chemically switchable "molecular brake", **5**, and the design of an unsuccessful "molecular ratchet", **6**.

in the "spokes". The next iteration toward controlling the motion involved construction of a potential "molecular ratchet", **6** [16], in which it was proposed that a helicene "pawl" might direct the rotation of an attached triptycene "wheel" in one direction owing to the pawl's chiral helical structure. Although the calculated energetics for rotation showed the energy barrier to be asymmetric (Fig. 7.4), ^{1}H nuclear magnetic resonance (NMR) experiments showed rotation to be occurring equally in both directions.

This result serves as a reminder that the Second Law of Thermodynamics cannot be escaped – although thermal motion can be used to move a system toward its equilibrium state, it cannot be used to power work [17]. The rate of any chemical transformation – clockwise and anticlockwise rotation in **6** included – depends on the energy of the transition state passed over and the temperature. However, state functions such as enthalpy and free energy do not depend on a system's history [18]. Thus, the energy of the transition state in Fig. 7.4, $\Delta H^{\ddagger}$, is independent of whether it is approached from the left (clockwise rotation of the triptycene as drawn) or the right (anticlockwise rotation) and the rates of rotation are constrained to be equal in either direction.

Figure 7.4 Schematic representation of the calculated enthalpy changes for rotation around the single degree of internal rotational freedom in proposed ratchet **6**.

The temptation to relate **6** to a macroscopic ratchet and pawl, which does use an asymmetric energy-profile to produce unidirectional motion from a randomly fluctuating input force, is clear – what is different about the molecular-level system? Smoluchowski and, more famously, Feynman have considered the problem of a molecular-level ratchet, showing that the effect of thermal fluctuations on the molecular-level "pawl" would cause it to disengage and the overall result would be random motion in either direction – the Second Law survives [19]. Feynman went on to show that for directional motion to occur energy must be added to the system in some way, to remove it from equilibrium [20]. The requirements for unidirectional motion at the molecular-level therefore are: (1) a randomizing force which enables the system to seek its equilibrium position; (2) some form of asymmetry in the structure, surroundings, or information within the system which differentiates two potential directions of travel; and, crucially, (3) an energy input which drives the system away from equilibrium.

In **6**, criteria 1 and 2 are clearly already present in the form of thermally powered random fluctuations around the single degree of internal rotational freedom and the helical chirality of the structure, respectively. Yet to achieve *unidirectional* motion the system must be driven away from equilibrium – to break "detailed balance" – so that the Second Law of Thermodynamics is not contravened. Kelly therefore proposed **7a** (Fig. 7.5), a modified version of the ratchet structure in which a chemical reaction is

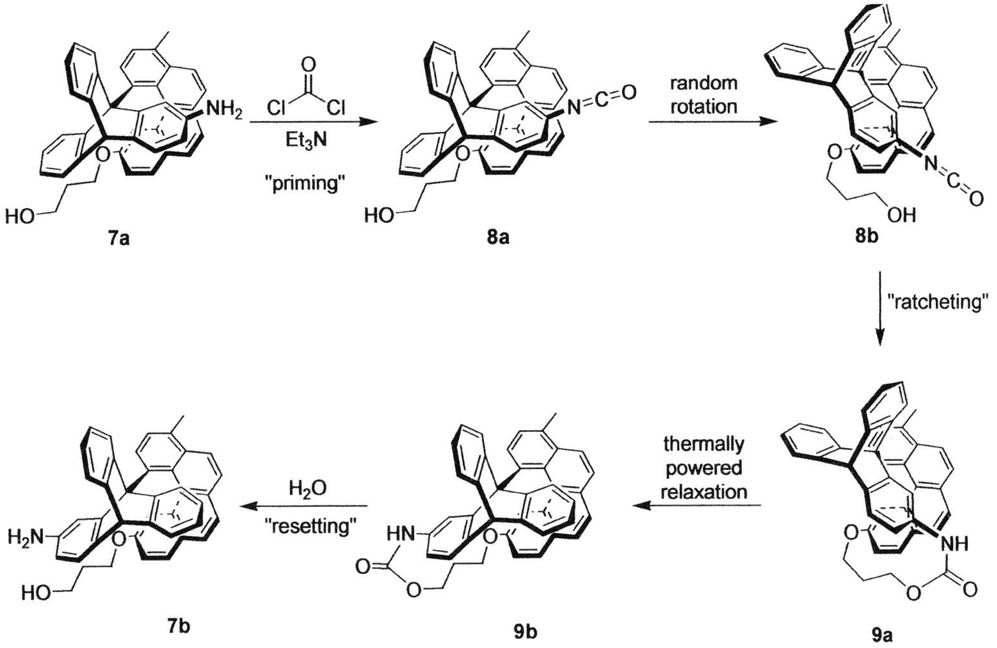

Figure 7.5 A chemically powered unidirectional rotary motor in action. Priming of the motor in its initial state with phosgene (**7a** → **8a**) enables a chemical reaction to occur when the helicene rotates far enough up its potential well toward the blocking triptycene arm (**8b**). This gives a tethered state **9a** for which rotation over the barrier to **9b** is an exoergic process and occurs under thermal control. Finally, the urethane linker can be cleaved to give the original rotor rotated by 120° (**7b**).

used as a source of energy [21]. Ignoring the amino group, all three possible positions for the helicene relative to the triptycene "spokes" are identical – the energy profile for 360° rotation would appear as three equal energy minima, separated by equal barriers. As the helicene oscillates back and forth in a trough, however, sometimes it will come close enough to the amine for chemical reaction to occur (as in **8b**). Priming the system with chemical "fuel" (phosgene in this example to give the isocyanate **8a**) results in "ratcheting" of the motion some way up the energy barrier (**9a**). Continuation of the rotation in the same direction, over the energy barrier, can occur under thermal control and is now an exoergic process (giving **9b**) before cleavage of the urethane gives the 120° rotated system (**7b**). Although the current system can only perform one third of a full rotation, it demonstrates the principles required for a fully operating and cyclable rotary system under chemical control and is a significant advance from the original restricted C–C bond-rotation systems.

7.3
Controlling Configurational Changes

Changes in configuration [22] – in particular cis–trans isomerization of double bonds – have been widely studied from theoretical, chemical and biological perspectives [8, 23]. Although such small-amplitude motions are not usually suitable for direct machine-like exploitation, they can provide photoswitchable control mechanisms within more complex devices (see below) and, in some instances, can even be harnessed to perform a significant mechanical task. Such systems are some of the first examples of molecular-level motion which can be controlled by application of an external stimulus.

The utility of isomerizations in mechanical systems is widely exploited in Nature. The photoisomerization of retinal (Fig. 7.6a) induces a conformational change in the visual pigment rhodopsin, enabling it to bind to protein-G, thereby triggering the visual signal-transduction process [24]. Equally remarkable are the photoactive yellow proteins (PYP) found in some eubacteria [25]. The photoinduced trans-to-cis isomerization of the olefin in a *p*-coumaric acid chromophore (Fig. 7.6b) activates a photocycle in which this small configurational change is amplified through the protein scaffold resulting in large conformational changes which ultimately cause the bacterium to swim away from harmful blue light. Despite being formally single bonded, the delocalization in amide C–N bonds enables peptide linkages to exist in distinct *Z* and *E* forms. Although the *E* forms are normally more stable, for tertiary amides there are only small thermodynamic differences between the two isomers. Xaa–Pro bonds therefore play an important role in controlling protein conformation – the existence of just one such bond as its *Z* isomer can have drastic effects on the secondary structure of the protein. Enzymes which catalyze *E* to *Z* isomerization – peptidyl–prolyl cis–trans isomerases (PPIases) – play important roles in directing protein folding and controlling the subsequent biological activity of many proline-containing proteins [26].

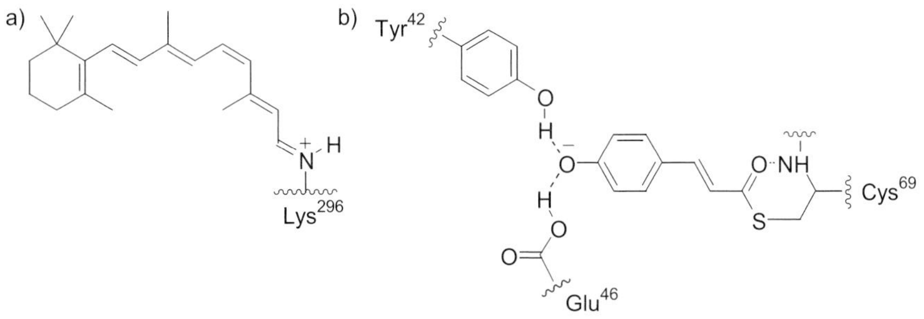

Figure 7.6 (a) Chemical structure of the rhodopsin chromophore – an iminium derivative of 11-*cis*-retinal. (b) Chemical structure of the PYP chromophore – a thioester of *p*-coumaric acid.

As a prototypical example of the reaction class, the photoisomerization of stilbenes has been extensively studied theoretically and preparatively for well over 50 years [27]. As organic photochemistry began to develop, other important photochromic systems were discovered (Fig. 7.7) [28]: the photoisomerization of azobenzenes [29], the interconversion of spiropyrans with merocyanine (and the associated spirooxazine/merocyanine system) [30], the photochromic reactions of fulgides [31], and chiroptical systems such as overcrowded alkenes (for example **10**, Fig. 7.9) [32]. All these processes are accompanied by marked changes in both physical and chemical properties, for example color, charge, and stereochemistry, and a wide range of switching applications have both been proposed and realized, from liquid-crystal display technology to variable-tint spectacles and optical recording media [33].

Although photochromism in modern diarylethene systems [34] and in fulgides is based on a photocyclization process (rendering the configurational cis–trans isomerization an undesirable side-reaction), most applications of these photochromic systems simply rely on the intrinsic electronic and spectroscopic changes on interconversion between the two species. As such, these systems do not normally fit our definition of molecular machines. There are some systems, however, which harness small configurational changes in a more mechanical fashion so that the resulting devices can, indeed, be described as machine-like.

A variety of configurational changes (but in particular azobenzene photoisomerization) have been used to alter peptide structures, creating semi-synthetic biomaterials whose activity can be controlled photonically [35]. These systems, which amplify the small configurational change of the synthetic unit to cause a significant change in the peptide secondary structure, to some extent mimic the combined role of proline and PPIases in controlling protein activity in the cell.

Combining photochromism with liquid crystal systems can lead to a variety of photoswitched changes in the properties of the liquid crystal [36]. The trans → cis photoisomerization of stilbene or azobenzene dopants, for example, can lead to reversible disruption of the nematic mesophase, effectively turning off liquid crystalline properties. More subtle control over the alignment of liquid crystals can also be achieved in systems in which the mesophase structure is maintained throughout.

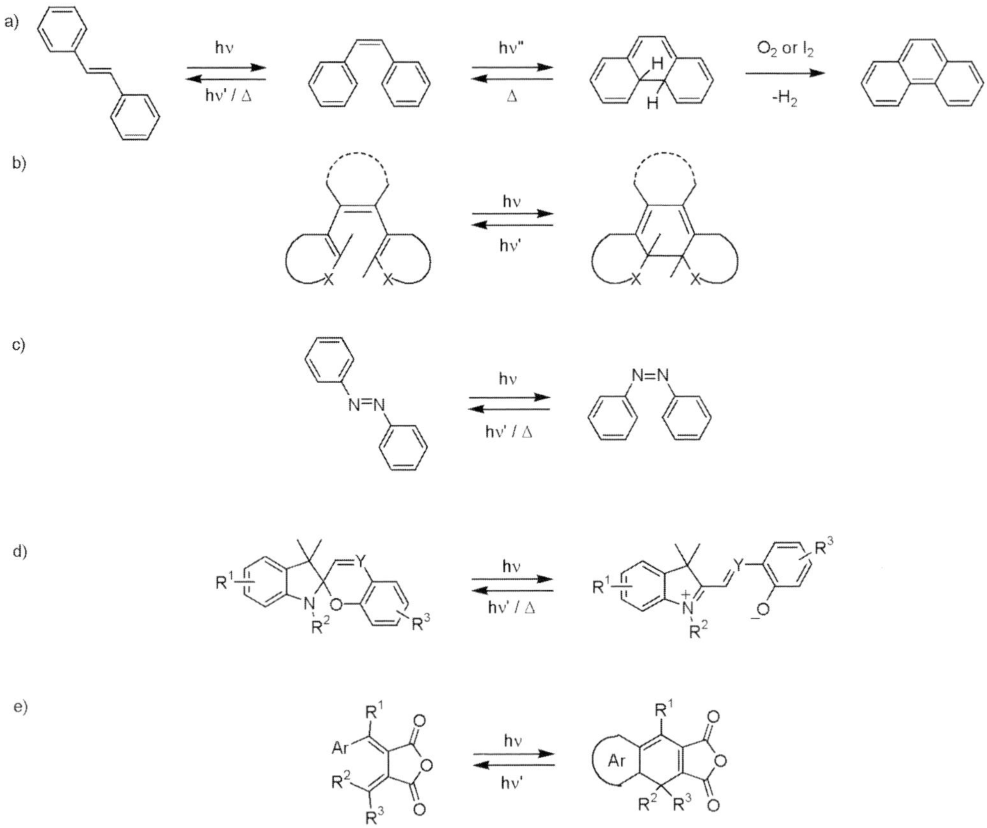

a)

b)

c)

d)

e)

Figure 7.7 Some photochromic systems which can potentially give rise to a mechanical response. (a) The cis–trans isomerization of stilbene can be accomplished photochemically. A competing process is the 6π-electron cyclization to give dihydrophenanthrene, which reverts to cis-stilbene in the dark or, in the presence of a suitable oxidant, irreversibly loses H_2 to give phenanthrene. (b) Other diarylethene systems exploit the photocyclization process by blocking the cis $\rightarrow$ trans and hydrogen-elimination processes through substitution, while replacing the phenyl rings with heteroaromatics stabilizes the cyclized form ($X=C$, NH, O, S). (c) The cis–trans isomerization of azobenzene. (d) The interconversion of spiropyrans ($Y = C$) or spirooxazines ($Y = N$) with their merocyanine forms. (e) The interconversion of ring-open and cyclized forms of fulgides, both of which are thermally stable (cis–trans isomerization around either bond can occur, but is undesirable).

Use of plane-polarized light to stimulate repetitive cis–trans isomerization of photochromic dopant molecules can lead to photoorientation of these species by axis-selective photoconversion. This directional orientation triggers concomitant alignment of the whole liquid crystal phase. A second vector of polarization can then produce a different aligned state. Potential applications of this photo-induced mechanical molecular-level motion include optical recording media and light-addressable displays [36].

In such systems, quite remarkable amplification of a relatively small photoinduced configurational change causes a global change in orientation in the bulk. It is even possible to create surface-controlled systems in which a monolayer of photochromic species (a so-called "command surface") controls the alignment of a whole liquid-crystal layer in contact with it. The same principles can be applied to the photonic control of the organization in other systems such as gels and Langmuir–Blodgett films [37].

Photoinduced mechanical changes have also been observed at the single-molecule level by use of atomic-force microscopy (AFM) [38]. One end of a polyazopeptide (a polymer of azobenzene units linked by dipeptide spacers) was isolated on a glass slide while the other was attached to a gold-coated AFM tip via a thioether–Au bond (Fig. 7.8). At zero applied force, repeated shortening and lengthening of the polymer was observed on photoisomerization of the azobenzene units to cis and trans respectively. This experimental arrangement was, furthermore, used to demonstrate the transduction of light energy into mechanical work. A "load" was applied to one of the polyazopeptides – in its fully trans form – by increasing the force applied by the AFM tip from 80 pN to 200 pN, resulting in stretching of the polymer. Photoisomerization, to yield what is assumed to be the fully cis polymer, resulted in contraction even against this external force ("raising the load"). Removal of the "load" by reducing the applied force, followed by re-isomerization to the trans form, restored the polymer to its original length – ready to lift another load. The work done by this single molecule device is ~5 × 10^{-20} J.

While investigating the use of overcrowded alkenes as chiroptical molecular switches, Feringa and coworkers created molecular brake **10** in which the speed of rotation around an arene–arene bond can be varied by changing the alkene configuration (Fig. 7.9) [39]. Counter-intuitively from the two-dimensional structures in Fig. 7.9, the rate of rotation around the indicated bond is faster in *cis*-**10** than in *trans*-**10** (demonstrated by the respective free energies of activation, $\Delta G^{\ddagger}$, for the process). Unfortunately, the photochemical interconversion between cis and trans could not be accomplished efficiently for this molecule, yet the system stands as proof that a change in configuration can alter not just optical properties or bulk orientations, but can also control large-amplitude submolecular motions.

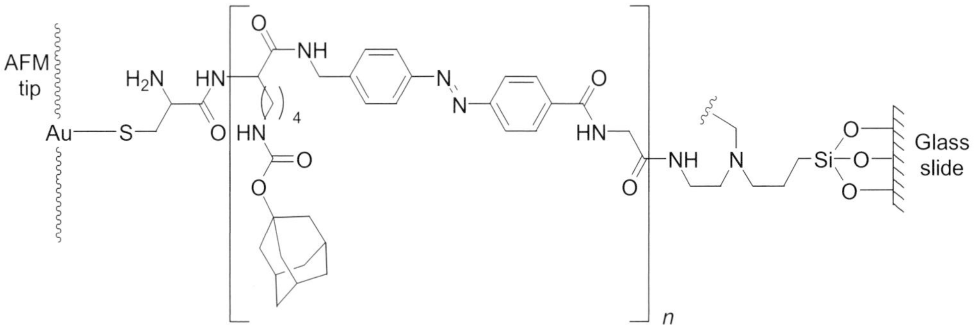

Figure 7.8 Experimental arrangement for observation of single-molecule extension and contraction using AFM (azobenzene unit shown as the extended *Z* isomer).

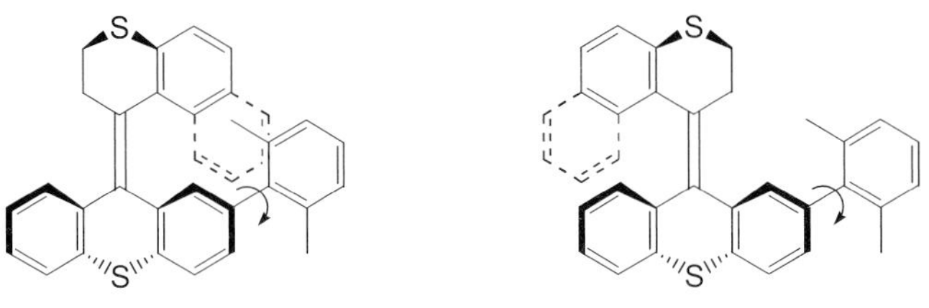

cis-**10**

$\Delta G^{\ddagger}_{303} = 19.0 \pm 0.2$ kcal mol^{-1}

trans-**10**

$\Delta G^{\ddagger}_{303} = 19.7 \pm 0.2$ kcal mol^{-1}

Figure 7.9 Counterintuitive variation of kinetic barrier to rotation around an aryl-aryl bond by isomerization of the double bond in an overcrowded alkene. Values for the free energy of activation ($\Delta G^{\ddagger}$) for the rotary motion in each configuration are shown.

The chiral helicity of molecules such as **10** causes the photochemically induced trans–cis isomerization to occur unidirectionally according to the handedness of the helix. Accordingly, these systems already have many of the requirements necessary for a unidirectional molecular rotor and, indeed, the incorporation of another stereogenic center enabled realization of this potential and creation of the first synthetic molecular rotor capable of achieving a full and repetitive 360° unidirectional rotation (Fig. 7.10) [40]. Irradiation ($\lambda > 250$ nm) of this extraordinary molecule, ($3R,3'R$)-(P,P)-trans-**11**, causes chiral helicity-directed clockwise rotation of the upper half relative to the lower portion (as drawn), at the same time switching the configuration of the double bond and inverting the helicity to give ($3R,3'R$)-(M,M)-cis-**11**. Unlike previously investigated systems however [32b], this form is not stable at temperatures above –55 °C, because the cyclohexyl ring methyl substituents are placed in an unfavorable equatorial position. At ambient temperatures, therefore, the system relaxes via a second, thermally activated helix inversion thus continuing to rotate in the same direction, giving ($3R,3'R$)-(P,P)-cis-**11**. Irradiation of this new species ($\lambda > 250$ nm) results again in photoisomerization and helix inversion to give ($3R,3'R$)-(M,M)-trans-**11**. Once more the methyl substituents are in an equatorial position and thermal relaxation (this time temperatures >60 °C are required) completes the 360° rotation giving the starting species ($3R,3'R$)-(P,P)-trans-**11**. As each different step in the cycle involves a change in helicity, the unidirectional process can be easily observed by the change in the circular dichroism (CD) spectrum at each stage. The four different states can be populated depending on the precise choice of wavelength and temperatures, and irradiation at temperatures above 60 °C results in continuous 360° rotation [41].

Because the photochemically induced isomerization process in these systems is known to be extremely fast (<300 ps), the rate-limiting step in the operation of **11** is the slowest of the thermally activated isomerizations. The effect of structure on the rate of these steps was investigated in a series of second-generation motors (Fig. 7.11) [42]. With a view to future applications, the "base" and "rotor" in these systems were differentiated, so that only one directing methyl group is present. This, however, is still sufficient for unidirectional rotation. For all examples in Fig. 7.11, the slowest thermal isomerization step has a lower kinetic barrier than that in **11**. It was,

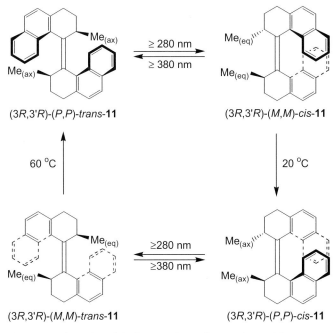

(3R,3'R)-(P,P)-*trans*-**11**   (3R,3'R)-(M,M)-*cis*-**11**

(3R,3'R)-(M,M)-*trans*-**11**   (3R,3'R)-(P,P)-*cis*-**11**

Figure 7.10 Operation of the first, continuously operating, unidirectional 360° rotor, (3R,3'R)-**11**.

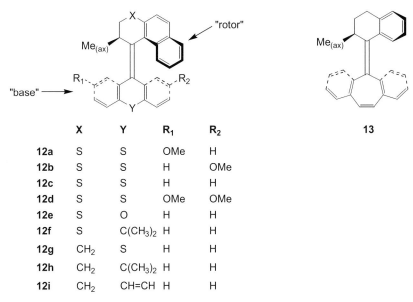

	X	Y	R₁	R₂
12a	S	S	OMe	H
12b	S	S	H	OMe
12c	S	S	H	H
12d	S	S	OMe	OMe
12e	S	O	H	H
12f	S	C(CH₃)₂	H	H
12g	CH₂	S	H	H
12h	CH₂	C(CH₃)₂	H	H
12i	CH₂	CH=CH	H	H

13

Figure 7.11 Second-generation light-driven unidirectional rotors **12a–i** and **13**.

furthermore, observed that smaller bridging groups at Y and, in particular, at X lowered the activation barrier to this process so that $\Delta G^{\ddagger}_{293} = 21.90$ kcal mol^{-1} for compound **12g**. Not all structural changes turned out to be intuitive, however. Rotor **13** contains an even smaller upper portion yet the free energy of activation for the thermal isomerization process in this molecule is actually higher than that of its phenanthrene analog **12i** [43]. The reason is that the reduced steric hindrance in **13** lowers the ground-state energy of the "unstable" isomer more than it lowers the transition-state energy for the thermal isomerization process.

A further increase in rate of rotation was, however, achieved by creation of five-membered homolog **14** (Fig. 7.12) [44]. Despite the greater conformational flexibility of the cyclopentyl ring, a significant energy difference between the pseudoequatorial and pseudoaxial positions of the appended methyl group still exists and unidirectional rotation occurs. As well as being accessible in higher synthetic yields, one of the thermal isomerizations in this molecule is extremely rapid and the other is competitive with the previous systems.

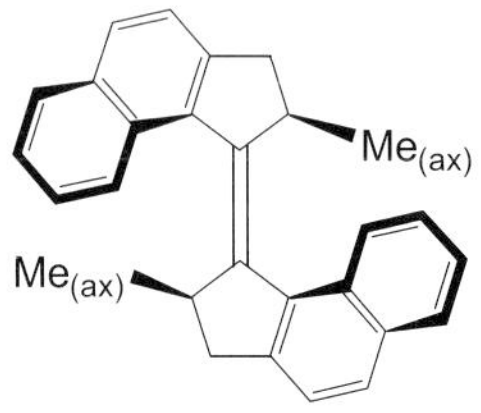

Me(ax)

Me(ax)

(2*R*,2'*R*)-(*P*,*P*)-*trans*-**14**

Figure 7.12 Third-generation unidirectional rotor, **14**. In this structure "ax" denotes pseudoaxial orientation of substituents on the cyclopentyl ring.

A potential drawback of third-generation molecule **14** is the reduced extent of isomerization at the photostationary states. Yet, owing to the "ratcheted" mechanism which does not allow backwards movement of the rotor, this does not reduce the integrity of the unidirectional process, it simply reduces the photoisomerization quantum yield.

7.4
Controlling Motion in Supramolecular Systems

We have seen how, by judicial choice of structure, it is possible to restrict submolecular motion such as bond rotation whereas configurational changes can provide an external control element for molecular geometry, orientation, and submolecular motion. Yet these generally enable only relatively crude control over molecular-level motion (the Feringa "motors" are an obvious exception to this caveat). Molecular recognition [45] and the control of noncovalent interactions provides the opportunity to *selectively, specifically,* and *strongly* position one molecule or molecular component relative to another.

Furthermore, a wide range of recognition systems are now known which can be modulated in strength – or even effectively switched "on" or "off" – using a range of external stimuli. We must be careful to remember, however, that a net mechanical task requires control of molecular-level motion. This means that the putative "machine" assembly must be *kinetically associated* (i.e. it cannot exchange with the bulk) over the timeframe of operation of the machine and that a stimulus-induced molecular recognition event is neither a sufficient nor necessary condition for construction of such a machine.

7.4.1
Switchable Host–Guest Systems

Using an external stimulus to modulate the binding affinity of a host for a guest is the simplest expression of controllable molecular recognition. A wide variety of stimuli can be used to precipitate changes – not just in geometric configuration but also electronic arrangement and environmental influences – which modulate noncovalent interactions. Switchable host–guest systems teach us much about the nature of noncovalent interactions and their manipulation, yet the requirement for kinetic association of components in a molecular machine rules out simple host–guest complexation where binding does not bring about a change in the conformation of either species or where transport of the guest between sites within the host is slow relative to exchange with the bulk. Although all complexation events clearly involve movement of the components relative to each other, the physics of complexation depend only on binding affinity and concentration, not factors associated with mechanical task performance (net position, mass, etc.). This is why myosin must move along a track to which it is kinetically associated over a series of sequential binding events to bring about muscle contraction; no mechanical task occurs through simple "on"/"off" binding to the track by myosin molecules from the bulk [46]. In other words, simple host–guest/supramolecular systems cannot function as nanoscale mechanical machines unless restrictions on the motion of the unbound species apply or the binding event brings about a mechanical (i.e. conformational) change in one of the molecular components.

Whilst this means that many switchable host–guest systems do not have the potential to act as mechanical machines, many still do. A good example of a molecular system on the borderline of machine-like behavior is given by the "tail-biting" crown ethers *E/Z*-**15**•H$^+$ (Fig. 7.13) developed by Shinkai and coworkers [47]. Nominally

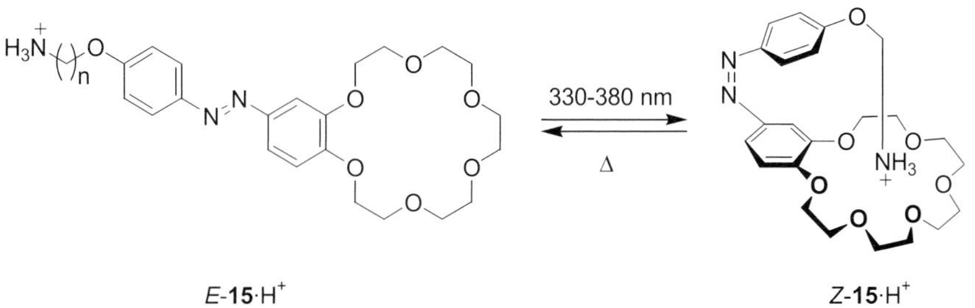

330–380 nm

Δ

E-**15**·H$^+$ *Z*-**15**·H$^+$

Figure 7.13 Isomerization of "tail-biting" crown ethers.

these systems are simply switched molecular receptors, not molecular machines. In the Z-**15**•H$^+$ forms, intramolecular binding of the ammonium ion to the crown ether prevents recognition of other cations in the medium whereas thermal isomerization to give E-**15**•H$^+$ restores the alkali metal cation binding properties of the crown ether. It is observed, however, that the rate of $Z \rightarrow E$ isomerization is lowered (by 1.6 to 2.2 fold) for Z-**15**•H$^+$ when compared with the deprotonated control Z-**15**. Crucially, the rate of this reaction increases with increasing concentration of K$^+$ ions in solution, so that the system can indeed be viewed as a primitive molecular machine in which a binding event at the crown ether affects the $Z \rightarrow E$ isomerization process of the azobenzene moiety *with which it is kinetically associated.*

7.4.2
Intramolecular Ion Translocation

Metal–ligand binding interactions are often kinetically stable and careful structural design can produce systems in which intramolecular motion is significantly favored over intermolecular exchange [48]. System **16**$^{4+}$ (Fig. 7.14) consists of a coordinatively unsaturated Cu^{II} center covalently linked to a redox-active Ni^{II}-cyclam unit [49]. In this form chloride anions bind strongly to the copper, filling the vacant coordination site ([**16**•Cl]$^{3+}$). Electrochemical oxidation of the nickel center to Ni^{III} dramatically increases its affinity for anions so that the chloride translocates to this new, more energetically favorable site. The motion is completely reversible on reduction of the nickel. There is, of course, the distinct possibility that the switching of anion position is an intermolecular process involving either free anion in solution or more than one molecule of **16**. However, the process was demonstrated not to be concentration-dependent, in contrast with the analogous noncovalently linked three-component system (Cu receptor + Ni receptor + Cl$^-$) for which the behavior is strongly concentration dependent. Further thermodynamic comparisons of the two systems suggest that the dominant mechanism in **16** is intramolecular translocation, brought about by folding of the ditopic receptor so that the motion pathway of the chloride ion is defined.

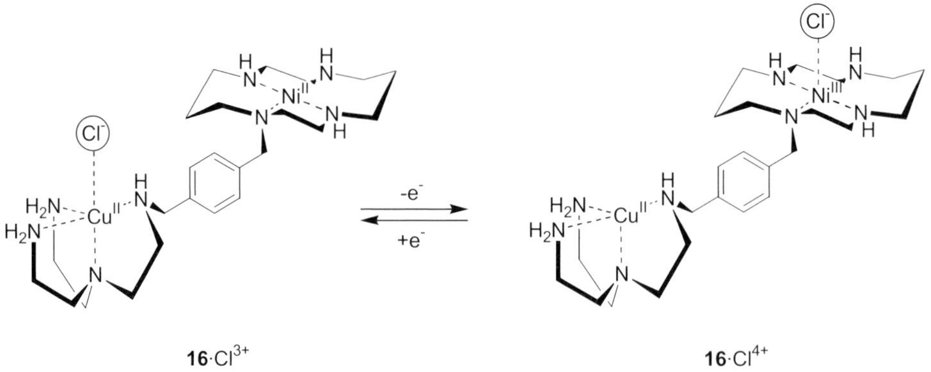

16·Cl^{3+} **16**·Cl^{4+}

Figure 7.14 Redox-driven intramolecular anion translocation.

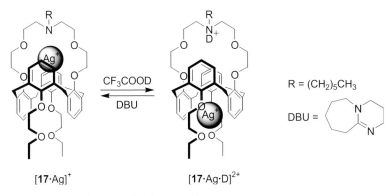

$[\mathbf{17 \cdot Ag}]^+$ $[\mathbf{17 \cdot Ag \cdot D}]^{2+}$

$R = (CH_2)_5 CH_3$

DBU =

Figure 7.15 Translocation of a silver cation between two sites in a calix[4]arene cavity – a "molecular syringe".

Calix[4]arene **17** performs reversible translocation of metal cations by protonation of the tertiary nitrogen. Dynamic ^1H NMR experiments indicate a high kinetic barrier to dissociation of the metal from the deprotonated receptor, suggesting that on protonation, translocation occurs in an intramolecular fashion through the cavity defined by the aromatic rings, in a manner resembling the operation of a syringe. In the protonated form, however, intermolecular exchange of the cation is much faster, so the return stroke, at least, can involve a significant amount of undefined intermolecular motion (Fig. 7.15) [50].

The first redox-driven cation translocation process was reported by Shanzer and coworkers in a helical complex $[\mathbf{18 \cdot Fe^{III}}]$ for which there is strong evidence for a fully intramolecular process [51]. The system works by exploiting the preference of Fe^{III} for hard ligands and Fe^{II} for a softer coordination sphere. Chemical reduction of $[\mathbf{18 \cdot Fe^{III}}]$ with ascorbic acid gives $[\mathbf{18 \cdot Fe^{II} \cdot 3H}]^{2+}$ which has spectral properties characteristic of the $Fe^{II}(bipyridyl)_3$ coordination sphere (Fig. 7.16). Subsequent re-oxidation (ammonium persulfate) returns the system to its original state. The equivalent intermolecular process between two monotopic ligands failed to occur under the same conditions, and even the translocation in **18** is relatively kinetically slow, suggesting an intramolecular process.

Pseudorotaxanes – supramolecular complexes in which a macrocyclic host encapsulates a linear, thread-like guest – are subject to the same issues with regard to kinetic stability as other host–guest complexes [52–54]. While these complexes are model systems for kinetically associated systems such as interlocked rotaxanes (see Sect. 7.5), their basic structure is similar to that of other supramolecular systems. Many interesting pseudorotaxane devices have been created with functions including reversible formation and de-threading of the interlocked species using a number of stimuli [55], switching between different preferred guests [56], and control of intramolecular electron transfer reactions [57]. Only for kinetically stable pseudorotaxanes, however – systems in which the components do not exchange with the bulk over the operation of the machine – are there sufficient restrictions on the motions of the unbound species for them to feature controlled mechanical behavior and otherwise they are best considered supramolecular devices not mechanical machines [58].

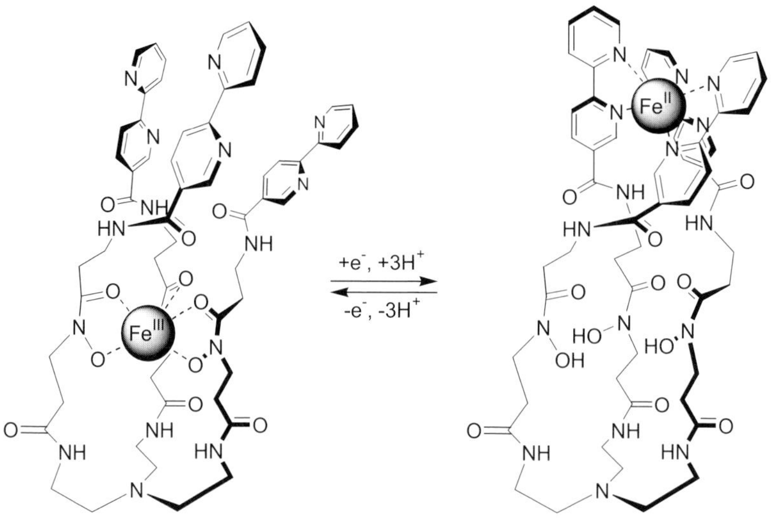

Figure 7.16 Redox-switched intramolecular cation translocation in a triple-stranded helical complex. The reductant is ascorbic acid and the oxidant $(NH_4)_2S_2O_8$.

7.5
Controlling Motion in Interlocked Systems

7.5.1
Basic Features

Catenanes are chemical structures in which two or more macrocycles are interlocked; in rotaxanes one or more macrocycles are mechanically prevented from de-threading from linear chains by bulky "stoppers" (Fig. 7.17) [59]. Although the components are not covalently connected, catenanes and rotaxanes are molecules – not supramolecular complexes – because covalent bonds must be broken to separate the constituent parts. In these kinetically associated species [60, 61] the mechanical bond restricts the degrees of freedom for relative movement of the components while often permitting extraordinarily large amplitude motion in the allowed vectors. This is directly analogous to the restriction of movement imposed on biological motors by a structural track [46] and is one reason interlocked structures have played a central role in the development of synthetic molecular machines [62].

Large-amplitude submolecular motion in catenanes and rotaxanes can be divided into two classes (Fig. 7.17) – pirouetting of the macrocycle around the thread (rotaxanes) or the other ring (catenanes), and translation of the macrocycle along the thread (rotaxanes) or around the other ring (catenanes). By analogy with conformational changes within classical molecules, the relative movements between interlocked species are termed co-conformational changes [22]. Motion in these systems is most

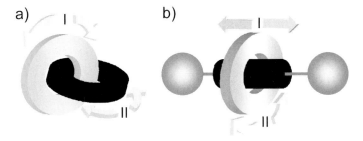

Figure 7.17 Cartoon representations of (a) a [2]catenane and (b) a [2]rotaxane [63]. Arrows show possible large-amplitude modes of movement for one component relative to another.

easily described by assuming one component to be a stationary frame of reference around or along which the other units move. Often the choice of this frame of reference is based on the relative size or complexity of the components. The distinction can become ambiguous, however, as discussed below.

For a long time these non-classical architectures were no more than academic curiosities, particularly as synthetic approaches relied on inefficient statistical or time-consuming covalently-directed approaches [64]. Introduction of supramolecular chemistry, however, enabled chemists to apply an understanding of noncovalent interactions to synthesis, resulting in various template methods to catenanes and rotaxanes [65]. In such syntheses noncovalent binding interactions between the components often "live-on" in the interlocked products. These interactions enable restriction of the relative motion of components, further to that defined by their architecture, and they can ultimately be manipulated to affect positional displacements. Much attention has been devoted both to intrinsic submolecular motion within these structures and control of this motion by manipulation of recognition events.

7.5.2
Inherent Dynamics: Ring Pirouetting in Rotaxanes

Pirouetting is the random rotation of the macrocycle about the axis defined by the thread. While this motion is often observed to be occurring, it can be difficult to study in detail because of the symmetry of the components. Benzylic amide macrocycle-based rotaxanes possess a useful characteristic in this respect. In many examples of these molecules, the benzylic amide macrocycle adopts a chair-like conformation meaning that for each pair of benzylic protons (H_E, Fig. 7.18), one is in an equatorial environment, while the other is axial [66]. For a macrocycle on a symmetric thread, two ^1H NMR signals would therefore be expected for the 8 benzylic protons in the molecule. Rotation of the macrocycle around the axis shown by 180° must, however, result in a chair-chair flip so as to maintain the hydrogen bonding network between the macrocyclic amide protons and carbonyl oxygen atoms on the thread. This co-conformational change therefore interconverts the axial and equatorial sets of protons twice during a full 360° revolution. It is possible to study chemical exchange process-

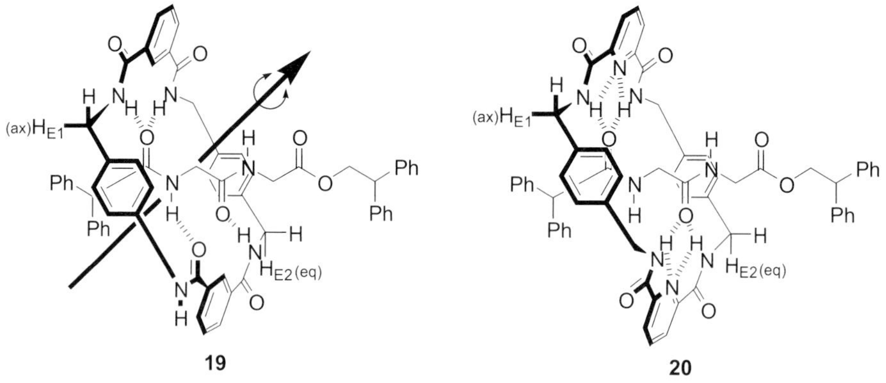

Figure 7.18 Peptidorotaxanes **19** and **20**. The arrow on **19** indicates the axis about which macrocycle pirouetting occurs.

es such as this by variable temperature (VT) NMR techniques, including the coalescence method [67], or spin polarization transfer by selective inversion recovery (SPT-SIR) [68].

For example, the room temperature ^{1}H NMR spectrum in CDCl$_3$ of the glycyl-glycine-based [2]rotaxane **19** contains the fewest possible signals for the macrocycle protons, indicating rapid pirouetting (820 s^{-1}) of the ring at ambient temperature [66]. The pyridyl-2,6-dicarbonyl-based macrocycle in **20** has a different, overall slightly stronger, hydrogen-bonding network between the components, producing a concomitant reduction in pirouetting rate to 100 s^{-1}.

As the rate of pirouetting is directly related to the strength with which the macrocycle is held to the thread, this motion can be a useful probe for evaluating the effect of structural changes on the strength of intercomponent interactions in rotaxanes. For the series of fumaric acid-based [2]rotaxanes shown in Fig. 7.19, yields of the rotaxane forming process (a five-component clipping reaction in which formation of the benzylic amide macrocycle is templated by the fumaric acid residue) increase in the order **23** < **22** < **21** [69]. As expected, the rates of macrocycle pirouetting in the products were shown to follow exactly the opposite order (Fig. 7.19), confirmation that the efficiency of the templated synthesis is directly related to the hydrogen-bonding ability of the thread templates.

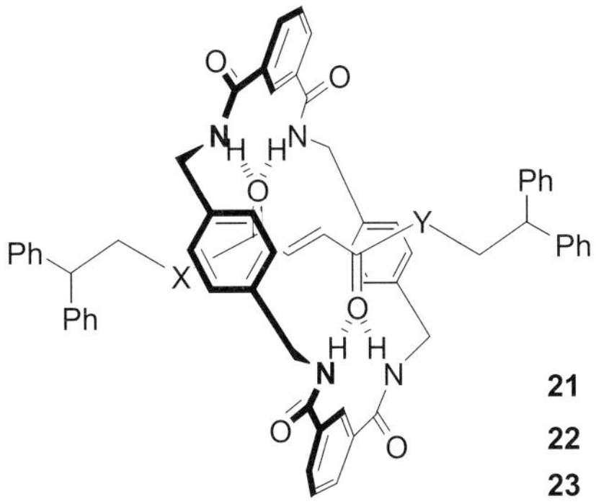

	X	Y	Yield	$\Delta G^{\ddagger}_{298}$ / kcal mol^{-1}
21	NH	NH	97%	11.5
22	NH	O	35%	9.3
23	O	O	3%	7.2

Figure 7.19 Variation of yield in rotaxane-forming reactions for a series of simple fumaric acid based templates is correlated with the strength of intercomponent interactions in the products, as evidenced by $\Delta G^{\ddagger}_{298}$, the energy barrier to pirouetting in CD_2Cl_2.

7.5.3
Inherent Dynamics: Ring Pirouetting in Catenanes

Pirouetting in catenanes is not quite so easily defined. Spinning of one macrocycle around an axis defined by its interlocked partner is directly analogous to the pirouetting motion in rotaxanes – the "thread" is simply cyclic in this case (e.g. motion I, Fig. 7.17a is pirouetting of the gray ring around the "stationary" black one). This same motion could, on the other hand, be regarded as rotation of the second macrocycle around the first in a manner analogous to shuttling in rotaxanes (i.e. motion I, Fig. 7.17a could also be regarded as rotation of the black ring around the "stationary" gray ring). Likewise, motion II can be considered to be pirouetting or rotation depending on which ring is taken as the stationary frame of reference. In practice, for the simplest case of homocircuit catenanes (when gray = black), the distinction is unnecessary, as motions I and II are identical, the overall result often being termed circumvolution (also called "circumrotation") if a full 360° rotation is involved.

7.5.3.1 Metal-based Homocircuit Catenates and Catenands
As a result of the degeneracy of the pirouetting and rotational motions, homocircuit catenanes are the simplest systems in which the intrinsic dynamics of interlocked architectures can be studied. The transition metal templated [2]catenates produced by the Sauvage group (e.g. 24, Fig. 7.20) were the first interlocked molecules to be produced by template synthesis [70–72]. The catenates have both rings coordinated to a metal ion and so are co-conformationally locked and no large-amplitude submolecular motion is observed. Removal of the metal template forms the corresponding catenand. The lack of intercomponent interactions in catenands usually means that the relative motion of the two rings is extremely fast and are difficult to study.

24

Figure 7.20 Chemical structure of the first transition metal templated catenate, **24**.

7.5.3.2 Amide-based Homocircuit Catenanes

Amide-based catenanes, in which hydrogen bonds template interlocking of the two rings, were first reported in 1992 by Hunter (**25**) [73], prompting the publication of similar work by Vögtle (**26**) (Fig. 7.21) [74]. In these systems, the steric bulk of the cyclohexyl groups prevents full circumvolution of the macrocycles. A 90° rotational motion which exchanges the non-equivalent isophthaloyl rings (labeled "1" and "2" in

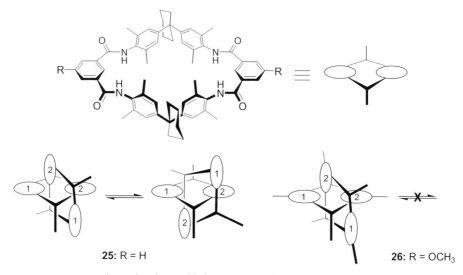

25: R = H   **26: R = OCH$_3$**

Figure 7.21 Hydrogen bond-assembled [2]catenanes **25** and **26** reported by Hunter and Vögtle respectively, showing the large-amplitude motion possible in **25**, but not in **26**.

Fig. 7.21) is possible, but even this is blocked on substitution at the 5-position of the isophthaloyl ring (e.g. in **26**).

The amide-based homocircuit catenane **27** (Fig. 7.22), discovered by chance in 1995, has more interesting and versatile dynamics [75]. The catenane has the simplest possible ^{1}H NMR spectrum for this kind of structure, only the six different constitutional types of proton are apparent in DMSO-d$_6$, the same number seen for the parent macrocycle [76]. Despite the smaller macrocycle cavity size of **27** compared with **25** and **26**, both the isophthaloyl and p-xylylene components can rotate through the cavity of the other macrocycle and the circumvolution process is rapid on the NMR timescale in polar solvents at room temperature. The catenane-formation process tolerates a number of aromatic 1,3-dicarbonyl and benzylic amide precursors, so a diverse range of analogs could be prepared and the effect of structure on the dynamic processes assessed [77].

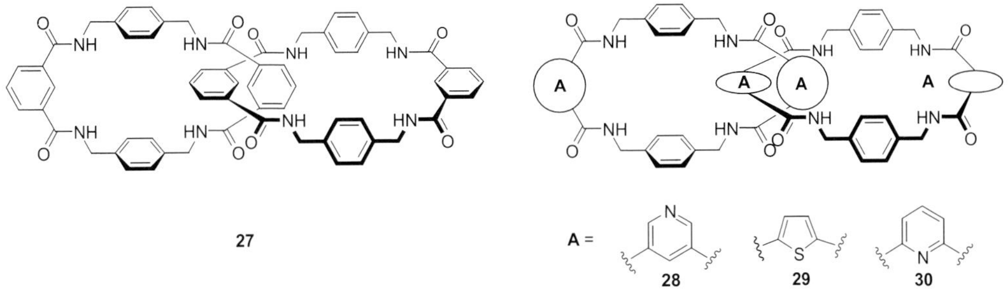

Figure 7.22 The original hydrogen bond-assembled benzylic amide [2]catenane **27** and structural analogs **28–30** with different aromatic-1,3-dicarbonyl units.

Any substitution on the isophthaloyl ring (at either the 5- or 4-position) prevents complete circumvolution of the macrocycles, thereby destroying the plane of symmetry bisecting the isophthaloyl units in each ring. A series of coalescence temperature measurements and SPT-SIR NMR experiments [77, 78] revealed circumvolution to be slow on the NMR timescale for the pyridine-2,6-dicarbonyl derivative (**30**) – just as its pirouetting rates are slower than other macrocycles in rotaxanes (Sect. 7.5.2). The thiophene derivative (**29**) was shown to rotate the fastest – in fact 3.2 million times faster than **30** in C$_2$D$_2$Cl$_4$ at room temperature.

This quite remarkable variation in dynamic behavior resulting from relatively simple structural changes can also be allied to the more subtle effects on rate of solvent composition. Hydrogen-bond-disrupting solvents such as CD$_3$OD and, to an even greater extent, (CD$_3$)$_2$SO increase the rate of circumvolution by competing for the hydrogen bonding groups in the macrocycle. This has the effect of weakening the ground-state interactions and stabilizing the intermediate co-conformations during circumvolution, thus lowering the energy barriers for the process. Variation of solvent composition enables fine-tuning of the circumvolution rate after selection of the approximate value by appropriate choice of structure [77, 78].

Both these solvent and structural effects on rate suggest that a key process in the circumvolution mechanism is rupture and formation of intercomponent hydrogen bonds. An isolated molecule molecular mechanics approach has been used to simulate the dynamics in these catenane systems, enabling simulation of the molecular shape at the transition states [78, 79]. Together with a full low-dimensional quantum-mechanical description of circumvolution [80], these studies enable the theoretical and practical understanding of the intrinsic motion in a catenane system necessary for development of molecular devices and materials in which the frequency of motion is a key characteristic.

7.5.3.3 Heterocircuit Catenanes

The very first catenanes in which dynamic processes were studied were not homo-circuit systems but rather molecules comprising two different interlocked rings. Most heterocircuit catenanes (gray ≠ black in Fig. 7.17a) consist of a larger ring containing two or more recognition sites for a smaller macrocycle. It is, therefore, usually convenient to consider such systems analogously to rotaxanes with the larger component the stationary frame of reference around which the smaller unit(s) move. [2]Catenane **31**$^{4+}$ is composed of a π-electron-deficient cyclophane (cyclobis(paraquat-*p*-phenylene or CBPQT^{4+}, Fig. 7.23) interlocked with an electron-rich bis-dioxyarene crown-ether macrocycle [81, 82]. To maximize intercomponent charge-transfer and hydrogen-bonding interactions, the cyclophane must circumscribe one of the hydroquinone units in the crown ether. Given this constraint, the two large-amplitude dynamic processes can be portrayed as pirouetting of the cyclophane around the occupied dioxyarene unit (process II, Fig. 7.23, alternatively viewed as rotation of the crown ether around the cyclophane) and rotation of the cyclophane around the crown ether from one dioxyarene station to another (process I, Fig. 7.23, alternatively viewed as pirouetting of the crown ether through the cyclophane). Both processes are observable in ^{1}H and ^{13}C VT NMR studies, enabling calculation of activation barriers for the two processes; process I is the higher in energy.

Enlargement of the electron-rich macrocycle in **32**$^{4+}$, which contains four dioxybenzene stations, renders the distinction between pirouetting and rotation less equivocal; the larger ring is intuitively regarded as stationary with the smaller one moving around it. By employing high pressure methods the related [3]catenane **33**$^{8+}$ can also be obtained [83]. Interestingly, two non-degenerate translational isomers are now possible for the rotational motion of the two cyclophanes – the two tetracations occupying adjacent stations (*proximal* isomer) or opposite stations (*distal* isomer, shown). The former situation is not observed [84], with the tetracationic cyclophanes preferring to minimize electrostatic repulsion by maximizing the distance between themselves at all times. More recently, the Stoddart group have reported an analogous, hydrogen bonded, [3]catenane which contains four secondary ammonium recognition sites in the large macrocycle as stations for two uncharged polyether macrocycles [85]. In this case, both translational isomers *are* observed, in approximately equal proportions. Given that, statistically, a 2:1 ratio of *proximal:distal* isomers would be expected, the observed ratio suggests that while clearly no large electrostatic repulsion exists, there is still some small (probably steric) interaction destabilizing the *proximal* isomer [86].

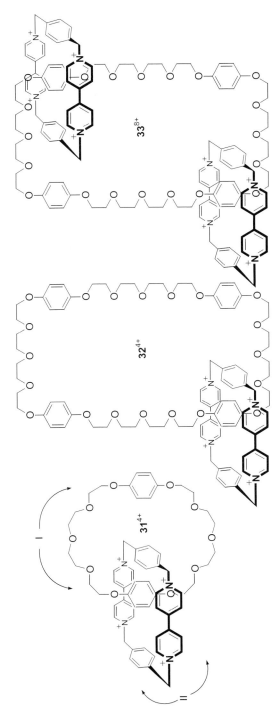

Figure 7.23 Examples of the first catenanes in which dynamic processes were studied. [3]Catenane **33**⁸⁺ is shown in the preferred *distal* translational isomer.

7.5.4
Inherent Dynamics: Shuttling in Rotaxanes

Shuttling is the movement of a macrocycle back and forth along the linear thread component of a rotaxane (or one macrocycle around another in a catenane). This motion takes the form of a Brownian motion-powered random walk, constrained to one dimension by the thread and to translational displacement boundaries by the bulky stoppers. By virtue of the template synthesis methods employed in interlocked molecule synthesis [65], rotaxanes without sites of attractive interaction between macrocycle and thread are relatively rare [87]. It is more common that the thread consists of one or more recognition elements, or "stations" for the macrocycle(s); shuttling therefore becomes the movement on, off or between such stations and, just as for the pirouetting motions, the dynamics depend on the strength of the intercomponent interactions.

7.5.4.1 Observation of Shuttling in Degenerate Shuttles
In the first [2]rotaxane for which the dynamics were studied, Stoddart and coworkers demonstrated shuttling behavior in 34^{4+} (which uses the same π–π interactions as the catenanes discussed in Sect. 7.5.3.3) [88]. In ^{1}H NMR experiments the behavior of proton signals from both components was consistent with the macrocycle moving between the two identical hydroquinol stations in a temperature-dependent fashion (Fig. 7.24). Directly analogous situations exist for several similar molecules [89].

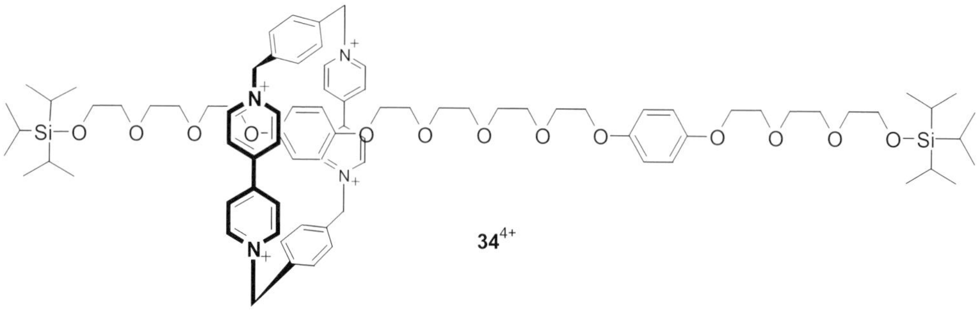

34^{4+}

Figure 7.24 The first "molecular shuttle".

Following the observation that hydrogen-bond-disrupting solvents such as $(CD_3)_2SO$ destroy the interactions between the benzylic amide macrocycle and a glycylglycine thread in [2]rotaxane **19** (Fig. 7.18) [66], a series of peptide-based molecular shuttles were described in which two glycylglycine stations for the benzylic amide macrocycle are separated by aliphatic linkers (Fig. 7.25) [90]. In both **35** and **36**, in $CDCl_3$, the macrocycle shuttles rapidly between the two degenerate peptide stations at room temperature, evidenced by the single set of signals for the two peptide stations in the ^{1}H NMR spectrum being resolved into two sets (for the occupied and unoccupied stations) on cooling the sample and freezing out the motion on the NMR timescale.

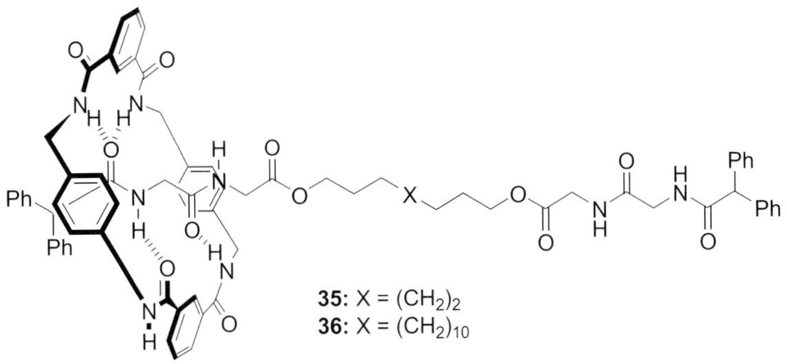

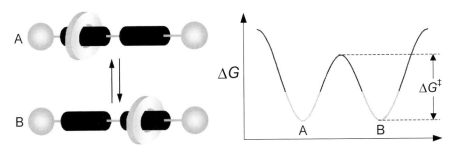

35: X = (CH₂)₂
36: X = (CH₂)₁₀

Figure 7.25 Peptide-based degenerate molecular shuttles **35** and **36**.

Just as for the pirouetting motions, the shuttling mechanism must require at least partial rupture of the intercomponent interactions at one station before formation of new interactions at the new station. The thermodynamics of the degenerate process in each of **34–36** therefore involves passage over an activation barrier from one energy well to another, identical, minimum (Fig. 7.26). Just as for the analogous benzylic amide catenanes, solvent composition has a profound effect on the shuttling rate in **35** and **36** – as little as 5 % CD₃OD in halogenated solutions of peptide-based rotaxanes leads to rate increases in excess of two orders of magnitude. The hydrogen-bond-disrupting methanol weakens the intercomponent interactions, effectively loosening the macrocycle from its station (i.e. reducing $\Delta G^{\ddagger}$) with a concomitant increase in the shuttling rate [90].

This increase in rate does not, however, continue indefinitely with increasing solvent hydrogen bond basicity. An additional feature of the rotaxanes compared with the analogous benzylic amide catenanes discussed in Sect. 7.5.3.2 is that a major change in solvent polarity (changing from halogenated solvents to (CD₃)₂SO) stops the macrocycle shuttling between the peptide stations and causes it to preferentially

A

ΔG

B

$\Delta G^{\ddagger}$

A   B

Figure 7.26 Idealized free-energy profile for movement between two identical stations in a degenerate molecular shuttle. The height of the barrier $\Delta G^{\ddagger}$ contains two components –

the energy required to break the noncovalent interactions holding it to the station and a distance-dependent diffusional component.

sit over the hydrophobic thread instead, hiding the thread from the unfavorably polar environment and enabling maximum hydrogen bonding between the peptide stations and solvent [90].

7.5.4.2 A Physical Model of Degenerate Molecular Shuttles

An interesting structural effect is observed on increasing the length of the spacer in degenerate shuttles. The rate of shuttling in [2]rotaxane **36** was compared to that in **35** (Fig. 7.25). Although ostensibly not involved in any interactions with the macrocycle, extension of the alkyl chain results in an experimentally measured reduction in the rate of shuttling which would correspond to an increase in activation energy for the process of 1.2 kcal mol^{-1} – an effect solely of the increased distance the macrocycle must travel [80].

All these effects are most clearly understood by considering the macrocycle as a particle moving along a one-dimensional potential energy (rather than free energy) surface (Fig. 7.27). At any point on the potential energy surface the gradient of the line gives the force exerted on the macrocycle by the thread. When the macrocycle is in the vicinity of a station, hydrogen bonds and/or other attractive noncovalent inter-

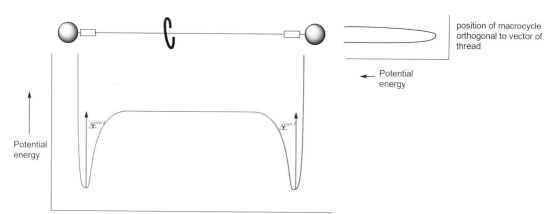

Figure 7.27 Idealized potential energy of the macrocycle in a two-station degenerate molecular shuttle. The potential energy surface shows the effect of the interaction between macrocycle and thread on the energy of the macrocycle (ignoring any complicating factors such as folding). Chemical potential energies (ΔE) usually follow trends similar to those of free energies (ΔG, Fig. 7.26) but there are some important differences; for example, the activation free energy of shuttling, $\Delta G^{\ddagger}$, corresponds to the energy required for the macrocycle to move all the way to the new binding site (i.e. includes a contribution for the distance the ring has to move along the track to reach the other station) whereas the $\Delta E^{\ddagger}$, shown here, is the energy required for the macrocycle to escape the forces exerted through noncovalent binding interactions at a station. The main plot shows ΔE in terms of the position of the macrocycle along the vector of the thread; the minor plot shows the ΔE in terms of the position of the macrocycle orthogonal to the vector of the thread, illustrating that the thread genuinely behaves as a one-dimensional potential-energy surface for the macrocycle.

actions exert large forces (typically varying as a high power of the inverse distance, r^{-n}) on the macrocycle, opposing its motion. When the macrocycle is on the thread *between* the stations, the hydrogen bonds and other noncovalent binding interactions are broken and no forces are exerted on the macrocycle by the thread (i.e. the gradient of the line is zero).

It is thermal energy which enables the macrocycle to escape these energy wells and explore the full length of the thread. For a population of shuttles at equilibrium, therefore, the macrocycles reside on the different stations and the thread according to a Boltzmann distribution and for a single molecule these populations correspond to the amount of time the macrocycle spends on each site. In a degenerate shuttle, in which the binding energy to each station is the same, and no significant interaction with the thread occurs, the macrocycle spends a negligible amount of time on the thread and splits its time equally between the two stations.

The rate at which the macrocycles escape from each station is given by a standard Arrhenius equation, depending on the depth of the energy well and the temperature. This is not the only factor involved in determining the rate at which macrocycles move between the two stations, however, there must also be some distance-dependent diffusional factor in the function describing rate of shuttling – easily understood from consideration of Fig. 7.27.

This phenomenon has been studied further in an attempt to link the experimentally determined values of $\Delta G^{\ddagger}$ with a quantum-mechanical description of the shuttling mechanism [91]. Calculation of the wavefunctions for the system shows that an increase in distance between the stations does not, of course, change the activation energy for cleavage of hydrogen bonds but rather, the effect is to widen the free energy potential well (Fig. 7.26). As this well widens, it has a higher density of states per unit energy (just like the simple "particle in a box" model). The closer to one another the levels are, the more readily thermally populated they are or, in other words, the larger is the partition function and hence $\Delta G^{\ddagger}$.

The nature of the wavefunctions at energies close to the top of the barrier is intriguing, because under this energy regime the maximum probability of finding the macrocycle is, in fact, over the aliphatic spacer. The shuttling process can therefore be thought of in terms of a function of free energy (Fig. 7.26) as long as we remember that the height of the $\Delta G^{\ddagger}$ barrier is affected by *both* binding strength and distance between the stations. The behavior of the macrocycle is similar to a cart moving along a roller coaster track shaped like the double potential in Fig. 7.26. At low temperatures, the cart mostly resides on the stations (oscillating with small amplitude in the troughs). As energy approaches the value of the barrier, the cart spends most of its time passing over the barrier. At higher energies still, the cart is most likely to be found at the extremes of its translational motion.

7.5.5
Controlling Translational Motion: Molecular Shuttles

With increasing understanding of the nature of the inherent restriction in degrees of freedom in interlocked architectures came the realization that control of inter-component positioning was achievable [92]. Switching on and off degenerate shuttling in a rotaxane or varying the frequency of circumvolution in a catenane by solvent effects, structural effects or temperature demonstrates control of motion at the most basic level and only hints at the possibilities which have now been realized.

In shuttles such as **34**$^{4+}$, considered in Sect. 7.5.4.1, there are two identical recognition sites (stations) for the macrocycle so that it is equally likely to reside on either – they are degenerate. As we have seen, the rate at which the macrocycle moves between the stations can be regulated by the temperature or, occasionally, solvent composition. A different kind of control, however, can be achieved in stimulus-responsive shuttles, in which the net location of the macrocycle can be varied by applying an external stimulus.

7.5.5.1 Single-station Switchable Shuttles

A limited amount of control over shuttling can be introduced into single-station rotaxanes if the affinity of the macrocycle recognition site can be affected by a stimulus. [2]Rotaxane **37**$^{4+}$ (Fig. 7.28) is closely related to those produced by the Stoddart group and features attractive interactions between a cyclobis(paraquat-*p*-phenylene) cyclophane and an electron-rich dialkoxybenzene station in the thread to template its formation [93]. In **37**$^{4+}$, however, the bulky stoppers are redox-active ferrocenyl residues. A laser flash photolysis pulse within the charge-transfer band of **37**$^{4+}$ induces electron transfer between the two interacting components, giving an intimate radical–ion pair (RIP). In closely related systems, decay of the RIP by charge recombination is rapid compared with any competing processes such as solvent penetration or spatial separation of the radical ions. The flexible thread, however, enables close contact of the electron rich ferrocenyl stoppers with the cyclophane by means of a secondary π-stacking interaction. This electronic coupling enables some of the radical ion pairs (~25 %) to undergo a secondary electron transfer step in which the hole on the hydroquinone residue is transferred to one of the ferrocenyl stoppers. This transfer simultaneously breaks the interactions between the cyclophane and both the stopper and dialkoxybenzene, so the system is free to undergo conformational changes involving unraveling of the thread and shuttling of the cyclophane away from the oxidized stopper to reduce electrostatic repulsion. The lifetime of this "long-lived" species is approximately 550 ns – a timescale on which some shuttling of the cyclophane could occur before charge recombination restores the initial state; there is,

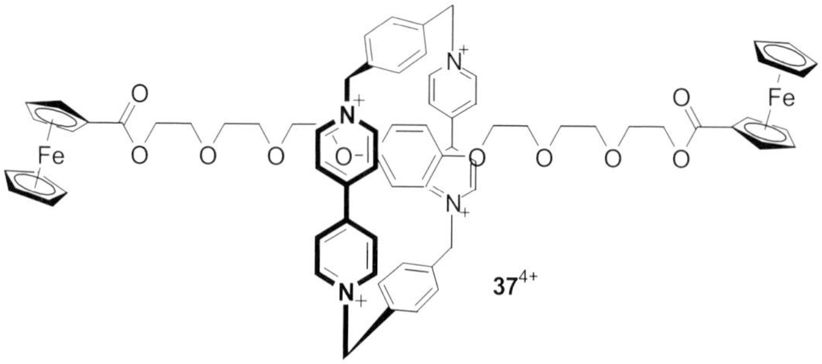

Figure 7.28 [2]Rotaxane **37**$^{4+}$ – a "single station" molecular shuttle.

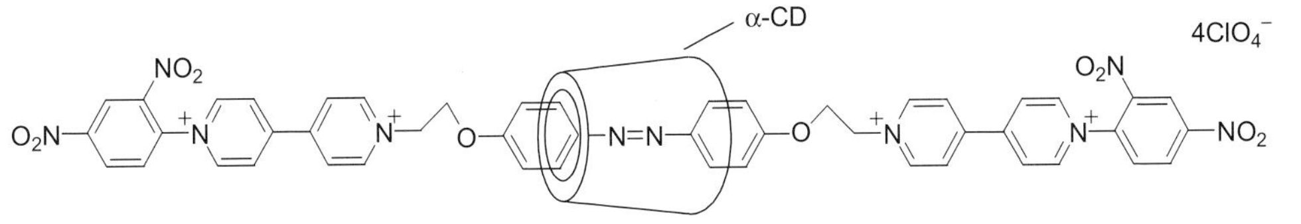

Figure 7.29 Photo-responsive single station shuttles **38** and **39**, based on azobenzene and stilbene units, respectively.

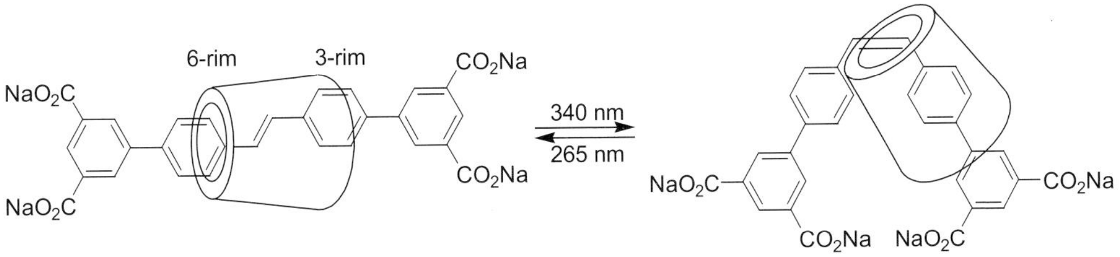

however, no direct evidence for this transient shuttling, nor means to control the position of the cyclophane in the spatially-remote charge-separated state.

It has also been possible to control the position of a cyclodextrin (CD) "macrocycle" on azobenzene and stilbene containing threads in two related systems [94, 95]. In the E isomers of [2]rotaxanes **38** and **39** (Fig. 7.29) in aqueous media the cyclodextrin spends most of its time over the central aromatic units. Irradiation at suitable wavelengths results in photoisomerization of the N=N and C=C double bonds in **38** and **39**, respectively, to the Z forms. The steric requirements of this "kinking" in the thread demands shuttling of the cyclodextrin away from the central units. Interestingly, in **39** [95], this motion is unidirectional, with the narrower 6-rim of the cyclodextrin always closest to the Z olefin. It was also demonstrated, however, that even in the E isomers the cyclodextrin only spends approximately two-thirds of its time over the central unit, and it is probably during its vacation of this site that photoisomerization occurs.

7.5.5.2 A Physical Model of Two-station, Stimuli-responsive, Molecular Shuttles

Rotaxanes in which the macrocycle can be translocated between two or more well-separated stations in response to an external signal should, in principle, provide a greater level of control for machines. As we have seen (Sect. 7.5.4), in any rotaxane the macrocycle distributes itself between the available binding sites according to the difference between the macrocycle binding energies and the temperature. If a suitably large difference in macrocycle affinity between two stations exists, the macrocycle resides overwhelmingly in one positional isomer or co-conformation [22]. In stimuli-responsive molecular shuttles, an external trigger is used to chemically modify the system and alter the noncovalent intercomponent interactions such that the second macrocycle binding site becomes energetically more favored, causing translocation of the macrocycle along the thread to the second station (Fig. 7.30). This might be

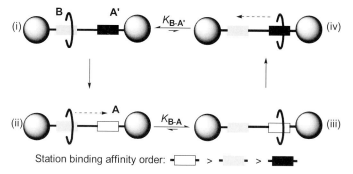

Figure 7.30 Translational submolecular motion in a stimulus-responsive molecular shuttle: (i) the macrocycle initially resides on the preferred station (B); (ii) a reaction occurs (A' → A) changing the relative binding potentials of the two stations such that, (iii), the macrocycle "shuttles" to the now-preferred station (A). If the reverse reaction (A → A') now occurs, (iv), the components return to their original positions.

achieved by addressing either of the stations (destabilizing the initially preferred site or increasing the binding affinity of the originally weaker station). The system can be returned to its original state by using a second chemical modification to restore the initial order of station-binding affinities. Performed consecutively these two steps enable the "machine" to perform a complete cycle of shuttling motion.

The physical basis for this motion is again best understood by consideration of the potential energy of the macrocycle as a function of its position along the thread (Fig. 7.31). It is important to appreciate that the external stimulus does not induce directional motion of the macrocycle *per se*, rather, by increasing the binding strength of the less populated station and/or destabilizing the initially preferred binding site, the system is put out of co-conformational equilibrium [96]. Relaxation towards the new global energy minimum subsequently occurs by *thermally activated* motion of the components, a phenomenon which we recognize as biased Brownian motion. In other words, biased Brownian motion arises from a difference in the activation energies for movement in different directions, *not* from the difference in energy minima. This results in net directional transport (a directional flux) of macrocycles when these barriers are suddenly changed, putting the system out of equilibrium.

Given this mode of action, a key requirement is finding ways of generating sufficiently large, long-lived binding energy differences between pairs of positional isomers. A Boltzmann distribution at 298 K requires a $\Delta\Delta E$ (or $\Delta\Delta G$) between translational co-conformers of ~2 kcal mol^{-1} for 95 % occupancy of one station. Achieving such discrimination in *two* states to form a positionally bistable shuttle (i.e. both $\Delta\Delta E_{\text{B-A'}}$ and $\Delta\Delta E_{\text{B-A}} \geq 2$ kcal mol^{-1}) by modifying only intrinsically weak, noncovalent binding modes is thus a significant challenge.

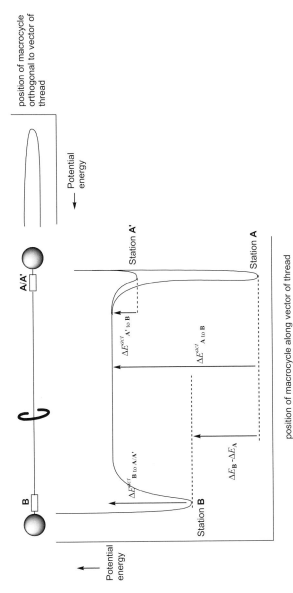

Figure 7.31 Idealized potential energy of the macrocycle in a stimulus-responsive molecular shuttle in which one station changes (A' → A) in response to the stimulus, and complicating factors such as folding are ignored. As before (Fig. 7.27), the potential energy surface shows the effect of the interaction between macrocycle and thread on the energy of the macrocycle. The main plot shows ΔE in terms of the position of the macrocycle along the vector of the thread; the minor plot shows ΔE in terms of the position of the macrocycle orthogonal to the vector of the thread.

7.5.5.3 Adding and Removing Protons to Control Shuttling

In fact, the very first bistable switchable molecular shuttle, reported by Stoddart and Kaifer in 1994, was a two-station design of this type [97]. The biphenol and benzidine units in the thread of [2]rotaxane 40^{4+} (Fig. 7.32) are both potential π-electron donor stations for the CBPQT^{4+} cyclophane, and at room temperature rapid shuttling of the macrocycle occurs, as in the related degenerate shuttles described above (Sect. 7.5.4.1). Cooling the sample to 229 K enables observation (by NMR and UV–visible absorption spectroscopy) of the two non-degenerate translational isomers in a ratio of 21:4 in favor of encapsulation of the benzidine station. Protonation of the basic benzidine residue with CF_3CO_2D results in electrostatic repulsion between the positively charged station and cyclophane so that the CBPQT^{4+} resides exclusively on the biphenol station. The system can subsequently be restored to its initial state on neutralization with pyridine-d_5.

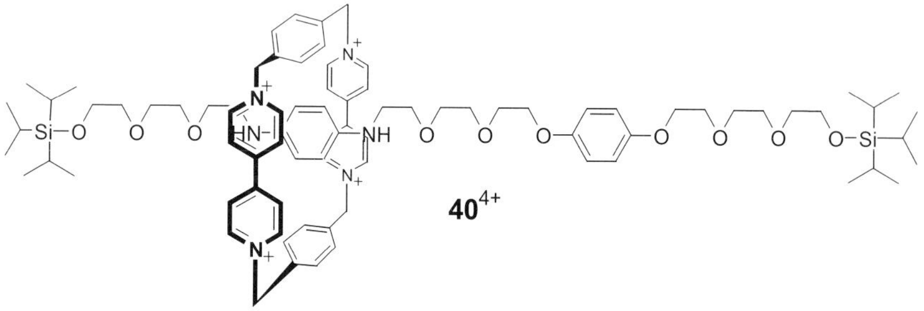

Figure 7.32 The first switchable molecular shuttle 40^{4+}, shown in the preferred co-conformation in the neutral state.

Although this system has many of the attributes of a fully controllable molecular shuttle, it has modest positional integrity in the non-protonated state – a much larger binding difference between the two stations was required. Inspired by the complexation of ammonium ions by crown ethers, Stoddart and coworkers developed a new series of systems based on threads containing secondary alkylammonium species and crown ether macrocycles [98]. [2]Rotaxane $41 \cdot H^{3+}$ (Fig. 7.33) was the first switchable molecular shuttle reported using these units [99]. The dibenzo-crown ether is bound to the ammonium cation by means of [N$^+$–H$\cdots$O] hydrogen bonds and well as weaker [C–H$\cdots$O] bonds from methylene groups in the α-position to the nitrogen. ^{1}H NMR spectroscopy in CD_3COCD_3 shows no shuttling motion, with the crown ether sitting overwhelmingly (but *not* exclusively [113]) over the cation, to the limits of detection set by this experimental technique even at room temperature. Deprotonation of the ammonium center with diisopropylethylamine turns off the interactions holding the macrocycle to this station, so it is free to shuttle on to the alternative bipyridinium station (41^{2+}). Despite no observation of binding between the crown ether and bipyridinium units in a non-interlocked pseudorotaxane model system, NMR studies on the deprotonated rotaxane show the system again adopting a co-conformation with high of positional integrity, sitting over the

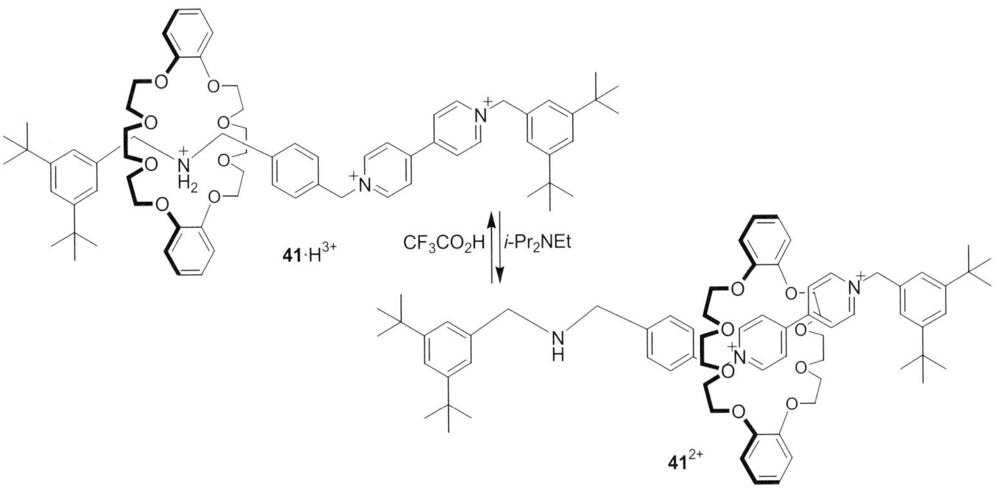

Figure 7.33 A pH-responsive bistable molecular shuttle with excellent positional integrity in both chemical states.

bipyridinium unit in an asymmetric fashion; a corresponding yellow charge-transfer interaction between the catechol rings and bipyridinium station is observed [100].

Although excellent macrocycle positional integrity in both chemical states is observed for this example, the low binding constant between the crown ether and bipyridinium moieties in the deprotonated state might be a limitation in more complex systems, especially if the macrocycle is afforded a greater degree of translational freedom (the distance between the stations in the current system is ~7 Å). Accordingly, Stoddart and coworkers have prepared a combination of three shuttle units in parallel using this system [101]. The molecule (**42·3H⁹⁺**) consists of three thread-like components connected at one end and encircled by three catechol–polyether macrocycles connected in a platform-like fashion (Fig. 7.34). Essentially, the shuttling action works precisely as in **41**, just over three equivalents of base are required to move the platform from ammonium ("top") to bipyridinium ("bottom") positions. The combined effect of three binding sites is excellent positional integrity in both positions.

The first pH switched shuttle to exploit [N-H···anion] hydrogen bonding interactions has also been reported [102]. In [2]rotaxane **43·H** (Fig. 7.35), formation of the benzylic amide macrocycle is templated by a succinamide station in the thread. The thread however also contains a cinnamate derivative (related to the *p*-coumaric acid chromophore of PYP, see Sect. 7.3). In the neutral form, the cinnamate phenol is a relatively poor hydrogen-bonding group and the macrocycle resides on the succinamide station >95 % of the time. Deprotonation to give **43⁻**, however, results in the macrocycle binding to the phenolate anion. Reprotonation of the phenol returns the system to its original state. Although a wide range of bases (with a variety of counterions) proved efficacious, shuttling was found to be extremely solvent-dependent. Hydrogen-bond-mediated systems usually perform best in "non-competing" solvents – those with low hydrogen bond basicity. Yet when deprotonation of **43·H** is cona)

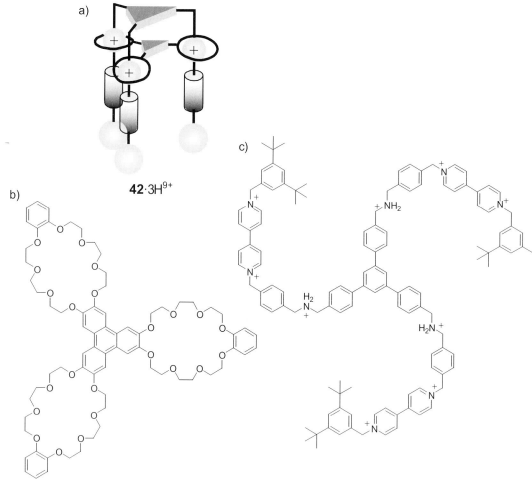

42·3H⁹⁺

b)

c)

Figure 7.34 (a) Schematic representation of a "molecular elevator" **42·3H⁹⁺**, with the "platform" on the "top" dialkylammonium station.

(b) Chemical structure of the macrocyclic "platform" component of **42**. (c) Chemical structure of "rig" component of **42**.

ducted in CDCl₃ or CD₂Cl₂, a change of position of the macrocycle does not occur. Rather an intramolecular folding event occurs to enable the phenolate to hydrogen bond with the macrocycle while it remains on the succinamide station. The interaction between macrocycle and succinamide station in **43·H** is so strong, however, that even in DMF-d₇ the excellent positional integrity is maintained. On deprotonation in this solvent, shuttling does indeed occur and in **43⁻** the only co-conformation detectable by NMR has the succinamide station vacated with the macrocycle hydrogen-bonding to the phenolate. This solvent dependence can be understood when we consider that the phenolate can only satisfy the hydrogen-bonding requirements of one isophthalamide unit in the macrocycle. The presence of a hydrogen-bond-accepting solvent such as DMF can therefore compensate for this loss of stabilization. The

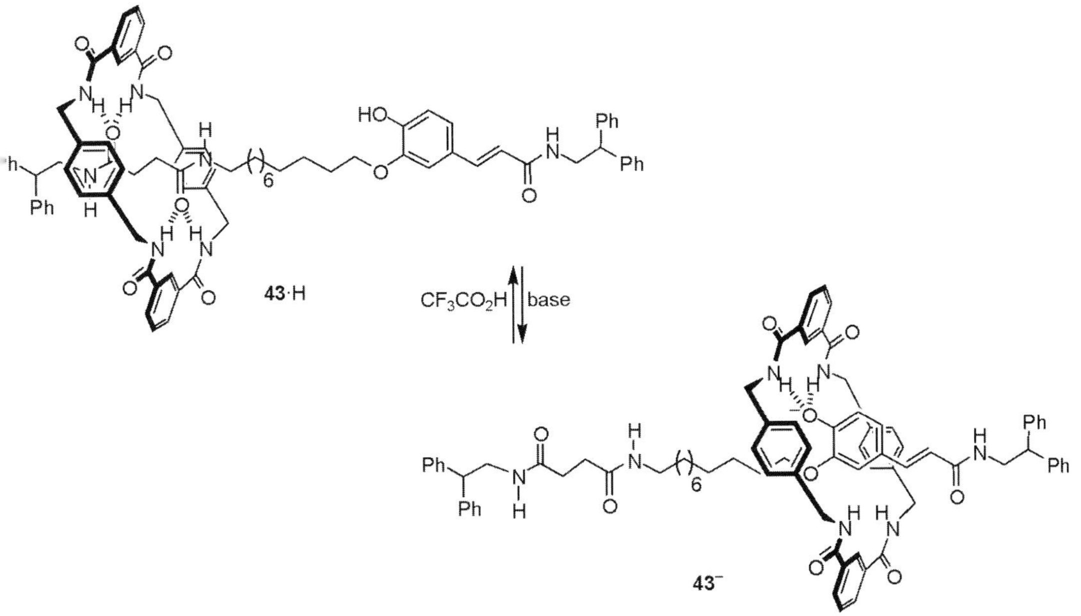

Figure 7.35 pH-switched anion shuttling in a hydrogen bonded [2]rotaxane. Bases used: LiOH, NaOH, KOH, CsOH, Bu$_4$NOH, tBuOK, DBU, Schwesinger's phosphazine P$_1$ base.

strength of binding to the anion is illustrated by the fact that the shuttling process continues to occur even in CD$_3$CN – only a moderate hydrogen-bond acceptor and weaker than the amide groups of the thread.

7.5.5.4 Adding and Removing Electrons to Control Shuttling

With the original switchable shuttle **40**$^{4+}$ it was found that the change of position could be achieved in a reagent-free manner by electrochemical oxidation of the benzidine station – shuttling away from this station occurring after oxidation to the radical cation. Unfortunately, several attempts to create analogous redox-switched shuttles with improved positional integrity failed to give the desired distribution of translational isomers in the ground state [103] and intrinsic shuttling at ambient temperatures always remained rapid [104, 105].

[2]Rotaxane **44** contains two potential hydrogen-bonding stations for the benzylic amide macrocycle – a succinamide (*succ*) station and a redox-active 3,6-di-*tert*-butyl-1,8-naphthalimide(*ni*) station – separated by a C$_{12}$ aliphatic spacer (Fig. 7.36) [106].

While the ability of the *succ* station to template formation of the macrocycle is well established, the neutral naphthalimide moiety is a poor hydrogen-bond acceptor. To minimize its free energy the macrocycle in **44** must therefore sit over the succinamide station in non-hydrogen-bonding solvents, so that co-conformation *succ*-**44** predominates (Fig. 7.37). In fact the difference between macrocycle-binding affinity is so great that *succ*-**44** is the only translational isomer detectable by ^{1}H NMR in CD-

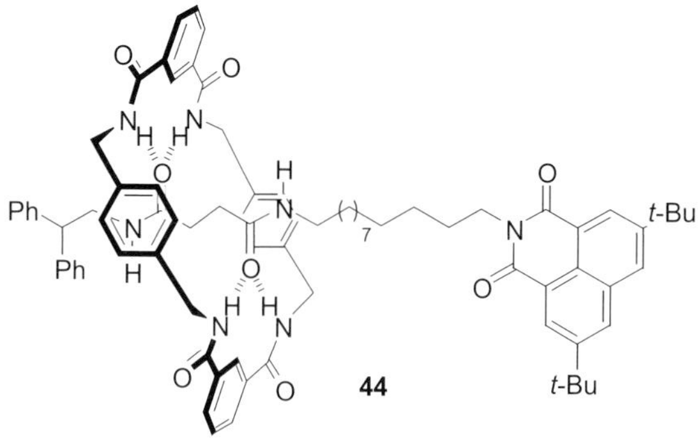

Figure 7.36 A photochemically and electrochemically addressable molecular shuttle.

Cl_3, CD_3CN and THF-d_8, and even in the strongly hydrogen-bond-disrupting $(CD_3)_2SO$ the macrocycle sits over the *succ* station approximately half the time. One-electron reduction of naphthalimides to the corresponding radical anion, however, results in a substantial increase in electron charge density on the imide carbonyls and a concomitant increase in hydrogen-bond-accepting ability. In **44**, this change in oxidation state reverses the relative hydrogen bonding abilities of the two thread stations so that co-conformation *ni*-**44**⁻ is preferred in the reduced state. Subsequent re-oxidation to the neutral state restores the original order of binding site affinities and the shuttle returns to its initial state as co-conformation *succ*-**44**. This process can be stimulated and observed in cyclic voltammetry experiments [106b] or, alternatively, photochemistry can be employed to initiate (through excitation of the naphthalimide group by a nanosecond laser pulse at 355 nm followed by electron transfer from a regenerable external electron donor) and observe (using transient absorption spectroscopy) the change of position [106a]. A number of control experiments were able to prove unequivocally that the dynamic process observed is reversible shuttling of the macrocycle between the stations rather than any other conformational or co-conformational change [106b].

Although many metal–ligand interactions can be rather kinetically stable, the availability of different oxidation states with different properties can be harnessed to control intercomponent motion in transition metal-based systems. [2]Rotaxane **45** (Fig. 7.38) exploits the unique properties (among the first row transition metals at least) of copper – in particular, the strong stereoelectronic requirements for the mono- and divalent cations [107]. The preference of Cu^I for four-coordinate tetrahedral complexes has been widely exploited to synthesize interlocked architectures (see Sect. 7.5.3.1), because it enforces the orthogonal arrangement of two bidentate ligands which can subsequently be cyclized to give interlocked species. In **45-I-N₄**, a macrocyclic ring containing a bidentate 2,9-diphenyl-1,10-phenanthroline (dpp) unit, is locked around a thread component which also contains one phenanthroline as well as one tridentate 2,2':6',2''-terpyridine (terpy) unit. The preference of Cu^I for tetrahedral geometries requires that the ring sits around the phenanthroline station in the

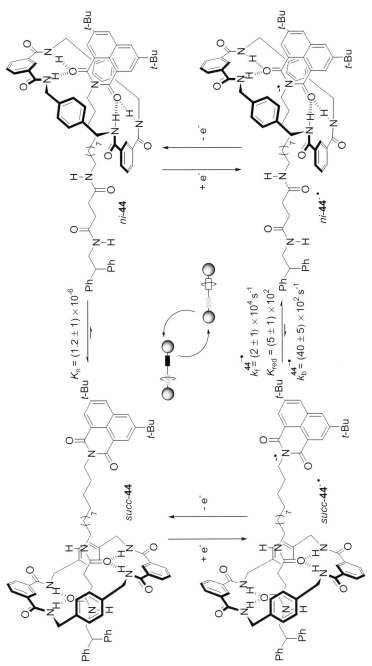

$K_n = (1.2 \pm 1) \times 10^{-6}$

$k_f^{44^{-\bullet}} = (2 \pm 1) \times 10^4 \ s^{-1}$

$K_{red}^{44^{-\bullet}} = (5 \pm 1) \times 10^2$

$k_b^{44^{-\bullet}} = (40 \pm 5) \times 10^2 \ s^{-1}$

ni-**44**

ni-**44**⁻•

succ-**44**

succ-**44**⁻•

Figure 7.37 The photochemically and electrochemically switchable, hydrogen-bonded molecular shuttle **44**. In the neutral state, the translational co-conformation *succ*-**44** is predominant because the *ni* station is a poor hydrogen-bond acceptor ($K_n = (1.2 \pm 1) \times 10^{-6}$). On reduction, the equilibrium between *succ*-**44**⁻• and *ni*-**44**⁻• is altered ($K_{red} = (5 \pm 1) \times 10^2$) because *ni*⁻• is a powerful hydrogen-bond acceptor and the macrocycle moves through biased Brownian motion. On reoxidation the macrocycle shuttles back to its original position. Repeated reduction and oxidation causes the macrocycle to shuttle forwards and backwards between the two stations. All the values shown refer to cyclic voltammetry experiments in anhydrous THF at 298 K with tetrabutylammonium hexafluorophosphate as the supporting electrolyte. Similar values were determined on photoexcitation and reduction of the ensuing triplet excited state by an external electron donor.

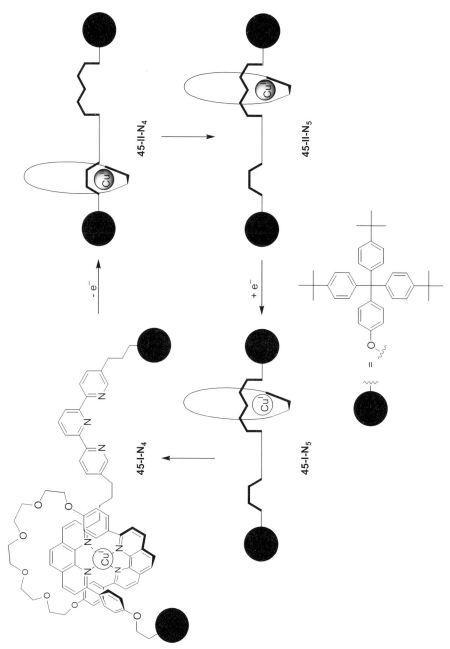

Figure 7.38 Redox-switched shuttling in a metal-templated [2]rotaxane, **45**.

thread with the metal ion coordinated between the two components. Electrochemical reduction gives **45-II-N$_4$**. As Cu^{II} prefers higher coordination numbers, thermally activated relaxation to the thermodynamically preferred **45-II-N$_5$** product occurs. Even the metastable **45-II-N$_4$** state is relatively kinetically inert, however, and this shuttling step takes several hours ($k = 1.5 \times 10^{-4}$ s^{-1}). Electrochemical reduction of the divalent species gives **45-I-N$_5$** which slowly converts to the starting material **45-I-N$_4$** ($10^{-4} \leq k \leq 10^{-2}$ s^{-1}). The shuttling process can also be accomplished by a photochemically-triggered oxidation. The reverse reductive step cannot be performed photochemically but proceeds successfully with a chemical reductant (ascorbic acid).

7.5.5.5 Changing Configuration to Control Shuttling

As with the control of submolecular motion in covalently bound systems, isomerization processes are an attractive means of controlling shuttling in rotaxanes. Shuttle *E/Z*-**46** (Fig. 7.39) employs the interconversion between fumaramide (trans) and maleamide (cis) isomers of the olefinic unit [108]. Fumaramide moieties are excellent binding sites for benzylic amide macrocycles – the trans olefin fixes the two strongly hydrogen-bond-accepting amide carbonyls in a close-to-ideal spatial arrangement for interaction with the amide protons from the macrocycle. Strong interactions between macrocycle and thread, in the guise of two sets of bifurcated hydrogen bonds, are the result. Although a similar hydrogen-bonding surface is presented to the macrocycle, binding to the succinamide station results in both a loss in entropy (because of loss of bond rotation) and also one less intracomponent hydrogen bond which cannot be reproduced in the rigid fumaramide station. The result is that only one positional isomer of *E*-**46** is observed in NMR studies at room temperature. Photoisomerization of the fumaramide station by irradiation at 254 nm reduces the number of possible intercomponent hydrogen bonds at this station from four to two so that a new co-conformational energy minimum now exists; the macrocycle now sits overwhelmingly on the succinamide station. Unlike the succinamide/naphthalimide system (Fig. 7.36, Sect. 7.5.5.4), this new state is indefinitely stable until a further

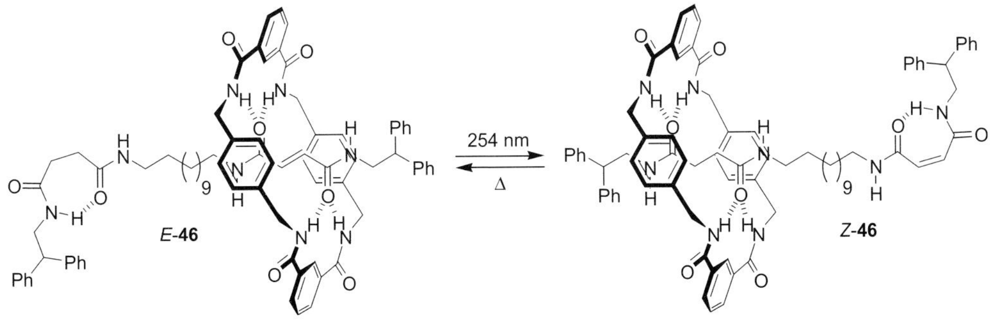

Figure 7.39 Bistable molecular shuttle *E/Z*-**46** in which self-binding of the "low affinity" station in each state is a major factor in producing excellent positional discrimination.

stimulus is applied – namely thermal or chemical re-isomerization of the maleamide unit back to fumaramide, thus restoring the shuttle to its initial state [109].

7.5.5.6 Entropy-driven Shuttling

Most of the shuttles which exhibit excellent positional discrimination can be switched using stimuli such as pH, light, polarity of the environment or electrochemistry to modify the enthalpy of macrocycle binding to one or both stations. In general the effect of temperature is simply to alter the extent of discrimination the macrocycle expresses for the different stations, not to alter the station preference [110]. In [2]rotaxane **47**, however, the macrocycle can be switched between stations simply by changing the temperature. In fact, **47** is a *tristable* molecular shuttle – a rotaxane in which the ring can be switched between three different positions on the thread (Fig. 7.40).

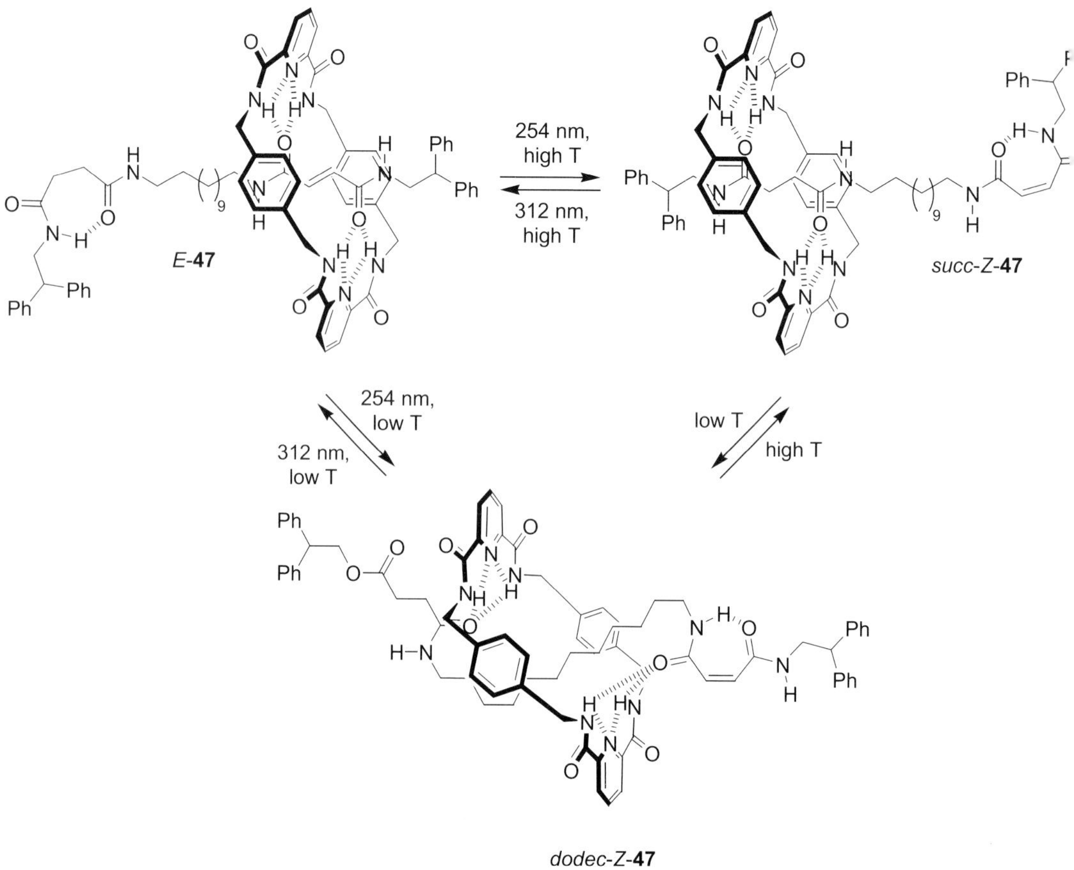

Figure 7.40 A tristable molecular shuttle **47**.

Structurally, **47** is closely related to **46**, the difference being substitution of the isophthaloyl unit in the macrocycle for a pyridine-2,6-dicarbonyl moiety. In the *E*-**47** form, the macrocycle resides over the strong fumaramide station at all temperatures investigated, as expected. Photoisomerization of *E*-**47** gave the maleamide *Z*-**47** isomer. The ^{1}H NMR spectra of this product showed clearly that shuttling away from the maleamide station had occurred, but the nature of the product was highly temperature-dependent. At elevated temperatures (308 K) the expected *succ*-*Z*-**47** co-conformation was observed; the macrocycle spends nearly all its time over the succinamide station. At lower temperatures, however, the macrocycle occupies neither the succinamide nor the maleamide units – the alkyl chain exhibits spectroscopic shifts indicative of encapsulation by the ring. This suggests that the thread adopts an S-shaped conformation with the macrocycle binding to one amide of each station (*dodec*-*Z*-**42**, Fig. 7.40). Unlike [2]rotaxane **46**, photochemical cis → trans isomerization was effective (irradiation of *Z*-**47** at 312 nm gave a photostationary state of >95:5 *E*:*Z* compared with ~45:55 under the same conditions for **46**) so no heating is necessary for this step and all the interconversions shown in Fig. 7.40 are possible.

The origin of this temperature-switchable effect is presumably the large difference in entropy of binding ($\Delta S_{binding}$) to the succinamide and alkyl chain stations, which enables the $T\Delta S_{binding}$ term to have a significant overall impact on $\Delta G_{binding}$ as temperature is varied. In the *succ*-*Z*-**47** co-conformation, the macrocycle forms two strong hydrogen bonds with an amide carbonyl and two, significantly weaker, bonds to the ester carbonyl. The *dodec*-*Z*-**47** co-conformation, however, enables formation of four strong hydrogen bonds to amide carbonyls making it enthalpically favored by ~2 kcal mol^{-1}. At low temperatures, when the effects of the entropy term are less significant, the molecule therefore adopts the *dodec*-*Z*-**42** co-conformation. At higher temperatures, the increased contribution from the $T\Delta S_{binding}$ term requires that the molecule adopts the more entropically favorable *succ*-*Z*-**42** co-conformation to minimize its energy [111].

This effect seems to be quite structure specific – no other shuttles in this series have shown temperature-dependent co-conformational preference. If suitable systems can be successfully designed, however, entropy-driven temperature control of positional isomerism could prove a useful addition to the expanding strategies for controlling submolecular motions [112].

Many of the shuttling examples described show remarkable degrees of control over submolecular fragment positioning and dynamics. They utilize a number of different stimuli-induced processes to effect macrocycle shuttling over large amplitudes (up to ~15 Å in **43**, **46**, and **47**) and operate over a range of timescales (a complete shuttling cycle in **44** is over in ~100 μs whereas in **46** both states are indefinitely stable and the time for a full cycle is infinitely variable). It must, however, be remembered that all such shuttles exist as an equilibrium of co-conformations and it is simply the position of the equilibrium that is varied [113].

7.5.6
Controlling the Motion: Ring Pirouetting in Rotaxanes

Control of macrocycle pirouetting in rotaxanes presents two challenges – frequency of random pirouetting and directionality (Fig. 7.41). The former has been achieved by means of the effects of temperature, structure, electric fields, light, and solvent; the latter has not yet been demonstrated.

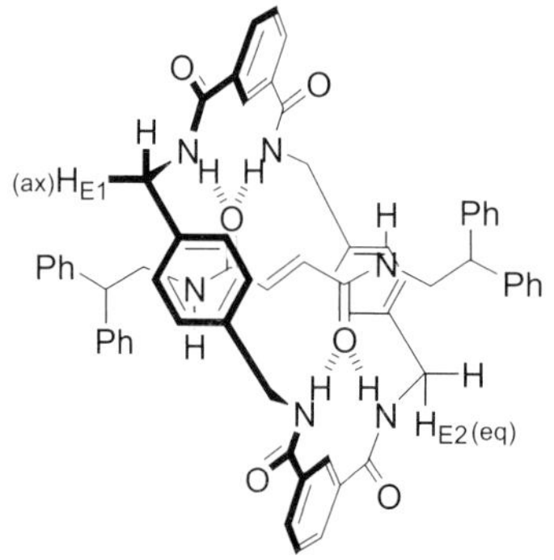

21

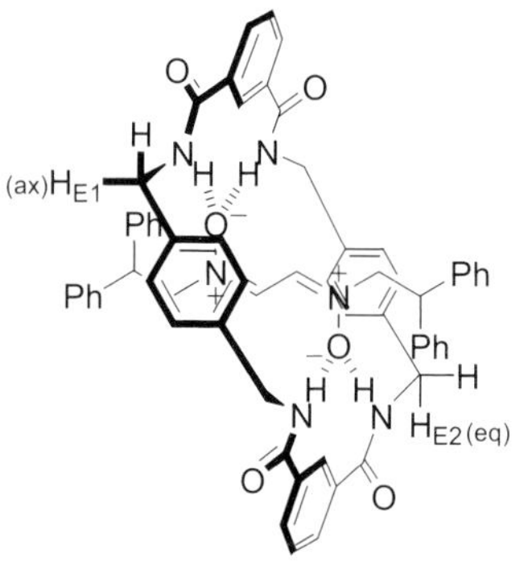

48

Figure 7.41 [2]Rotaxanes **21** and **48** in which the rate of pirouetting can be controlled by application of an alternating current electric field.

Alternating-current (a.c.) electric fields would be an ideal stimulus for control of submolecular dynamics in many applications. In two of the benzylic amide-based, hydrogen-bonded [2]rotaxanes (**21** and **48**), it was observed that application of a.c fields of approximately 50 Hz resulted in unusual Kerr-effect responses unique to the interlocked architecture (i.e. not observed for either of the components alone) [114]. Reducing the field strength had the same effect as increasing temperature – enhancement of the response – indicating that the underlying phenomenon is addressable by these two stimuli. Both VT NMR experiments and molecular modeling confirmed that macrocycle pirouetting is the only possible dynamic process on this timescale in these structures. The experimental and theoretical studies also predict the slightly more complex Kerr effect response observed experimentally for **21** compared with **48**. Application of an a.c. electric field therefore attenuates macrocycle pirouetting in **21** and **48**. The extent of the dampening can be varied with the strength of the applied field – even modest fields of ~1 V cm^{-1} produce rate reductions a large as 2–3 orders of magnitude.

An alternative strategy for affecting pirouetting rates would be to apply some stimulus which directly alters the structure or electronics of the thread or macrocycle, to adjust the strength of intercomponent interactions. This has been achieved for the fumaramide-based [2]rotaxanes **49–51** (Fig. 7.42), which are closely analogous to **21** above [115]. The decrease in intercomponent binding affinity on photoisomerization

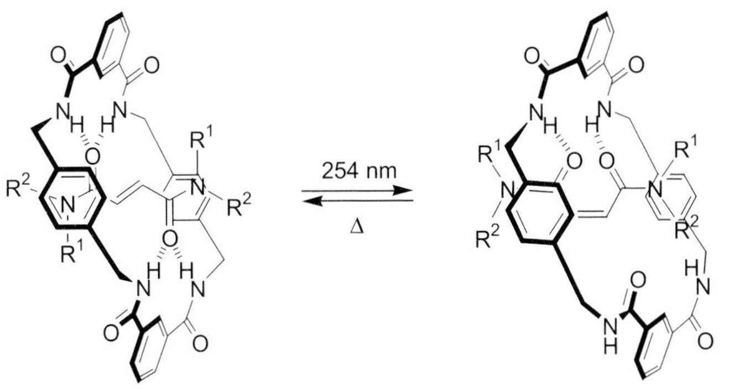

E-49	$R^1 = R^2 = CH_2CO_2CH_2Ph$	Z-49
E-50	R^1 Me, $R_2 = CH_2CHPh_2$	Z-50
E-51	$R^1 = H$, $R^2 = CH_2CHPh_2$	Z-51

Figure 7.42 Photoisomerization of [2]rotaxanes **49–51** which results in pirouetting rate enhancements of up to six orders of magnitude. The reverse process Z → E isomerization can be effected by heating a 0.02 M solution of the Z rotaxanes at 400 K or generating bromine radicals (cat. Br₂, *hv* 400 nm) or by reversible Michael addition of piperidine (RT, 1 h).

of the fumaramide units in **49–51**, to the *cis*-maleamide isomers, gives a huge increase in rate of pirouetting of more than six orders of magnitude. The switching process is also reversible; subjecting the maleamide rotaxanes to heat or a variety of chemical stimuli results in re-isomerization to the more thermally stable trans olefin isomers, with accompanying reinstatement of the strong hydrogen-bonding network [116, 117].

7.5.7
Controlling the Motion: Switchable Catenanes – the Issue of Directionality

The fundamental principles of controlling shuttling in rotaxanes and rotation in catenanes are the same. For example, homocircuit [2]catenane **52**, acts somewhat like the two-station degenerate shuttle **35** (Sect. 7.5.4.1) [118]. In halogenated solvents such as CDCl₃ the two macrocycles interact by hydrogen bonding between their aromatic 1,3-diamide groups, resulting in a "host–guest" relationship in which each (constitutionally identical) ring adopts a different conformation and exists in a different chemical environment (**52**-a, Fig. 7.43, also the structure observed in the solid state). Pirouetting of the two rings interconverts this host–guest relationship and is fast on the NMR timescale. In a hydrogen bond-disrupting solvent such as (CD₃)₂SO, however, the preferred co-conformation has the amides exposed on the surface, where they can interact with the surrounding medium, while the hydrophobic alkyl chains are buried in the middle of the molecule (**52**-b, Fig. 7.43). Of course, with disruption of the main noncovalent interactions between the two rings, the frequency of movements is also greatly increased in polar media.

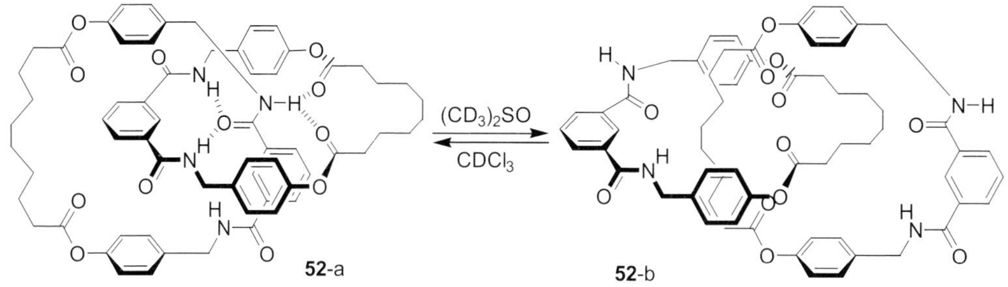

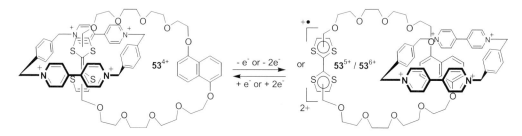

Figure 7.43 Translational isomerism in an amphiphilic benzylic amide [2]catenane **52**.

Heterocircuit [2]catenane **53**[4+] (Fig. 7.44) behaves similarly to analogous two-station non-degenerate [2]rotaxanes [119]. Initially the tetracationic cyclophane encircles the more electron-rich tetrathiafulvalene station (TTF), as evidenced by ^{1}H NMR and UV–visible absorption spectroscopy studies. Oxidation of the TTF to either its radical cation or dication can be achieved chemically or electrochemically. This oxidation results in complete movement of the cyclophane away from the cationic residue to the hydroxynaphthalene unit. This process has been observed using cyclic voltammetry measurements and ^{1}H NMR and UV–visible absorption spectroscopy and is accompanied by a green to maroon color change. The process can be reversed by either chemical or electrochemical reduction of the TTF unit back to its neutral state. There is, however, no control over which direction the motion occurs; the cyclophane has a choice of two identical routes between the stations and half the molecules will rotate one way, the other half the other way [120].

Figure 7.44 Chemically and electrochemically driven translational isomer switching of [2]catenane **53**.

Sauvage has demonstrated both electrochemical and photochemical control over ring motion in hetero-[2]catenate **54** (Fig. 7.45) which is directly related to [2]rotaxane **45** (Sect. 7.5.5.4) [121]. The observed behavior of the catenate is essentially the same as that of the rotaxane analog, although the 4-coordinate to 5-coordinate (dpp → terpy) shuttling process is slower in the catenate and the reverse terpy → dpp step is faster. These discrepancies are explained by the easier access to the metal for solvent

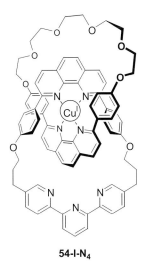

54-I-N₄

Figure 7.45 Heterocircuit switchable catenate **54**.

or ionic species in the rotaxane, thus stabilizing transition states on the way to higher-coordination-number species [122].

The behavior of the related homocircuit [2]catenate **55** (Fig. 7.46) – in which each ring now contains a bidentate dpp unit and tridentate terpy site – is more complicated [123]. In **55-I-N₄**, the Cu^I template coordinates to the two dpp units in the expected tetrahedral arrangement. Oxidation of the metal center to its divalent state (by either chemical or electrochemical means) reverses the order of preference for coordination numbers: $CN = 6 > 5 > 4$ is the preferred order for Cu^{II}. The result is circumvolution of the rings to give the preferred hexacoordinated species. It was demonstrated that this process occurs by revolution of one ring to give an intermediate five-coordinate species (**55-II-N₅**). In comparison with the other metal templated systems, this process is relatively fast, the translational process being on the timescale of tens of seconds. The process is completely reversible, via the same five-coordinate geometry, on reduction to Cu^I (i.e. via **55-I-N₅**).

As such, this catenate can exist as three translational "isomers" (six different states when the oxidation state of the metal ion is considered). The intermediate five-coordinate states are, however, only transient and cannot be isolated. A system which does have three distinct, stable states is [2]catenane **56**. Furthermore, **56** – and the related [3]catenane **57** – extends the fumaramide photoisomerization strategy seen in [2]rotaxanes **46**, **47** and **49–51** to create the first examples of stimuli-driven sequential and unidirectional rotation in interlocked molecules, thereby illustrating the key difference between shuttling in two-station rotaxanes and rotation in catenanes – the issue of directionality [124].

Sequential movement of one macrocycle between three stations on another ring requires independent switching of the affinity for two of the units to change the relative order of binding affinity, as shown schematically in Fig. 7.47.

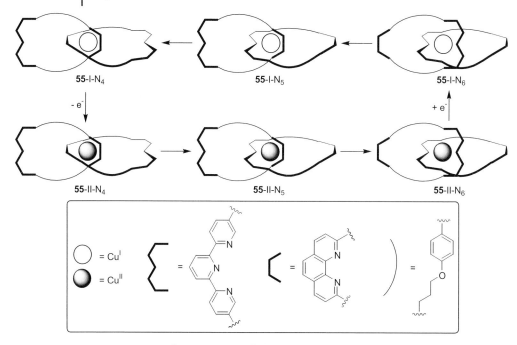

Figure 7.46 Oxidation state-controlled switching of [2]catenate **55** between three distinct co-conformations.

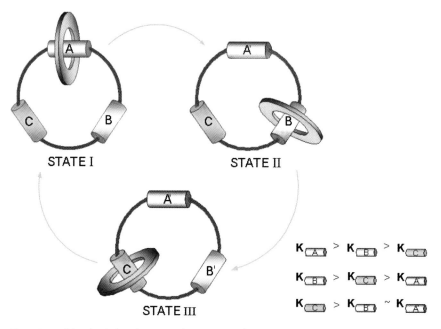

Figure 7.47 Stimulus-induced sequential movement of a macrocycle between three different binding sites in a [2]catenane.

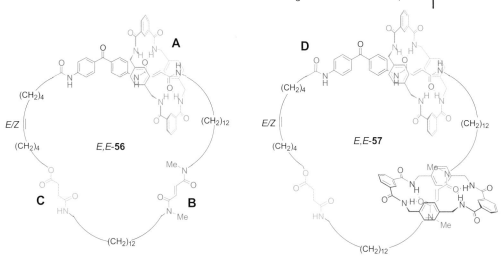

Figure 7.48 [2]Catenane **56** and [3]catenane **57**, shown as their *E,E* isomers.

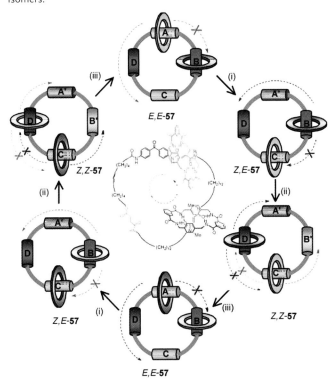

Figure 7.49 Stimulus-induced unidirectional rotation in a four station [3]catenane, **57**. (i) 350 nm, CH$_2$Cl$_2$, 5 min, 67 %; (ii) 254 nm, CH$_2$Cl$_2$, 20 min, 50 %; (iii) Δ, 100 °C, C$_2$H$_2$Cl$_4$, 24 h, ~100 %, or catalytic ethylene-diamine, 50 °C, 48 h, 65 %; or catalytic Br$_2$, 400–670 nm, CH$_2$Cl$_2$, −78 °C, 10 min. ~100 %.

In **56** (Fig. 7.48), this is achieved by using two fumaramide stations with different macrocycle binding affinity, one of which (station A, green) is located next to a benzophenone unit. This enables selective, photosensitized isomerization of station A by irradiation at 350 nm before direct photoisomerization of the other fumaramide station (station B, red) at 254 nm. Station B, being a methylated fumaramide residue, has lower affinity for the macrocycle than station A. The third station (station C, orange) – a succinic amide ester – is not photoactive and is intermediate in macrocycle-binding affinity between the two fumaramide stations and their maleamide counterparts. A fourth station, an isolated amide group (shown as D in *E,E*-**57**) which can make fewer intercomponent hydrogen bonding contacts than A, B, or C, is also present but only plays a significant role in the behavior of the [3]catenane.

Consequently, in the initial state (state I, Fig. 7.47), the small macrocycle resides on the green, non-methylated fumaramide station. Isomerization of this station (irradiation at 350 nm, green → blue) destabilizes the system and the macrocycle finds its new energy minimum on the red station (state II). Subsequent photoisomerization of this station (irradiation at 254 nm, red → pink) means the macrocycle must now move on to the succinic amide ester unit (orange, state III). Finally, heating the catenane (or treating it with photo-generated bromine radicals or piperidine) results in isomerization of both the *Z* olefins back to their *E* forms (pink → red and blue → green) so the original order of binding affinity is restored and the macrocycle returns to its original position on the green station.

The ^{1}H NMR spectra of each diastereomer show the positional integrity of the small macrocycle to be excellent at all stages of the process, but the rotation is undeniably not unidirectional – over the complete sequence of reactions an equal number of macrocycles go from A, though B and C, and back to A again in each direction.

To bias the direction the macrocycle takes from station to station, temporary barriers are required at each stage to restrict Brownian motion in one particular direction and so bias the path taken by the macrocycle from station to station. Such a situation is intrinsically present in [3]catenane **57** (Fig. 7.48). Irradiation at 350 nm of *E,E*-**57** causes counterclockwise (as drawn) rotation of the light blue macrocycle to the succinic amide ester (orange) station to give *Z,E*-**57**. Isomerization (254 nm) of the remaining fumaramide group causes the other (purple) macrocycle to relocate to the single amide (dark green) station (*Z,Z*-**57**) and, again, this occurs counterclockwise because the clockwise route is blocked by the other (light blue) macrocycle. This "follow-the-leader" process, each macrocycle in turn moving and then blocking a direction of passage for the other macrocycle, is repeated throughout the sequence of transformations shown in Fig. 7.49. After three diastereomer interconversions, *E,E*-**57** is again formed but 360° rotation of each of the small rings has not yet occurred, they have only swapped places. Complete unidirectional rotation of both small rings occurs only after the synthetic sequence (i) to (iii) has been completed twice. Somewhat counterintuitively, the small macrocycles in the [2]- and [3]catenanes move from station to station in an opposite sequence in response to the same series of chemical reactions.

7.6
From Laboratory to Technology: Toward Useful Molecular Machines

7.6.1
Current Challenges

The creation of novel molecular architectures – both classical covalent and mechanically interlocked structures – which in some way restrict the thermally-powered movements of their constituent parts has been central to developing control over motion at the molecular level. These structural constraints have been successfully combined with external switching of molecular structure and/or electronics to create systems with a remarkable level of control over both the relative position of units and the frequency of their motion. Whatever the application, it is clear that the machine and its components must be able to interact with the macroscopic world, either directly or through further interactions with other molecular-scale devices. This problem of "wiring" molecular machines also has implications for the physical construction of molecular devices which might very probably also require the creation of ordered arrays of functional molecules. In the following sections, we shall briefly examine some of these challenges and current progress towards surmounting them.

7.6.2
Reporting Motion: Switches and Memories

In direct analogy to the more established molecular switches which do not employ (significant) submolecular motion, any change in properties accompanied by the submolecular motion in a molecular machine could be applied to some sort of switching or information storage device. The fundamental requirements for such systems are the same as their non-mechanical counterparts and include: (1) a useful and detectable difference between two states; (2) fatigue resistance; (3) stability of each state under the operating conditions; (4) rapid response times; and particularly for memory applications, (5) non-destructive read-out [32a, 33].

7.6.2.1 Solid-state Molecular Electronic Devices
In a series of ground-breaking experiments, molecular shuttles (Sect. 7.5.5) have been employed as the active molecular component in solid-state molecular electronic devices. In the first example of such devices to utilize interlocked molecules, a monolayer of redox-active, degenerate two-station rotaxanes (58^{4+}, Fig. 7.50) was sandwiched between electrodes made from titanium and aluminum oxide [125]. The molecules were aligned during the manufacturing process, due to designed hydrophobic and hydrophilic regions in the thread portions. On application of a negative voltage to the resulting junctions (up to –2 V), a current flows, owing to resonant tunneling through the single-molecule layer, and the variation of current with increasing voltage is strongly non-linear. Application of a small, oxidizing voltage (typically +0.7 V, for several minutes) irreversibly reduces the current response by a factor of 60 to 80. It was shown, however, that the observed response is characteristic of

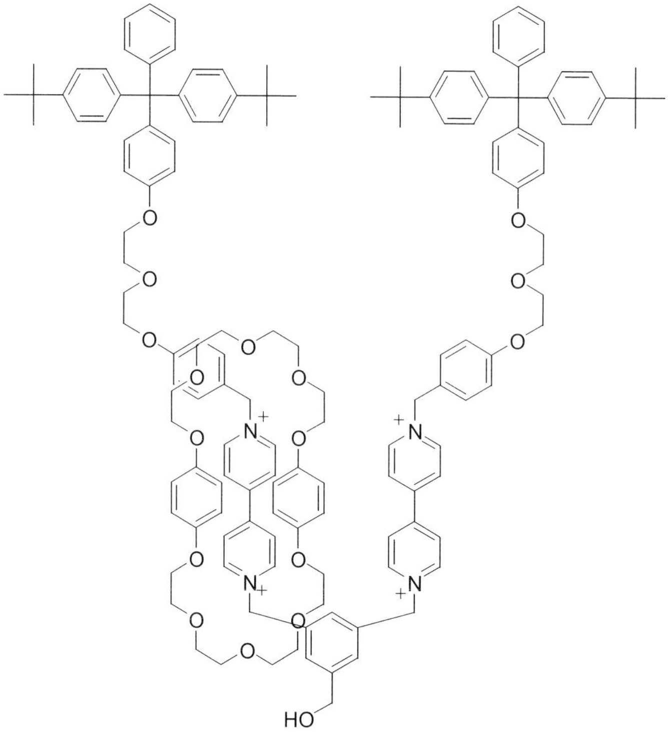

58⁴⁺

Figure 7.50 Chemical structure of amphiphilic [2]rotaxane **58⁴⁺** used in the first interlocked-molecule-based electronic switches. Alignment of the molecules was achieved by exploiting to the hydrophobic and hydrophilic portions, respectively, at the top and bottom of the molecule as drawn.

the thread portion, being essentially the same for monolayers of [2]rotaxane **58⁴⁺** and its related thread and [3]rotaxane analogs, so that for this species mechanical interlocking and motion do not seem to play a significant role. Although these devices each contained several million rotaxane or thread molecules, the electronic properties of the organic molecules in the solid state correlated well with the known solution-state characteristics, suggesting that scaling down toward single-molecule dimensions would not change the mode of operation. Furthermore, connection of such switches in linear arrays enabled creation of wired-logic gates performing OR and AND functions and with much more pronounced voltage responses compared with resistor-based circuits.

In a second generation of devices, bistable interlocked molecules were employed and reversible switching was achieved. Such a *molecular switchable tunnel junction* (MSTJ) was first demonstrated using a monolayer of bistable [2]catenanes **53⁴⁺** (Sect. 7.5.7) sandwiched between silicon and titanium/aluminum electrodes [126]. With these devices a moderate (a factor of ~2) but reversible change in resistance is observed on application of oxidizing (opening the switch) and reducing (closing the switch) voltages. The "read" voltage was in the region 0.1 to 0.3 V, and so did not interfere with switch configuration, while the devices withstood many switching cycles. A rational design process then led to devices made from a related [2]pseudorotaxane (**59**$^{4+}$, Fig. 7.51), for which the on/off current ratio was much larger (~10^2) and the "on" currents larger than for the [2]catenane-based system [127]. The output currents and switching voltages were, however, found to vary widely between both cycles and devices. On further investigation, these problems were attributed to the formation of molecular domains, with the result that device characteristics were no longer determined by single-molecule properties.

This led to further refinement of shuttle structure, producing kinetically stable pseudo[2]rotaxanes **60**$^{4+}$ and, ultimately, **61**$^{4+}$. MSTJ comprising these rotaxanes had stable switching voltages of approximately –2 and +2 V with reasonable on/off ratios and switch-closed currents [128]. Such favorable characteristics enabled preparation of nanometer-scale devices with properties similar to those of the original micrometer analogs, demonstrating that operation is occurring on the molecular level. Furthermore, these devices were successfully connected in a circuit architecture known as a 2D crossbar. The resulting circuit could be used as a reliable 64-bit random access memory (RAM). The even more demanding task of creating a logic circuit could also be demonstrated by hard-wiring 1D circuits. A true 2D MSTJ-based crossbar logic circuit, however, will require junctions with features yet to be achieved, for example diode character and gain.

Throughout these investigations several control molecules were tested to probe the origin of the switching effect [126–128]. These have demonstrated that some form of bistability, together with the presence of *both* interlocked components, is necessary for the switching properties. It is not clear, however, whether relative motions of the components are responsible. In particular, resistance switching is still observed for devices made from a related single-station [2]rotaxane, albeit with an on/off ratio significantly lower than for the two station analogs **60**$^{4+}$ and **61**$^{4+}$. It has also been demonstrated that the switching process is thermally activated – at lower temperatures, higher voltages are required to operate the switch. Although this might suggest that some form of submolecular motion is required for operation of the device, it can also be understood in terms of reduced polarizability of the silicon electrode at reduced temperatures [127].

Although these devices have high potential as possible components for truly single-molecule electronic devices, a significant barrier to achieving this goal will be the physical connections used to wire the device. At very small dimensions it is expected that metal wires will have the best conductance characteristics but, unfortunately, devices similar to those described above, but with two metal electrodes, did not have switching properties dependent on the nature of the organic monolayer [129]. Devices which use single-walled carbon nanotubes in place of the silicon electrode have, however, been successfully created and this might provide a valuable alternative in the creation of real-world devices [130].

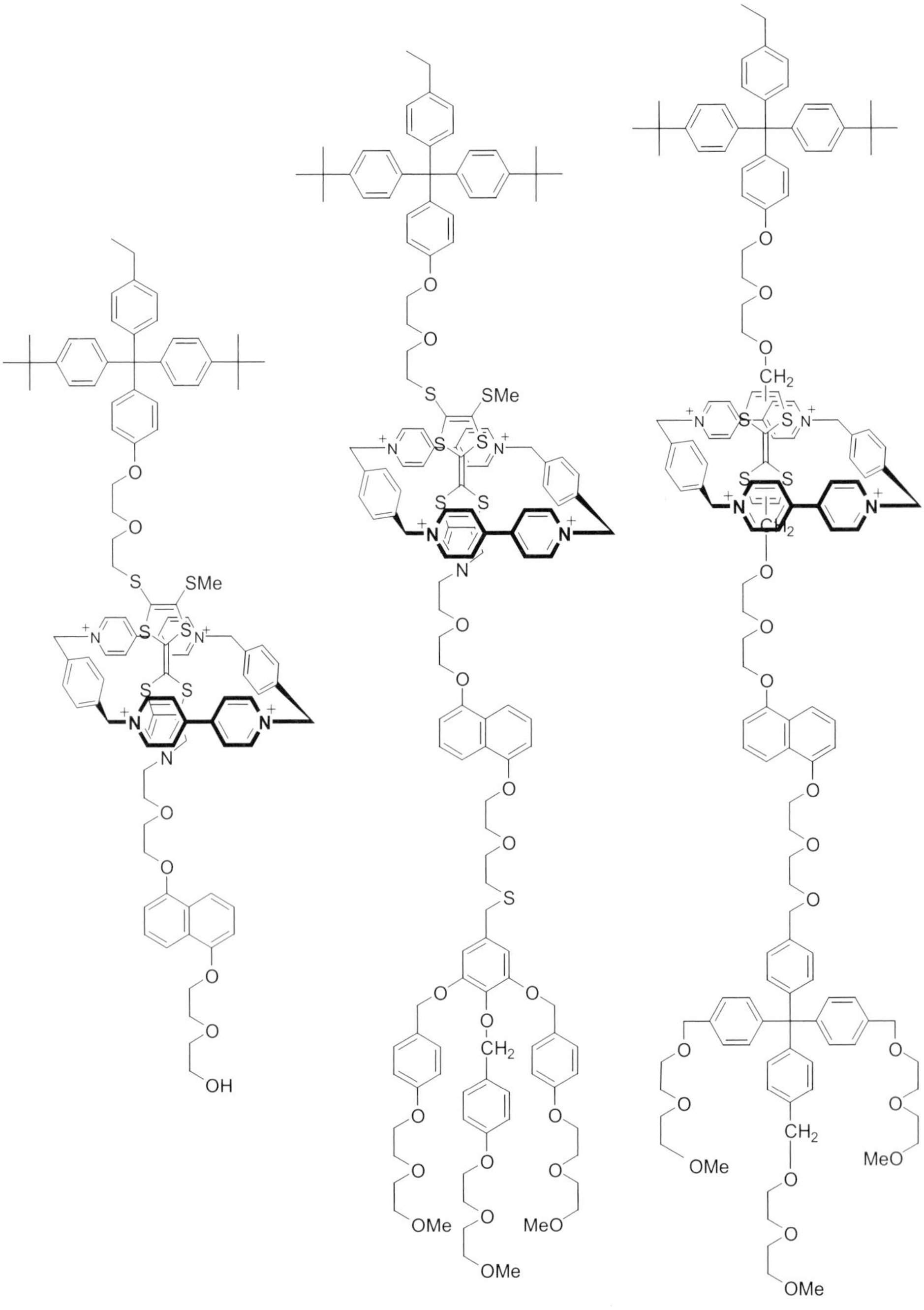

59^{4+} 60^{4+} 61^{4+}

Figure 7.51 Chemical structures of [2]pseudorotaxane **59**$^{4+}$ and [2]rotaxanes **60**$^{4+}$ and **61**$^{4+}$ used as the active components in molecular switched tunnel junctions (MSTJ).

7.6.2.2   A Molecular Muscle

The principles of shuttling in metal-templated rotaxanes have been extended to create a dimeric system in which the submolecular motion results in lengthening and contraction of the molecule in a manner reminiscent of the operation of the actin-myosin complex which is the basis of natural muscle [131]. Each monomer unit for the construction of the rotaxane dimer (Fig. 7.52) consists of a bidentate dpp site embedded in a macrocyclic ring, as in **45** (see Sect. 7.5.5.4), but this ring is now connected to a thread portion which also contains a phenanthroline (a 2,9-dimethylphenanthroline, dmp) ligand and a tridentate terpyridine site. The Cu^I ions used to template formation of the dimer coordinate preferentially – once again in a tetrahedral geometry – to the dpp of one unit and the dmp of another component, resulting in the threaded dimer structure $[62 \cdot 2Cu]^{2+}$ (Fig. 7.53). Unfortunately, in this example electrochemical oxidation to the Cu^{II} species did not trigger motion – even this unfavorable geometry for divalent copper is kinetically too stable. Instead, demetallation (excess KCN, room temperature, 3 h) to give the free ligand system **62**, followed by insertion of Zn^{II} ions ($Zn(NO_3)_2$, room temperature, 1 h) gives the contracted form $[62 \cdot 2Zn]^{4+}$, as evidenced by ^1H NMR studies. The molecule can then be returned to its original length simply by treatment with excess $[Cu(CH_3CN)_4] \cdot PF_6$.

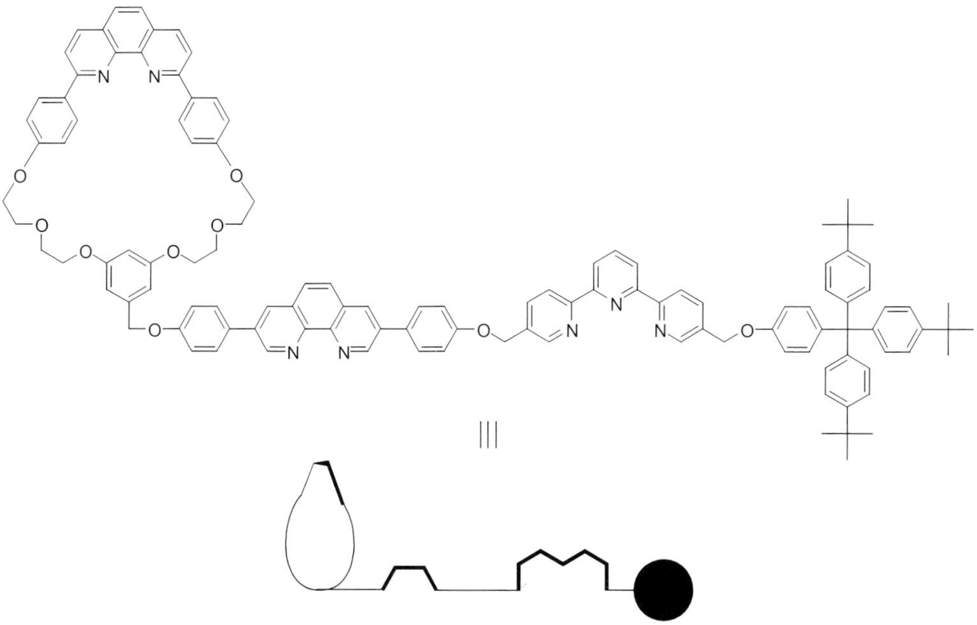

Figure 7.52   Monomer unit for the construction of artificial molecular muscle rotaxane dimer **62**.

In this molecule, therefore, the shuttling of metal templates between two sites on the linear portions results in the filaments gliding over each other – as the distance between the metal centers is increased, the overall length of the molecule decreases.

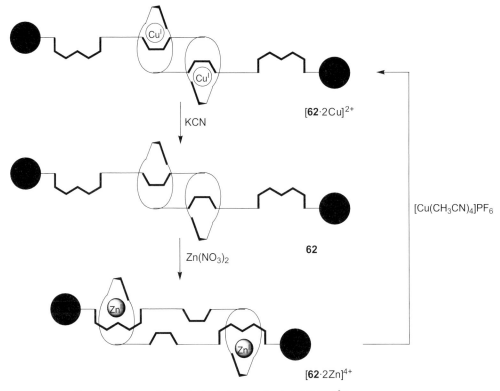

Figure 7.53 Reversible switching between extended ([**62**•2Cu]$^{2+}$)
and contracted ([**62**•2Zn]$^{4+}$) forms in a chemically-switched artificial
molecular muscle.

During this process, the length of the molecule changes from approximately 85 Å to
65 Å, a reduction of 24 %.

7.6.2.3 Shuttle-based Mechanical Switches

The coupling of submolecular motion in a shuttle with a change in molecular properties can lead to mechanical switches operating at the molecular level and which
could perform a range of functions.

In a study of chiral dipeptide [2]rotaxanes it was found that the presence of an intrinsically achiral benzylic amide macrocycle near the chiral center could induce an
asymmetric response in the aromatic ring absorption bands. This induced circular
dichroism (ICD) effect was strongest in apolar solvents when intercomponent interactions are maximized and it was shown that the chirality is transmitted from the
amino acid asymmetric center on the thread, via the achiral macrocycle to the aromatic rings of the achiral C-terminal stopper on the thread [132]. These observations
led to the design of chiroptical switch *E/Z*-**63** which utilizes the same fumaramide–maleamide function as [2]rotaxane shuttles **46** and **47** in Sect. 7.5.5, above
(Fig. 7.54) [133]. Unlike chiroptical switches in which the presence of or handedness

of chirality is intrinsically altered [32], *E/Z*-**63** remains chiral and non-racemic, with the same handedness throughout; it is the *expression* of chirality that is altered. In the *E*-**63** form, the macrocycle is held over the excellent fumaramide template and thus far from the chiral center in the peptidic station. Correspondingly, the CD response is flat, as observed for the free thread. In the *Z*-**63** isomer, in which the macrocycle resides on the peptide station, close to the L-Leu residue, a strong (-13k deg cm^2 dmol^{-1}), negative CD response is observed. Preparatively, the $E \rightarrow Z$ isomerization is most efficiently carried out by irradiation at 312 nm in the presence of a benzophenone sensitizer (photostationary state 70:30 *Z*:*E*), while the $Z \rightarrow E$ transformation can be achieved almost quantitatively by irradiation at 400–670 nm in the presence of catalytic Br$_2$. In the context of any switching application, however, addition of any external reagent is obviously undesirable and a more modest difference between photostationary states can be achieved by irradiation at 254 nm (photostationary state 56:44 *Z*:*E*) and 312 nm (photostationary state 49:51 *Z*:*E*). Even under these conditions a large net change (>1500 deg cm^2 dmol^{-1}) in the elliptical polarization response is observed and this is reproducible over several cycles.

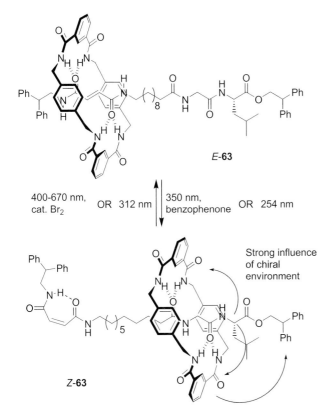

Figure 7.54 Chiroptical switching in [2]rotaxane-based molecular shuttle *E/Z*-**63**.

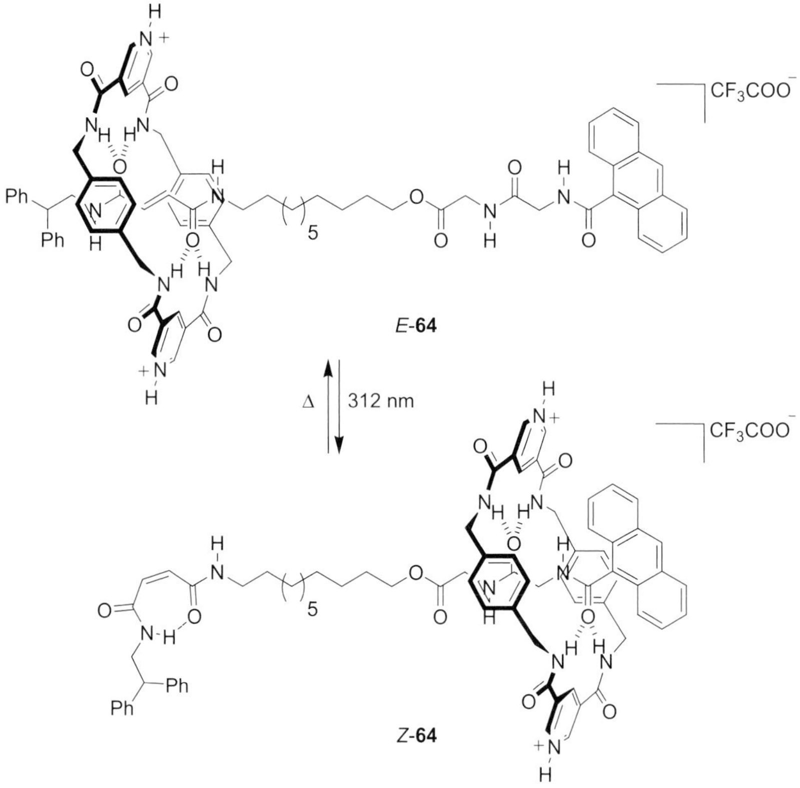

Figure 7.55 A fluorescent molecular switch based on [2]rotaxane molecular shuttle *E/Z*-**64**.

Encouraged by these results, bistable molecular shuttle *E/Z*-**64** was created (Fig. 7.55) [134]. This system also relies on the photoswitchable fumaramide/ maleamide station, but attached to the intermediate, dipeptide station is an anthracene fluorophore, and the macrocycle now contains pyridinium units – known to quench anthracene fluorescence by electron transfer. In both the free thread and *E*-**64**, strong fluorescence (λ_{exc} = 365 nm) is observed, and shuttling of the macrocycle on to the glycylglycine station in *Z*-**64** almost completely quenches this emission. At the maximum of *E*-**64** emission (λ_{max} = 417 nm) there is a remarkable 200:1 difference in intensity between the two states – strikingly visible to the naked eye.

These two examples demonstrate a generic approach which could be taken to create mechanical molecular switches for a variety of distance-dependent properties. The fumaramide/maleamide unit provides a means of changing the position of the macrocycle using a number of olefin isomerization procedures (the strategy is equally relevant for any other bistable stimuli-responsive shuttle system for which there is good macrocycle positional integrity between the two stations in both states). The glycylglycine is a non-reactive station of intermediate affinity and the alkyl spacer can be

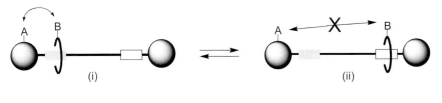

Station binding affinity order: ▭ > ▬ > ■

Figure 7.56 Exploiting a well-defined, large amplitude positional change to trigger property changes. (i) A and B interact to produce a physical response (fluorescence quenching, specific dipole, or magnetic moment, NLO properties, color, creation/concealment of a binding site or reactive/catalytic group, hydrophobic/hydrophilic region, etc.); (ii) moving A and B far apart mechanically switches off the interaction and the corresponding property effect.

varied in length to suit the distance dependency of the property in question. Suitable functionalization of the macrocycle and one end of the thread should therefore lead to molecular switches that could perform several functions (Fig. 7.56).

A similar strategy has been used to achieve fluorescence switching in [2]rotaxane E/Z-**65** (Fig. 7.57) [135]. In E-**65**•2H, the α-cyclodextrin ring sits preferentially over the *trans*-stilbene unit and this co-conformation is further stabilized by strong hydrogen-bonding interactions between hydroxyl groups on the 3-rim of the cyclodextrin and the isophthaloyl stopper group. As for the previous α-CD based systems (Fig. 7.29, Sect. 7.5.5.1), photoisomerization of the stilbene unit can occur only when thermal motion has moved the ring away from the binding site. The strength of the combined hydrophobic and hydrogen-bonding interactions (even in water) for E-**65**•2H, however, means that the shuttle is conformationally *locked* – irradiation at 355 nm, which should isomerize the stilbene moiety, results in no change to the system. Formation of the disodium salt of the isophthaloyl group (giving E-**65**•2Na) breaks the hydrogen-bonding network and although NMR studies show that the cyclodextrin continues to sit over the stilbene station, enough thermal motion is now present to enable photoisomerization, to give Z-**65**•2Na (photostationary state 63:37 Z:E). In this state, the cyclodextrin ring is forced to reside over the biphenyl group. The shuttling motion is accompanied by an increase of approximately 46 % in the fluorescence intensity of the 4-aminonaphthalimide stopper (λ_{max} = 530 nm). This change is attributed to restriction of vibrational and rotational movement of the methylene and biaryl linkages thus disfavoring non-radiative relaxation processes. The shuttling and fluorescence changes are completely reversible on re-isomerization to E-**65**•2Na at 280 nm.

The same researchers have elaborated this concept in [2]rotaxane E/Z-**66** (Fig. 7.58) which contains two fluorophores; the intensity of each of these can be selectively enhanced depending on the position of the ring [136]. In this instance the configurational switch is an azobenzene group but the operation is essentially the same as above. In E-**66**, the α-CD sits over the azobenzene moiety, thus enhancing the fluorescence of the adjacent 4-aminonaphthalimide stopper (λ_{max} = 520 nm). Photoiso-

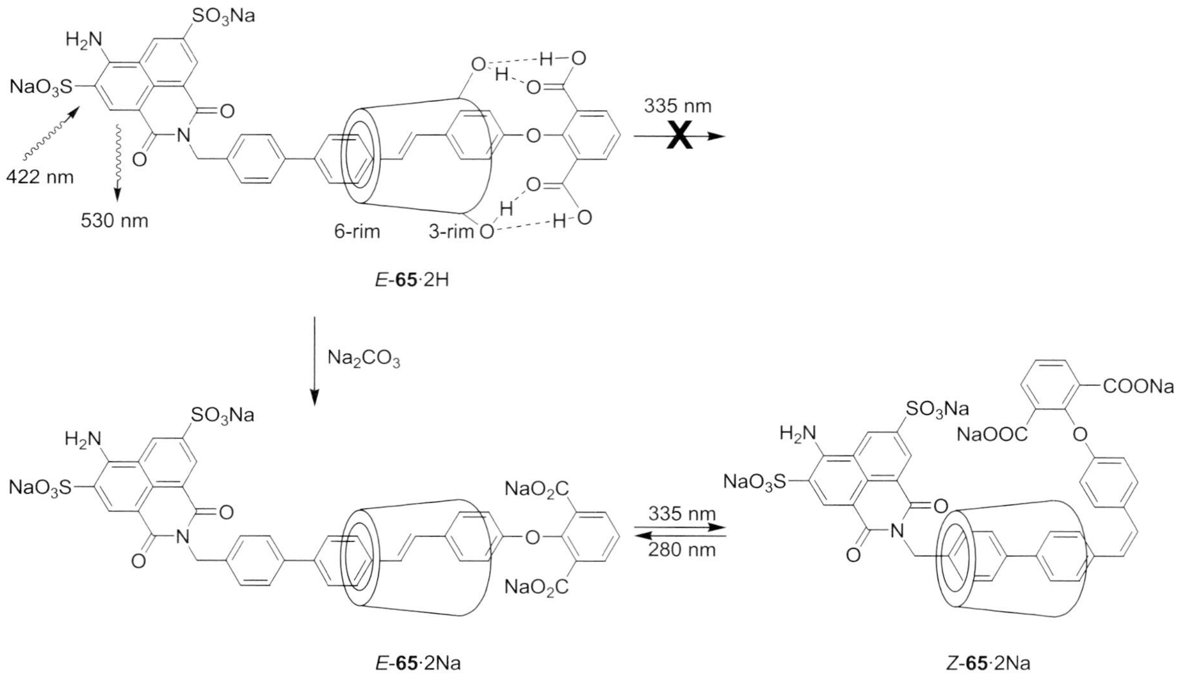

Figure 7.57 Photoswitched shuttling in [2]rotaxane *E*/*Z*-**65**·2Na which results in fluorescence enhancement of the 4-aminonaphthalimide group. The photoisomerization process is prevented when the free carboxylic acids of the isophthaloyl stopper are present (*E*-**65**·2H).

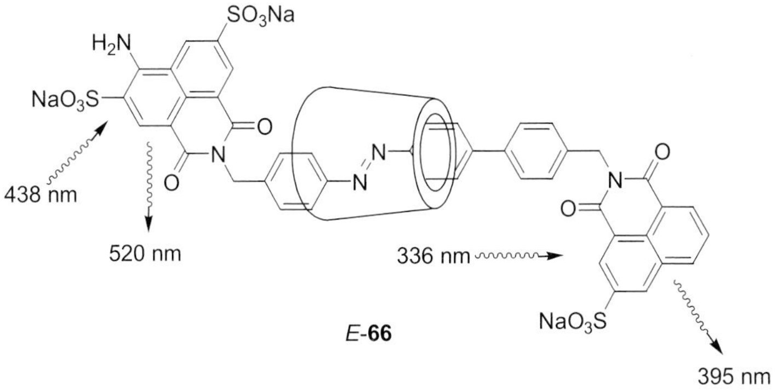

Figure 7.58 [2]Rotaxane *E*-**66** in which the F$_{520}$ band is selectively enhanced. Photoisomerization of the azobenzene unit (360 nm) results in shuttling of the α-cyclodextrin ring to the biaryl site and concomitant reduction of the F$_{520}$ emission together with enhancement of the F$_{395}$ band.

merization to *Z*-**66** (irradiation at 360 nm; photostationary state 59:41 *Z*:*E*) results in shuttling of the ring to the biaryl station, simultaneously reducing the intensity of the F$_{520}$ emission while enhancing the fluorescence of the right-hand fluorophore (λ_{max} = 395 nm). The original state (100 % *E*-**66**) can be achieved on irradiation at 430 nm and the process can be repeated over several cycles.

7.6.3
The Interface with Real-world Technology

It has already been mentioned that the implementation of a technology based on molecular machines faces several problems with regard to the actual engineering of devices and their interconnection with each other and with the macroscopic world. Although light-operated systems such as those discussed in Sect. 7.6.2.3 might have potential in some respects, the generally non-directional nature of emission processes would make specific communication between nanoscale devices difficult, and the focus resolution of laser light that might be used to address such systems is limited by the wavelength. On the other hand, systems designed to perform mechanical tasks, for example rotaxane dimer **62** described in Sect. 7.6.2.2, might require anchoring to a massive object, and correlation of the efforts of many molecular machines will probably be required to produce a desired macroscopic effect. The devices and circuits outlined in Sect. 7.6.2.1 highlight the utility of electrical signals for selective addressing of small-scale devices which seem to operate at the single-molecule level, yet the device dimensions are limited by the nature of the electrodes required.

It is recognized that transfer of molecular machine technology to solid substrates might be a key step in the development of many potential applications. There are already several examples of switchable molecular recognition events (including threading and de-threading of pseudorotaxanes) which occur at surfaces functionalized with either host or guest components [137]. In one example the reversible recognition event is also observed in a nanoporous silicate matrix [137f]. The state of the art has been demonstrated by Stoddart and coworkers who have harnessed the photochemically triggered de-threading of a surface-mounted pseudorotaxane to release molecules trapped in nanoscopic pores on the same surface [138]. In addition, switched conformational motions of immobilized, fully-covalent molecules have been used to alter the properties of surfaces [139].

The problem of achieving controlled movement within surface-mounted or solid-state interlocked molecules is more demanding, however, with both components constrained by the solid support at all times. The first evidence for externally stimulated submolecular motion of a catenane in the solid state was observed when a vacuum-evaporated thin film of benzylic amide [2]catenane **27** was subjected to oscillating electric fields, producing an unexpected second-order nonlinear optical effect [140].

Electrochemically-induced shuttling has been observed in self-assembled monolayers (SAM) on gold wire of an alkanethiol derivative of [2]rotaxane 61^{4+} [141]. Shuttling of the macrocycle on oxidation and subsequent reduction of the TTF unit was observed by voltammetry measurements in an analogous fashion to solution-state studies. Although this motion is occurring at the solid–liquid interface, the characteristics of the shuttling process, which involves a relatively long-lived metastable state, provides some backing for the occurrence of similar submolecular motion in the solid-state devices discussed in Sect. 7.6.2.1.

A different redox-active molecular shuttle could be orientated perpendicular to the surface of a titanium dioxide nanoparticle, by attachment of a tripodal phosphonate unit at one end of the thread [142]. The use of nanoparticles enabled characterization of the dynamic processes in the surface-mounted species by both ^{1}H NMR and cyclic voltammetry. Redox-triggered shuttling of the macrocycle was indeed observed, although in this example, positional integrity was found to be poor.

A potentially useful way of detecting and subsequently communicating the state of any type of molecular switch is to translate its state into an electrical signal [143]. This has been accomplished for a molecular shuttle attached to the surface of a gold electrode [144]. In this case, photochemically induced shuttling of a cyclodextrin ring closer to the electrode surface was detectable as an increase in the rate of oxidation of a redox-active ferrocene unit attached to the mobile ring.

Probe microscopy techniques such as scanning–tunnelling microscopy (STM) and AFM have been instrumental in the development of many fields over the last two decades, in particular where atomic-level movements are to be effected or observed [145], yet their use in the study and operation of synthetic molecular machines is still relatively underdeveloped. Recently, however, this technology has been used to move components in interlocked molecules. In one report, an STM tip was used to reversibly move one or two adjacent cyclodextrin rings along the poly(ethylene glycol)-derived thread in a polyrotaxane [146]. It has also been demonstrated that an AFM tip can be used to create a regular array of deformations on thin films of benzylic amide-based rotaxanes [147]. This effect is unique to films made from interlocked molecules (compared with their non-interlocked components) and is a result of coupled nucleation recrystallization being favored by the ease of intercomponent mobility (even in the solid state) for these molecules. The characteristics of these features are readily variable with the nature of the film and show potential for application in information-storage technology.

The high resolution achievable with atomically sharp microscope probes, together with the range of possible functions (e.g. mechanical, electrical, or thermal perturbations and observations are all possible) suggests that such techniques may prove key in the future development of useful devices based on synthetic molecular machines.

7.7
Summary and Outlook

The last few years have witnessed dramatic advances in the ability to control submolecular motion in synthetic systems. Although these new-found skills continue to be refined, attention is now turning to the application of synthetic structures to perform useful tasks and create true molecular machines. Inspiration and assistance is continually offered as a result of the accumulating understanding of biological motor proteins and information processing, and the fantastic achievements of modern-day microelectronics and mechanical engineering, yet we must be sure to remember that the devices created by the synthetic chemist will have fundamentally different characteristics, strengths, and weaknesses to their counterparts from other regimes. Consequently, it is difficult to predict what the eventual applications of molecular machines will be. In the short term, changing surface properties, switching, and memory devices look particularly attractive. A greater challenge, however, would be to harness the motion of submolecular components to perform useful work at the molecular-level – that is, to transduce the energy input from an external stimulus into directed motion of some cargo or motion against an energy gradient. In each of Sects. 7.2, 7.3, and 7.5.7 we have seen the first examples of unidirectional rotors based on rotation around single, double, and mechanical bonds respectively. Although the physical principles underlying unidirectional motion at the molecular-level were perhaps not fully appreciated by chemists before the creation of this first wave of systems, these molecules demonstrate the basic principles required to perform work at the molecular level. The development of molecular-level systems which perform functions by responding to stimuli as a result of Boolean logic operations [148] has clear relevance to the future development of compartmentalized molecular machines which are more sophisticated than this current generation of mechanical switches and simple motors. Although current efforts in molecular logic have concentrated largely on photonic supramolecular systems and have not been applied to systems which rely on mechanical motion, their successful operation suggests a way in which existing but relatively simple molecular machine components might be connected to produce useful devices at the next level of complexity.

As with many areas of chemistry, mimicry of biological systems continues to be an important driving-force in this field. There is a synergistic relationship between the two subject areas, with biology providing the inspiration and operating principles while the comparatively simplistic synthetic systems provide fully characterizable models for the more complicated natural processes. An elegant example is the toroidal, processive epoxidation catalyst described by Rowan, Nolte, and coworkers [149]. Although, one day, such systems might provide catalysts for post-polymerization modification of polymers, the mechanically linked nature of the catalyst–substrate complex should also provide important insights into the operation of the processive enzymes which are so central to fundamental processes in the cell.

Traditional covalent and noncovalent synthetic chemistry techniques are not the only route to molecular-level devices and machines. The unusual mechanical and electrical properties of nanotubes and other fullerenes constructed from boron nitride or carbon has enabled the downwards extension of current-day microelectromechanical systems (MEMS; the current state of the art, and perhaps limit, of silicon-based devices) toward true nanometer-scale electromechanical systems (NEMS) [150]. This is an example of "hard nanotechnology", but another area of intense interest involves so-called "soft" nanomaterials. This area encompasses functional materials created by self-assembly processes giving organized arrays of molecules on "hard" templates. Smaller length-scales and milder conditions are achievable compared with silicon-based lithography techniques [151]. Such processes might prove useful in the organization of synthetic molecular machines to create useful devices. Also included in this area however are several materials, for example polymers and gels, in which macroscopic mechanical motion can be generated in response to an external stimulus. Traditionally, these effects have been non-specific processes such as the uptake or expulsion of counterions during redox processes [152], yet there has been a recent move toward the creation of polymeric materials which stretch and contract as a result of specific molecular interactions [153] or nuclear arrangements [154].

Yet another field of interest turns to biology for the components to construct "artificial" molecular devices. Although some impressive successes have been achieved by harnessing biological motors to power motion outside the cell [155], others have borrowed one of Nature's most useful construction materials – nucleic acids – to build novel constructs capable of molecular-level motion [156].

Although the field is in its infancy, it is clear that the concept of harnessing molecular-level motion to perform useful tasks is exciting researchers in many disciplines. Creating entirely synthetic molecular machines should enable us to exploit the full potential of synthesis to create systems with currently unrealized characteristics – to use controlled mechanical motion to drive chemical reactions, make artificial machines which can "sort" ions into different types, or drive themselves and their cargo energetically uphill along tracks. Equally important will be the experimentally-derived insight into motion at the molecular level which will tell us how biological systems work. For synthetic molecular machines, the best is yet to come.

References

1 For an introduction to Brownian motors, see: R. D. Astumian, P. Hänggi, *Phys. Today* **2002**, *55 (11)*, 33–39.

2 For reviews on Brownian ratchet mechanisms, see: (a) F. Jülicher, A. Ajdari, J. Prost, *Rev. Mod. Phys.* **1997**, *69*, 1269–1281; (b) P. Reimann, *Phys. Rep.* **2002**, *361*, 57–265; (c) *Appl. Phys. A* **2002**, *75*, 167–352 (special issue on Ratchets and Brownian motors: basics, experiments and applications); (d) B. J. Gabrys, K. Pesz, S. J. Bartkiewicz, *Physica A* **2004**, *336*, 112–122.

3 For selected examples, see ref. [2c] and: (a) R. D. Astumian, M. Bier, *Eur. Biophys. J.* **1996**, *70*, 637–653; (b) R. D. Astumian, *Science* **1997**, *276*, 917–922; (c) R. D. Astumian, I. Derényi, *Eur. Biophys. J. Biophys. Lett.* **1998**, *27*, 474–489; (d) I. Derényi, T. Vicsek, *Physica A* **1998**, *249*, 397–406; (e) R. F. Fox, *Phys. Rev. E* **1998**, *57*, 2177–2203; (f) F. Jülicher, *arXiv:physics* **1999**, /9908054; (g) R. D. Astumian, I. Derényi, *Biophys. J.* **1999**, *77*, 993–1002; (h) R. D. Astumian, *Philos. Trans. R. Soc. Lond. Ser. B* **2000**, *355*, 511–522; (i) M. Kurzynski, P. Chełminiak, *Physica A* **2004**, *336*, 123–132.

4 *Molecular Motors* (Ed.: M. Schliwa), Wiley-VCH, Weinheim, **2003**.

5 H. Hess, G. D. Bachand, V. Vogel, *Chem. Eur. J.* **2004**, *10*, 2110–2116.

6 (a) V. Balzani, A. Credi, F. M. Raymo, J. F. Stoddart, *Angew. Chem. Int. Ed.* **2000**, *39*, 3349–3391; (b) J. F. Stoddart, *Acc. Chem. Res.* **2001**, *34*, 410–411; (c) R. Ballardini, V. Balzani, A. Credi, M. T. Gandolfi, M. Venturi, *Acc. Chem. Res.* **2001**, *34*, 445–455; (d) V. Balzani, A. Credi, M. Venturi, *Chem. Eur. J.* **2002**, *8*, 5524–5532; (e) V. Balzani, A. Credi, M. Venturi, *Molecular Devices and Machines. A Journey into the Nanoworld*, Wiley–VCH, Weinheim, **2003**; (f) J. D. Crowley, A. J. Goshe, I. M. Steele, B. Bosnich, *Chem. Eur. J.* **2004**, *10*, 1944–1955; (g) C. J. Easton, S. F. Lincoln, L. Barr, H. Onagi, *Chem. Eur. J.* **2004**, *10*, 3120–3128.

7 E. M. Purcell, *Am. J. Phys.* **1977**, *45*, 3–11.

8 E. L. Eliel, S. H. Wilen, *Stereochemistry of Organic Compounds*, Wiley–Interscience, New York, **1994**.

9 J. D. Kemp, K. S. Pitzer, *J. Chem. Phys.* **1936**, *4*, 749.

10 M. Ōki, *Angew. Chem. Int. Ed. Engl.* **1976**, *15*, 87–93.

11 (a) H. Iwamura, K. Mislow, *Acc. Chem. Res.* **1988**, *21*, 175–182; (b) K. Mislow, *Chemtracts: Org. Chem.* **1989**, *2*, 151–174; (c) J. Vacek, J. Michl, *New J. Chem.* **1997**, *21*, 1259–1268.

12 T. C. Bedard, J. S. Moore, *J. Am. Chem. Soc.* **1995**, *117*, 10662–10671.

13 (a) J. Clayden, J. H. Pink, *Angew. Chem. Int. Ed.* **1998**, *37*, 1937–1939; (b) A. Ahmed, R. A. Bragg, J. Clayden, L. W. Lai, C. McCarthy, J. H. Pink, N. Westlund, S. A. Yasin, *Tetrahedron* **1998**, *54*, 13277–13294; (c) R. A. Bragg, J. Clayden, *Org. Lett.* **2000**, *2*, 3351–3354; (d) R. A. Bragg, J. Clayden, G. A. Morris, J. H. Pink, *Chem. Eur. J.* **2002**, *8*, 1279–1289.

14 (a) T. R. Kelly, *Acc. Chem. Res.* **2001**, *34*, 514–522; (b) J. P. Sestelo, T. R. Kelly, *Appl. Phys. A* **2002**, *75*, 337–343.

15 T. R. Kelly, M. C. Bowyer, K. V. Bhaskar, D. Bebbington, A. Garcia, F. R. Lang, M. H. Kim, M. P. Jette, *J. Am. Chem. Soc.* **1994**, *116*, 3657–3658.

16 (a) T. R. Kelly, I. Tellitu, J. P. Sestelo, *Angew. Chem. Int. Ed. Engl.* **1997**, *36*, 1866–1868; (b) T. R. Kelly, J. P. Sestelo, I. Tellitu, *J. Org. Chem.* **1998**, *63*, 3655–3665.

17 (a) A. P. Davis, *Angew. Chem. Int. Ed.* **1998**, *37*, 909–910; (b) G. Musser, *Sci. Am.* **1999**, *280 (2)*, 24.

18 P. W. Atkins, *Physical Chemistry*, 6th ed., Oxford University Press, Oxford, **1998**.

19 (a) M. von Smoluchowski, *Physik. Z.* **1912**, *13*, 1069–1080; (b) R. P. Feynman, R. B. Leighton, M. Sands, *The Feynman Lectures on Physics, Vol. 1 (Chap. 46)*, Addison–Wesley, Reading, MA, **1963**.

20 (a) For a modern critique of Feynman's analysis, see: J. M. R. Parrondo, P. Espanol, *Am. J. Phys.* **1996**, *64*, 1125–1130; (b) M. O. Magnasco, G. Stolovitzky, *J. Stat. Phys.* **1998**, *93*, 615–632.

21 (a) T. R. Kelly, H. De Silva, R. A. Silva, *Nature* **1999**, *401*, 150–152; (b) T. R. Kelly, R. A. Silva, H. De Silva, S. Jasmin, Y. J. Zhao, *J. Am. Chem. Soc.* **2000**, *122*, 6935–6949.

22 There are differences between the technical definition of the terms "configuration" and "conformation" and their common usage. While formally configurational differences involve differences in bond angles and conformational differences suggest changes in torsion angles, here "configuration" is applied to *cis–trans* isomerism around double bonds as convention suggests. The same categorization can be extended to include isomerism around the C–N bond of amides, although this is often considered as a conformational change around a formal single bond as in **4**. For further discussion, see ref. [8].
The arrangement in space of the components of interlocked molecules and supramolecular complexes requires definition of another stereochemical term; "coconformation". This refers to the relative positions of the noncovalently linked components with respect to each other, see: M. C. T. Fyfe, P. T. Glink, S. Menzer, J. F. Stoddart, A. J. P. White, D. J. Williams, *Angew. Chem. Int. Ed. Engl.* **1997**, *36*, 2068–2070.

23 For a comprehensive review, see: C. Dugave, L. Demange, *Chem. Rev.* **2003**, *103*, 2475–2532.

24 (a) T. Yoshizawa in *CRC Handbook of Organic Photochemistry and Photobiology* (Eds.: W. M. Horspool, P.-S. Song), CRC Press, Boca Raton, **1995**, pp. 1493–1499; (b) R. Needleman, *ibid.*, pp. 1508–1515; (c) T. G. Ebrey, J. Liang, *ibid.* pp. 1500–1507; (d) B. Borhan, M. L. Souto, H. Imai, Y. Shichida, K. Nakanishi, *Science* **2000**, *288*, 2209–2212.

25 (a) B. Perman, V. Srajer, Z. Ren, T. Y. Teng, C. Pradervand, T. Ursby, D. Bourgeois, F. Schotte, M. Wulff, R. Kort, K. Hellingwerf, K. Moffat, *Science* **1998**, *279*, 1946–1950; (b) I. Antes, W. Thiel, W. F. van Gunsteren, *Eur. Biophys. J.* **2002**, *31*, 504–520; (c) K. J. Hellingwerf, J. Hendriks, T. Gensch, *J. Phys. Chem. A* **2003**, *107*, 1082–1094.

26 (a) G. Fischer, *Angew. Chem. Int. Ed. Engl.* **1994**, *33*, 1415–1436; (b) G. Fischer, *Chem. Soc. Rev.* **2000**, *29*, 119–127; (c) J. Balbach, F. X. Schmid in *Mechanisms of Protein Folding: Frontiers in Molecular Biology, Vol. 32*, 2nd ed. (Ed.: R. H. Pain), Oxford University Press, Oxford, **2000**, pp. 212–249.

27 (a) D. H. Waldeck, *Chem. Rev.* **1991**, *91*, 415–436; (b) H. Meier, *Angew. Chem. Int. Ed. Engl.* **1992**, *31*, 1399–1420; (c) J. Saltiel, D. F. Sears Jr., D.-H. Ko, K.-M. Park in *CRC Handbook of Organic Photochemistry and Photobiology* (Eds.: W. M. Horspool, P.-S. Song), CRC Press, Boca Raton, **1995**, pp. 3–15.

28 A photochromic process is one in which a light-induced interconversion between two species results in a change in its absorption spectrum.

29 H. Suginome in *CRC Handbook of Organic Photochemistry and Photobiology* (Eds.: W. M. Horspool, P.-S. Song), CRC Press, Boca Raton, **1995**, pp. 824–840.

30 G. Berkovic, V. Krongauz, V. Weiss, *Chem. Rev.* **2000**, *100*, 1741–1753.

31 (a) H. G. Heller in *CRC Handbook of Organic Photochemistry and Photobiology* (Eds.: W. M. Horspool, P.-S. Song), CRC Press, Boca Raton, **1995**, pp. 173–183; (b) Y. Yokoyama, *Chem. Rev.* **2000**, *100*, 1717–1739; (c) Y. Yokoyama in *Molecular Switches* (Ed.: B. L. Feringa), Wiley–VCH, Weinheim, **2001**, pp. 107–121.

32 (a) B. L. Feringa, N. P. M. Huck, A. M. Schoevaars, *Adv. Mater.* **1996**, *8*, 681–684; (b) B. L. Feringa, R. A. van Delden, N. Koumura, E. M. Geertsema, *Chem. Rev.* **2000**, *100*, 1789–1816; (c) B. L. Feringa, R. A. van Delden, M. K. J. ter Wiel in *Molecular Switches* (Ed.: B. L. Feringa), Wiley–VCH, Weinheim, **2001**, pp. 123–163.

33 For reviews, see: (a) B. L. Feringa, W. F. Jager, B. Delange, *Tetrahedron* **1993**, *49*, 8267–8310; (b) *Chem. Rev.* **2000**, *100*, 1683–1890 (special issue on Photochromism: Memories and Switches); (c) *Molecular Switches* (Ed.: B. L. Feringa), Wiley–VCH, Weinheim, **2001**.

34 M. Irie, *Chem. Rev.* **2000**, *100*, 1685–1716.

35 (a) I. Willner, *Acc. Chem. Res.* **1997**, *30*, 347–356; (b) I. Willner, B. Willner in *Molecular Switches* (Ed.: B. L. Feringa), Wiley–VCH, Weinheim, **2001**, pp. 165–218.

36 (a) K. Ichimura, *Chem. Rev.* **2000**, *100*, 1847–1873; (b) N. Tamaoki, *Adv. Mater.* **2001**, *13*, 1135–1147; (c) T. Ikeda, A. Kanazawa in *Molecular Switches* (Ed.: B. L. Feringa), Wiley–VCH, Weinheim, **2001**, pp. 363–397.

37 See, for example: (a) G. S. Kumar, D. C. Neckers, *Chem. Rev.* **1989**, *89*, 1915–1925; (b) J. A. Delaire, K. Nakatani, *Chem. Rev.* **2000**, *100*, 1817–1845; (c) F. Ciardelli, O. Pieroni in *Molecular Switches* (Ed.: B. L. Feringa), Wiley–VCH, Weinheim, **2001**, pp. 399–441.

38 T. Hugel, N. B. Holland, A. Cattani, L. Moroder, M. Seitz, H. E. Gaub, *Science* **2002**, *296*, 1103–1106.

39 A. M. Schoevaars, W. Kruizinga, R. W. J. Zijlstra, N. Veldman, A. L. Spek, B. L. Feringa, *J. Org. Chem.* **1997**, *62*, 4943–4948.

40 (a) B. L. Feringa, N. Koumura, R. A. van Delden, M. K. J. ter Wiel, *Appl. Phys. A* **2002**, *75*, 301–308; (b) B. L. Feringa, *Acc. Chem. Res.* **2001**, *34*, 504–513.

41 N. Koumura, R. W. J. Zijlstra, R. A. van Delden, N. Harada, B. L. Feringa, *Nature* **1999**, *401*, 152–155.

42 N. Koumura, E. M. Geertsema, M. B. van Gelder, A. Meetsma, B. L. Feringa, *J. Am. Chem. Soc.* **2002**, *124*, 5037–5051.

43 E. M. Geertsema, N. Koumura, M. K. J. ter Wiel, A. Meetsma, B. L. Feringa, *Chem. Commun.* **2002**, 2962–2963.

44 M. K. J. ter Wiel, R. A. van Delden, A. Meetsma, B. L. Feringa, *J. Am. Chem. Soc.* **2003**, *125*, 15076–15086.

45 (a) J.-M. Lehn, *Supramolecular Chemistry: Concepts and Perspectives*, VCH, Weinheim, **1995**; (b) *Comprehensive Supramolecular Chemistry* (Eds.: J.-M. Lehn, J. L. Atwood, J. E. D. Davies, D. D. MacNicol, F. Vögtle), Pergamon, Oxford, **1996**.

46 Myosin requires an actin track to move along: (a) S. M. Block, *Cell* **1996**, *87*, 151–157; while kinesin processes along the pathway provided by microtubule filaments: (b) S. M. Block, *Cell* **1998**, *93*, 5–8; even ion pumps operate through precisely defined structural channels: (c) J. C. Skou, *Angew. Chem. Int. Ed.* **1998**, *37*, 2321–2328.

47 S. Shinkai, M. Ishihara, K. Ueda, O. Manabe, *J. Chem. Soc.–Perkin Trans. 2* **1985**, 511–518.

48 V. Amendola, L. Fabbrizzi, C. Mangano, P. Pallavicini, *Acc. Chem. Res.* **2001**, *34*, 488–493.

49 L. Fabbrizzi, F. Gatti, P. Pallavicini, E. Zambarbieri, *Chem. Eur. J.* **1999**, *5*, 682–690.

50 A. Ikeda, T. Tsudera, S. Shinkai, *J. Org. Chem.* **1997**, *62*, 3568–3574.

51 L. Zelikovich, J. Libman, A. Shanzer, *Nature* **1995**, *374*, 790–792.

52 For representative examples of pseudorotaxane architectures based on dialkylammonium linear components and polyether macrocycles, see: (a) P. R. Ashton, P. J. Campbell, E. J. T. Chrystal, P. T. Glink, S. Menzer, D. Philp, N. Spencer, J. F. Stoddart, P. A. Tasker, D. J. Williams, *Angew. Chem. Int. Ed. Engl.* **1995**, *34*, 1865–1869; (b) P. R. Ashton, E. J. T. Chrystal, P. T. Glink, S. Menzer, C. Schiavo, J. F. Stoddart, P. A. Tasker, D. J. Williams, *Angew. Chem. Int. Ed. Engl.* **1995**, *34*, 1869–1871; (c) P. R. Ashton, E. J. T. Chrystal, P. T. Glink, S. Menzer, C. Schiavo, N. Spencer, J. F. Stoddart, P. A. Tasker, A. J. P. White, D. J. Williams, *Chem. Eur. J.* **1996**, *2*, 709–728; (d) P. R. Ashton, M. C. T. Fyfe, P. T. Glink, S. Menzer, J. F. Stoddart, A. J. P. White, D. J. Williams, *J. Am. Chem. Soc.* **1997**, *119*, 12514–12524; (e) P. R. Ashton, I. Baxter, M. C. T. Fyfe, F. M. Raymo, N. Spencer, J. F. Stoddart, A. J. P. White, D. J. Williams, *J. Am. Chem. Soc.* **1998**, *120*, 2297–2307; (f) P. R. Ashton, M. C. T. Fyfe, M. V. Martinez-Diaz, S. Menzer, C. Schiavo, J. F. Stoddart, A. J. P. White, D. J. Williams, *Chem. Eur. J.* **1998**, *4*, 1523–1534; (g) T. Clifford, A. Abushamleh, D. H. Busch, *Proc. Natl. Acad. Sci. USA* **2002**, *99*, 4830–4836; (h) H. W. Gibson, N. Yamaguchi, J. W. Jones, *J. Am. Chem. Soc.* **2003**, *125*, 3522–3533.

53 For an example of a pseudorotaxane architecture based on π-donor guest and π-acceptor cyclophane interactions, see: P. L. Anelli, P. R. Ashton, N. Spencer, A. M. Z. Slawin, J. F. Stoddart, D. J. Williams, *Angew. Chem. Int. Ed. Engl.* **1991**, *30*, 1036–1039.

54 For examples of other types of pseudorotaxane architectures, see: (a) A. Mirzoian, A. E. Kaifer, *Chem. Eur. J.* **1997**, *3*, 1052–1058; (b) K. M. Huh, T. Ooya, S. Sasaki, N. Yui, *Macromolecules* **2001**, *34*, 2402–2404; (c) J. A. Wisner, P. D. Beer, N. G. Berry, B. Tomapatanaget, *Proc. Natl. Acad. Sci. USA* **2002**, *99*, 4983–4986; (d) F. H. Huang, F. R. Fronczek, H. W. Gibson, *J. Am. Chem. Soc.* **2003**, *125*, 9272–9273; (e) F. H. Huang, H. W. Gibson, W. S. Bryant, D. S. Nagvekar, F. R. Fronczek, *J. Am. Chem. Soc.* **2003**, *125*, 9367–9371; (f) Y. Inoue, T. Kanbara, T. Yamamoto, *Tetrahedron Lett.* **2004**, *45*, 4603–4606.

55 (a) R. Ballardini, V. Balzani, M. T. Gandolfi, L. Prodi, M. Venturi, D. Philp, H. G. Ricketts, J. F. Stoddart, *Angew. Chem. Int. Ed. Engl.* **1993**, *32*, 1301–1303; (b) R. Ballardini, V. Balzani, A. Credi, M. T. Gandolfi, S. J. Langford, S. Menzer, L. Prodi, J. F. Stoddart, M. Venturi, D. J. Williams, *Angew. Chem. Int. Ed. Engl.* **1996**, *35*, 978–981; (c) P. R. Ashton, R. Ballardini, V. Balzani, S. E. Boyd, A. Credi, M. T. Gandolfi, M. GomezLopez, S. Iqbal, D. Philp, J. A. Preece, L. Prodi, H. G. Ricketts, J. F. Stoddart, M. S. Tolley, M. Venturi, A. J. P. White, D. J. Williams, *Chem. Eur. J.* **1997**, *3*, 152–170; (d) M. Asakawa, P. R. Ashton, V. Balzani, A. Credi, G. Mattersteig, O. A. Matthews, M. Montalti, N. Spencer, J. F. Stoddart, M. Venturi, *Chem. Eur. J.* **1997**, *3*, 1992–1996; (e) A. Credi, V. Balzani, S. J. Langford, J. F. Stoddart, *J. Am. Chem. Soc.* **1997**, *119*, 2679–2681; (f) P. R. Ashton, R. Ballardini, V. Balzani, M. GomezLopez, S. E. Lawrence, M. V. Martinez-Diaz, M. Montalti, A. Piersanti, L. Prodi, J. F. Stoddart, D. J. Williams, *J. Am. Chem. Soc.* **1997**, *119*, 10641–10651; (g) M. Montalti, L. Prodi, *Chem. Commun.* **1998**, 1461–1462; (h) P. R. Ashton, V. Balzani, O. Kocian, L. Prodi, N. Spencer, J. F. Stoddart, *J. Am. Chem. Soc.* **1998**, *120*, 11190–11191; (i) V. Balzani, A. Credi, F. Marchioni, J. F. Stoddart, *Chem. Commun.* **2001**, 1860–1861; (j) T. Fujimoto, A. Nakamura, Y. Inoue, Y. Sakata, T. Kaneda, *Tetrahedron Lett.* **2001**, *42*, 7987–7989.

56 A. Credi, M. Montalti, V. Balzani, S. J. Langford, F. M. Raymo, J. F. Stoddart, *New J. Chem.* **1998**, *22*, 1061–1065.

57 (a) E. Ishow, A. Credi, V. Balzani, F. Spadola, L. Mandolini, *Chem. Eur. J.* **1999**, *5*, 984–989; (b) R. Ballardini, V. Balzani, M. Clemente-Leon, A. Credi, M. T. Gandolfi, E. Ishow, J. Perkins, J. F. Stoddart, H. R. Tseng, S. Wenger, *J. Am. Chem. Soc.* **2002**, *124*, 12786–12795.

58 For examples of kinetically stable pseudorotaxane systems, see: (a) J.-P. Collin, P. Gaviña, J.-P. Sauvage, *Chem. Commun.* **1996**, 2005–2006; (b) J. W. Lee, K. P.

Kim, K. Kim, *Chem. Commun.* **2001**, 1042–1043.

59 (a) G. Schill, *Catenanes, Rotaxanes and Knots*, Academic Press, New York, **1971**; (b) D. M. Walba, *Tetrahedron* **1985**, *41*, 3161–3212; (c) D. B. Amabilino, J. F. Stoddart, *Chem. Rev.* **1995**, *95*, 2725–2828; (d) D. A. Leigh, A. Murphy, *Chem. Ind.* **1999**, 178–183; (e) G. A. Breault, C. A. Hunter, P. C. Mayers, *Tetrahedron* **1999**, *55*, 5265–5293; (f) *Molecular Catenanes Rotaxanes and Knots* (Eds.: J.-P. Sauvage, C. O. Dietrich-Buchecker), Wiley–VCH, Weinheim, **1999**.

60 There are some examples of catenanes and rotaxanes constructed under thermodynamic control using dynamic processes such as olefin metathesis or imine bond formation (see for example: J. S. Hannam, T. J. Kidd, D. A. Leigh, A. J. Wilson, *Org. Lett.* **2003**, *5*, 1907–1910 and references therein). Such systems cannot act as molecular machines if they exchange components with the bulk quicker than the timescale of their stimuli-induced motion.

61 Another synthetic strategy, 'slippage', theoretically provides a route to rotaxanes under thermodynamic control. However, the chemical assemblies thus far prepared by this route are misclassified in the literature. They do not satisfy the structural requirements of a rotaxane since the components can be separated from each other without breaking covalent bonds. Rather they are pseudo-rotaxanes – threaded host–guest complexes which are kinetically stable at room temperature. This incorrect classification is not normally an issue because the physical behavior of kinetically stable pseudo-rotaxanes generally mirrors that of rotaxanes. However, it increasingly appears to be confusing non-specialists into thinking that rotaxanes are supramolecular species or that pseudo-rotaxanes that are not kinetically stable can act as molecular machines. Slippage *could* be used as a strategy to form rotaxanes if the stabilization gained from the ring binding on the thread was sufficient to mean that covalent bond breaking is less energetically demanding than de-threading of the rings over the stoppers. To date no structures of this type have been reported.

62 For reviews on the development of interlocked molecules as molecular machines, see refs. [6a, 6c, 6e] and: (a) V. Balzani, M. Gomez-Lopez, J. F. Stoddart, *Acc. Chem. Res.* **1998**, *31*, 405–414; (b) J.-P. Sauvage, *Acc. Chem. Res.* **1998**, *31*, 611–619; (c) J.-C. Chambron, J.-P. Sauvage, *Chem. Eur. J.* **1998**, *4*, 1362–1366; (d) J.-P. Collin, P. Gaviña, V. Heitz, J.-P. Sauvage, *Eur. J. Inorg. Chem.* **1998**, 1–14; (e) M.-J. Blanco, M. C. Jimenez, J.-C. Chambron, V. Heitz, M. Linke, J.-P. Sauvage, *Chem. Soc. Rev.* **1999**, *28*, 293–305; (f) Z. Asfari, J. Vicens, *J. Inclusion Phenom. Macro. Chem.* **2000**, *36*, 103–118; (g) *Molecular Machines and Motors: Structure and Bonding, Vol. 99* (Ed.: J.-P. Sauvage), Springer, Berlin, **2001**; (h) A. R. Pease, J. O. Jeppesen, J. F. Stoddart, Y. Luo, C. P. Collier, J. R. Heath, *Acc. Chem. Res.* **2001**, *34*, 433–444; (i) C. A. Schalley, K. Beizai, F. Vögtle, *Acc. Chem. Res.* **2001**, *34*, 465–476; (j) J.-P. Collin, C. Dietrich-Buchecker, P. Gaviña, M. C. Jimenez-Molero, J.-P. Sauvage, *Acc. Chem. Res.* **2001**, *34*, 477–487; (k) B. X. Colasson, C. Dietrich-Buchecker, M. C. Jimenez-Molero, J.-P. Sauvage, *J. Phys. Org. Chem.* **2002**, *15*, 476–483.

63 Numbers in square brackets preceding the names of interlocked compounds indicate the number of mechanically interlocked species, e.g. a [2]rotaxane consists of one macrocycle and one thread component mechanically interlocked; two macrocycles both locked around the same thread would constitute a [3]rotaxane.

64 The concept of constructing a molecular catenane was first postulated by Willstätters before 1912 (see refs. [59a, 59b]). The first statistical synthesis of a catenane was achieved by Wasserman in 1960: (a) E. Wasserman, *J. Am. Chem. Soc.* **1960**, *82*, 4433–4434; and the first directed synthesis by Lüttringhaus and Schill in 1964: (b) G. Schill, A. Lüttringhaus, *Angew. Chem. Int. Ed. Engl.* **1964**, *3*, 546–547. The first rotaxane was synthesized by Harrison and Harrison in 1967: (c) I. T. Harrison, S. Harrison, *J. Am. Chem. Soc.* **1967**, *89*, 5723–5724; (d) I. T. Harrison, *J. Chem. Soc. Perkin Trans. 1* **1974**, 301–304.

65 For reviews on templated synthesis, see: (a) C. O. Dietrich-Buchecker, J.-P. Sauvage, *Chem. Rev.* **1987**, *87*, 795–810; (b) D. H. Busch, N. A. Stephenson, *Coord. Chem. Rev.* **1990**, *100*, 119–154; (c) D. Philp, J. F. Stoddart, *Synlett* **1991**, 445–458; (d) S. Anderson, H. L. Anderson, J. K. M. Sanders, *Acc. Chem. Res.* **1993**, *26*, 469–475; (e) R. Cacciapaglia, L. Mandolini, *Chem. Soc. Rev.* **1993**, *22*, 221–231; (f) R. Hoss, F. Vögtle, *Angew. Chem. Int. Ed. Engl.* **1994**, *33*, 375–384; (g) M. Fujita, K. Ogura, *Coord. Chem. Rev.* **1996**, *148*, 249–264; (h) M. C. T. Fyfe, J. F. Stoddart, *Acc. Chem. Res.* **1997**, *30*, 393–401; (i) R. Jager, F. Vögtle, *Angew. Chem. Int. Ed. Engl.* **1997**, *36*, 930–944; (j) *Templated Organic Synthesis* (Eds.: F. Diederich, P. J. Strang), Wiley–VCH, Weinheim, **1999**; (k) M. Fujita, *Acc. Chem. Res.* **1999**, *32*, 53–61; (l) T. J. Hubin, D. H. Busch, *Coord. Chem. Rev.* **2000**, *200*, 5–52; (m) J. F. Stoddart, H. R. Tseng, *Proc. Natl. Acad. Sci. USA* **2002**, *99*, 4797–4800; (n) R. Vilar, *Angew. Chem. Int. Ed.* **2003**, *42*, 1460–1477.

66 D. A. Leigh, A. Murphy, J. P. Smart, A. M. Z. Slawin, *Angew. Chem. Int. Ed. Engl.* **1997**, *36*, 728–732.

67 J. Sandström, *Dynamic NMR Spectroscopy*, Academic Press, London, **1982**.

68 F. W. Dahlquist, K. J. Longmur, R. B. Du Vernet, *J. Magn. Reson.* **1975**, *17*, 406–413.

69 F. G. Gatti, D. A. Leigh, S. A. Nepogodiev, A. M. Z. Slawin, S. J. Teat, J. K. Y. Wong, *J. Am. Chem. Soc.* **2001**, *123*, 5983–5989.

70 (a) C. O. Dietrich-Buchecker, J.-P. Sauvage, J. P. Kintzinger, *Tetrahedron Lett.* **1983**, *24*, 5095–5098; (b) C. O. Dietrich-Buchecker, J.-P. Sauvage, J.-M. Kern, *J. Am. Chem. Soc.* **1984**, *106*, 3043–3045; (c) C. Dietrich-Buchecker, J.-P. Sauvage, *Tetrahedron* **1990**, *46*, 503–512.

71 For reviews, see ref. [65a] and: (a) J.-P. Sauvage, *Acc. Chem. Res.* **1990**, *23*, 319–327; (b) J. C. Chambron, J. P. Collin, V. Heitz, D. Jouvenot, J. M. Kern, P. Mobian, D. Pomeranc, J. P. Sauvage, *Eur. J. Org. Chem.* **2004**, 1627–1638.

72 Metal-coordinated catenanes are sometimes known as catenates. Demetallation to give the uncomplexed, but still interlocked, ligands gives a catenand. No similar nomenclature is used for metal-coordinated rotaxane-like structures however, which are named as their organic counterparts. In recent times, the use of the terms "catenate" and "catenand" has largely been superseded in the literature by the more general "catenane".

73 C. A. Hunter, *J. Am. Chem. Soc.* **1992**, *114*, 5303–5311.

74 F. Vögtle, S. Meier, R. Hoss, *Angew. Chem. Int. Ed. Engl.* **1992**, *31*, 1619–1622.

75 A. G. Johnston, D. A. Leigh, R. J. Pritchard, M. D. Deegan, *Angew. Chem. Int. Ed. Engl.* **1995**, *34*, 1209–1212.

76 This previously inaccessible species was obtained by trapping round an acyclic, stoppered benzylic amide thread to create a [2]rotaxane, followed by controlled cleavage of the thread and resultant de-threading: A. G. Johnston, D. A. Leigh, A. Murphy, J. P. Smart, M. D. Deegan, *J. Am. Chem. Soc.* **1996**, *118*, 10662–10663.

77 A. G. Johnston, D. A. Leigh, L. Nezhat, J. P. Smart, M. D. Deegan, *Angew. Chem. Int. Ed. Engl.* **1995**, *34*, 1212–1216.

78 D. A. Leigh, A. Murphy, J. P. Smart, M. S. Deleuze, F. Zerbetto, *J. Am. Chem. Soc.* **1998**, *120*, 6458–6467.

79 M. S. Deleuze, D. A. Leigh, F. Zerbetto, *J. Am. Chem. Soc.* **1999**, *121*, 2364–2379.

80 D. A. Leigh, A. Troisi, F. Zerbetto, *Chem. Eur. J.* **2001**, *7*, 1450–1454.

81 P. L. Anelli, P. R. Ashton, R. Ballardini, V. Balzani, M. Delgado, M. T. Gandolfi, T. T. Goodnow, A. E. Kaifer, D. Philp, M. Pietraszkiewicz, L. Prodi, M. V. Reddington, A. M. Z. Slawin, N. Spencer, J. F. Stoddart, C. Vicent, D. J. Williams, *J. Am. Chem. Soc.* **1992**, *114*, 193–218.

82 P. R. Ashton, T. T. Goodnow, A. E. Kaifer, M. V. Reddington, A. M. Z. Slawin, N. Spencer, J. F. Stoddart, C. Vicent, D. J. Williams, *Angew. Chem. Int. Ed. Engl.* **1989**, *28*, 1396–1399.

83 P. R. Ashton, C. L. Brown, E. J. T. Chrystal, K. P. Parry, M. Pietraszkiewicz, N. Spencer, J. F. Stoddart, *Angew. Chem. Int. Ed. Engl.* **1991**, *30*, 1042–1045.

84 The distal isomer may still be involved transiently in the circumvolution mechanism.

85 S.-H. Chiu, A. M. Elizarov, P. T. Glink, J. F. Stoddart, *Org. Lett.* **2002**, *4*, 3561–3564.

86 For further studies on the effect of structure on the dynamic properties of similar catenanes, see: (a) D. B. Amabilino, P. R. Ashton, M. S. Tolley, J. F. Stoddart, D. J. Williams, *Angew. Chem. Int. Ed. Engl.* **1993**, *32*, 1297–1301; (b) P. R. Ashton, M. A. Blower, S. Iqbal, C. H. McLean, J. F. Stoddart, M. S. Tolley, D. J. Williams, *Synlett* **1994**, 1059–1062; (c) D. B. Amabilino, P. R. Ashton, G. R. Brown, W. Hayes, J. F. Stoddart, M. S. Tolley, D. J. Williams, *J. Chem. Soc. Chem. Commun.* **1994**, 2479–2482; (d) D. B. Amabilino, P. L. Anelli, P. R. Ashton, G. R. Brown, E. Cordova, L. A. Godinez, W. Hayes, A. E. Kaifer, D. Philp, A. M. Z. Slawin, N. Spencer, J. F. Stoddart, M. S. Tolley, D. J. Williams, *J. Am. Chem. Soc.* **1995**, *117*, 11142–11170; (e) M. Asakawa, P. R. Ashton, S. E. Boyd, C. L. Brown, R. E. Gillard, O. Kocian, F. M. Raymo, J. F. Stoddart, M. S. Tolley, A. J. P. White, D. J. Williams, *J. Org. Chem.* **1997**, *62*, 26–37; (f) R. Ballardini, V. Balzani, A. Credi, C. L. Brown, R. E. Gillard, M. Montalti, D. Philp, J. F. Stoddart, M. Venturi, A. J. P. White, B. J. Williams, D. J. Williams, *J. Am. Chem. Soc.* **1997**, *119*, 12503–12513; (g) B. Cabezon, J. G. Cao, F. M. Raymo, J. F. Stoddart, A. J. P. White, D. J. Williams, *Angew. Chem. Int. Ed.* **2000**, *39*, 148–151.

87 With the notable exception of the earliest statistically constructed [2]rotaxanes (see refs. [64c, 64d]), there are very few examples of rotaxanes without any recognition elements in the thread, see however: (a) C. Heim, A. Affeld, M. Nieger, F. Vögtle, *Helv. Chim. Acta* **1999**, *82*, 746–759; (b) J. S. Hannam, S. M. Lacy, D. A. Leigh, C. G. Saiz, A. M. Z. Slawin, S. Stitchell, *Angew. Chem. Int. Ed.* **2004**, *43*, 3260–3264.

88 P. L. Anelli, N. Spencer, J. F. Stoddart, *J. Am. Chem. Soc.* **1991**, *113*, 5131–5133.

89 (a) P. R. Ashton, D. Philp, N. Spencer, J. F. Stoddart, *J. Chem. Soc. Chem. Commun.* **1992**, 1124–1128; (b) P. R. Ashton, M. R. Johnston, J. F. Stoddart, M. S. Tolley, J. W. Wheeler, *J. Chem. Soc. Chem. Commun.* **1992**, 1128–1131.

90 A. S. Lane, D. A. Leigh, A. Murphy, *J. Am. Chem. Soc.* **1997**, *119*, 11092–11093.

91 D. A. Leigh, A. Troisi, F. Zerbetto, *Angew. Chem. Int. Ed.* **2000**, *39*, 350–353.

92 Even in 1991, when reporting the first degenerate molecular shuttle (see ref. [88]), Stoddart noted that: "The opportunity now exists to desymmetrize the molecular shuttle by inserting nonidentical 'stations' along the polyether 'thread' in such a

manner that these different 'stations' can be addressed selectively by chemical, electrochemical, or photochemical means and so provide a mechanism to drive the 'bead' to and fro between 'stations' along the 'thread'.

93 (a) A. C. Benniston, A. Harriman, *Angew. Chem. Int. Ed. Engl.* **1993**, *32*, 1459–1461; (b) A. C. Benniston, A. Harriman, V. M. Lynch, *J. Am. Chem. Soc.* **1995**, *117*, 5275–5291.

94 H. Murakami, A. Kawabuchi, K. Kotoo, M. Kunitake, N. Nakashima, *J. Am. Chem. Soc.* **1997**, *119*, 7605–7606.

95 C. A. Stanier, S. J. Alderman, T. D. W. Claridge, H. L. Anderson, *Angew. Chem. Int. Ed.* **2002**, *41*, 1769–1772.

96 As long as the chemical reaction is fast with respect to the movement of the macrocycle.

97 R. A. Bissell, E. Cordova, A. E. Kaifer, J. F. Stoddart, *Nature* **1994**, *369*, 133–137.

98 See ref. [52] and: (a) A. G. Kolchinski, D. H. Busch, N. W. Alcock, *J. Chem. Soc. Chem. Commun.* **1995**, 1289–1291; (b) P. T. Glink, C. Schiavo, J. F. Stoddart, D. J. Williams, *Chem. Commun.* **1996**, 1483–1490; (c) P. R. Ashton, P. T. Glink, J. F. Stoddart, P. A. Tasker, A. J. P. White, D. J. Williams, *Chem. Eur. J.* **1996**, *2*, 729–736.

99 (a) M. V. Martinez-Diaz, N. Spencer, J. F. Stoddart, *Angew. Chem. Int. Ed. Engl.* **1997**, *36*, 1904–1907; (b) P. R. Ashton, R. Ballardini, V. Balzani, I. Baxter, A. Credi, M. C. T. Fyfe, M. T. Gandolfi, M. Gomez-Lopez, M. V. Martinez-Diaz, A. Piersanti, N. Spencer, J. F. Stoddart, M. Venturi, A. J. P. White, D. J. Williams, *J. Am. Chem. Soc.* **1998**, *120*, 11932–11942.

100 For an example of another pH switched shuttle from the Stoddart group, see: A. M. Elizarov, S.-H. Chiu, J. F. Stoddart, *J. Org. Chem.* **2002**, *67*, 9175–9181.

101 J. D. Badjić, V. Balzani, A. Credi, S. Silvi, J. F. Stoddart, *Science* **2004**, *303*, 1845–1849.

102 C. M. Keaveney, D. A. Leigh, *Angew. Chem. Int. Ed.* **2004**, *43*, 1222–1224.

103 (a) P. R. Ashton, R. A. Bissell, N. Spencer, J. F. Stoddart, M. S. Tolley, *Synlett* **1992**, 914–918; (b) P. R. Ashton, R. A. Bissell, R. Gorski, D. Philp, N. Spencer, J. F. Stoddart, M. S. Tolley, *Synlett* **1992**,

919–922; (c) P. R. Ashton, R. A. Bissell, N. Spencer, J. F. Stoddart, M. S. Tolley, *Synlett* **1992**, 923–926; (d) P. L. Anelli, M. Asakawa, P. R. Ashton, R. A. Bissell, G. Clavier, R. Gorski, A. E. Kaifer, S. J. Langford, G. Mattersteig, S. Menzer, D. Philp, A. M. Z. Slawin, N. Spencer, J. F. Stoddart, M. S. Tolley, D. J. Williams, *Chem. Eur. J.* **1997**, *3*, 1113–1135; (e) D. B. Amabilino, P. R. Ashton, S. E. Boyd, M. GomezLopez, W. Hayes, J. F. Stoddart, *J. Org. Chem.* **1997**, *62*, 3062–3075.

104 Shuttle **41** also shows a certain degree of electrochemical switching in the deprotonated state when electrochemical reduction of the bipyridinium unit causes movement of the macrocycle back towards the amine moiety. In this state however, the electron-rich macrocycle has no significant favorable interactions with the thread and its position is not well-defined, see ref. [99a].

105 For more recent examples of electrochemically switched shuttles from the Stoddart group, see: H. R. Tseng, S. A. Vignon, P. C. Celestre, J. Perkins, J. O. Jeppesen, A. Di Fabio, R. Ballardini, M. T. Gandolfi, M. Venturi, V. Balzani, J. F. Stoddart, *Chem. Eur. J.* **2004**, *10*, 155–172; and for an example of light- and electrochemically-triggered shuttling from the same group, see: P. R. Ashton, R. Ballardini, V. Balzani, A. Credi, K. R. Dress, E. Ishow, C. J. Kleverlaan, O. Kocian, J. A. Preece, N. Spencer, J. F. Stoddart, M. Venturi, S. Wenger, *Chem. Eur. J.* **2000**, *6*, 3558–3574.

106 (a) A. M. Brouwer, C. Frochot, F. G. Gatti, D. A. Leigh, L. Mottier, F. Paolucci, S. Roffia, G. W. H. Wurpel, *Science* **2001**, *291*, 2124–2128; (b) A. Altieri, F. G. Gatti, E. R. Kay, D. A. Leigh, D. Martel, F. Paolucci, A. M. Z. Slawin, J. K. Y. Wong, *J. Am. Chem. Soc.* **2003**, *125*, 8644–8654.

107 N. Armaroli, V. Balzani, J.-P. Collin, P. Gaviña, J.-P. Sauvage, B. Ventura, *J. Am. Chem. Soc.* **1999**, *121*, 4397–4408.

108 A. Altieri, G. Bottari, F. Dehez, D. A. Leigh, J. K. Y. Wong, F. Zerbetto, *Angew. Chem. Int. Ed.* **2003**, *42*, 2296–2300.

109 For an example of shuttling triggered by azobenzene photoisomerisation, see: M. Asakawa, P. R. Ashton, V. Balzani, C. L. Brown, A. Credi, O. A. Matthews, S. P.

Newton, F. M. Raymo, A. N. Shipway, N. Spencer, A. Quick, J. F. Stoddart, A. J. P. White, D. J. Williams, *Chem. Eur. J.* **1999**, *5*, 860–875.

110   Heat has been used to effect cis–trans isomerization in **46** resulting in a concomitant net change of position of the ring (see ref. [108]), while a temperature increase has also been used to overcome a significant kinetic barrier to shuttling, following a chemical change in a kinetically stable pseudorotaxane (see Ref. [58b]). In neither of these cases is the process reversible on cooling the material without some other stimulus being applied.

111   A C_{12} chain has >500 000 (3^{12}) possible C–C rotamers and a significant number of these degrees of freedom must be lost upon forming the *dodec-Z-***47** structure.

112   A series of theoretical designs for entropy-driven mechanical motion in mechanically interlocked molecules have been suggested, see: A. Hanke, R. Metzler, *Chem. Phys. Lett.* **2002**, *359*, 22–26.

113   The use of terms such as "all or nothing" or "quantitative" are inappropriate descriptions of an equilibrium.

114   V. Bermudez, N. Capron, T. Gase, F. G. Gatti, F. Kajzar, D. A. Leigh, F. Zerbetto, S. W. Zhang, *Nature* **2000**, *406*, 608–611.

115   F. G. Gatti, S. Leon, J. K. Y. Wong, G. Bottari, A. Altieri, M. A. F. Morales, S. J. Teat, C. Frochot, D. A. Leigh, A. M. Brouwer, F. Zerbetto, *Proc. Natl. Acad. Sci. USA* **2003**, *100*, 10–14.

116   For examples where the ring in a rotaxane is pirouetted between two distinct orientations, using the same principles as molecular shuttling (Section 7.5.5), see: (a) L. Raehm, J.-M. Kern, J.-P. Sauvage, *Chem. Eur. J.* **1999**, *5*, 3310–3317; (b) I. Poleschak, J.-M. Kern, J.-P. Sauvage, *Chem. Commun.* **2004**, 474–476.

117   It has recently been suggested that unidirectional pirouetting could result from simply derivatizing rotaxanes with chiral or knotted stoppers (see: O. Lukin, T. Kubota, Y. Okamoto, F. Schelhase, A. Yoneva, W. M. Müller, U. Müller, F. Vögtle, *Angew. Chem. Int. Ed.* **2003**, *42*, 4542–4545). As discussed in Section 7.2, this is incorrect. Directional rotation can only result from an external energy source being used to drive the system temporarily away from equilibrium.

118   D. A. Leigh, K. Moody, J. P. Smart, K. J. Watson, A. M. Z. Slawin, *Angew. Chem. Int. Ed. Engl.* **1996**, *35*, 306–310.

119   M. Asakawa, P. R. Ashton, V. Balzani, A. Credi, C. Hamers, G. Mattersteig, M. Montalti, A. N. Shipway, N. Spencer, J. F. Stoddart, M. S. Tolley, M. Venturi, A. J. P. White, D. J. Williams, *Angew. Chem. Int. Ed.* **1998**, *37*, 333–337.

120   For other examples of switchable motion in [2]catenanes from the Stoddart group, see ref. [109] and: (a) P. R. Ashton, L. Perezgarcia, J. F. Stoddart, A. J. P. White, D. J. Williams, *Angew. Chem. Int. Ed. Engl.* **1995**, *34*, 571–574; (b) P. R. Ashton, R. Ballardini, V. Balzani, A. Credi, M. T. Gandolfi, S. Menzer, L. Perezgarcia, L. Prodi, J. F. Stoddart, M. Venturi, A. J. P. White, D. J. Williams, *J. Am. Chem. Soc.* **1995**, *117*, 11171–11197; (c) M. Asakawa, P. R. Ashton, V. Balzani, S. E. Boyd, A. Credi, G. Mattersteig, S. Menzer, M. Montalti, F. M. Raymo, C. Ruffilli, J. F. Stoddart, M. Venturi, D. J. Williams, *Eur. J. Org. Chem.* **1999**, 985–994; (d) V. Balzani, A. Credi, G. Mattersteig, O. A. Matthews, F. M. Raymo, J. F. Stoddart, M. Venturi, A. J. P. White, D. J. Williams, *J. Org. Chem.* **2000**, *65*, 1924–1936; (e) P. R. Ashton, V. Baldoni, V. Balzani, A. Credi, H. D. A. Hoffmann, M. V. Martinez-Diaz, F. M. Raymo, J. F. Stoddart, M. Venturi, *Chem. Eur. J.* **2001**, *7*, 3482–3493.

121   (a) A. Livoreil, C. O. Dietrich-Buchecker, J.-P. Sauvage, *J. Am. Chem. Soc.* **1994**, *116*, 9399–9400; (b) F. Baumann, A. Livoreil, W. Kaim, J.-P. Sauvage, *Chem. Commun.* **1997**, 35–36; (c) A. Livoreil, J.-P. Sauvage, N. Armaroli, V. Balzani, L. Flamigni, B. Ventura, *J. Am. Chem. Soc.* **1997**, *119*, 12114–12124.

122   For an example of an electrochemically switched [2]catenane based on the redox reactions of metal centers embedded in one of the rings, see: B. Korybut-Daszkiewicz, A. Więckowska, R. Bilewicz, S. Domagala, K. Woźniak, *Angew. Chem. Int. Ed.* **2004**, *43*, 1668–1672.

123   D. J. Cardenas, A. Livoreil, J.-P. Sauvage, *J. Am. Chem. Soc.* **1996**, *118*, 11980–11981.

124 D. A. Leigh, J. K. Y. Wong, F. Dehez, F. Zerbetto, *Nature* **2003**, *424*, 174–179.

125 (a) C. P. Collier, E. W. Wong, M. Běhloradský, F. M. Raymo, J. F. Stoddart, P. J. Kuekes, R. S. Williams, J. R. Heath, *Science* **1999**, *285*, 391–394; (b) E. W. Wong, C. P. Collier, M. Běhloradský, F. M. Raymo, J. F. Stoddart, J. R. Heath, *J. Am. Chem. Soc.* **2000**, *122*, 5831–5840.

126 C. P. Collier, G. Mattersteig, E. W. Wong, Y. Luo, K. Beverly, J. Sampaio, F. M. Raymo, J. F. Stoddart, J. R. Heath, *Science* **2000**, *289*, 1172–1175.

127 C. P. Collier, J. O. Jeppesen, Y. Luo, J. Perkins, E. W. Wong, J. R. Heath, J. F. Stoddart, *J. Am. Chem. Soc.* **2001**, *123*, 12632–12641.

128 Y. Luo, C. P. Collier, J. O. Jeppesen, K. A. Nielsen, E. Delonno, G. Ho, J. Perkins, H. R. Tseng, T. Yamamoto, J. F. Stoddart, J. R. Heath, *ChemPhysChem* **2002**, *3*, 519–525.

129 (a) Y. Chen, D. A. A. Ohlberg, X. M. Li, D. R. Stewart, R. S. Williams, J. O. Jeppesen, K. A. Nielsen, J. F. Stoddart, D. L. Olynick, E. Anderson, *Appl. Phys. Lett.* **2003**, *82*, 1610–1612; (b) Y. Chen, G. Y. Jung, D. A. A. Ohlberg, X. M. Li, D. R. Stewart, J. O. Jeppesen, K. A. Nielsen, J. F. Stoddart, R. S. Williams, *Nanotechnology* **2003**, *14*, 462–468; (c) D. R. Stewart, D. A. A. Ohlberg, P. A. Beck, Y. Chen, R. S. Williams, J. O. Jeppesen, K. A. Nielsen, J. F. Stoddart, *Nano Lett.* **2004**, *4*, 133–136.

130 M. R. Diehl, D. W. Steuerman, H. R. Tseng, S. A. Vignon, A. Star, P. C. Celestre, J. F. Stoddart, J. R. Heath, *ChemPhysChem* **2003**, *4*, 1335–1339.

131 (a) M. C. Jimenez, C. Dietrich-Buchecker, J.-P. Sauvage, *Angew. Chem. Int. Ed.* **2000**, *39*, 3284–3287; (b) M. C. Jimenez-Molero, C. Dietrich-Buchecker, J.-P. Sauvage, *Chem. Eur. J.* **2002**, *8*, 1456–1466; (c) M. C. Jimenez-Molero, C. Dietrich-Buchecker, J.-P. Sauvage, *Chem. Commun.* **2003**, 1613–1616.

132 M. Asakawa, G. Brancato, M. Fanti, D. A. Leigh, T. Shimizu, A. M. Z. Slawin, J. K. Y. Wong, F. Zerbetto, S. W. Zhang, *J. Am. Chem. Soc.* **2002**, *124*, 2939–2950.

133 G. Bottari, D. A. Leigh, E. M. Pérez, *J. Am. Chem. Soc.* **2003**, *125*, 13360–13361.

134 E. M. Pérez, D. T. F. Dryden, D. A. Leigh, G. Teobaldi, F. Zerbetto, **2004**, *J. Am. Chem. Soc.,* **2004**, *126*, 12210–12211.

135 Q.-C. Wang, D.-H. Qu, J. Ren, K. Chen, H. Tian, *Angew. Chem. Int. Ed.* **2004**, *43*, 2661–2665.

136 D.-H. Qu, Q.-C. Wang, J. Ren, H. Tian, *Org. Lett.* **2004**, *6*, 2085–2088.

137 See, for example: (a) A. K. Boal, V. M. Rotello, *J. Am. Chem. Soc.* **1999**, *121*, 4914–4915; (b) Y. Ge, D. K. Smith, *Anal. Chem.* **2000**, *72*, 1860–1865; (c) G. Cooke, F. M. A. Duclairoir, V. M. Rotello, J. F. Stoddart, *Tetrahedron Lett.* **2000**, *41*, 8163–8166; (d) A. Labande, D. Astruc, *Chem. Commun.* **2000**, 1007–1008; (e) M. C. Daniel, J. Ruiz, S. Nlate, J. Palumbo, J. C. Blais, D. Astruc, *Chem. Commun.* **2001**, 2000–2001; (f) S. Y. Chia, J. G. Cao, J. F. Stoddart, J. I. Zink, *Angew. Chem. Int. Ed.* **2001**, *40*, 2447–2451; (g) M. R. Bryce, G. Cooke, F. M. A. Duclairoir, P. John, D. F. Perepichka, N. Polwart, V. M. Rotello, J. F. Stoddart, H. R. Tseng, *J. Mater. Chem.* **2003**, *13*, 2111–2117; (h) K. Kim, W. S. Jeon, J.-K. Kang, J. W. Lee, S. Y. Jon, T. Kim, K. Kim, *Angew. Chem. Int. Ed.* **2003**, *42*, 2293–2296.

138 R. Hernandez, H. R. Tseng, J. W. Wong, J. F. Stoddart, J. I. Zink, *J. Am. Chem. Soc.* **2004**, *126*, 3370–3371.

139 (a) J. Lahann, S. Mitragotri, T. N. Tran, H. Kaido, J. Sundaram, I. S. Choi, S. Hoffer, G. A. Somorjai, R. Langer, *Science* **2003**, *299*, 371–374; (b) X. M. Wang, A. B. Kharitonov, E. Katz, I. Willner, *Chem. Commun.* **2003**, 1542–1543.

140 T. Gase, D. Grando, P. A. Chollet, F. Kajzar, A. Murphy, D. A. Leigh, *Adv. Mater.* **1999**, *11*, 1303–1305.

141 H. R. Tseng, D. M. Wu, N. X. L. Fang, X. Zhang, J. F. Stoddart, *ChemPhysChem* **2004**, *5*, 111–116.

142 B. Long, K. Nikitin, D. Fitzmaurice, *J. Am. Chem. Soc.* **2003**, *125*, 15490–15498.

143 (a) I. Willner, B. Willner, *J. Mater. Chem.* **1998**, *8*, 2543–2556; (b) A. N. Shipway, I. Willner, *Acc. Chem. Res.* **2001**, *34*, 421–432.

144 I. Willner, V. Pardo-Yissar, E. Katz, K. T. Ranjit, *J. Electroanal. Chem.* **2001**, *497*, 172–177.

145 See for example the observation of single-molecule stretching and contraction described in Section 7.2 and ref. [38], and additionally: (a) D. M. Eigler, C. P. Lutz, W. E. Rudge, *Nature* **1991**, *352*, 600–603; (b) T. A. Jung, R. R. Schlittler, J. K. Gimzewski, H. Tang, C. Joachim, *Science* **1996**, *271*, 181–184; (c) J. K. Gimzewski, C. Joachim, R. R. Schlittler, V. Langlais, H. Tang, I. Johannsen, *Science* **1998**, *281*, 531–533.

146 H. Shigekawa, K. Miyake, J. Sumaoka, A. Harada, M. Komiyama, *J. Am. Chem. Soc.* **2000**, *122*, 5411–5412.

147 M. Cavallini, F. Biscarini, S. Leon, F. Zerbetto, G. Bottari, D. A. Leigh, *Science* **2003**, *299*, 531–531.

148 (a) P. Ball, *Nature* **2000**, *406*, 118–120; (b) A. P. de Silva, N. D. McClenaghan, C. P. McCoy in *Molecular Switches* (Ed.: B. L. Feringa), Wiley–VCH, Weinheim, **2001**, pp. 339–361; (c) F. M. Raymo, *Adv. Mater.* **2002**, *14*, 401–414; (d) G. J. Brown, A. P. de Silva, S. Pagliari, *Chem. Commun.* **2002**, 2461–2463.

149 (a) P. Thordarson, E. J. A. Bijsterveld, A. E. Rowan, R. J. M. Nolte, *Nature* **2003**, *424*, 915–918; (b) P. Thordarson, R. J. M. Nolte, A. E. Rowan, *Aust. J. Chem.* **2004**, *57*, 323–327; (c) P. R. Carlier, *Angew. Chem. Int. Ed.* **2004**, *43*, 2602–2605.

150 See, for example: (a) S. J. Tans, A. R. M. Verschueren, C. Dekker, *Nature* **1998**, *393*, 49–52; (b) H. Park, J. Park, A. K. L. Lim, E. H. Anderson, A. P. Alivisatos, P. L. McEuen, *Nature* **2000**, *407*, 57–60 (nanoscale transistors); (c) A. M. Fennimore, T. D. Yuzvinsky, W. Q. Han, M. S. Fuhrer, J. Cumings, A. Zettl, *Nature* **2003**, *424*, 408–410 (nanoscale rotational actuator).

151 For reviews, see: (a) G. M. Whitesides, M. Boncheva, *Proc. Natl. Acad. Sci. USA* **2002**, *99*, 4769–4774; (b) I. W. Hamley, *Angew. Chem. Int. Ed.* **2003**, *42*, 1692–1712.

152 T. F. Otero, J. M. Sansinena, *Adv. Mater.* **1998**, *10*, 491–494 and references therein.

153 (a) M. Barboiu, J.-M. Lehn, *Proc. Natl. Acad. Sci. USA* **2002**, *99*, 5201–5206; (b) H. J. Schneider, T. J. Liu, N. Lomadze, *Angew. Chem. Int. Ed.* **2003**, *42*, 3544–3546.

154 M. J. Marsella, R. J. Reid, S. Estassi, L. S. Wang, *J. Am. Chem. Soc.* **2002**, *124*, 12507–12510.

155 See ref. [5] and, for example: (a) G. Steinberg-Yfrach, J. L. Rigaud, E. N. Durantini, A. L. Moore, D. Gust, T. A. Moore, *Nature* **1998**, *392*, 479–482; (b) C. Montemagno, G. Bachand, *Nanotechnology* **1999**, *10*, 225–231; (c) R. K. Soong, G. D. Bachand, H. P. Neves, A. G. Olkhovets, H. G. Craighead, C. D. Montemagno, *Science* **2000**, *290*, 1555–1558.

156 (a) B. Yurke, A. J. Turberfield, A. P. Mills, F. C. Simmel, J. L. Neumann, *Nature* **2000**, *406*, 605–608; (b) D. S. Liu, S. Balasubramanian, *Angew. Chem. Int. Ed.* **2003**, *42*, 5734–5736; (c) W. U. Dittmer, A. Reuter, F. C. Simmel, *Angew. Chem. Int. Ed.* **2004**, *43*, 3550–3553; (d) Y. Chen, M. Wang, C. Mao, *Angew. Chem. Int. Ed.* **2004**, *43*, 3554–3557.

8
Replicable Nanoscaffolded Multifunctionality – A Chemical Perspective

Wolf-Matthias Pankau, Sergey Antsypovich, Lars Eckardt, Johanna Stankiewicz, Sven Mönninghoff, Jan Zimmermann, Maya Radeva, and Günter von Kiedrowski

Abstract

Self-replicating nanorobots were foreseen in technological dreams and visions the scientific basis of which attracted solid criticism from chemical and physical reasoning. If, however, one views such constructs as three-dimensionally defined noncovalent nanoscaffolding of a multitude of modular functions whose array is replicable nonautonomously, many pieces of the technology needed for their implementation has recently became available. Gold cluster-labeled molecules have been remotely controlled by GHz radio frequencies causing local and selective inductive heating, and monoconjugable thermostable gold clusters will soon become commercially available. Charged molecules have been electrophoretically steered and manipulated on the surface of microelectrode array chips. Surface-promoted replication and exponential amplification of DNA analogs (SPREAD) might find applications in the cloning and copying of informational nanostructures on the surface of such chips. Synthetic tris-oligonucleotidyl junctions have been reported as covalent building blocks for noncovalent DNA nanostructures and it has been shown that kinetic control during noncovalent synthesis favors small and defined nanostructures instead of polymeric networks. Very recently it was demonstrated that functionalized DNA nanoscaffolds with stiff tensegrity, for example tetrahedra, self-assemble from maximally instructed sets of 3- or [3+1]-arm junctions and that connectivity information in such nanoscaffolds can be copied. The implications of these developments are discussed here with respect to a possible implementation scheme.

8.1
Introduction

The paradigm of material production in chemistry is synthesis; the paradigm of material production in biology is growth based on self-replication and molecular infor-

Functional Synthetic Receptors. Edited by T. Schrader, A. D. Hamilton

ISBN: 3-527-30655-2

mation processing. Material properties in physics and chemistry arise by design and engineering, material properties in biology emerge in the process of natural evolution. The gap between synthesis and design on the chemical and physical side and growth and evolution on the biological side calls for a new generation of technologies that aim to combine the advantages of both man-made and natural strategies. This is not only a challenge but also a opportunity for technology in the nanosciences. In the industrial context of material sciences there is a strong argument against the mass-production of highly complex and intelligent materials based on nonstandard polymers. The argument is simply based on the cost of synthesis. It is, however, getting weaker, as production technologies become available that are either based on biological systems or mimic the latter systems in their elementary ability to replicate and to grow. Replicable nanomachinery [1], known by the public as nanorobots, were seen as the holy grail of nanotechnology, and medical applications of the latter were envisaged as yielding a new generation of drugs with the functionality to perform complex tasks in the human body [2]. The scientific concepts behind the schemes proposed for implementation of nanorobots attracted solid criticism from chemical and physical reasoning, however [3, 4]. Central to the criticism is the concept of a nanomechanical assembler that enables the manufacturing of nanomachines, including itself, in an atom-by-atom fashion. If, however, nanomechanical fabrication of nanomachines is replaced by their self-assembly from modular parts in the sense of a molecular self-instruction process, the creation of replicable nanomachines seem to be feasible today.

8.2
A Manifesto for Nanorobot Implementation

We propose here a scheme for the implementation of a prototype generation of replicable nanomachines, that is based on:

1. noncovalent informational nanoscaffolding and arraying of a multitude of modular functions expressed by organic, inorganic, or biological components in a defined 3D arrangement;
2. the self-assembly of such objects from synthetic three-arm junctions where each arm is a composite/conjugate from an informational molecule such as DNA, RNA, or a synthetic mimic of RNA and an individual functional module;
3. tensegrity and maximum instruction as the keys for nanoarchitecture control;
4. the copying of connectivity information in such junctions as the key for nanomachine replication;
5. the controlled cloning and copying of junctions on the surface of electronically addressable chips by means of a chip variant of the procedure published as "surface promoted replication and exponential amplification of DNA analogs" (SPREAD);
6. the use of linear conjugates between oligonucleotides and single modular functions as basic construction elements for fabrication of functionalized three-arm junctions and the self-assembly of the latter into multifunctionalized nanoscaffolded machines;

7. the directed evolution of replicable nanomachines by a variant of the SELEX procedure starting from random connectivity information;
8. the use of such nanomachines to probe nanoepitopes on the surfaces of biological
   cells; and
9. the external control of the operation of such nanomachines by GHz radio-fre-
   quency magnetic field inductive heating of metal clusters attached to such constructs.

Although several key steps of the proposed technology have already been realized
in laboratories, including ours, the full potential of replicable nanoscaffolded, multifunctional materials and machines can only be exploited if the pieces of technology
finally fit together. This is why we believe that a kind of open project in chemistry
might be the most suitable form here.

8.2.1
Noncovalent Informational Nanoscaffolding

This as been demonstrated in a number of ways by using *linear* DNA as informational
templates [5]. Conjugates of oligonucleotides with nanoparticles, dyes, proteins, and
other functions were hybridized with the oligonucleotide part to the complementary
stretch of the linear template. The array of function modules was studied with a variety of techniques including TEM, STM, and AFM. Defined three-dimensional
nanoarchitectures are, however, difficult to achieve by following this strategy, because
linear assemblies can fold in many different ways.

8.2.2
Self-assembly From Synthetic Three-arm Junctions

Nadrian Seeman has demonstrated the feasibility of synthesizing three-dimensional
nanoarchitectures such as cubes [6] from DNA as the construction material. Seeman's nanoobjects are covalent structures, the synthesis of which started from sets
of three linear oligonucleotides that self-assembled into noncovalent three-way junctions (Fig. 8.1). These junctions had sticky ends which were joined and then ligated
to form covalent nanostructures. Recently, technology has become available which
enables the automated synthesis of covalent tris-oligo junctions bearing either identical or nonidentical sequences [7–10]. Using this technology it is now possible to
work on the reverse strategy, namely, self-assembly of nanostructures from sets of covalent tris-oligo junctions. Complementary tris-oligo junctions with three identical
sequences have been shown to self-assemble into nanoobjects with the topology of
acetylene and cyclobutadiene if one understands a DNA double-strand as the noncovalent topological analogy of a C–C bond (Fig. 8.2) [9]. It was also shown that kinetic
control – applied by means of rapid cooling during hybridization – favors the formation of small and defined nanoobjects instead of large "polymeric" networks of noncovalent superstructures [9]. Very recently, a feasibility study on the issue (2) of this
proposal proved successful. It was shown that a tetrahedral nanoobject self-assem-

Seeman's strategy

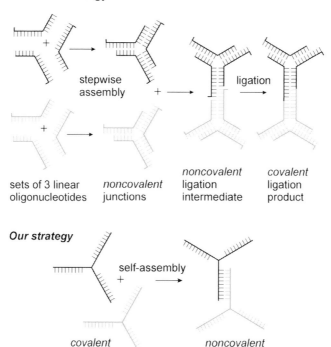

sets of 3 linear
oligonucleotides

noncovalent
junctions

noncovalent
ligation
intermediate

covalent
ligation
product

Our strategy

covalent
trisoligo junctions

noncovalent
product

Figure 8.1 Two strategies for the construction of DNA-nanoobjects. Seeman's strategy is based on the synthesis of covalent objects from noncovalent junctions. Our strategy involves self assembly of covalent junctions to form noncovalent objects.

bles from four tris-oligo junctions each having three individual sequences to give a total of $4 \times 3 = 12$ noncovalent interaction elements (Fig. 8.3) [10]. Pairing of these sequences yielded six doublestranded "bonds" encoding the edges of the tetrahedron. Gel electrophoresis studies on the assembly and enzymatic digestion of the tetrahe-

Figure 8.2 Self-assembly of two complementary tris-oligo junctions (left) yields objects with the topology of acetylene (middle) and cyclobutadiene (right). Kinetic control during self-assembly was shown to favor the acetylene topology.

dron and its noncovalent intermediates confirmed that all six planned bonds were formed. Circular dichroism studies indicated a B-DNA conformation of the duplex bonds and UV–melting studies revealed the expected melting cooperativity. Distance probing between the nodes of the tetrahedron using fluorescence resonance energy transfer (FRET) gave evidence of the equal bond length expected for tetrahedral geometry. The latter studies also showed that the tetrahedron can act as a noncovalent nanoscaffold for arraying of functional modules, for example dyes, in a defined 3D arrangement [10].

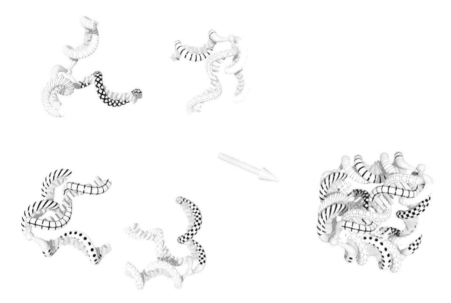

Figure 8.3 Self-assembly of a tetrahedral nanoobject from four tris-oligo junctions.

8.2.3
Tensegrity and Maximum Instruction as the Keys for Nanoarchitecture Control

A tetrahedral scaffold is the smallest object where the topology enforces a "three-dimensional" geometry on the nanometer scale. We usually understand tris-oligo nanoscaffolds as constructs having rigid DNA doublestranded "sticks" that are connected by flexible linker "threads". As long as the flexible units are small compared with the rigid units, such nanoscaffolds are either stiff, in the sense that topology/connectivity determines geometry, or not stiff. A cube-shaped scaffold, for example, is not stiff, because it can fold in many different ways. In the macroscopic world a cube-shaped scaffold can even collapse under the influence of gravity, whereas a tetrahedral scaffold (Fig. 8.4) does not. Stiff scaffolds maintain their geometric shape and will even stiffen if a force is applied. R. Buckminster Fuller introduced the term "tensegrity" to describe the phenomenon of (geometrical) integrity under ten-

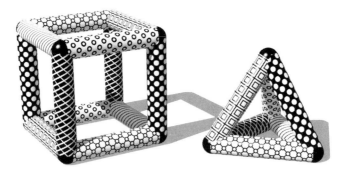

Figure 8.4 The tetrahedron, unlike the cube is an example of a nano-object with (a) stiff tensegrity and (b) maximum instruction. Each texture in the objects indicates an individual doublestranded sequence.

sion and it was recently reported that this concept has implications for the molecular world [11]. Another issue of nanoscaffold engineering is based on Lehn's *instructed mixture paradigm* [12]. Lehn coined this term for a mixture of interacting molecules for which the overall supramolecular structure/s is/are instructed by a number of selective interactions in the set of building blocks. There are, however, different ways of instructing the mixture. One proposal is to refer to "*maximum instruction*" if one deals with a set of building blocks whose self-assembly leads to the target super-structure as the smallest member in a family of conceivable alternatives. It has already been shown that noncovalent synthesis can be steered to favor the smallest superstructure by rapid cooling as a means of applying kinetic control [9]. For any given tris-oligo nanoscaffold the maximally instructed set is that where each corner belongs to an individual tris-oligo and each edge to an individual doublestranded sequence. The maximally instructed set for a tetrahedral nanoscaffold consists of four tris-oligos, each with three different sequences pairing to encode six double-stranded bonds. If pairing of all sequences in the set occurs, the noncovalent products can be composed of four molecules (tetrahedron), eight (cube), twelve (truncated tetrahedron), and so on.

8.2.4
Chemical Copying of Connectivity Information (CCC) as the Key for Nanomachine Replication

We consider connectivity information as a central concept in the design of replicable nanomachines. For systems with stiff tensegrity, the whole scaffold nanoarchitecture is encoded in the connectivity information, as a result of which each junction "knows" where to integrate during the self-assembly process. Very recently it was shown that the connectivity information of one kind of such junctions can be copied by chemical means (Fig. 8.5) [13]. Copying of connectivity information is based on a template-directed tris-linking reaction in which the tris-oligo junction binds three linear complements whose 5'-ends come into close spatial proximity and thus can be

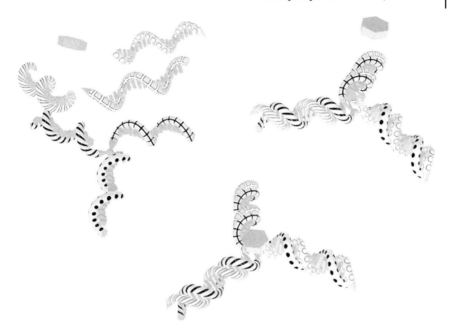

Figure 8.5 Illustration of chemical connectivity copying (CCC) by template-directed tris-linking. Left: The copying process involves a 3'-tris-oligonucleotidyl with three individually defined sequences, three linear complements whose 5'-ends are drawn as blunt ends, and a tris-linker shown as a honeycomb cap. Right: Hybridization leads to a quartermolecular complex in which the blunt ends come into close spatial proximity. Bottom: The tris-linker has reacted to connect the 5'-ends of the copy.

connected by a suitable tris-linking reagent. At this point it should be mentioned that there are two kinds of tris-oligo junction, so-called 3'-tris-oligos with the 3'-ends connected and so-called 5'-tris-oligos with a connection between the 5'-ends. Currently we know that 3'-tris-oligo junctions can be used as templates for fabrication of 5'-tris-oligo junctions. The proof of replicability of a nanoscaffolded machinery requires also demonstration of the reverse process, which has not yet been implemented. Currently, however, it seems it is just a matter of a short period of time before a full copying cycle will be demonstrated.

8.2.5
Cloning and Copying on Surfaces Using eSPREAD

Surface-promoted replication and exponential amplification of DNA analogs (SPREAD) was introduced as a general procedure for replication of DNA-like materials [14]. The procedure, originally invented to overcome the issue of product inhibition in artificial self-replicating systems, has special potential for non-natural modes of copying such as CCC, for which no biotechnological tools, for example polymerases or ligases, are available. As such, and because of its generality, we see this

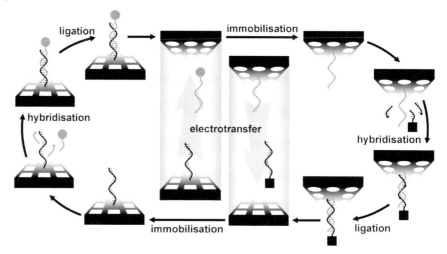

Figure 8.6 The eSPREAD-procedure makes use of two planar, parallel electrodes to which permeation layers enabling two orthogonal immobilization chemistries (circles/squares) are attached. A template is bound to the lower electrode. Hybridization and ligation of complementary fragments yields a copy which is transferred to an adjacent site of the counterelectrode with the help of an electric field. The copy is immobilized to become a template for production of its complement, which is transferred to the lower electrode by reversing the electric field. eSPREAD thus proceeds with the conservation of sequential and spatial information, enabling the copying and cloning of template molecules by means of a flip-flop between the two electrodes.

procedure as a key method for nanomachine replication. An electronic chip variant of the procedure (Fig. 8.6) became the issue of a patent [15]. We propose to develop the eSPREAD procedure (also called as cloning and amplification technology, CAT) for the cloning and copying of nanomachines on the surface of such electrophoresis chips.

8.2.6
Linear Conjugates as Building Blocks for Junction eSPREADing and Nanoscaffolded Multifunctionality

We propose modular monoconjugates as construction elements for junction replication and amplification using eSPREAD (introduced in step 4 of Fig. 8.7). The scheme proceeds from triple-function 3'-tris-oligo templates where one of the three functions is needed for tris-oligo template immobilization. The copy product of steps 1–3 is a 5'-tris-oligo that is used as a template in the second half of the cycle. Any modular function that is (a) conjugable to an linear oligonucleotide and (b) compatible with the implementation procedure of eSPREAD can be used here. As such, peptides, cyclopeptides, proteins, saccharides and oligosaccharides, dyes, organic conductors and semiconductors, lipids, steroids, dendrimers, fullerenes, nanotubes, and inorganic nanocrystals, clusters, and nanoparticles might become the issue of nanoscaffolded multifunctionality. Moreover, because CCC is a purely chemical process, the informational material from which the scaffold is self-assembled does not necessarily need to be based on oligonucleotides. Any functional mimic, for example "spiegelmers" [16], PNA [17], pRNA [18], and many more might replace the oligonucleotide part of the scaffold. Fig. 8.8 illustrates the self-assembly process using a single triple-function tris-oligo. Not only three but all of the twelve sequences in the four tris-oligos of a tetrahedral nanoscaffold can bear an individual function module.

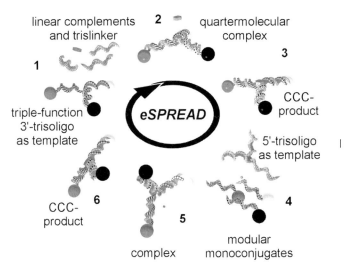

Figure 8.7 eSPREADing the connectivity information of a triple-function 3'-tris-oligo template (1). The copying and amplification process involves modular monoconjugates (4) as basic construction elements. The three functions per junction are indicated by balls with different shades of gray.

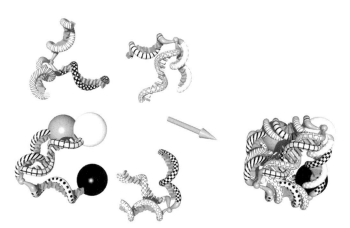

Figure 8.8 Incorporation of a triple-function 3'-tris-oligo during the self-assembly process.

8.2.7
Directed Evolution of Replicable Nanomachines

When eSPREADing of connectivity information has led to successful procedures for junction amplification, it is only a small step to the directed in-vitro evolution of multifunctional nanoscaffolded machinery. Starting from modular monoconjugates as basic construction elements, random connectivity information could be seeded by statistical, i.e. nontemplated, synthesis of triple-functionalized junctions. Statistically, starting from 12 monoconjugates there are 12 possible homotrimers, 396 possible heterotrimers with two different arms, and 7920 tris-oligos with three different arms. As such, statistics clearly favors multifunctionality. If one asks for the total number of possible arrangements on a tetrahedral scaffold with six distinguishable duplex bonds, there are six possibilities to make the first bond, but in two different orientations. For the second bond there are five possibilities again in two orientations. The third bond comes up with four possibilities in two orientations and so on. So the total number of arrangements is $6! \times 2^6 = 46,080$. In general, a spherically closed ob-

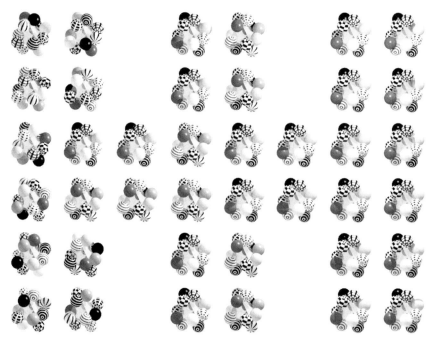

Figure 8.9 Illustration of a directed evolution process of replicable nanomachines based on a tetrahedral scaffold. A population of twelve arrangements (left) is probed to suit a given complex function, e.g. to bind to a cluster of receptors on the surface of a cell. The constructs (two here) which fulfill the task are selected from the initial population by chemical means, e.g. by affinity chromatography, and then amplified to arrive at a new population in which the functional constructs are enriched (middle). Another round of selection and amplification using more stringent conditions finally leads to a (more ore less) uniform population of nanomachines with an optimum array of function modules.

ject with n bonds has $2^n \times n!$ possible arrangements. A nanoscaffold with dodeca-hedrane topology ($n = 30$), for example, could exist in as many as 2.8×10^{41} configurations. One can argue that the functional diversity of such constructs is lower for reasons of symmetry. This arguments, however, are valid only for ideal geometries with equal bond lengths. When a set of tris-oligo building blocks with different arm lengths is assembled statistically, geometrically distorted objects will result which might have the full repertoire of multifunctionality. Multifunction probing might proceed similarly to the SELEX procedure [19], in which a random population of linear RNA and DNA molecules undergoes a series of selection and amplification steps finally leading to a rather uniform population of molecules which fulfills the requested function optimally. An illustration of the procedure is shown in Fig. 8.9.

8.2.8
Probing the Existence of Nanoepitopes on the Surface of Biological Cells

Among the possible tasks for such constructs are catalysis, binding, and sensing tasks. Although the gross geometry of a nanoscaffold with stiff tensegrity seems to be predictable, the detailed orientation and positioning of the modules might be more difficult to control. Depending on the nature of the module itself, the linker between the module and the scaffold, and the lengths of the junction arms, modules might prefer to orient themselves into the interior or exterior of the scaffold. Although catalysis and small molecule-binding tasks might benefit from endopresentation of modules, binding to objects larger than the scaffold will benefit from their exopresentation. A rather appealing target for application of such constructs is the surface of biological cells. Little is known about the extent to which the arrangement of individual cell-surface molecules and receptors such as integrins, G-protein coupled receptors, lectins, cell-adhesion molecules, and others are spatially organized by the cytoskeleton and to what extent these molecules are presented in a statistical arrangement. If the cytoskeleton "imprints" 2D-spatial information on the outside of the cell, this arrangement might vary with the state of cell cycle and as a function of cell differentiation. Nanoepitope screening by arraying of the ligands of the cell surface molecules via an artificial nanoscaffold might shed new light on the classification of cell types and tissues. If cell types, including cancer cells, can, indeed, be recognized by cooperative binding to several different modules arrayed on the scaffolds, it seems conceivable to load the scaffolds with additional "therapy" modules. In particular, modules with the function of alerting the immune system or metallic nanoparticles used to dissipate heat upon receipt of external radiation seem promising candidates for stepwise implementation of medical nanomachines.

8.2.9
External Control of the Operation of Such Nanomachines by GHz Radio-frequency Magnetic Field Inductive Heating of Metal Clusters Attached to Such Constructs

Remote control of the operation of nanomachines might be based on induced conformation change. This might affect, e.g., the activation of modules for a remotely controlled action, or the arrangement of modules (exo versus endo presentation). External control might be achieved by means of a change of temperature, light, pH, ionic strength, and the concentration of specific factors (e.g. heavy metal ions) for which selective recognition modules might be loaded on the scaffolds. If, however, one considers a potential use of such nanomachines inside the human body all of these means of control will have limited use only. Most elegant in this context is, however, the possibility of switching conformations by dissipation of heat from inductive heating of metallic or magnetic nanoparticles attached to such constructs. Very recently it was shown that a gold nanocrystal (1.4 nm) can act as a nanoscale antenna for the receipt of the magnetic component of GHz radio frequency radiation [20]. It was demonstrated that irradiation of a gold-labeled molecular beacon leads to its opening, whereas no opening occurred in the absence of the gold cluster label. The results indicated that this technique can be used to heat single molecules [20]. The technique of single-molecule heating may soon go beyond its proof-of-concept stage, when thermostable gold-cluster labels become available. A new generation of biocompatible gold clusters, tailored for nanotechnological, biomolecular, and nanomedical applications, will be introduced soon. The material (called as RUBiGold) is based on a robotic grip design for packaging of the gold-cluster core (Fig. 8.10) [21]; its thermostability is orders of magnitude higher than that of the material currently available. Current research in this field is devoted to the development of bioconjugable nanowires, nanocircles, and nanocoils of precisely defined size. The length of such constructs is expected to control the extent of single-molecule heating.

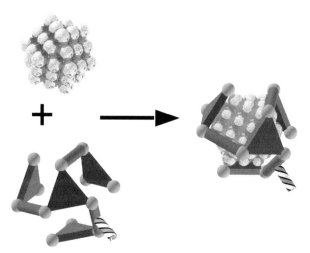

Figure 8.10   RUBiGold is based on a single dodecadentate thioether grip carrying a functional moiety (candystick texture) that enables the monoconjugation of gripped gold cluster (1.4 nm) to biomolecules and other materials.

8.3
Conclusion

The vision of self-replicating nanorobots as introduced by Drexler is based on the concept of a general-purpose nanomechanical assembler that builds any molecule and material including itself in a atom-by-atom fashion. It underestimates the individuality and diversity of chemical behavior and implies an unfruitful lack of appreciation of chemistry as a science and of chemical synthesis as an art. Being able to use an AFM tip to move around xenon atoms on the surface of a single crystal to write a company's logo is indeed a remarkable achievement, but from here to the nanomechanical synthesis of objects shaped in the diamond lattice with atomic precision is a very long way. Nevertheless, Drexler's vision should not be abandoned, because it holds a number of promising ideas of where to go and what to do in the future. Drexler's core idea of self-replicating nanorobots, however, might be challenged by alternative and more natural concepts of implementation such as those described in this manifesto. Chemistry has been looking in this direction for several years – the first demonstration of a chemical self-replicating system [22] and even a cautiously hidden claim to go for self-reproducing molecular robots [23] were reported in the year Drexler's book was published. In a technical report published by Iben Browning as early as 1956 and reprinted in 1978, future chemistry was seen as a central science facing two challenges – deciphering of the origin-of-life problem and creation of artificial self-reproducing molecular machinery. It seems that this future has arrived today.

Acknowledgment

This work was supported by Deutsche Forschungsgemeinschaft (SFB-452), Bundesministerium für Bildung und Forschung (Bionanotechnologie), and Fonds der Chemischen Industrie. We thank Michael Wüstefeld for DNA synthesis, Rolf Breuckmann for MALDI and ESI analysis, and Beate Materne and Carsten Lodwig for technical assistance.

References

1 Winkless, N. and Browning, E. *Robots on Your Doorstep*. Robotics Press, Portland, **1978**.

2 Drexler, K.E. *Engines of Creation: The Coming Era of Nanotechnology*. Anchor Press, Doubleday, **1986**.

3 Smalley, R.E. Of chemistry, love, and nanobots. *Scientific American* **2001**, *285*(3), 76–77.

4 Whitesides, G.M. The once and future nanomachine. *Scientific American* **2001**, *285*(3), 78–83.

5 Overview: Niemeyer, C. M. Nanoparticles, proteins, and nucleic acids: biotechnology meets materials science. *Angew. Chem. Int. Ed.*, **2001**, *40*, 4128–4158.

6 Chen, J. and Seeman, N.C. The synthesis from DNA of a molecule with the connectivity of a cube, *Nature* **1991**, *350*, 631–633.

7 Jordan, S. *Synthesis of oligonucleotide building blocks for self-replication and self-assembly experiments* (in German), thesis, Göttingen, **1993**.

8 Shchepinov, M.S., Mir, K.U., Elder, J.K., Frank-Kamenetskii, M.D., and Southern, E.M. *Nucleic Acids Res.* **1999**, *27*, 3035–3041.

9 Scheffler, M., Dorenbeck, A., Jordan, S., Wüstefeld, M., and von Kiedrowski, G. Self-Assembly of Trisoligonucleotidyls: The Case for Nano-Acetylene and Nano-Cyclobutadiene. *Angew. Chem. Int. Ed.* **1999**, *38*, 3311–3315.

10 Dorenbeck, A., Scheffler, M., Wüstefeld, M., and von Kiedrowski, G. Noncovalent synthesis of a tetrahedral nanoscaffold from 3'-trisoligonucleotidyls with individually defined sequences. *Angew. Chem.*, in press.

11 Ingber, D.E., *Ann. Rev. Physiol.* **1997**, *59*, 575–599.

12 Lehn, J.-M., Supramolecular Chemistry – Concepts and Perspectives, VCH, Weinheim, **1994**.

13 Eckardt, L., Naumann, K., Pankau, W.M., Rein, M., Schweitzer, M., Windhab, N., and von Kiedrowski, G. Chemical copying of connectivity. *Nature*, **2002**, *420*, 286.

14 Luther, A., Brandsch, R., and von Kiedrowski, G. Surface-promoted replication and exponential amplification of DNA analogues. *Nature* **1998**, *396*, 245–248.

15 Fürste, J.P., Klussman, S., Klein, T., and von Kiedrowski, G. Cloning and copying on surfaces, EP 1135527, WO 0032809, US 2002022276 .

16 Nolte, A., Klussmann, S. Bald, R., Erdmann, V.A., and Fürste, J.P. Mirror-design of L-oligonucleotide ligands binding to L-arginine. *Nature Biotech.* **1996**, *14*, 1116–1119.

17 Nielsen, P.E., Egholm, M., Berg, R.H., and Buchard, O. Sequence-selective recognition of DNA by strand displacement with a thymidine-substituted polyamide. *Science* **1991**, *254*, 1497–1500.

18 Pitsch, S., Wendeborn, S. Jaun, B., and Eschenmoser, A. Why pentose- and not hexose-nucleic acids? Part VII: Pyranosyl-RNA. *Helv. Chim. Acta* **1993**, *76*, 2161–2183.

19 Tuerk, C. and Gold, L. **Systematic evolution of ligands by exponential enrichment: RNA ligands to bacteriophage T4 DNA polymerase.** Science **1990**, *249*, 505–510.

20 Hamad-Schifferli, K., Schwartz, J.J., Santos, A.T., Zhang, S., and Jacobson, J.M. Remote electronic control of DNA hybridisation through inductive coupling to an attached nanocrystal antenna. *Nature* **2002**, *415*, 152–155.

21 Pankau, M., Mönninghoff, S., and von Kiedrowski, G. Thermostable and mono-conjugable gold cluster by gripping with a dodekadentate thioether ligand. Submitted, patent pending.

22 von Kiedrowski, G. A self-replicating hexadeoxynucleotide. *Angew. Chem. Intl. Ed. Engl.* **1986**, *25*, 932–935.

23 von Kiedrowski, G. A Self-Reproducing Molecular Robot. *Orig. Life Evol. Biosph.* **1986**, *16*, 468.

Index

Functional Synthetic Receptors. Edited by T. Schrader, A. D. Hamilton
Copyright © 2005 WILEY-VCH Verlag GmbH & Co. KGaA, Weinheim
ISBN: 3-527-30655-2